D0148699

THE THEORY OF THE

RIEMANN ZETA-FUNCTION

THE THEORY OF THE RIEMANN ZETA-FUNCTION

BY

E. C. TITCHMARSH

F.R.S.

FORMERLY SAVILIAN PROFESSOR OF GEOMETRY IN THE UNIVERSITY OF OXFORD

SECOND EDITION

REVISED BY

D. R. HEATH-BROWN

FELLOW OF MAGDALEN COLLEGE, UNIVERSITY OF OXFORD

CLARENDON PRESS · OXFORD

Oxford University Press, Walton Street, Oxford OX2 6DP
Oxford New York Toronto
Delhi Bombay Calcutta Madras Karachi
Kuala Lumpur Singapore Hong Kong Tokyo
Nairobi Dar es Salaam Cape Town
Melbourne Auckland Madrid
and associated companies in
Berlin Ibadan

Oxford is a trade mark of Oxford University Press

Published in the United States
by Oxford University Press Inc., New York

First published 1951
Second edition 1986
Reprinted 1988, 1994

British Library Cataloguing in Publication Data
Titchmarsh, E. C.
The theory of the Riemann zeta-function.
1. Calculus 2. Functions, Zeta
3. Riemann–Hilbert problems
I. Title
515.9′82 QA320

Library of Congress Cataloging in Publication Data
Titchmarsh, E. C. (Edward Charles), 1899–
The theory of the Riemann zeta-function.
Bibliography: p.
1. Functions, Zeta. I. Heath-Brown, D. R.
II. Title.
QA246.T44 1986 512′.73 86–12520
ISBN 0 19 853369 1 (Pbk)

Printed and bound in Great Britain by
Biddles Ltd, Guildford and King's Lynn

PREFACE TO THE SECOND EDITION

SINCE the first edition was written, a vast amount of further work has been done. This has been covered by the end-of-chapter notes. In most instances, restrictions on space have prohibited the inclusion of full proofs, but I have tried to give an indication of the methods used wherever possible. (Proofs of quite a few of the recent results described in the end of chapter notes may be found in the book by Ivić [**3**].) I have also corrected a number of minor errors, and made a few other small improvements to the text. A considerable number of recent references have been added.

In preparing this work I have had help from Professors J. B. Conrey, P. D. T. A. Elliott, A. Ghosh, S. M. Gonek, H. L. Montgomery, and S. J. Patterson. It is a pleasure to record my thanks to them.

OXFORD
1986

D. R. H.-B.

PREFACE TO FIRST EDITION

THIS book is a successor to my Cambridge Tract *The Zeta-Function of Riemann*, 1930, which is now out of print and out of date. It seems no longer practicable to give an account of the subject in such a small space as a Cambridge Tract, so that the present work, though on exactly the same lines as the previous one, is on a much larger scale. As before, I do not discuss general prime-number theory, though it has been convenient to include some theorems on primes.

Most of this book was compiled in the 1930's, when I was still researching on the subject. It has been brought partly up to date by including some of the work of A. Selberg and of Vinogradov, though a great deal of recent work is scantily represented.

The manuscript has been read by Dr. S. H. Min and by Prof. D. B. Sears, and my best thanks are due to them for correcting a large number of mistakes. I must also thank Prof. F. V. Atkinson and Dr. T. M. Fleet for their kind assistance in reading the proof-sheets.

OXFORD
1951

E. C. T.

CONTENTS

I

THE FUNCTION $\zeta(s)$ AND THE DIRICHLET SERIES RELATED TO IT

1.1. Definition of $\zeta(s)$. The Riemann zeta-function $\zeta(s)$ has its origin in the identity expressed by the two formulae

$$\zeta(s) = \sum_{n=1}^{\infty} \frac{1}{n^s}, \tag{1.1.1}$$

where n runs through all integers, and

$$\zeta(s) = \prod_p \left(1-\frac{1}{p^s}\right)^{-1}, \tag{1.1.2}$$

where p runs through all primes. Either of these may be taken as the definition of $\zeta(s)$; s is a complex variable, $s = \sigma+it$. The Dirichlet series (1.1.1) is convergent for $\sigma > 1$, and uniformly convergent in any finite region in which $\sigma \geqslant 1+\delta$, $\delta > 0$. It therefore defines an analytic function $\zeta(s)$, regular for $\sigma > 1$.

The infinite product is also absolutely convergent for $\sigma > 1$; for so is

$$\sum_p \left|\frac{1}{p^s}\right| = \sum_p \frac{1}{p^\sigma},$$

this being merely a selection of terms from the series $\sum n^{-\sigma}$. If we expand the factor involving p in powers of p^{-s}, we obtain

$$\prod_p \left(1+\frac{1}{p^s}+\frac{1}{p^{2s}}+\ldots\right).$$

On multiplying formally, we obtain the series (1.1.1), since each integer n can be expressed as a product of prime-powers p^m in just one way. The identity of (1.1.1) and (1.1.2) is thus an analytic equivalent of the theorem that the expression of an integer in prime factors is unique.

A rigorous proof is easily constructed by taking first a finite number of factors. Since we can multiply a finite number of absolutely convergent series, we have

$$\prod_{p\leqslant P} \left(1+\frac{1}{p^s}+\frac{1}{p^{2s}}+\ldots\right) = 1+\frac{1}{n_1^s}+\frac{1}{n_2^s}+\ldots,$$

where n_1, n_2,..., are those integers none of whose prime factors exceed P.

Since all integers up to P are of this form, it follows that, if $\zeta(s)$ is defined by (1.1.1),

$$\left|\zeta(s)-\prod_{p\leqslant P}\left(1-\frac{1}{p^s}\right)^{-1}\right| = \left|\zeta(s)-1-\frac{1}{n_1^s}-\frac{1}{n_2^s}-\ldots\right|$$
$$\leqslant \frac{1}{(P+1)^\sigma}+\frac{1}{(P+2)^\sigma}+\ldots.$$

This tends to 0 as $P\to\infty$, if $\sigma>1$; and (1.1.2) follows.

This fundamental identity is due to Euler, and (1.1.2) is known as Euler's product. But Euler considered it for particular values of s only, and it was Riemann who first considered $\zeta(s)$ as an analytic function of a complex variable.

Since a convergent infinite product of non-zero factors is not zero, we deduce that *$\zeta(s)$ has no zeros for $\sigma>1$.* This may be proved directly as follows. We have for $\sigma>1$

$$\left(1-\frac{1}{2^s}\right)\left(1-\frac{1}{3^s}\right)\ldots\left(1-\frac{1}{P^s}\right)\zeta(s) = 1+\frac{1}{m_1^s}+\frac{1}{m_2^s}+\ldots,$$

where m_1, m_2,..., are the integers all of whose prime factors exceed P. Hence

$$\left|\left(1-\frac{1}{2^s}\right)\ldots\left(1-\frac{1}{P^s}\right)\zeta(s)\right| \geqslant 1-\frac{1}{(P+1)^\sigma}-\frac{1}{(P+2)^\sigma}-\ldots>0$$

if P is large enough. Hence $|\zeta(s)|>0$.

The importance of $\zeta(s)$ in the theory of prime numbers lies in the fact that it combines two expressions, one of which contains the primes explicitly, while the other does not. The theory of primes is largely concerned with the function $\pi(x)$, the number of primes not exceeding x. We can transform (1.1.2) into a relation between $\zeta(s)$ and $\pi(x)$; for if $\sigma>1$,

$$\log\zeta(s) = -\sum_p \log\left(1-\frac{1}{p^s}\right) = -\sum_{n=2}^{\infty}\{\pi(n)-\pi(n-1)\}\log\left(1-\frac{1}{n^s}\right)$$
$$= -\sum_{n=2}^{\infty}\pi(n)\left\{\log\left(1-\frac{1}{n^s}\right)-\log\left(1-\frac{1}{(n+1)^s}\right)\right\}$$
$$= \sum_{n=2}^{\infty}\pi(n)\int_n^{n+1}\frac{s}{x(x^s-1)}\,dx = s\int_2^{\infty}\frac{\pi(x)}{x(x^s-1)}\,dx. \qquad (1.1.3)$$

The rearrangement of the series is justified since $\pi(n)\leqslant n$ and

$$\log(1-n^{-s}) = O(n^{-\sigma}).$$

Again
$$\frac{1}{\zeta(s)} = \prod_p \left(1-\frac{1}{p^s}\right),$$
and on carrying out the multiplication we obtain
$$\frac{1}{\zeta(s)} = \sum_{n=1}^{\infty} \frac{\mu(n)}{n^s} \quad (\sigma > 1), \tag{1.1.4}$$
where $\mu(1) = 1$, $\mu(n) = (-1)^k$ if n is the product of k different primes, and $\mu(n) = 0$ if n contains any factor to a power higher than the first. The process is easily justified as in the case of $\zeta(s)$.

The function $\mu(n)$ is known as the Möbius function. It has the property
$$\sum_{d|q} \mu(d) = 1 \ (q = 1), \quad 0 \ (q > 1), \tag{1.1.5}$$
where $d \mid q$ means that d is a divisor of q. This follows from the identity
$$1 = \sum_{m=1}^{\infty} \frac{1}{m^s} \sum_{n=1}^{\infty} \frac{\mu(n)}{n^s} = \sum_{q=1}^{\infty} \frac{1}{q^s} \sum_{d|q} \mu(d).$$
It also gives the 'Möbius inversion formula'
$$g(q) = \sum_{d|q} f(d), \tag{1.1.6}$$
$$f(q) = \sum_{d|q} \mu\left(\frac{q}{d}\right) g(d), \tag{1.1.7}$$
connecting two functions $f(n)$, $g(n)$ defined for integral n. If f is given and g defined by (1.1.6), the right-hand side of (1.1.7) is
$$\sum_{d|q} \mu\left(\frac{q}{d}\right) \sum_{r|d} f(r).$$
The coefficient of $f(q)$ is $\mu(1) = 1$. If $r < q$, then $d = kr$, where $k \mid q/r$. Hence the coefficient of $f(r)$ is
$$\sum_{k|q/r} \mu\left(\frac{q}{kr}\right) = \sum_{k'|q/r} \mu(k') = 0$$
by (1.1.5). This proves (1.1.7). Conversely, if g is given, and f is defined by (1.1.7), then the right-hand side of (1.1.6) is
$$\sum_{d|q} \sum_{r|d} \mu\left(\frac{d}{r}\right) g(r),$$
and this is $g(q)$, by a similar argument. The formula may also be

derived formally from the obviously equivalent relations

$$F(s)\zeta(s) = \sum_{n=1}^{\infty} \frac{g(n)}{n^s}, \qquad F(s) = \frac{1}{\zeta(s)} \sum_{n=1}^{\infty} \frac{g(n)}{n^s},$$

where

$$F(s) = \sum_{n=1}^{\infty} \frac{f(n)}{n^s}.$$

Again, on taking logarithms and differentiating (1.1.2), we obtain, for $\sigma > 1$,

$$\begin{aligned} \frac{\zeta'(s)}{\zeta(s)} &= -\sum_{p} \frac{\log p}{p^s}\left(1-\frac{1}{p^s}\right)^{-1} \\ &= -\sum_{p} \log p \sum_{m=1}^{\infty} \frac{1}{p^{ms}} \\ &= -\sum_{n=2}^{\infty} \frac{\Lambda(n)}{n^s}, \end{aligned} \tag{1.1.8}$$

where $\Lambda(n) = \log p$ if n is p or a power of p, and otherwise $\Lambda(n) = 0$. On integrating we obtain

$$\log \zeta(s) = \sum_{n=2}^{\infty} \frac{\Lambda_1(n)}{n^s} \quad (\sigma > 1), \tag{1.1.9}$$

where $\Lambda_1(n) = \Lambda(n)/\log n$, and the value of $\log \zeta(s)$ is that which tends to 0 as $\sigma \to \infty$, for any fixed t.

1.2. Various Dirichlet series connected with $\zeta(s)$. In the first place

$$\zeta^2(s) = \sum_{n=1}^{\infty} \frac{d(n)}{n^s} \quad (\sigma > 1), \tag{1.2.1}$$

where $d(n)$ denotes the number of divisors of n (including 1 and n itself). For

$$\zeta^2(s) = \sum_{\mu=1}^{\infty} \frac{1}{\mu^s} \sum_{\nu=1}^{\infty} \frac{1}{\nu^s} = \sum_{n=1}^{\infty} \frac{1}{n^s} \sum_{\mu\nu=n} 1,$$

and the number of terms in the last sum is $d(n)$. And generally

$$\zeta^k(s) = \sum_{n=1}^{\infty} \frac{d_k(n)}{n^s} \quad (\sigma > 1), \tag{1.2.2}$$

where $k = 2, 3, 4,\ldots$, and $d_k(n)$ denotes the number of ways of expressing n as a product of k factors, expressions with the same factors in a different order being counted as different. For

$$\zeta^k(s) = \sum_{\nu_1=1}^{\infty} \frac{1}{\nu_1^s} \cdots \sum_{\nu_k=1}^{\infty} \frac{1}{\nu_k^s} = \sum_{n=1}^{\infty} \frac{1}{n^s} \sum_{\nu_1 \ldots \nu_k = n} 1,$$

and the last sum is $d_k(n)$.

Since we have also

$$\zeta^2(s) = \prod_p \left(1-\frac{1}{p^s}\right)^{-2} = \prod_p \left(1+\frac{2}{p^s}+\frac{3}{p^{2s}}+\ldots\right), \tag{1.2.3}$$

on comparing the coefficients in (1.2.1) and (1.2.3) we verify the elementary formula

$$d(n) = (m_1+1)\ldots(m_r+1) \tag{1.2.4}$$

for the number of divisors of

$$n = p_1^{m_1} p_2^{m_2} \ldots p_r^{m_r}. \tag{1.2.5}$$

Similarly from (1.2.2)

$$d_k(n) = \frac{(k+m_1-1)!}{m_1!\,(k-1)!} \ldots \frac{(k+m_r-1)!}{m_r!\,(k-1)!}. \tag{1.2.6}$$

We next note the expansions

$$\frac{\zeta(s)}{\zeta(2s)} = \sum_{n=1}^{\infty} \frac{|\mu(n)|}{n^s} \quad (\sigma > 1), \tag{1.2.7}$$

where $\mu(n)$ is the coefficient in (1.1.4);

$$\frac{\zeta^2(s)}{\zeta(2s)} = \sum_{n=1}^{\infty} \frac{2^{\nu(n)}}{n^s} \quad (\sigma > 1), \tag{1.2.8}$$

where $\nu(n)$ is the number of different prime factors of n;

$$\frac{\zeta^3(s)}{\zeta(2s)} = \sum_{n=1}^{\infty} \frac{d(n^2)}{n^s} \quad (\sigma > 1), \tag{1.2.9}$$

and

$$\frac{\zeta^4(s)}{\zeta(2s)} = \sum_{n=1}^{\infty} \frac{\{d(n)\}^2}{n^s} \quad (\sigma > 1). \tag{1.2.10}$$

To prove (1.2.7), we have

$$\frac{\zeta(s)}{\zeta(2s)} = \prod_p \frac{1-p^{-2s}}{1-p^{-s}} = \prod_p \left(1+\frac{1}{p^s}\right),$$

and this differs from the formula for $1/\zeta(s)$ only in the fact that the signs are all positive. The result is therefore clear. To prove (1.2.8), we have

$$\frac{\zeta^2(s)}{\zeta(2s)} = \prod_p \frac{1-p^{-2s}}{(1-p^{-s})^2} = \prod_p \frac{1+p^{-s}}{1-p^{-s}}$$

$$= \prod_p (1+2p^{-s}+2p^{-2s}+\ldots),$$

and the result follows. To prove (1.2.9),

$$\frac{\zeta^3(s)}{\zeta(2s)} = \prod_p \frac{1-p^{-2s}}{(1-p^{-s})^3} = \prod_p \frac{1+p^{-s}}{(1-p^{-s})^2}$$
$$= \prod_p \{(1+p^{-s})(1+2p^{-s}+3p^{-2s}+...)\}$$
$$= \prod_p \{1+3p^{-s}+...+(2m+1)p^{-ms}+...\},$$

and the result follows, since, if n is (1.2.5),

$$d(n^2) = (2m_1+1)...(2m_r+1).$$

Similarly

$$\frac{\zeta^4(s)}{\zeta(2s)} = \prod_p \frac{1-p^{-2s}}{(1-p^{-s})^4} = \prod_p \frac{1+p^{-s}}{(1-p^{-s})^3}$$
$$= \prod_p (1+p^{-s})\{1+3p^{-s}+...+\tfrac{1}{2}(m+1)(m+2)p^{-ms}+...\}$$
$$= \prod_p \{1+4p^{-s}+...+(m+1)^2 p^{-ms}+...\},$$

and (1.2.10) follows.

Other formulae are

$$\frac{\zeta(2s)}{\zeta(s)} = \sum_{n=1}^{\infty} \frac{\lambda(n)}{n^s} \quad (\sigma > 1), \tag{1.2.11}$$

where $\lambda(n) = (-1)^r$ if n has r prime factors, a factor of degree k being counted k times;

$$\frac{\zeta(s-1)}{\zeta(s)} = \sum_{n=1}^{\infty} \frac{\phi(n)}{n^s} \quad (\sigma > 2), \tag{1.2.12}$$

where $\phi(n)$ is the number of numbers less than n and prime to n; and

$$\frac{1-2^{1-s}}{1-2^{-s}}\zeta(s-1) = \sum_{n=1}^{\infty} \frac{a(n)}{n^s} \quad (\sigma > 2), \tag{1.2.13}$$

where $a(n)$ is the greatest odd divisor of n. Of these, (1.2.11) follows at once from

$$\frac{\zeta(2s)}{\zeta(s)} = \prod_p \left(\frac{1-p^{-s}}{1-p^{-2s}}\right) = \prod_p \left(\frac{1}{1+p^{-s}}\right) = \prod_p (1-p^{-s}+p^{-2s}-...).$$

Also

$$\frac{\zeta(s-1)}{\zeta(s)} = \prod_p \left(\frac{1-p^{-s}}{1-p^{1-s}}\right) = \prod_p \left\{\left(1-\frac{1}{p^s}\right)\left(1+\frac{p}{p^s}+\frac{p^2}{p^{2s}}+...\right)\right\}$$
$$= \prod_p \left\{1+\left(1-\frac{1}{p}\right)\left(\frac{p}{p^s}+\frac{p^2}{p^{2s}}+...\right)\right\},$$

and (1.2.12) follows, since, if $n = p_1^{m_1}...p_r^{m_r}$,

$$\phi(n) = n\left(1-\frac{1}{p_1}\right)...\left(1-\frac{1}{p_r}\right).$$

Finally

$$\frac{1-2^{1-s}}{1-2^{-s}}\zeta(s-1) = \frac{1-2^{1-s}}{1-2^{-s}}\prod_p \frac{1}{1-p^{1-s}}$$

$$= \frac{1}{1-2^{-s}}\,\frac{1}{1-3^{1-s}}\,\frac{1}{1-5^{1-s}}...$$

$$= \left(1+\frac{1}{2^s}+\frac{1}{2^{2s}}+...\right)\left(1+\frac{3}{3^s}+\frac{3^2}{3^{2s}}+...\right)...,$$

and (1.2.13) follows.

Many of these formulae are, of course, simply particular cases of the general formula

$$\sum_{n=1}^{\infty}\frac{f(n)}{n^s} = \prod_p\left\{1+\frac{f(p)}{p^s}+\frac{f(p^2)}{p^{2s}}+...\right\},$$

where $f(n)$ is a multiplicative function, i.e. is such that, if $n = p_1^{m_1}p_2^{m_2}...$, then

$$f(n) = f(p_1^{m_1})f(p_2^{m_2})....$$

Again, let $f_k(n)$ denote the number of representations of n as a product of k factors, each greater than unity when $n > 1$, the order of the factors being essential. Then clearly

$$\sum_{n=2}^{\infty}\frac{f_k(n)}{n^s} = \{\zeta(s)-1\}^k \quad (\sigma > 1). \tag{1.2.14}$$

Let $f(n)$ be the number of representations of n as a product of factors greater than unity, representations with factors in a different order being considered as distinct; and let $f(1) = 1$. Then

$$f(n) = \sum_{k=1}^{\infty} f_k(n).$$

Hence

$$\sum_{n=1}^{\infty}\frac{f(n)}{n^s} = 1+\sum_{k=1}^{\infty}\{\zeta(s)-1\}^k = 1+\frac{\zeta(s)-1}{1-\{\zeta(s)-1\}}$$

$$= \frac{1}{2-\zeta(s)}. \tag{1.2.15}$$

It is easily seen that $\zeta(s) = 2$ for $s = \alpha$, where α is a real number greater than 1; and $|\zeta(s)| < 2$ for $\sigma > \alpha$, so that (1.2.15) holds for $\sigma > \alpha$.

1.3. Sums involving $\sigma_a(n)$. Let $\sigma_a(n)$ denote the sum of the ath powers of the divisors of n. Then

$$\zeta(s)\zeta(s-a) = \sum_{\mu=1}^{\infty}\frac{1}{\mu^s}\sum_{\nu=1}^{\infty}\frac{\nu^a}{\nu^s} = \sum_{n=1}^{\infty}\frac{1}{n^s}\sum_{\mu\nu=n}\nu^a,$$

i.e. $$\zeta(s)\zeta(s-a) = \sum_{n=1}^{\infty}\frac{\sigma_a(n)}{n^s} \quad (\sigma > 1,\ \sigma > \mathbf{R}(a)+1). \tag{1.3.1}$$

Since the left-hand side is, if $a \neq 0$,

$$\prod_p \left(1+\frac{1}{p^s}+\frac{1}{p^{2s}}+\ldots\right)\left(1+\frac{p^a}{p^s}+\frac{p^{2a}}{p^{2s}}+\ldots\right)$$
$$= \prod_p \left(1+\frac{1+p^a}{p^s}+\frac{1+p^a+p^{2a}}{p^s}+\ldots\right) = \prod_p \left(1+\frac{1-p^{2a}}{1-p^a}\frac{1}{p^s}+\ldots\right)$$

we have $$\sigma_a(n) = \frac{1-p_1^{(m_1+1)a}}{1-p_1^a}\cdots\frac{1-p_r^{(m_r+1)a}}{1-p_r^a}, \tag{1.3.2}$$

if n is (1.2.5), as is also obvious from elementary considerations.

The formula†

$$\frac{\zeta(s)\zeta(s-a)\zeta(s-b)\zeta(s-a-b)}{\zeta(2s-a-b)} = \sum_{n=1}^{\infty}\frac{\sigma_a(n)\sigma_b(n)}{n^s} \tag{1.3.3}$$

is valid for $\sigma > \max\{1, \mathbf{R}(a)+1, \mathbf{R}(b)+1, \mathbf{R}(a+b)+1\}$. The left-hand side is equal to

$$\prod_p \frac{1-p^{-2s+a+b}}{(1-p^{-s})(1-p^{-s+a})(1-p^{-s+b})(1-p^{-s+a+b})}.$$

Putting $p^{-s} = z$, the partial-fraction formula gives

$$\frac{1-p^{a+b}z^2}{(1-z)(1-p^az)(1-p^bz)(1-p^{a+b}z)}$$
$$= \frac{1}{(1-p^a)(1-p^b)}\left\{\frac{1}{1-z}-\frac{p^a}{1-p^az}-\frac{p^b}{1-p^bz}+\frac{p^{a+b}}{1-p^{a+b}z}\right\}$$
$$= \frac{1}{(1-p^a)(1-p^b)}\sum_{m=0}^{\infty}(1-p^{(m+1)a}-p^{(m+1)b}+p^{(m+1)(a+b)})z^m$$
$$= \frac{1}{(1-p^a)(1-p^b)}\sum_{m=0}^{\infty}(1-p^{(m+1)a})(1-p^{(m+1)b})z^m.$$

† Ramanujan (2), B. M. Wilson (1).

Hence

$$\frac{\zeta(s)\zeta(s-a)\zeta(s-b)\zeta(s-a-b)}{\zeta(2s-a-b)} = \prod_p \sum_{m=0}^{\infty} \frac{1-p^{(m+1)a}}{1-p^a} \frac{1-p^{(m+1)b}}{1-p^b} \frac{1}{p^{ms}},$$

and the result follows from (1.3.2). If $a = b = 0$, (1.3.3) reduces to (1.2.10).

Similar formulae involving $\sigma_a^{(q)}(n)$, the sum of the ath powers of those divisors of n which are qth powers of integers, have been given by Crum (1).

1.4. It is also easily seen that, if $f(n)$ is multiplicative, and

$$\sum_{n=1}^{\infty} \frac{f(n)}{n^s}$$

is a product of zeta-functions such as occurs in the above formulae, and k is a given positive integer, then

$$\sum_{n=1}^{\infty} \frac{f(kn)}{n^s}$$

can also be summed. An example will illustrate this point. The function $\sigma_a(n)$ is 'multiplicative', i.e. if m is prime to n

$$\sigma_a(mn) = \sigma_a(m)\sigma_a(n).$$

Hence

$$\sum_{n=1}^{\infty} \frac{\sigma_a(n)}{n^s} = \prod_p \sum_{m=0}^{\infty} \frac{\sigma_a(p^m)}{p^{ms}},$$

and, if $k = \prod p^l$,

$$\sum_{n=1}^{\infty} \frac{\sigma_a(kn)}{n^s} = \prod_p \sum_{m=0}^{\infty} \frac{\sigma_a(p^{l+m})}{p^{ms}}.$$

Hence

$$\sum_{n=1}^{\infty} \frac{\sigma_a(kn)}{n^s} = \zeta(s)\zeta(s-a) \prod_{p|k} \left\{ \sum_{m=0}^{\infty} \frac{\sigma_a(p^{l+m})}{p^{ms}} \Big/ \sum_{m=0}^{\infty} \frac{\sigma_a(p^m)}{p^{ms}} \right\}.$$

Now if $a \neq 0$,

$$\sum_{m=0}^{\infty} \frac{\sigma_a(p^{l+m})}{p^{ms}} = \sum_{m=0}^{\infty} \frac{1-p^{(l+m+1)a}}{(1-p^a)p^{ms}} = \frac{1-p^{a-s}-p^{(l+1)a}+p^{(l+1)a-s}}{(1-p^a)(1-p^{-s})(1-p^{a-s})}.$$

Hence

$$\sum_{n=0}^{\infty} \frac{\sigma_a(kn)}{n^s} = \zeta(s)\zeta(s-a) \prod_{p|k} \frac{1-p^{a-s}-p^{(l+1)a}+p^{(l+1)a-s}}{1-p^a}. \tag{1.4.1}$$

Making $a \to 0$,

$$\sum_{n=0}^{\infty} \frac{d(kn)}{n^s} = \zeta^2(s) \prod_{p|k} (l+1-lp^{-s}). \tag{1.4.2}$$

1.5. Ramanujan's sums.† Let

$$c_k(n) = \sum_h e^{-2nh\pi i/k} = \sum_h \cos\frac{2nh\pi}{k}, \tag{1.5.1}$$

where h runs through all positive integers less than and prime to k. Many formulae involving these sums were proved by Ramanujan.

We shall first prove that

$$c_k(n) = \sum_{d|k,d|n} \mu\left(\frac{k}{d}\right) d. \tag{1.5.2}$$

The sum
$$\eta_k(n) = \sum_{m=0}^{k-1} e^{-2nm\pi i/k}$$

is equal to k if $k \mid n$ and 0 otherwise. Denoting by (r, d) the highest common factor of r and d, so that $(r, d) = 1$ means that r is prime to d,

$$\sum_{d|k} c_d(n) = \sum_{d|k} \sum_{(r,d)=1,r<d} e^{-2nr\pi i/d} = \eta_k(n).$$

Hence by the inversion formula of Möbius (1.1.7)

$$c_k(n) = \sum_{d|k} \mu\left(\frac{k}{d}\right) \eta_d(n),$$

and (1.5.2) follows. In particular

$$c_k(1) = \mu(k). \tag{1.5.3}$$

The result can also be written

$$c_k(n) = \sum_{dr=k,d|n} \mu(r)d.$$

Hence
$$\frac{c_k(n)}{k^s} = \sum_{dr=k,d|n} \frac{\mu(r)}{r^s} d^{1-s}.$$

Summing with respect to k, we remove the restriction on r, which now assumes all positive integral values. Hence‡

$$\sum_{k=1}^{\infty} \frac{c_k(n)}{k^s} = \sum_{r,d|n} \frac{\mu(r)}{r^s} d^{1-s} = \frac{\sigma_{1-s}(n)}{\zeta(s)}, \tag{1.5.4}$$

the series being absolutely convergent for $\sigma > 1$ since $|c_k(n)| \leqslant \sigma_1(n)$, by (1.5.2).

We have also

$$\sum_{n=1}^{\infty} \frac{c_k(n)}{n^s} = \sum_{n=1}^{\infty} \frac{1}{n^s} \sum_{d|k\, d|n} \mu\left(\frac{k}{d}\right) d$$

$$= \sum_{d|k} \mu\left(\frac{k}{d}\right) d \sum_{m=1}^{\infty} \frac{1}{(md)^s} = \zeta(s) \sum_{d|k} \mu\left(\frac{k}{d}\right) d^{1-s}. \tag{1.5.5}$$

† Ramanujan (3), Hardy (5).
‡ Two more proofs are given by Hardy, *Ramanujan*, 137–41.

We can also sum series of the form†

$$\sum_{n=1}^{\infty} \frac{c_k(n)f(n)}{n^s},$$

where $f(n)$ is a multiplicative function. For example,

$$\sum_{n=1}^{\infty} \frac{c_k(n)d(n)}{n^s} = \sum_{n=1}^{\infty} \frac{d(n)}{n^s} \sum_{\delta|k,\delta|n} \delta\mu\left(\frac{k}{\delta}\right)$$

$$= \sum_{\delta|k} \delta\mu\left(\frac{k}{\delta}\right) \sum_{m=1}^{\infty} \frac{d(m\delta)}{(m\delta)^s}$$

$$= \zeta^2(s) \sum_{\delta|k} \delta^{1-s}\mu\left(\frac{k}{\delta}\right) \prod_{p|\delta} (l+1-lp^{-s})$$

if $\delta = \prod p^l$. If $k = \prod p^\lambda$ the sum is

$$k^{1-s} \prod_{p|k} (\lambda+1-\lambda p^{-s}) - \sum_{p|k} \left(\frac{k}{p}\right)^{1-s} \{\lambda-(\lambda-1)p^{-s}\} \prod_{\substack{p'|k \\ p'\neq p}} (\lambda+1-\lambda p'^{-s}) +$$

$$+ \sum_{pp'|k} \left(\frac{k}{pp'}\right)^{1-s} \{\lambda-(\lambda-1)p^{-s}\}\{\lambda-(\lambda-1)p'^{-s}\} \prod_{\substack{p''|k \\ p''\neq p,p'}} (\lambda+1-\lambda p''^{-s}) - \dots$$

$$= k^{1-s} \prod_{p|k} \left\{(\lambda+1-\lambda p^{-s}) - \frac{1}{p^{1-s}}\{\lambda-(\lambda-1)p^{-s}\}\right\}$$

$$= k^{1-s} \prod_{p|k} \left\{1-\frac{1}{p}+\lambda\left(1-\frac{1}{p^s}\right)\left(1-\frac{1}{p^{1-s}}\right)\right\}.$$

Hence

$$\sum_{n=1}^{\infty} \frac{c_k(n)d(n)}{n^s} = \zeta^2(s)k^{1-s} \prod_{p|k} \left\{1-\frac{1}{p}+\lambda\left(1-\frac{1}{p^s}\right)\left(1-\frac{1}{p^{1-s}}\right)\right\}. \tag{1.5.6}$$

We can also sum $$\sum_{n=1}^{\infty} \frac{c_k(qn)f(n)}{n^s}.$$

For example, in the simplest case $f(n) = 1$, the series is

$$\sum_{n=1}^{\infty} \frac{1}{n^s} \sum_{\delta|k,\delta|qn} \delta\mu\left(\frac{k}{\delta}\right).$$

For given δ, n runs through those multiples of δ/q which are integers. If δ/q in its lowest terms is δ_1/q_1, these are the numbers δ_1, $2\delta_1$,.... Hence the sum is

$$\sum_{\delta|k} \delta\mu\left(\frac{k}{\delta}\right) \sum_{r=1}^{\infty} \frac{1}{(r\delta_1)^s} = \zeta(s) \sum_{\delta|k} \delta\mu\left(\frac{k}{\delta}\right)\delta_1^{-s}.$$

† Crum (1).

Since $\delta_1 = \delta/(q,\delta)$, the result is

$$\sum_{n=1}^{\infty} \frac{c_k(qn)}{n^s} = \zeta(s) \sum_{\delta|k} \delta^{1-s}\mu\left(\frac{k}{\delta}\right)(q,\delta)^s. \tag{1.5.7}$$

1.6. There is another class of identities involving infinite series of zeta-functions. The simplest of these is†

$$\sum_p \frac{1}{p^s} = \sum_{n=1}^{\infty} \frac{\mu(n)}{n} \log\zeta(ns). \tag{1.6.1}$$

We have
$$\log\zeta(s) = \sum_m \sum_p \frac{1}{mp^{ms}} = \sum_{m=1}^{\infty} \frac{P(ms)}{m},$$

where $P(s) = \sum p^{-s}$. Hence

$$\sum_{n=1}^{\infty} \frac{\mu(n)}{n} \log\zeta(ns) = \sum_{n=1}^{\infty} \frac{\mu(n)}{n} \sum_{m=1}^{\infty} \frac{P(mns)}{m} = \sum_{r=1}^{\infty} \frac{P(rs)}{r} \sum_{n|r} \mu(n),$$

and the result follows from (1.1.5).

A closely related formula is

$$\sum_{n=1}^{\infty} \frac{\nu(n)}{n^s} = \zeta(s) \sum_{n=1}^{\infty} \frac{\mu(n)}{n} \log\zeta(ns), \tag{1.6.2}$$

where $\nu(n)$ is defined under (1.2.8). This follows at once from (1.6.1) and the identity

$$\sum_{n=1}^{\infty} \frac{\nu(n)}{n^s} = \sum_{m=1}^{\infty} \frac{1}{m^s} \sum_p \frac{1}{p^s}.$$

Denoting by $b(n)$ the number of divisors of n which are primes or powers of primes, another identity of the same class is

$$\sum_{n=1}^{\infty} \frac{b(n)}{n^s} = \zeta(s) \sum_{n=1}^{\infty} \frac{\phi(n)}{n} \log\zeta(ns), \tag{1.6.3}$$

where $\phi(n)$ is defined under (1.2.12). For the left-hand side is equal to

$$\sum_{m=1}^{\infty} \frac{1}{m^s} \sum_p \left(\frac{1}{p^s} + \frac{1}{p^{2s}} + \frac{1}{p^{3s}} + \ldots\right),$$

and the series on the right is

$$\sum_{n=1}^{\infty} \frac{\phi(n)}{n} \sum_{m=1}^{\infty} \sum_p \frac{1}{mp^{mns}} = \sum_p \sum_\nu \frac{1}{\nu p^{\nu s}} \sum_{n|\nu} \phi(n).$$

Since
$$\sum_{n|\nu} \phi(n) = \nu,$$

the result follows.

† See Landau and Walfisz (1), Estermann (1), (2).

II

THE ANALYTIC CHARACTER OF $\zeta(s)$, AND THE FUNCTIONAL EQUATION

2.1. Analytic continuation and the functional equation, first method. Each of the formulae of Chapter I is proved on the supposition that the series or product concerned is absolutely convergent. In each case this restricts the region where the formula is proved to be valid to a half-plane. For $\zeta(s)$ itself, and in all the fundamental formulae of § 1.1, this is the half-plane $\sigma > 1$.

We have next to inquire whether the analytic function $\zeta(s)$ can be continued beyond this region. The result is

THEOREM 2.1. *The function $\zeta(s)$ is regular for all values of s except $s = 1$, where there is a simple pole with residue 1. It satisfies the functional equation*

$$\zeta(s) = 2^s\pi^{s-1}\sin\tfrac{1}{2}s\pi\Gamma(1-s)\zeta(1-s). \tag{2.1.1}$$

This can be proved in a considerable variety of different ways, some of which will be given in later sections. We shall first give a proof depending on the following summation formula.

Let $\phi(x)$ be any function with a continuous derivative in the interval $[a,b]$. Then, if $[x]$ denotes the greatest integer not exceeding x,

$$\sum_{a<n\leqslant b}\phi(n) = \int_a^b \phi(x)\,dx + \int_a^b (x-[x]-\tfrac{1}{2})\phi'(x)\,dx + \\ +(a-[a]-\tfrac{1}{2})\phi(a)-(b-[b]-\tfrac{1}{2})\phi(b). \tag{2.1.2}$$

Since the formula is plainly additive with respect to the interval $(a,b]$ it suffices to suppose that $n \leqslant a < b \leqslant n+1$. One then has

$$\int_a^b (x-n-\tfrac{1}{2})\phi'(x)\,dx = (b-n-\tfrac{1}{2})\phi(b)-(a-n-\tfrac{1}{2})\phi(a)-\int_a^b \phi(x)dx,$$

on integrating by parts. Thus the right hand side of (2.1.2) reduces to $([b]-n)\phi(b)$. This vanishes unless $b = n+1$, in which case it is $\phi(n+1)$, as required.

In particular, let $\phi(n) = n^{-s}$, where $s \neq 1$, and let a and b be positive integers. Then

$$\sum_{n=a+1}^{b} \frac{1}{n^s} = \frac{b^{1-s}-a^{1-s}}{1-s} - s\int_a^b \frac{x-[x]-\frac{1}{2}}{x^{s+1}}\,dx + \tfrac{1}{2}(b^{-s}-a^{-s}). \quad (2.1.3)$$

First take $\sigma > 1$, $a = 1$, and make $b \to \infty$. Adding 1 to each side, we obtain

$$\zeta(s) = s\int_1^\infty \frac{[x]-x+\frac{1}{2}}{x^{s+1}}\,dx + \frac{1}{s-1} + \frac{1}{2}. \quad (2.1.4)$$

Since $[x]-x+\frac{1}{2}$ is bounded, this integral is convergent for $\sigma > 0$, and uniformly convergent in any finite region to the right of $\sigma = 0$. It therefore defines an analytic function of s, regular for $\sigma > 0$. The right-hand side therefore provides the analytic continuation of $\zeta(s)$ up to $\sigma = 0$, and there is clearly a simple pole at $s = 1$ with residue 1.

For $0 < \sigma < 1$ we have

$$\int_0^1 \frac{[x]-x}{x^{s+1}}\,dx = -\int_0^1 x^{-s}\,dx = \frac{1}{s-1}, \qquad \frac{s}{2}\int_1^\infty \frac{dx}{x^{s+1}} = \frac{1}{2},$$

and (2.1.4) may be written

$$\zeta(s) = s\int_0^\infty \frac{[x]-x}{x^{s+1}}\,dx \quad (0 < \sigma < 1). \quad (2.1.5)$$

Actually (2.1.4) gives the analytic continuation of $\zeta(s)$ for $\sigma > -1$; for if

$$f(x) = [x]-x+\tfrac{1}{2}, \qquad f_1(x) = \int_1^x f(y)\,dy,$$

then $f_1(x)$ is also bounded, since, as is easily seen,

$$\int_k^{k+1} f(y)\,dy = 0$$

for any integer k. Hence

$$\int_{x_1}^{x_2} \frac{f(x)}{x^{s+1}}\,dx = \left[\frac{f_1(x)}{x^{s+1}}\right]_{x_1}^{x_2} + (s+1)\int_{x_1}^{x_2} \frac{f_1(x)}{x^{s+2}}\,dx,$$

which tends to 0 as $x_1 \to \infty$, $x_2 \to \infty$, if $\sigma > -1$. Hence the integral in (2.1.4) is convergent for $\sigma > -1$. Also it is easily verified that

$$s\int\limits_0^1 \frac{[x]-x+\frac{1}{2}}{x^{s+1}}\,dx = \frac{1}{s-1}+\frac{1}{2} \quad (\sigma < 0).$$

Hence
$$\zeta(s) = s\int\limits_0^\infty \frac{[x]-x+\frac{1}{2}}{x^{s+1}}\,dx \quad (-1 < \sigma < 0). \tag{2.1.6}$$

Now we have the Fourier series

$$[x]-x+\tfrac{1}{2} = \sum_{n=1}^\infty \frac{\sin 2n\pi x}{n\pi}, \tag{2.1.7}$$

where x is not an integer. Substituting in (2.1.6), and integrating term by term, we obtain

$$\begin{aligned}\zeta(s) &= \frac{s}{\pi}\sum_{n=1}^\infty \frac{1}{n}\int\limits_0^\infty \frac{\sin 2n\pi x}{x^{s+1}}\,dx \\ &= \frac{s}{\pi}\sum_{n=1}^\infty \frac{(2n\pi)^s}{n}\int\limits_0^\infty \frac{\sin y}{y^{s+1}}\,dy \\ &= \frac{s}{\pi}(2\pi)^s\{-\Gamma(-s)\}\sin\tfrac{1}{2}s\pi\,\zeta(1-s),\end{aligned}$$

i.e. (2.1.1). This is valid primarily for $-1 < \sigma < 0$. Here, however, the right-hand side is analytic for all values of s such that $\sigma < 0$. It therefore provides the analytic continuation of $\zeta(s)$ over the remainder of the plane, and there are no singularities other than the pole already encountered at $s = 1$.

We have still to justify the term-by-term integration. Since the series (2.1.7) is boundedly convergent, term-by-term integration over any finite range is permissible. It is therefore sufficient to prove that

$$\lim_{\lambda\to\infty}\sum_{n=1}^\infty \frac{1}{n}\int\limits_\lambda^\infty \frac{\sin 2n\pi x}{x^{s+1}}\,dx = 0 \quad (-1 < \sigma < 0).$$

Now
$$\begin{aligned}\int\limits_\lambda^\infty \frac{\sin 2n\pi x}{x^{s+1}}\,dx &= \left[-\frac{\cos 2n\pi x}{2n\pi x^{s+1}}\right]_\lambda^\infty - \frac{s+1}{2n\pi}\int\limits_\lambda^\infty \frac{\cos 2n\pi x}{x^{s+2}}\,dx \\ &= O\left(\frac{1}{n\lambda^{\sigma+1}}\right)+O\left(\frac{1}{n}\int\limits_\lambda^\infty \frac{dx}{x^{\sigma+2}}\right) = O\left(\frac{1}{n\lambda^{\sigma+1}}\right),\end{aligned}$$

and the desired result clearly follows.

The functional equation (2.1.1) may be written in a number of different ways. Changing s into $1-s$, it is

$$\zeta(1-s) = 2^{1-s}\pi^{-s}\cos\tfrac{1}{2}s\pi\Gamma(s)\zeta(s). \tag{2.1.8}$$

It may also be written

$$\zeta(s) = \chi(s)\zeta(1-s), \tag{2.1.9}$$

where
$$\chi(s) = 2^s\pi^{s-1}\sin\tfrac{1}{2}s\pi\Gamma(1-s) = \pi^{s-\frac{1}{2}}\frac{\Gamma(\frac{1}{2}-\frac{1}{2}s)}{\Gamma(\frac{1}{2}s)}, \tag{2.1.10}$$

and
$$\chi(s)\chi(1-s) = 1. \tag{2.1.11}$$

Writing
$$\xi(s) = \tfrac{1}{2}s(s-1)\pi^{-\frac{1}{2}s}\Gamma(\tfrac{1}{2}s)\zeta(s), \tag{2.1.12}$$

it is at once verified from (2.1.8) and (2.1.9) that

$$\xi(s) = \xi(1-s). \tag{2.1.13}$$

Writing
$$\Xi(z) = \xi(\tfrac{1}{2}+iz) \tag{2.1.14}$$

we obtain
$$\Xi(z) = \Xi(-z). \tag{2.1.15}$$

The functional equation is therefore equivalent to the statement that $\Xi(z)$ is an even function of z.

The approximation near $s = 1$ can be carried a stage farther; we have

$$\zeta(s) = \frac{1}{s-1}+\gamma+O(|s-1|), \tag{2.1.16}$$

where γ is Euler's constant. For by (2.1.4)

$$\begin{aligned}
\lim_{s\to 1}\left\{\zeta(s)-\frac{1}{s-1}\right\} &= \int_1^\infty \frac{[x]-x+\frac{1}{2}}{x^2}\,dx+\tfrac{1}{2} \\
&= \lim_{n\to\infty}\int_1^n \frac{[x]-x}{x^2}\,dx+1 \\
&= \lim_{n\to\infty}\left\{\sum_{m=1}^{n-1} m\int_m^{m+1}\frac{dx}{x^2}-\log n+1\right\} \\
&= \lim_{n\to\infty}\left\{\sum_{m=1}^{n-1}\frac{1}{m+1}+1-\log n\right\} = \gamma.
\end{aligned}$$

2.2. A considerable number of variants of the above proof of the functional equation have been given. A similar argument was applied by Hardy,† not to $\zeta(s)$ itself, but to the function

$$\sum_{n=1}^\infty \frac{(-1)^{n-1}}{n^s} = (1-2^{1-s})\zeta(s). \tag{2.2.1}$$

† Hardy (6).

This Dirichlet series is convergent for all real positive values of s, and so, by a general theorem on the convergence of Dirichlet series, for all values of s such that $\sigma > 0$. Here, of course, the pole of $\zeta(s)$ at $s = 1$ is cancelled by the zero of the other factor. These facts enable us to simplify the discussion in some respects.

Hardy's proof runs as follows. Let

$$f(x) = \sum_{n=0}^{\infty} \frac{\sin(2n+1)x}{2n+1}.$$

This series is boundedly convergent and

$$f(x) = (-1)^m \tfrac{1}{4}\pi \quad \text{for} \quad m\pi < x < (m+1)\pi \quad (m = 0, 1, \ldots).$$

Multiplying by x^{s-1} $(0 < s < 1)$, and integrating over $(0, \infty)$, we obtain

$$\tfrac{1}{4}\pi \sum_{m=0}^{\infty} (-1)^m \int_{m\pi}^{(m+1)\pi} x^{s-1}\,dx = \Gamma(s)\sin\tfrac{1}{2}s\pi \sum_{n=0}^{\infty} \frac{1}{(2n+1)^{s+1}}$$
$$= \Gamma(s)\sin\tfrac{1}{2}s\pi(1-2^{-s-1})\zeta(s+1).$$

The term-by-term integration may be justified as in the previous proof. The series on the left is

$$\frac{\pi^s}{s}\Big[1+\sum_{m=1}^{\infty}(-1)^m\{(m+1)^s-m^s\}\Big].$$

This series is convergent for $s < 1$, and, as a little consideration of the above argument shows, uniformly convergent for $\mathbf{R}(s) \leqslant 1-\delta < 1$. Its sum is therefore an analytic function of s, regular for $\mathbf{R}(s) < 1$. But for $s < 0$ it is

$$2(1^s-2^s+3^s-\ldots) = 2(1-2^{s+1})\zeta(-s).$$

Its sum is therefore the same analytic function of s for $\mathbf{R}(s) < 1$. Hence, for $0 < s < 1$,

$$\frac{\pi^{s+1}}{2s}(1-2^{s+1})\zeta(-s) = \Gamma(s)\sin\tfrac{1}{2}s\pi(1-2^{-s-1})\zeta(s+1),$$

and the functional equation again follows.

2.3. Still another proof is based on Poisson's summation formula

$$\sum_{n=-\infty}^{\infty} f(n) = \sum_{-\infty}^{\infty} \int_{-\infty}^{\infty} f(u)\cos 2\pi nu\,du. \tag{2.3.1}$$

If we put $f(x) = |x|^{-s}$ and ignore all questions of convergence, we obtain the result formally at once. The proof may be established in various ways. If we integrate by parts to obtain integrals involving $\sin 2\pi nu$, we obtain a proof not fundamentally distinct from the first proof given here.† The formula can also be used to give a proof depending‡ on $(1-2^{1-s})\zeta(s)$.

Actually cases of Poisson's formula enter into several of the following proofs; (2.6.3) and (2.8.2) are both cases of Poisson's formula.

2.4. Second method. The whole theory can be developed in another way, which is one of Riemann's methods. Here the fundamental formula is

$$\zeta(s) = \frac{1}{\Gamma(s)}\int\limits_0^\infty \frac{x^{s-1}}{e^x-1}\,dx \quad (\sigma > 1). \tag{2.4.1}$$

To prove this, we have for $\sigma > 0$

$$\int\limits_0^\infty x^{s-1}e^{-nx}\,dx = \frac{1}{n^s}\int\limits_0^\infty y^{s-1}e^{-y}\,dy = \frac{\Gamma(s)}{n^s}.$$

Hence

$$\Gamma(s)\zeta(s) = \sum_{n=1}^\infty \int\limits_0^\infty x^{s-1}e^{-nx}\,dx = \int\limits_0^\infty x^{s-1}\sum_{n=1}^\infty e^{-nx}\,dx = \int\limits_0^\infty \frac{x^{s-1}}{e^x-1}\,dx$$

if the inversion of the order of summation and integration can be justified; and this is so by absolute convergence if $\sigma > 1$, since

$$\sum_{n=1}^\infty \int\limits_0^\infty x^{\sigma-1}e^{-nx}\,dx = \Gamma(\sigma)\zeta(\sigma)$$

is convergent for $\sigma > 1$.

Now consider the integral

$$I(s) = \int\limits_C \frac{z^{s-1}}{e^z-1}\,dz,$$

where the contour C starts at infinity on the positive real axis, encircles the origin once in the positive direction, excluding the points $\pm 2i\pi$, $\pm 4i\pi$,..., and returns to positive infinity. Here z^{s-1} is defined as

$$e^{(s-1)\log z}$$

when the logarithm is real at the beginning of the contour; thus $\mathbf{I}(\log z)$ varies from 0 to 2π round the contour.

We can take C to consist of the real axis from ∞ to ρ $(0 < \rho < 2\pi)$, the circle $|z| = \rho$, and the real axis from ρ to ∞. On the circle,

$$|z^{s-1}| = e^{(\sigma-1)\log|z|-t\arg z} \leqslant |z|^{\sigma-1}e^{2\pi|t|},$$

$$|e^z-1| > A|z|.$$

† Mordell (2). ‡ Ingham, *Prime Numbers*, 46.

Hence the integral round this circle tends to zero with ρ if $\sigma > 1$. On making $\rho \to 0$ we therefore obtain

$$I(s) = -\int_0^\infty \frac{x^{s-1}}{e^x-1}\,dx + \int_0^\infty \frac{(xe^{2\pi i})^{s-1}}{e^x-1}\,dx$$

$$= (e^{2\pi is}-1)\Gamma(s)\zeta(s)$$

$$= \frac{2i\pi e^{i\pi s}}{\Gamma(1-s)}\zeta(s).$$

Hence
$$\zeta(s) = \frac{e^{-i\pi s}\Gamma(1-s)}{2\pi i} \int_C \frac{z^{s-1}}{e^z-1}\,dz. \tag{2.4.2}$$

This formula has been proved for $\sigma > 1$. The integral $I(s)$, however, is uniformly convergent in any finite region of the s-plane, and so defines an integral function of s. Hence the formula provides the analytic continuation of $\zeta(s)$ over the whole s-plane. The only possible singularities are the poles of $\Gamma(1-s)$, viz. $s = 1, 2, 3,\ldots$. We know already that $\zeta(s)$ is regular at $s = 2, 3,\ldots$, and in fact it follows at once from Cauchy's theorem that $I(s)$ vanishes at these points. Hence the only possible singularity is a simple pole at $s = 1$. Here

$$I(1) = \int_C \frac{dz}{e^z-1} = 2\pi i,$$

and
$$\Gamma(1-s) = -\frac{1}{s-1}+\ldots.$$

Hence the residue at the pole is 1.

If s is any integer, the integrand in $I(s)$ is one-valued, and $I(s)$ can be evaluated by the theorem of residues. Since

$$\frac{z}{e^z-1} = 1-\tfrac{1}{2}z+B_1\frac{z^2}{2!}-B_2\frac{z^4}{4!}+\ldots,$$

where B_1, B_2,... are Bernoulli's numbers, we find the following values of $\zeta(s)$:

$$\zeta(0) = -\tfrac{1}{2}, \qquad \zeta(-2m) = 0, \qquad \zeta(1-2m) = \frac{(-1)^m B_m}{2m} \quad (m = 1, 2,\ldots). \tag{2.4.3}$$

To deduce the functional equation from (2.4.2), take the integral along the contour C_n consisting of the positive real axis from infinity to $(2n+1)\pi$, then round the square with corners $(2n+1)\pi(\pm 1 \pm i)$, and then back to infinity along the positive real axis. Between the contours

C and C_n the integrand has poles at the points $\pm 2i\pi,\ldots, \pm 2in\pi$. The residues at $2mi\pi$ and $-2mi\pi$ are together

$$(2m\pi e^{\frac{1}{2}i\pi})^{s-1}+(2m\pi e^{\frac{3}{2}i\pi})^{s-1} = (2m\pi)^{s-1}e^{i\pi(s-1)}2\cos\tfrac{1}{2}\pi(s-1)$$
$$= -2(2m\pi)^{s-1}e^{i\pi s}\sin\tfrac{1}{2}\pi s.$$

Hence by the theorem of residues

$$I(s) = \int\limits_{C_n} \frac{z^{s-1}}{e^z-1}\,dz+4\pi ie^{i\pi s}\sin\tfrac{1}{2}\pi s\sum_{m=1}^{n}(2m\pi)^{s-1}.$$

Now let $\sigma < 0$ and make $n \to \infty$. The function $1/(e^z-1)$ is bounded on the contours C_n, and $z^{s-1} = O(|z|^{\sigma-1})$. Hence the integral round C_n tends to zero, and we obtain

$$I(s) = 4\pi ie^{i\pi s}\sin\tfrac{1}{2}\pi s\sum_{m=1}^{\infty}(2m\pi)^{s-1}$$
$$= 4\pi ie^{i\pi s}\sin\tfrac{1}{2}\pi s(2\pi)^{s-1}\zeta(1-s).$$

The functional equation now follows again.

Two minor consequences of the functional equation may be noted here. The formula

$$\zeta(2m) = 2^{2m-1}\pi^{2m}\frac{B_m}{(2m)!} \quad (m = 1, 2,\ldots) \tag{2.4.4}$$

follows from the functional equation (2.1.1), with $s = 1-2m$, and the value obtained above for $\zeta(1-2m)$. Also

$$\zeta'(0) = -\tfrac{1}{2}\log 2\pi. \tag{2.4.5}$$

For the functional equation gives

$$-\frac{\zeta'(1-s)}{\zeta(1-s)} = -\log 2\pi-\tfrac{1}{2}\pi\tan\tfrac{1}{2}s\pi+\frac{\Gamma'(s)}{\Gamma(s)}+\frac{\zeta'(s)}{\zeta(s)}.$$

In the neighbourhood of $s = 1$

$$\tfrac{1}{2}\pi\tan\tfrac{1}{2}s\pi = -\frac{1}{s-1}+O(|s-1|), \qquad \frac{\Gamma'(s)}{\Gamma(s)} = \frac{\Gamma'(1)}{\Gamma(1)}+\ldots = -\gamma+\ldots,$$

and $$\frac{\zeta'(s)}{\zeta(s)} = \frac{-\{1/(s-1)^2\}+k+\ldots}{\{1/(s-1)\}+\gamma+k(s-1)+\ldots} = -\frac{1}{s-1}+\gamma+\ldots.$$

where k is a constant. Hence, making $s \to 1$, we obtain

$$-\frac{\zeta'(0)}{\zeta(0)} = -\log 2\pi,$$

and (2.4.5) follows.

2.5. Validity of (2.2.1) for all s. The original series (1.1.1) is naturally valid for $\sigma > 1$ only, on account of the pole at $s = 1$. The series (2.2.1) is convergent, and represents $(1-2^{1-s})\zeta(s)$, for $\sigma > 0$. This series ceases to converge on $\sigma = 0$, but there is nothing in the nature of the function represented to account for this. In fact if we use summability instead of ordinary convergence the equation still holds to the left of $\sigma = 0$.

THEOREM 2.5. *The series* $\sum_{n=1}^{\infty}(-1)^{n-1}n^{-s}$ *is summable* (A) *to the sum* $(1-2^{1-s})\zeta(s)$ *for all values of* s.

Let $0 < x < 1$. Then

$$\sum_{n=1}^{\infty}\frac{(-1)^{n-1}}{n^s}x^n = \sum_{n=1}^{\infty}\frac{(-1)^{n-1}x^n}{\Gamma(s)}\int_0^{\infty}e^{-nu}u^{s-1}\,du$$

$$=\frac{1}{\Gamma(s)}\int_0^{\infty}u^{s-1}\sum_{n=1}^{\infty}(-1)^{n-1}x^ne^{-nu}\,du = \frac{1}{\Gamma(s)}\int_0^{\infty}u^{s-1}\frac{xe^{-u}}{1+xe^{-u}}\,du.$$

This is justified by absolute convergence for $\sigma > 1$, and the result by analytic continuation for $\sigma > 0$.

We can now replace this by a loop-integral in the same way as (2.4.2) was obtained from (2.4.1). We obtain

$$\sum_{n=1}^{\infty}\frac{(-1)^{n-1}}{n^s}x^n = \frac{e^{-i\pi s}\Gamma(1-s)}{2\pi i}\int_C w^{s-1}\frac{xe^{-w}}{1+xe^{-w}}\,dw,$$

when C encircles the origin as before, but excludes all zeros of $1+xe^{-w}$, i.e. the points $w = \log x+(2m+1)i\pi$.

It is clear that, as $x \to 1$, the right-hand side tends to a limit, uniformly in any finite region of the s-plane excluding positive integers; and, by the theory of analytic continuation, the limit must be $(1-2^{1-s})\zeta(s)$. This proves the theorem except if s is a positive integer, when the proof is elementary.

Similar results hold for other methods of summation.

2.6. Third method. This is also one of Riemann's original proofs. We observe that if $\sigma > 0$

$$\int_0^{\infty}x^{\frac{1}{2}s-1}e^{-n^2\pi x}\,dx = \frac{\Gamma(\frac{1}{2}s)}{n^s\pi^{\frac{1}{2}s}}.$$

Hence if $\sigma > 1$

$$\frac{\Gamma(\frac{1}{2}s)\zeta(s)}{\pi^{\frac{1}{2}s}} = \sum_{n=1}^{\infty}\int_0^{\infty}x^{\frac{1}{2}s-1}e^{-n^2\pi x}\,dx = \int_0^{\infty}x^{\frac{1}{2}s-1}\sum_{n=1}^{\infty}e^{-n^2\pi x}\,dx,$$

the inversion being justified by absolute convergence, as in § 2.4.

Writing

$$\psi(x) = \sum_{n=1}^{\infty} e^{-n^2\pi x} \tag{2.6.1}$$

we therefore have

$$\zeta(s) = \frac{\pi^{\frac{1}{2}s}}{\Gamma(\frac{1}{2}s)} \int_0^\infty x^{\frac{1}{2}s-1}\psi(x)\, dx \quad (\sigma > 1). \tag{2.6.2}$$

Now it is known that, for $x > 0$,

$$\sum_{n=-\infty}^{\infty} e^{-n^2\pi x} = \frac{1}{\sqrt{x}} \sum_{n=-\infty}^{\infty} e^{-n^2\pi/x},$$

$$2\psi(x)+1 = \frac{1}{\sqrt{x}}\left\{2\psi\left(\frac{1}{x}\right)+1\right\}. \tag{2.6.3}$$

Hence (2.6.2) gives

$$\pi^{-\frac{1}{2}s}\Gamma(\tfrac{1}{2}s)\zeta(s) = \int_0^1 x^{\frac{1}{2}s-1}\psi(x)\, dx + \int_1^\infty x^{\frac{1}{2}s-1}\psi(x)\, dx$$

$$= \int_0^1 x^{\frac{1}{2}s-1}\left\{\frac{1}{\sqrt{x}}\psi\left(\frac{1}{x}\right)+\frac{1}{2\sqrt{x}}-\frac{1}{2}\right\} dx + \int_1^\infty x^{\frac{1}{2}s-1}\psi(x)\, dx$$

$$= \frac{1}{s-1}-\frac{1}{s}+\int_0^1 x^{\frac{1}{2}s-\frac{3}{2}}\psi\left(\frac{1}{x}\right) dx + \int_1^\infty x^{\frac{1}{2}s-1}\psi(x)\, dx$$

$$= \frac{1}{s(s-1)}+\int_1^\infty (x^{-\frac{1}{2}s-\frac{1}{2}}+x^{\frac{1}{2}s-1})\psi(x)\, dx.$$

The last integral is convergent for all values of s, and so the formula holds, by analytic continuation, for all values of s. Now the right-hand side is unchanged if s is replaced by $1-s$. Hence

$$\pi^{-\frac{1}{2}s}\Gamma(\tfrac{1}{2}s)\zeta(s) = \pi^{-\frac{1}{2}+\frac{1}{2}s}\Gamma(\tfrac{1}{2}-\tfrac{1}{2}s)\zeta(1-s), \tag{2.6.4}$$

which is a form of the functional equation.

2.7. Fourth method; proof by self-reciprocal functions. Still another proof of the functional equation is as follows. For $\sigma > 1$, (2.4.1) may be written

$$\zeta(s)\Gamma(s) = \int_0^1 \left(\frac{1}{e^x-1}-\frac{1}{x}\right)x^{s-1}\, dx + \frac{1}{s-1} + \int_1^\infty \frac{x^{s-1}\, dx}{e^x-1},$$

and this holds by analytic continuation for $\sigma > 0$. Also for $0 < \sigma < 1$

$$\frac{1}{s-1} = -\int_1^\infty \frac{x^{s-1}}{x}\, dx.$$

Hence $$\zeta(s)\Gamma(s) = \int_0^\infty \left(\frac{1}{e^x-1}-\frac{1}{x}\right)x^{s-1}\,dx \quad (0<\sigma<1). \tag{2.7.1}$$

Now it is known that the function

$$f(x) = \frac{1}{e^{x\sqrt{(2\pi)}}-1}-\frac{1}{x\sqrt{(2\pi)}} \tag{2.7.2}$$

is self-reciprocal for sine transforms, i.e. that

$$f(x) = \sqrt{\left(\frac{2}{\pi}\right)}\int_0^\infty f(y)\sin xy\,dy. \tag{2.7.3}$$

Hence, putting $x = \xi\sqrt{(2\pi)}$ in (2.7.1),

$$\zeta(s)\Gamma(s) = (2\pi)^{\frac{1}{2}s}\int_0^\infty f(\xi)\xi^{s-1}\,d\xi$$

$$= (2\pi)^{\frac{1}{2}s}\sqrt{\left(\frac{2}{\pi}\right)}\int_0^\infty \xi^{s-1}\,d\xi\int_0^\infty f(y)\sin\xi y\,dy.$$

If we can invert the order of integration, this is

$$2^{\frac{1}{2}s+\frac{1}{2}}\pi^{\frac{1}{2}s-\frac{1}{2}}\int_0^\infty f(y)\,dy\int_0^\infty \xi^{s-1}\sin\xi y\,d\xi$$

$$= 2^{\frac{1}{2}s+\frac{1}{2}}\pi^{\frac{1}{2}s-\frac{1}{2}}\int_0^\infty f(y)y^{-s}\,dy\int_0^\infty u^{s-1}\sin u\,du$$

$$= 2^{\frac{1}{2}s+\frac{1}{2}}\pi^{\frac{1}{2}s-\frac{1}{2}}(2\pi)^{\frac{1}{2}s-\frac{1}{2}}\Gamma(1-s)\zeta(1-s)\,\frac{\pi}{2\cos\frac{1}{2}\pi s\Gamma(1-s)},$$

and the functional equation again follows.

To justify the inversion, we observe that the integral

$$\int_0^\infty f(y)\sin\xi y\,dy$$

converges uniformly over $0<\delta\leqslant\xi\leqslant\Delta$. Hence the inversion of this part is valid, and it is sufficient to prove that

$$\lim_{\substack{\delta\to 0\\ \Delta\to\infty}}\int_0^\infty f(y)\,dy\left(\int_0^\delta+\int_\Delta^\infty\right)\xi^{s-1}\sin\xi y\,d\xi = 0.$$

Now $$\int_0^\delta \xi^{s-1}\sin\xi y\,d\xi = \int_0^\delta O(\xi^{\sigma-1}\,\xi y)\,d\xi = O(\delta^{\sigma+1}y)$$

and also $$= y^{-s}\int_0^{\delta y} u^{s-1}\sin u\,du = O(y^{-\sigma}).$$

Since $f(y) = O(1)$ as $y \to 0$, and $= O(y^{-1})$ as $y \to \infty$, we obtain

$$\int_0^\infty f(y)\,dy \int_0^\delta \xi^{s-1}\sin\xi y\,d\xi$$

$$= \int_0^1 O(\delta^{\sigma+1}y)\,dy + \int_1^{1/\delta} O(\delta^{\sigma+1})\,dy + \int_{1/\delta}^\infty O(y^{-\sigma-1})\,dy = O(\delta^\sigma) \to 0.$$

A similar method shows that the integral involving Δ also tends to 0.

2.8. Fifth method. The process by which (2.7.1) was obtained from (2.4.1) can be extended indefinitely. For the next stage, (2.7.1) gives

$$\Gamma(s)\zeta(s) = \int_0^1 \left(\frac{1}{e^x-1} - \frac{1}{x} + \frac{1}{2}\right)x^{s-1}\,dx - \frac{1}{2s} + \int_1^\infty \left(\frac{1}{e^x-1} - \frac{1}{x}\right)x^{s-1}\,dx,$$

and this holds by analytic continuation for $\sigma > -1$. But

$$\int_1^\infty \tfrac{1}{2}x^{s-1}\,dx = -\frac{1}{2s} \quad (-1 < \sigma < 0).$$

Hence

$$\Gamma(s)\zeta(s) = \int_0^\infty \left(\frac{1}{e^x-1} - \frac{1}{x} + \frac{1}{2}\right)x^{s-1}\,dx \quad (-1 < \sigma < 0). \tag{2.8.1}$$

Now

$$\frac{1}{e^x-1} = \frac{1}{x} - \frac{1}{2} + 2x\sum_{n=1}^\infty \frac{1}{4n^2\pi^2+x^2}. \tag{2.8.2}$$

Hence

$$\Gamma(s)\zeta(s) = \int_0^\infty 2x\sum_{n=1}^\infty \frac{1}{4n^2\pi^2+x^2}x^{s-1}\,dx = 2\sum_{n=1}^\infty \int_0^\infty \frac{x^s}{4n^2\pi^2+x^2}\,dx$$

$$= 2\sum_{n=1}^\infty (2n\pi)^{s-1}\frac{\pi}{2\cos\frac{1}{2}s\pi} = \frac{2^{s-1}\pi^s}{\cos\frac{1}{2}s\pi}\zeta(1-s),$$

the functional equation. The inversion is justified by absolute convergence if $-1 < \sigma < 0$.

2.9. Sixth method. The formula†

$$\zeta(s) = \frac{e^{i\pi s}}{2\pi i}\int_{c-i\infty}^{c+i\infty} \left\{\frac{\Gamma'(1+z)}{\Gamma(1+z)} - \log z\right\}z^{-s}\,dz \quad (-1 < c < 0) \tag{2.9.1}$$

is easily proved by the calculus of residues if $\sigma > 1$; and the integrand is $O(|z|^{-\sigma-1})$, so that the integral is convergent, and the formula holds by analytic continuation, if $\sigma > 0$.

† Kloosterman (1).

We may next transform this into an integral along the positive real axis after the manner of § 2.4. We obtain

$$\zeta(s) = -\frac{\sin \pi s}{\pi} \int_0^\infty \left\{\frac{\Gamma'(1+x)}{\Gamma(1+x)} - \log x\right\} x^{-s}\, dx \quad (0 < \sigma < 1). \tag{2.9.2}$$

To deduce the functional equation, we observe that†

$$\frac{\Gamma'(x)}{\Gamma(x)} = \log x - \frac{1}{2x} - 2\int_0^\infty \frac{t\, dt}{(t^2+x^2)(e^{2\pi t}-1)}.$$

Hence

$$\frac{\Gamma'(1+x)}{\Gamma(1+x)} - \log x = \frac{\Gamma'(x)}{\Gamma(x)} + \frac{1}{x} - \log x$$

$$= \frac{1}{2x} - 2\int_0^\infty \frac{t\, dt}{(t^2+x^2)(e^{2\pi t}-1)} = -2\int_0^\infty \frac{t}{t^2+x^2}\left(\frac{1}{e^{2\pi t}-1} - \frac{1}{2\pi t}\right) dt.$$

Hence (2.9.2) gives

$$\zeta(s) = \frac{2\sin \pi s}{\pi} \int_0^\infty x^{-s}\, dx \int_0^\infty \frac{t}{t^2+x^2}\left(\frac{1}{e^{2\pi t}-1} - \frac{1}{2\pi t}\right) dt$$

$$= \frac{2\sin \pi s}{\pi} \int_0^\infty \left(\frac{1}{e^{2\pi t}-1} - \frac{1}{2\pi t}\right) t\, dt \int_0^\infty \frac{x^{-s}}{t^2+x^2}\, dx$$

$$= \frac{\sin \pi s}{\cos \frac{1}{2}\pi s} \int_0^\infty \left(\frac{1}{e^{2\pi t}-1} - \frac{1}{2\pi t}\right) t^{-s}\, dt$$

$$= 2\sin \tfrac{1}{2}\pi s (2\pi)^{s-1} \int_0^\infty \left(\frac{1}{e^{u}-1} - \frac{1}{u}\right) u^{-s}\, du$$

$$= 2\sin \tfrac{1}{2}\pi s (2\pi)^{s-1} \Gamma(1-s)\zeta(1-s)$$

by (2.7.1). The inversion is justified by absolute convergence.

2.10. Seventh method. Still another method of dealing with $\zeta(s)$, due to Riemann, has been carried out in detail by Siegel.‡ It depends on the evaluation of the following infinite integral.

Let
$$\Phi(a) = \int_L \frac{e^{iw^2/(4\pi)+aw}}{e^w - 1}\, dw, \tag{2.10.1}$$

where L is a straight line inclined at an angle $\frac{1}{4}\pi$ to the real axis, and

† Whittaker and Watson, § 12.32, example. ‡ Siegel (2).

intersecting the imaginary axis between O and $2\pi i$. The integral is plainly convergent for all values of a.

We have

$$\Phi(a+1)-\Phi(a) = \int\limits_L \frac{e^{\frac{1}{4}iw^2/\pi}}{e^w-1}(e^{(a+1)w}-e^{aw})\,dw$$

$$= \int\limits_L e^{\frac{1}{4}iw^2/\pi+aw}\,dw$$

$$= \int\limits_L e^{\frac{1}{4}i(w-2i\pi a)^2/\pi+i\pi a^2}\,dw$$

$$= e^{i\pi a^2}\int e^{\frac{1}{4}iW^2/\pi}\,dW,$$

where $W = w-2i\pi a$. Here we may move the contour to the parallel line through the origin, so that the last integral is

$$e^{\frac{1}{4}i\pi}\int\limits_{-\infty}^{\infty} e^{-\frac{1}{4}\rho^2/\pi}\,d\rho = 2\pi e^{\frac{1}{4}i\pi}.$$

Hence
$$\Phi(a+1)-\Phi(a) = 2\pi e^{i\pi(a^2+\frac{1}{4})}. \tag{2.10.2}$$

Next let L' be the line parallel to L and intersecting the imaginary axis at a distance 2π below its intersection with L. Then by the theorem of residues

$$\int\limits_{L'} \frac{e^{\frac{1}{4}iw^2/\pi+aw}}{e^w-1}\,dw - \int\limits_L \frac{e^{\frac{1}{4}iw^2/\pi+aw}}{e^w-1}\,dw = 2\pi i.$$

But

$$\int\limits_{L'} \frac{e^{\frac{1}{4}iw^2/\pi+aw}}{e^w-1}\,dw = \int\limits_L \frac{e^{\frac{1}{4}i(w-2\pi i)^2/\pi+a(w-2\pi i)}}{e^w-1}\,dw$$

$$= \int\limits_L \frac{e^{\frac{1}{4}iw^2/\pi+w-i\pi+a(w-2\pi i)}}{e^w-1}\,dw = -e^{-2\pi ia}\Phi(a+1).$$

Hence
$$-e^{-2\pi ia}\Phi(a+1)-\Phi(a) = 2\pi i. \tag{2.10.3}$$

Eliminating $\Phi(a+1)$, we have

$$\Phi(a) = -\frac{2\pi i+2\pi e^{i\pi(a^2-2a+\frac{1}{4})}}{1+e^{-2\pi ia}}, \tag{2.10.4}$$

or

$$\Phi(a) = 2\pi\frac{\cos\pi(\frac{1}{2}a^2-a-\frac{1}{8})}{\cos\pi a}e^{i\pi(\frac{1}{2}a^2-\frac{5}{8})}. \tag{2.10.5}$$

If $a = \frac{1}{2}iz/\pi+\frac{1}{2}$, the result (2.10.4) takes the form

$$\int\limits_L \frac{e^{\frac{1}{4}iw^2/\pi+\frac{1}{2}izw/\pi+\frac{1}{2}w}}{e^w-1}\,dw = \frac{2\pi i}{e^z-1}-2\pi i\frac{e^{-\frac{1}{4}iz^2/\pi+\frac{1}{2}z}}{e^z-1}.$$

Multiplying by z^{s-1} ($\sigma > 1$), and integrating from 0 to $\infty e^{-\frac{1}{4}i\pi}$, we obtain

$$\int\limits_L \frac{e^{\frac{1}{4}iw^2/\pi+\frac{1}{2}w}}{e^w-1}\,dw \int\limits_0^{\infty e^{-\frac{1}{4}i\pi}} e^{\frac{1}{2}izw/\pi}z^{s-1}\,dz$$

$$= 2\pi i\Gamma(s)\zeta(s)-2\pi i \int\limits_0^{\infty e^{-\frac{1}{4}i\pi}} \frac{e^{-\frac{1}{4}iz^2/\pi+\frac{1}{2}z}}{e^z-1}z^{s-1}\,dz.$$

The inversion on the left-hand side is justified by absolute convergence; in fact

$$w = -c+\rho e^{\frac{1}{4}i\pi}, \qquad z = re^{-\frac{1}{4}i\pi},$$

where $c > 0$, so that $\qquad \mathbf{R}(izw) = -cr/\sqrt{2}.$

Now

$$\int\limits_0^{\infty e^{-\frac{1}{4}i\pi}} e^{\frac{1}{2}izw/\pi}z^{s-1}\,dz = e^{\frac{1}{4}i\pi s}\int\limits_0^{\infty} e^{-\frac{1}{2}yw/\pi}y^{s-1}\,dy = e^{\frac{1}{4}i\pi s}\left(\frac{w}{2\pi}\right)^{-s}\Gamma(s),$$

and

$$\int\limits_0^{\infty e^{-\frac{1}{4}i\pi}} \frac{e^{-\frac{1}{4}iz^2/\pi+\frac{1}{2}z}}{e^z-1}z^{s-1}\,dz = \frac{1}{1+e^{-is\pi}}\int\limits_{\bar{L}} \frac{e^{-\frac{1}{4}iz^2/\pi+\frac{1}{2}z}}{e^z-1}z^{s-1}\,dz,$$

where $\bar{L}$ is the reflection of L in the real axis. Hence

$$\zeta(s) = \frac{e^{\frac{1}{4}is\pi}(2\pi)^s}{2\pi i}\int\limits_L \frac{e^{\frac{1}{4}iw^2/\pi+\frac{1}{2}w}}{e^w-1}w^{-s}\,dw+\frac{1}{\Gamma(s)(1+e^{-is\pi})}\int\limits_{\bar{L}} \frac{e^{-\frac{1}{4}iz^2/\pi+\frac{1}{2}z}}{e^z-1}z^{s-1}\,dz,$$

or

$$\pi^{-\frac{1}{2}s}\Gamma(\tfrac{1}{2}s)\zeta(s) = e^{\frac{1}{4}i\pi(s-1)}2^{s-1}\pi^{\frac{1}{2}s-1}\Gamma(\tfrac{1}{2}s)\int\limits_L \frac{e^{\frac{1}{4}iw^2/\pi+\frac{1}{2}w}}{e^w-1}w^{-s}\,dw+$$

$$+e^{\frac{1}{4}i\pi s}2^{-s}\pi^{-\frac{1}{2}s-\frac{1}{2}}\Gamma(\tfrac{1}{2}-\tfrac{1}{2}s)\int\limits_{\bar{L}} \frac{e^{-\frac{1}{4}iz^2/\pi+\frac{1}{2}z}}{e^z-1}z^{s-1}\,dz. \qquad (2.10.6)$$

This formula holds by the theory of analytic continuation for all values of s.

If $s = \frac{1}{2}+it$, the two terms on the right are conjugates. Hence $f(s) = \pi^{-\frac{1}{2}s}\Gamma(\frac{1}{2}s)\zeta(s)$ is real on $\sigma = \frac{1}{2}$. Hence

$$f(s) = f(\sigma+it) = \overline{f(1-\sigma+it)} = f(1-\sigma-it) = f(1-s),$$

the functional equation.

2.11. A general formula involving $\zeta(s)$. It was observed by Müntz† that several of the formulae for $\zeta(s)$ which we have obtained are particular cases of a formula containing an arbitrary function.

We have formally

$$\int_0^\infty x^{s-1}\sum_{n=1}^\infty F(nx)\,dx = \sum_{n=1}^\infty \int_0^\infty x^{s-1}F(nx)\,dx$$

$$= \sum_{n=1}^\infty \frac{1}{n^s}\int_0^\infty y^{s-1}F(y)\,dy$$

$$= \zeta(s)\int_0^\infty y^{s-1}F(y)\,dy,$$

where $F(x)$ is arbitrary; and the process is justifiable if $F(x)$ is bounded in any finite interval, and $O(x^{-\alpha})$, where $\alpha > 1$, as $x \to \infty$. For then

$$\sum_{n=1}^\infty \left|\frac{1}{n^s}\right| \int_0^\infty |y^{s-1}F(y)|\,dy$$

exists if $1 < \sigma < \alpha$, and the inversion is justified.

Suppose next that $F'(x)$ is continuous, bounded in any finite interval, and $O(x^{-\beta})$, where $\beta > 1$, as $x \to \infty$. Then as $x \to 0$

$$\sum_{n=1}^\infty F(nx) - \int_0^\infty F(ux)\,du = x\int_0^\infty F'(ux)(u-[u])\,du$$

$$= x\int_0^{1/x} O(1)\,du + x\int_{1/x}^\infty O\{(ux)^{-\beta}\}\,du = O(1),$$

i.e.

$$\sum_{n=1}^\infty F(nx) = \frac{1}{x}\int_0^\infty F(v)\,dv + O(1) = \frac{c}{x} + O(1),$$

say. Hence

$$\int_0^\infty x^{s-1}\sum_{n=1}^\infty F(nx)\,dx$$

$$= \int_0^1 x^{s-1}\left\{\sum_{n=1}^\infty F(nx) - \frac{c}{x}\right\}dx + \frac{c}{s-1} + \int_1^\infty x^{s-1}\sum_{n=1}^\infty F(nx)\,dx,$$

and the right-hand side is regular for $\sigma > 0$ (except at $s = 1$). Also for $\sigma < 1$

$$\frac{c}{s-1} = -c\int_1^\infty x^{s-2}\,dx.$$

† Müntz (1).

Hence we have Müntz's formula

$$\zeta(s)\int_0^\infty y^{s-1}F(y)\,dy = \int_0^\infty x^{s-1}\left\{\sum_{n=1}^\infty F(nx)-\frac{1}{x}\int_0^\infty F(v)\,dv\right\}dx, \tag{2.11.1}$$

valid for $0<\sigma<1$ if $F(x)$ satisfies the above conditions.

If $F(x)=e^{-x}$ we obtain (2.7.1); if $F(x)=e^{-\pi x^2}$ we obtain a formula equivalent to those of § 2.6; if $F(x)=1/(1+x^2)$ we obtain a formula which is also obtained by combining (2.4.1) with the functional equation. If $F(x)=x^{-1}\sin\pi x$ we obtain a formula equivalent to (2.1.6), though this $F(x)$ does not satisfy our general conditions.

If $F(x)=1/(1+x)^2$, we have

$$\sum_{n=1}^\infty F(nx)-\frac{1}{x}\int_0^\infty F(v)\,dv = \sum_{n=1}^\infty \frac{1}{(1+nx)^2}-\frac{1}{x}$$

$$= \frac{1}{x^2}\left[\frac{d^2}{d\xi^2}\log\Gamma(\xi+1)\right]_{\xi=1/x}-\frac{1}{x}.$$

Hence $$\frac{(1-s)\pi}{\sin\pi s}\zeta(s) = \int_0^\infty \xi^{1-s}\left\{\frac{d^2}{d\xi^2}\log\Gamma(\xi+1)-\frac{1}{\xi}\right\}d\xi,$$

and on integrating by parts we obtain (2.9.2).

2.12. Zeros; factorization formulae.

THEOREM 2.12. *$\xi(s)$ and $\Xi(z)$ are integral functions of order* 1.

It follows from (2.1.12) and what we have proved about $\zeta(s)$ that $\xi(s)$ is regular for $\sigma>0$, $(s-1)\zeta(s)$ being regular at $s=1$. Since $\xi(s)=\xi(1-s)$, $\xi(s)$ is also regular for $\sigma<1$. Hence $\xi(s)$ is an integral function.

Also

$$|\Gamma(\tfrac{1}{2}s)| = \left|\int_0^\infty e^{-u}u^{\frac{1}{2}s-1}\,du\right| \leqslant \int_0^\infty e^{-u}u^{\frac{1}{2}\sigma-1}\,du = \Gamma(\tfrac{1}{2}\sigma) = O(e^{A\sigma\log\sigma}) \quad (\sigma>0), \tag{2.12.1}$$

and (2.1.4) gives for $\sigma\geqslant\frac{1}{2}$, $|s-1|>A$,

$$\zeta(s) = O\left(|s|\int_1^\infty\frac{du}{u^{\frac{3}{2}}}\right)+O(1) = O(|s|). \tag{2.12.2}$$

Hence (2.1.12) gives $$\xi(s) = O(e^{A|s|\log|s|}) \tag{2.12.3}$$

for $\sigma\geqslant\frac{1}{2}$, $|s|>A$. By (2.1.13) this holds for $\sigma\leqslant\frac{1}{2}$ also. Hence $\xi(s)$ is of order 1 at most. The order is exactly 1 since as $s\to\infty$ by real values $\log\zeta(s)\sim 2^{-s}$, $\log\xi(s)\sim\frac{1}{2}s\log s$.

Hence also $$\Xi(z) = O(e^{A|z|\log|z|}) \quad (|z| > A),$$

and $\Xi(z)$ is of order 1. But $\Xi(z)$ is an even function. Hence $\Xi(\sqrt{z})$ is also an integral function, and is of order $\frac{1}{2}$. It therefore has an infinity of zeros, whose exponent of convergence is $\frac{1}{2}$. Hence $\Xi(z)$ has an infinity of zeros, whose exponent of convergence is 1. The same is therefore true of $\xi(s)$. Let $\rho_1, \rho_2,...$ be the zeros of $\xi(s)$.

We have already seen that $\zeta(s)$ has no zeros for $\sigma > 1$. It then follows from the functional equation (2.1.1) that $\zeta(s)$ has no zeros for $\sigma < 0$ except for simple zeros at $s = -2, -4, -6,...$; for, in (2.1.1), $\zeta(1-s)$ has no zeros for $\sigma < 0$, $\sin\frac{1}{2}s\pi$ has simple zeros at $s = -2, -4,...$ only, and $\Gamma(1-s)$ has no zeros.

The zeros of $\zeta(s)$ at $-2, -4,...,$ are known as the 'trivial zeros'. They do not correspond to zeros of $\xi(s)$, since in (2.1.12) they are cancelled by poles of $\Gamma(\frac{1}{2}s)$. It therefore follows from (2.1.12) that $\xi(s)$ has no zeros for $\sigma > 1$ or for $\sigma < 0$. Its zeros $\rho_1, \rho_2,...$ therefore all lie in the strip $0 \leqslant \sigma \leqslant 1$; and they are also zeros of $\zeta(s)$, since $s(s-1)\Gamma(\frac{1}{2}s)$ has no zeros in the strip except that at $s = 1$, which is cancelled by the pole of $\zeta(s)$.

We have thus proved that $\zeta(s)$ has an infinity of zeros $\rho_1, \rho_2,...$ in the strip $0 \leqslant \sigma \leqslant 1$. Since

$$(1-2^{1-s})\zeta(s) = 1-\frac{1}{2^s}+\frac{1}{3^s}-... > 0 \quad (0 < s < 1) \tag{2.12.4}$$

and $\zeta(0) \neq 0$, $\zeta(s)$ has no zeros on the real axis between 0 and 1. The zeros $\rho_1, \rho_2,...$ are therefore all complex.

The remainder of the theory is largely concerned with questions about the position of these zeros. At this point we shall merely observe that they are in conjugate pairs, since $\zeta(s)$ is real on the real axis; and that, if ρ is a zero, so is $1-\rho$, by the functional equation, and hence so is $1-\bar{\rho}$. If $\rho = \beta+i\gamma$, then $1-\bar{\rho} = 1-\beta+i\gamma$. Hence the zeros either lie on $\sigma = \frac{1}{2}$, or occur in pairs symmetrical about this line.

Since $\xi(s)$ is an integral function of order 1, and $\xi(0) = -\zeta(0) = \frac{1}{2}$, Hadamard's factorization theorem gives, for all values of s,

$$\xi(s) = \tfrac{1}{2}e^{b_0 s}\prod_\rho \left(1-\frac{s}{\rho}\right)e^{s/\rho}, \tag{2.12.5}$$

where b_0 is a constant. Hence

$$\zeta(s) = \frac{e^{bs}}{2(s-1)\Gamma(\frac{1}{2}s+1)}\prod \left(1-\frac{s}{\rho}\right)e^{s/\rho}, \tag{2.12.6}$$

where $b = b_0+\frac{1}{2}\log\pi$. Hence also

$$\frac{\zeta'(s)}{\zeta(s)} = b-\frac{1}{s-1}-\frac{1}{2}\frac{\Gamma'(\frac{1}{2}s+1)}{\Gamma(\frac{1}{2}s+1)}+\sum_\rho\left(\frac{1}{s-\rho}+\frac{1}{\rho}\right). \tag{2.12.7}$$

Making $s \to 0$, this gives

$$\frac{\zeta'(0)}{\zeta(0)} = b+1-\frac{1}{2}\frac{\Gamma'(1)}{\Gamma(1)}.$$

Since $\zeta'(0)/\zeta(0) = \log 2\pi$ and $\Gamma'(1) = -\gamma$, it follows that

$$b = \log 2\pi-1-\tfrac{1}{2}\gamma. \tag{2.12.8}$$

2.13. In this section† we shall show that the only function which satisfies the functional equation (2.1.1), and has the same general characteristics as $\zeta(s)$, is $\zeta(s)$ itself.

Let $G(s)$ be an integral function of finite order, $P(s)$ a polynomial, and $f(s) = G(s)/P(s)$, and let

$$f(s) = \sum_{n=1}^{\infty}\frac{a_n}{n^s} \tag{2.13.1}$$

be absolutely convergent for $\sigma > 1$. Let

$$f(s)\Gamma(\tfrac{1}{2}s)\pi^{-\frac{1}{2}s} = g(1-s)\Gamma(\tfrac{1}{2}-\tfrac{1}{2}s)\pi^{-\frac{1}{2}(1-s)}, \tag{2.13.2}$$

where

$$g(1-s) = \sum_{n=1}^{\infty}\frac{b_n}{n^{1-s}},$$

the series being absolutely convergent for $\sigma < -\alpha < 0$. Then $f(s) = C\zeta(s)$, where C is a constant.

We have, for $x > 0$,

$$\phi(x) = \frac{1}{2\pi i}\int_{2-i\infty}^{2+i\infty} f(s)\Gamma(\tfrac{1}{2}s)\pi^{-\frac{1}{2}s}x^{-\frac{1}{2}s}\,ds$$

$$= \sum_{n=1}^{\infty}\frac{a_n}{2\pi i}\int_{2-i\infty}^{2+i\infty}\Gamma(\tfrac{1}{2}s)(\pi n^2x)^{-\frac{1}{2}s}\,ds$$

$$= 2\sum_{n=1}^{\infty}a_n e^{-\pi n^2 x}.$$

Also, by (2.13.2),

$$\phi(x) = \frac{1}{2\pi i}\int_{2-i\infty}^{2+i\infty} g(1-s)\Gamma(\tfrac{1}{2}-\tfrac{1}{2}s)\pi^{-\frac{1}{2}(1-s)}x^{-\frac{1}{2}s}\,ds.$$

We move the line of integration from $\sigma = 2$ to $\sigma = -1-\alpha$. We observe

† Hamburger (1)–(4), Siegel (1).

that $f(s)$ is bounded on $\sigma = 2$, and $g(1-s)$ is bounded on $\sigma = -1-\alpha$; since

$$\frac{\Gamma(\frac{1}{2}s)}{\Gamma(\frac{1}{2}-\frac{1}{2}s)} = O(|t|^{\sigma-\frac{1}{2}}),$$

it follows that $g(1-s) = O(|t|^{\frac{3}{2}})$ on $\sigma = 2$. We can therefore, by the Phragmén-Lindelöf principle, apply Cauchy's theorem, and obtain

$$\phi(x) = \frac{1}{2\pi i}\int\limits_{-\alpha-1-i\infty}^{-\alpha-1+i\infty} g(1-s)\Gamma(\tfrac{1}{2}-\tfrac{1}{2}s)\pi^{-\frac{1}{2}(1-s)}x^{-\frac{1}{2}s}\,ds+\sum_{\nu=1}^{m} R_\nu,$$

where R_1, R_2,..., are the residues at the poles, say s_1,..., s_m. Thus

$$\sum_{\nu=1}^{m} R_\nu = \sum_{\nu=1}^{m} x^{-\frac{1}{2}s_\nu}Q_\nu(\log x) = Q(x),$$

where the $Q_\nu(\log x)$ are polynomials in $\log x$. Hence

$$\phi(x) = \frac{1}{\sqrt{x}}\sum_{n=1}^{\infty}\frac{b_n}{2\pi i}\int\limits_{-\alpha-1-i\infty}^{-\alpha-1+i\infty}\Gamma(\tfrac{1}{2}-\tfrac{1}{2}s)(\pi n^2/x)^{-\frac{1}{2}+\frac{1}{2}s}\,ds+Q(x)$$

$$= \frac{2}{\sqrt{x}}\sum_{n=1}^{\infty} b_n e^{-\pi n^2/x}+Q(x).$$

Hence
$$\sum_{n=1}^{\infty} a_n e^{-\pi n^2 x} = \frac{1}{\sqrt{x}}\sum_{n=1}^{\infty} b_n e^{-\pi n^2/x}+\tfrac{1}{2}Q(x).$$

Multiply by $e^{-\pi t^2 x}$ $(t > 0)$, and integrate over $(0,\infty)$. We obtain

$$\sum_{n=1}^{\infty}\frac{a_n}{\pi(t^2+n^2)} = \sum_{n=1}^{\infty}\frac{b_n}{t}e^{-2\pi nt}+\tfrac{1}{2}\int\limits_0^{\infty} Q(x)e^{-\pi t^2 x}\,dx,$$

and the last term is a sum of terms of the form

$$\int\limits_0^{\infty} x^a \log^b x\, e^{-\pi t^2 x}\,dx,$$

where the b's are integers and $\mathbf{R}(a) > -1$; i.e. it is a sum of terms of the form $t^\alpha \log^\beta t$.

Hence
$$\sum_{n=1}^{\infty} a_n\left(\frac{1}{t+in}+\frac{1}{t-in}\right)-\pi t H(t) = 2\pi\sum_{n=1}^{\infty} b_n e^{-2\pi nt},$$

where $H(t)$ is a sum of terms of the form $t^\alpha \log^\beta t$.

Now the series on the left is a meromorphic function, with poles at $\pm in$. But the function on the right is periodic, with period i. Hence (by analytic continuation) so is the function on the left. Hence the residues at ki and $(k+1)i$ are equal, i.e. $a_k = a_{k+1}$ $(k = 1, 2,...)$. Hence $a_k = a_1$ for all k, and the result follows.

2.14. Some series involving $\zeta(s)$. We have†

$$\zeta(s)-\frac{1}{s-1} = 1-\tfrac{1}{2}s\{\zeta(s+1)-1\}-\frac{s(s+1)}{2.3}\{\zeta(s+2)-1\}-\ldots \quad (2.14.1)$$

for all values of s. For the right-hand side is

$$1-\frac{1}{s-1}\sum_{n=2}^{\infty}\frac{1}{n^{s-1}}\left\{\frac{(s-1)s}{1.2}\frac{1}{n^2}+\frac{(s-1)s(s+1)}{1.2.3}\frac{1}{n^3}+\ldots\right\}$$

$$= 1-\frac{1}{s-1}\sum_{n=2}^{\infty}\frac{1}{n^{s-1}}\left\{\left(1-\frac{1}{n}\right)^{1-s}-1-\frac{s-1}{n}\right\}$$

$$= 1-\frac{1}{s-1}\sum_{n=2}^{\infty}\left\{\frac{1}{(n-1)^{s-1}}-\frac{1}{n^{s-1}}-\frac{s-1}{n^s}\right\}$$

$$= \zeta(s)-\frac{1}{s-1}.$$

The inversion of the order of summation is justified for $\sigma > 0$ by the convergence of

$$\sum_{n=2}^{\infty}\frac{1}{n^{\sigma-1}}\sum_{k=0}^{\infty}\frac{|s|\ldots(|s|+k)}{(k+1)!}\frac{1}{n^{k+2}} = \sum_{n=2}^{\infty}\frac{1}{n^{\sigma}}\left\{\left(1-\frac{1}{n}\right)^{-|s|}-1\right\}.$$

The series obtained is, however, convergent for all values of s.

Another formula‡ which can be proved in a similar way is

$$(1-2^{1-s})\zeta(s) = s\frac{\zeta(s+1)}{2^{s+1}}+\frac{s(s+1)}{1.2}\frac{\zeta(s+2)}{2^{s+2}}+\ldots, \quad (2.14.2)$$

also valid for all values of s.

Either of these formulae may be used to obtain the analytic continuation of $\zeta(s)$ over the whole plane.

2.15. Some applications of Mellin's inversion formulae.§ Mellin's inversion formulae connecting the two functions $f(x)$ and $\mathfrak{F}(s)$ are

$$\mathfrak{F}(s) = \int_0^{\infty} f(x)x^{s-1}\,dx, \qquad f(x) = \frac{1}{2\pi i}\int_{\sigma-i\infty}^{\sigma+i\infty}\mathfrak{F}(s)x^{-s}\,ds. \quad (2.15.1)$$

The simplest example is

$$f(x) = e^{-x}, \qquad \mathfrak{F}(s) = \Gamma(s) \quad (\sigma > 0). \quad (2.15.2)$$

From (2.4.1) we derive the pair

$$f(x) = \frac{1}{e^x-1}, \qquad \mathfrak{F}(s) = \Gamma(s)\zeta(s) \quad (\sigma > 1), \quad (2.15.3)$$

† Landau, *Handbuch*, 272.

‡ Ramaswami (1).

§ See E. C. Titchmarsh, *Introduction to the Theory of Fourier Integrals*, §§ 1.5, 1.29, 2.1, 2.7, 3.17.

and from (2.6.2) the pair

$$f(x) = \psi(x), \qquad \mathfrak{F}(s) = \pi^{-s}\Gamma(s)\zeta(2s) \quad (\sigma > \tfrac{1}{2}). \tag{2.15.4}$$

The inverse formulae are thus

$$\frac{1}{2\pi i}\int_{\sigma-i\infty}^{\sigma+i\infty} \Gamma(s)\zeta(s)x^{-s}\,ds = \frac{1}{e^x-1} \quad (\sigma > 1) \tag{2.15.5}$$

and

$$\frac{1}{2\pi i}\int_{\sigma-i\infty}^{\sigma+i\infty} \pi^{-s}\Gamma(s)\zeta(2s)x^{-s}\,ds = \psi(x) \quad (\sigma > \tfrac{1}{2}). \tag{2.15.6}$$

Each of these can easily be proved directly by inserting the series for $\zeta(s)$ and integrating term-by-term, using (2.15.2).

As another example, (2.9.2), with s replaced by $1-s$, gives the Mellin pair

$$f(x) = \frac{\Gamma'(1+x)}{\Gamma(1+x)} - \log x, \qquad \mathfrak{F}(s) = -\frac{\pi\zeta(1-s)}{\sin \pi s} \quad (0 < \sigma < 1). \tag{2.15.7}$$

The inverse formula is thus

$$\frac{\Gamma'(1+x)}{\Gamma(1+x)} - \log x = -\frac{1}{2i}\int_{\sigma-i\infty}^{\sigma+i\infty} \frac{\zeta(1-s)}{\sin \pi s}x^{-s}\,ds. \tag{2.15.8}$$

Integrating with respect to x, and replacing s by $1-s$, we obtain

$$\log\Gamma(1+x) - x\log x + x = -\frac{1}{2i}\int_{\sigma-i\infty}^{\sigma+i\infty} \frac{\zeta(s)x^s}{s\sin \pi s}\,ds \quad (0 < \sigma < 1). \tag{2.15.9}$$

This formula is used by Whittaker and Watson to obtain the asymptotic expansion of $\log\Gamma(1+x)$.

Next, let $f(x)$ and $\mathfrak{F}(s)$ be related by (2.15.1), and let $g(x)$ and $\mathfrak{G}(s)$ be similarly related. Then we have, subject to appropriate conditions,

$$\frac{1}{2\pi i}\int_{c-i\infty}^{c+i\infty} \mathfrak{F}(s)\mathfrak{G}(w-s)\,ds = \int_0^\infty f(x)g(x)x^{w-1}\,dx. \tag{2.15.10}$$

Take for example $\mathfrak{F}(s) = \mathfrak{G}(s) = \Gamma(s)\zeta(s)$, so that

$$f(x) = g(x) = 1/(e^x-1).$$

Then, if $\mathbf{R}(w) > 2$, the right-hand side is

$$\int_0^\infty \frac{x^{w-1}}{(e^x-1)^2}\,dx = \int_0^\infty (e^{-2x}+2e^{-3x}+3e^{-4x}+\ldots)x^{w-1}\,dx$$

$$= \left(\frac{1}{2^w}+\frac{2}{3^w}+\frac{3}{4^w}+\ldots\right)\Gamma(w) = \Gamma(w)\{\zeta(w-1)-\zeta(w)\}.$$

Thus if $1 < c < \mathbf{R}(w)-1$

$$\frac{1}{2\pi i}\int\limits_{c-i\infty}^{c+i\infty} \Gamma(s)\Gamma(w-s)\zeta(s)\zeta(w-s)\,ds = \Gamma(w)\{\zeta(w-1)-\zeta(w)\}. \quad (2.15.11)$$

Similarly, taking $\mathfrak{F}(s) = \mathfrak{G}(s) = \Gamma(s)\zeta(2s)$, so that

$$f(x) = g(x) = \psi(x/\pi) = \sum_{n=1}^{\infty} e^{-n^2x},$$

the right-hand side of (2.15.10) is, if $\mathbf{R}(w) > 1$,

$$\int\limits_0^{\infty} \sum_{m=1}^{\infty}\sum_{n=1}^{\infty} e^{-(m^2+n^2)x}x^{w-1}\,dx = \Gamma(w)\sum_{m=1}^{\infty}\sum_{n=1}^{\infty}\frac{1}{(m^2+n^2)^w}.$$

This may also be written

$$\Gamma(w)\left\{\frac{1}{4}\sum_{n=1}^{\infty}\frac{r(n)}{n^w}-\zeta(2w)\right\},$$

where $r(n)$ is the number of ways of expressing n as the sum of two squares; or as

$$\Gamma(w)\{\zeta(w)\eta(w)-\zeta(2w)\},$$

where

$$\eta(w) = 1^{-w}-3^{-w}+5^{-w}-\dots.$$

Hence† if $\frac{1}{2} < c < \mathbf{R}(w)-\frac{1}{2}$

$$\frac{1}{2\pi i}\int\limits_{c-i\infty}^{c+i\infty} \Gamma(s)\Gamma(w-s)\zeta(2s)\zeta(2w-2s)\,ds = \Gamma(w)\{\zeta(w)\eta(w)-\zeta(2w)\}. \quad (2.15.12)$$

2.16. Some integrals involving $\Xi(t)$. There are some cases‡ in which integrals of the form

$$\Phi(x) = \int\limits_0^{\infty} f(t)\Xi(t)\cos xt\,dt$$

can be evaluated. Let $f(t) = |\phi(it)|^2 = \phi(it)\phi(-it)$, where ϕ is analytic. Writing $y = e^x$,

$$\Phi(x) = \tfrac{1}{2}\int\limits_{-\infty}^{\infty} \phi(it)\phi(-it)\Xi(t)y^{it}\,dt$$

$$= \tfrac{1}{2}\int\limits_{-\infty}^{\infty} \phi(it)\phi(-it)\xi(\tfrac{1}{2}+it)y^{it}\,dt$$

$$= \frac{1}{2i\sqrt{y}}\int\limits_{\frac{1}{2}-i\infty}^{\frac{1}{2}+i\infty} \phi(s-\tfrac{1}{2})\phi(\tfrac{1}{2}-s)\xi(s)y^s\,ds$$

$$= \frac{1}{2i\sqrt{y}}\int\limits_{\frac{1}{2}-i\infty}^{\frac{1}{2}+i\infty} \phi(s-\tfrac{1}{2})\phi(\tfrac{1}{2}-s)(s-1)\Gamma(1+\tfrac{1}{2}s)\pi^{-\frac{1}{2}s}\zeta(s)y^s\,ds.$$

† Hardy (4). A generalization is given by Taylor (1). ‡ Ramanujan (1).

Taking $\phi(s) = 1$, this is equal to

$$\frac{1}{i\sqrt{y}}\sum_{n=1}^{\infty}\int_{2-i\infty}^{2+i\infty}\{\Gamma(2+\tfrac{1}{2}s)-\tfrac{3}{2}\Gamma(1+\tfrac{1}{2}s)\}\left(\frac{y}{n\sqrt{\pi}}\right)^{s}ds$$

$$=\frac{1}{i\sqrt{y}}\sum_{n=1}^{\infty}\left\{2\int_{3-i\infty}^{3+i\infty}\Gamma(w)\left(\frac{y}{n\sqrt{\pi}}\right)^{2w-4}dw-3\int_{2-i\infty}^{2+i\infty}\Gamma(w)\left(\frac{y}{n\sqrt{\pi}}\right)^{2w-2}dw\right\}$$

$$=\frac{4\pi}{\sqrt{y}}\sum_{n=1}^{\infty}\left(\frac{y}{n\sqrt{\pi}}\right)^{-4}e^{-n^2\pi/y^2}-\frac{6\pi}{\sqrt{y}}\sum_{n=1}^{\infty}\left(\frac{y}{n\sqrt{\pi}}\right)^{-2}e^{-n^2\pi/y^2}.$$

Hence

$$\int_0^\infty \Xi(t)\cos xt\,dt = 2\pi^2\sum_{n=1}^{\infty}(2\pi n^4 e^{-9x/2}-3n^2e^{-5x/2})\exp(-n^2\pi e^{-2x}). \tag{2.16.1}$$

Again, putting $\phi(s) = 1/(s+\frac{1}{2})$, we have

$$\Phi(x) = -\frac{1}{2i\sqrt{y}}\int_{\frac{1}{2}-i\infty}^{\frac{1}{2}+i\infty}\frac{1}{s}\Gamma(1+\tfrac{1}{2}s)\pi^{-\frac{1}{2}s}\zeta(s)y^s\,ds$$

$$= -\frac{1}{4i\sqrt{y}}\int_{\frac{1}{2}-i\infty}^{\frac{1}{2}+i\infty}\Gamma(\tfrac{1}{2}s)\pi^{-\frac{1}{2}s}\zeta(s)y^s\,ds$$

$$= -\frac{\pi}{\sqrt{y}}\psi\left(\frac{1}{y^2}\right)+\tfrac{1}{2}\pi\sqrt{y}$$

in the notation of § 2.6. Hence

$$\int_0^\infty \frac{\Xi(t)}{t^2+\frac{1}{4}}\cos xt\,dt = \tfrac{1}{2}\pi\{e^{\frac{1}{2}x}-2e^{-\frac{1}{2}x}\psi(e^{-2x})\}. \tag{2.16.2}$$

The case $\phi(s) = \Gamma(\frac{1}{2}s-\frac{1}{4})$ was also investigated by Ramanujan, the result being expressed in terms of another integral.

2.17. The function $\zeta(s, a)$. A function which is in a sense a generalization of $\zeta(s)$ is the Hurwitz zeta-function, defined by

$$\zeta(s,a) = \sum_{n=0}^{\infty}\frac{1}{(n+a)^s} \quad (0 < a \leqslant 1,\ \sigma > 1).$$

This reduces to $\zeta(s)$ when $a = 1$, and to $(2^s-1)\zeta(s)$ when $a = \frac{1}{2}$. We shall obtain here its analytic continuation and functional equation, which are required later. This function, however, has no Euler product unless $a = \frac{1}{2}$ or $a = 1$, and so does not share the most characteristic properties of $\zeta(s)$.

As in § 2.4

$$\zeta(s,a) = \sum_{n=0}^{\infty} \frac{1}{\Gamma(s)} \int_0^{\infty} x^{s-1}e^{-(n+a)x}\,dx = \frac{1}{\Gamma(s)} \int_0^{\infty} \frac{x^{s-1}e^{-ax}}{1-e^{-x}}\,dx. \tag{2.17.1}$$

We can transform this into a loop integral as before. We obtain

$$\zeta(s,a) = \frac{e^{-i\pi s}\Gamma(1-s)}{2\pi i} \int_C \frac{z^{s-1}e^{-az}}{1-e^{-z}}\,dz. \tag{2.17.2}$$

This provides the analytic continuation of $\zeta(s,a)$ over the whole plane; it is regular everywhere except for a simple pole at $s = 1$ with residue 1.

Expanding the loop to infinity as before, the residues at $2m\pi i$ and $-2m\pi i$ are together

$$\begin{aligned}(2m\pi e^{\frac{1}{2}i\pi})^{s-1}e^{-2m\pi ia} + (2m\pi e^{\frac{3}{2}i\pi})^{s-1}e^{2m\pi ia} \\ &= (2m\pi)^{s-1}e^{i\pi(s-1)}2\cos\{\tfrac{1}{2}\pi(s-1)+2m\pi a\} \\ &= -2(2m\pi)^{s-1}e^{i\pi s}\sin(\tfrac{1}{2}\pi s+2m\pi a).\end{aligned}$$

Hence, if $\sigma < 0$,

$$\zeta(s,a) = \frac{2\Gamma(1-s)}{(2\pi)^{1-s}}\left\{\sin\tfrac{1}{2}\pi s\sum_{m=1}^{\infty}\frac{\cos 2m\pi a}{m^{1-s}} + \cos\tfrac{1}{2}\pi s\sum_{m=1}^{\infty}\frac{\sin 2m\pi a}{m^{1-s}}\right\}. \tag{2.17.3}$$

If $a = 1$, this reduces to the functional equation for $\zeta(s)$.

NOTES FOR CHAPTER 2

2.18. Selberg [1] has given a very general method for obtaining the analytic continuation and functional equation of certain types of zeta-function which arise as the 'constant terms' of Eisenstein series. We sketch a form of the argument in the classical case. Let $\mathscr{H} = \{z = x+iy: y > 0\}$ be the upper half plane and define

$$E(z, s) = \sum_{\substack{c,d=-\infty \\ (c,d)=1}}^{\infty} \frac{y^s}{|cz+d|^{2s}} \quad (z \in \mathscr{H},\ \sigma > 1)$$

and

$$B(z, s) = \zeta(2s)E(z, s) = \sum_{\substack{c,d=-\infty \\ (c,d)\neq(0,0)}}^{\infty} \frac{y^s}{|cz+d|^{2s}} \quad (z \in \mathscr{H},\ \sigma > 1),$$

these series being absolutely and uniformly convergent in any compact subset of the region $\mathbf{R}(s) > 1$. Here $E(z, s)$ is an Eisenstein series, while $B(z, s)$ is, apart from the factor y^s, the Epstein zeta-function for the lattice generated by 1 and z. We shall find it convenient to work with $B(z, s)$ in preference to $E(z, s)$.

We begin with two basic observations. Firstly one trivially has

$$B(z+1, s) = B(-1/z, s) = B(z, s). \tag{2.18.1}$$

(Thus, in fact, $B(z, s)$ is invariant under the full modular group.) Secondly, if Δ is the Laplace–Beltrami operator

$$\Delta = -y^2\left(\frac{\partial^2}{\partial x^2} + \frac{\partial^2}{\partial y^2}\right),$$

then

$$\Delta\left(\frac{y^s}{|cz+d|^{2s}}\right) = s(1-s)\frac{y^s}{|cz+d|^{2s}}, \tag{2.18.2}$$

whence

$$\Delta B(z, s) = s(1-s)B(z, s) \quad (\sigma > 1). \tag{2.18.3}$$

We proceed to obtain the Fourier expansion of $B(z, s)$ with respect to x. We have

$$B(z, s) = \sum_{-\infty}^{\infty} a_n(y, s)\, e^{2\pi i n x},$$

where

$$a_n(y, s) = y^s \sum_{c, d} \int_0^1 \frac{e^{-2\pi i n x}\, dx}{|cx+d+icy|^{2s}}$$

$$= 2\delta_n y^s \zeta(2s) + 2y^s \sum_{c=1}^{\infty} \sum_{d=-\infty}^{\infty} \int_0^1 \frac{e^{-2\pi i n x}\, dx}{|cx+d+icy|^{2s}},$$

with $\delta_n = 1$ or 0 according as $n = 0$ or not. The d summation above is

$$\sum_{k=1}^{c} \sum_{j=-\infty}^{\infty} \int_0^1 \frac{e^{-2\pi i n x}\, dx}{|c(x+j)+k+icy|^{2s}} = \sum_{k=1}^{c} \int_{-\infty}^{\infty} \frac{e^{-2\pi i n x}\, dx}{|cx+k+icy|^{2s}}$$

$$= c^{-2s} y^{1-2s} \int_{-\infty}^{\infty} \frac{e^{-2\pi i n y v}\, dv}{(v^2+1)^s} \sum_{k=1}^{c} e^{2\pi i n k/c},$$

and the sum over k is c or 0 according as $c|n$ or not. Moreover

$$\int_{-\infty}^{\infty} \frac{dv}{(v^2+1)^s} = \frac{\pi^{\frac{1}{2}}\Gamma(s-\frac{1}{2})}{\Gamma(s)},$$

and

$$\int_{-\infty}^{\infty} \frac{e^{-2\pi i n y v}}{(v^2+1)^s} dv = 2\pi^s(|n|y)^{s-\frac{1}{2}} \frac{K_{s-\frac{1}{2}}(2\pi|n|y)}{\Gamma(s)} \qquad (n \neq 0),$$

in the usual notation of Bessel functions†.

We now have

$$B(z, s) = \phi(s)y^s + \psi(s)y^{1-s} + B_0(z, s) \quad (\sigma > 1), \tag{2.18.4}$$

where

$$\phi(s) = 2\zeta(2s), \qquad \psi(s) = 2\pi^{\frac{1}{2}} \frac{\Gamma(s-\frac{1}{2})}{\Gamma(s)} \zeta(2s-1)$$

and

$$B_0(z, s) = 8\pi^s y^{\frac{1}{2}} \sum_{n=1}^{\infty} n^{s-\frac{1}{2}} \sigma_{1-2s}(n) \cos(2\pi n x) \frac{K_{s-\frac{1}{2}}(2\pi n y)}{\Gamma(s)}. \tag{2.18.5}$$

We observe at this point that

$$K_u(t) \ll t^{-\frac{1}{2}} e^{-t} \quad (t \to \infty)$$

for fixed u, whence the series (2.18.5) is convergent for all s, and so defines an entire function. Moreover we have

$$B_0(z, s) \ll e^{-y} \quad (y \to \infty) \tag{2.18.6}$$

for fixed s. Similarly one finds

$$\frac{\partial B_0(z, s)}{\partial y} \ll e^{-y} \quad (y \to \infty). \tag{2.18.7}$$

We proceed to derive the 'Maass–Selberg' formula. Let $D = \{z \in \mathscr{H}: |z| \geqslant 1, |\mathbf{R}(z)| \leqslant \frac{1}{2}\}$ be the standard fundamental region for the modular group, and let $D_Y = \{z \in D: \mathbf{I}(z) \leqslant Y\}$, where $Y \geqslant 1$. Let $\mathbf{R}(s)$, $\mathbf{R}(w) > 1$ and write, for convenience, $F = B(z, s)$, $G = B(z, w)$. Then, according to (2.18.3), we have

$$\begin{aligned} \{s(1-s) - w(1-w)\} \iint_{D_Y} FG \frac{dx\,dy}{y^2} &= \iint_{D_Y} (G\Delta F - F\Delta G) \frac{dx\,dy}{y^2} \\ &= \iint_{D_Y} (F\nabla^2 G - G\nabla^2 F)\, dx\,dy \\ &= \int_{\partial D_Y} (F\nabla G - G\nabla F) \cdot d\mathbf{n}, \end{aligned}$$

† see Watson, *Theory of Bessel functions* §6.16.

by Green's Theorem. The integrals along $x = \pm\frac{1}{2}$ cancel, since $F(z+1) = F(z)$, $G(z+1) = G(z)$ (see (2.18.1)). Similarly the integral for $|z| = 1$ vanishes, since $F(-1/z) = F(z)$, $G(-1/z) = G(z)$. Thus

$$\{s(1-s)-w(1-w)\}\iint_{D_Y} FG\frac{dxdy}{y^2} = \int_{-\frac{1}{2}}^{\frac{1}{2}}\left(F\frac{\partial G}{\partial y}(x, Y) - G\frac{\partial F}{\partial y}(x, Y)\right)dx. \tag{2.18.8}$$

The functions y^s and y^{1-s} also satisfy the eigenfunction equation (2.18.3) (by (2.18.2) with $c = 0$, $d = 1$) and thus, by (2.18.4) so too does $B_0(z, s)$. Consequently, if $Z \geqslant Y$, an argument analogous to that above yields

$$\{s(1-s)-w(1-w)\}\int_Y^Z\int_{-\frac{1}{2}}^{\frac{1}{2}} F_0 G_0 \frac{dx\,dy}{y^2}$$

$$= \int_{-\frac{1}{2}}^{\frac{1}{2}}\left(F_0\frac{\partial G_0}{\partial y}(x, Z) - G_0\frac{\partial F_0}{\partial y}(x, Z)\right)dx$$

$$- \int_{-\frac{1}{2}}^{\frac{1}{2}}\left(F_0\frac{\partial G_0}{\partial y}(x, Y) - G_0\frac{\partial F_0}{\partial y}(x, Y)\right)dx,$$

where $F_0 = B_0(z, s)$, $G_0 = B_0(z, w)$. Here we have used $F_0(z+1) = F_0(z)$ and $G_0(z+1) = G_0(z)$. (Note that we no longer have the corresponding relations involving $-1/z$.) We may now take $Z \to \infty$, using (2.18.6) and (2.18.7), so that the first integral on the right above vanishes. On adding the result to (2.18.8) we obtain the Maass–Selberg formula

$$[s(1-s)-w(1-w)]\iint_D \tilde{B}(z, s)\,\tilde{B}(z, w)\frac{dxdy}{y^2}$$

$$= \int_{-\frac{1}{2}}^{\frac{1}{2}}\left(F\frac{\partial G}{\partial y}(x, Y) - G\frac{\partial F}{\partial y}(x, Y)\right)dx$$

$$- \int_{-\frac{1}{2}}^{\frac{1}{2}}\left(F_0\frac{\partial G_0}{\partial y}(x, Y) - G_0\frac{\partial F_0}{\partial y}(x, Y)\right)dx$$

$$= (s-w)\{\psi(s)\,\psi(w)\,Y^{1-s-w} - \phi(s)\,\phi(w)\,Y^{s+w-1}\}$$

$$+ (1-s-w)\{\phi(s)\,\psi(w)\,Y^{s-w} - \psi(s)\,\phi(w)\,Y^{w-s}\}, \tag{2.18.9}$$

where

$$\tilde{B}(z, s) = \begin{cases} B(z, s) & (y \leqslant Y), \\ B_0(z, s) & (y > Y). \end{cases}$$

2.19. In the general case there are now various ways in which one can proceed in order to get the analytic continuation of ϕ and ψ. However one point is immediate: once the analytic continuation has been established one may take $w = 1 - s$ in (2.18.9) to obtain the relation

$$\phi(s)\,\phi(1-s) = \psi(s)\,\psi(1-s), \tag{2.19.1}$$

which can be thought of as a weak form of the functional equation.

The analysis we shall give takes advantage of certain special properties not available in the general case. We shall take $Y = 1$ in (2.18.9) and expand the integral on the left to obtain

$$(s-w)\alpha(s+w)\psi(s)\psi(w) + \beta(s, w)\psi(s) + \gamma(s, w)\psi(w) + \delta(s, w) = 0, \tag{2.19.2}$$

where

$$\alpha(u) = (1-u)\iint_{D_1} y^{-u}\,dx\,dy - 1 = -2\int_0^{\frac{1}{2}} (1-x^2)^{\frac{1}{2}(1-u)}\,dx$$

and β, γ, δ involve the functions ϕ and B_0, but not ψ. If we know that $\zeta(s)$ has a continuation to the half plane $\mathbf{R}(s) > \sigma_0$ then $\phi(s)$ has a continuation to $\mathbf{R}(s) > \frac{1}{2}\sigma_0$, so that $\alpha, \beta, \gamma, \delta$ are meromorphic there. If

$$(s-w)\alpha(s+w)\psi(w) + \beta(s, w) = 0 \tag{2.19.3}$$

identically for $\mathbf{R}(s)$, $\mathbf{R}(w) > 1$, then

$$\psi(w) = -\frac{\beta(s, w)}{(s-w)\alpha(s+w)}, \tag{2.19.4}$$

which gives the analytic continuation of $\psi(w)$ to $\mathbf{R}(w) > \frac{1}{2}\sigma_0$. Note that $(s-w)\alpha(s+w)$ does not vanish identically. If (2.19.3) does not hold for all s and w then (2.19.2) yields

$$\psi(s) = -\frac{\gamma(s, w)\psi(w) + \delta(s, w)}{(s-w)\alpha(s+w)\psi(w) + \beta(s, w)}, \tag{2.19.5}$$

which gives the analytic continuation of $\psi(s)$ to $\mathbf{R}(s) > \frac{1}{2}\sigma_0$, on choosing a suitable w in the region $\mathbf{R}(w) > 1$. In either case $\zeta(s)$ may be continued to $\mathbf{R}(s) > \sigma_0 - 1$. This process shows that $\zeta(s)$ has a meromorphic continuation to the whole complex plane.

Some information on possible poles comes from taking $w = \bar{s}$ in (2.18.9), so that $\tilde{B}(z, w) = \overline{\tilde{B}(z, s)}$. Then

$$(2\sigma-1)\iint_D |\tilde{B}(z, s)|^2 \frac{dx dy}{y^2} = \{|\phi(s)|^2 Y^{2\sigma-1} - |\psi(s)|^2 Y^{1-2\sigma}\}$$

$$+(2\sigma-1)\frac{\phi(s)\overline{\psi(s)}\, Y^{2it} - \psi(s)\overline{\phi(s)}\, Y^{-2it}}{2it}.$$

If $t \neq 0$ we may choose $Y \geqslant 1$ so that the second term on the right vanishes. It follows that

$$|\psi(s)|^2 Y^{1-2\sigma} \leqslant |\phi(s)|^2 Y^{2\sigma-1}$$

for $\sigma \geqslant \frac{1}{2}$. Thus ψ is regular for $\sigma \geqslant \frac{1}{2}$ and $t \neq 0$, providing that ϕ is. Hence $\zeta(s)$ has no poles for $\mathbf{R}(s) > 0$, except possibly on the real axis.

If we take $\frac{1}{2} < \mathbf{R}(s), \mathbf{R}(w) < 1$ in (2.19.5), so that $\phi(s)$ and $\phi(w)$ are regular, we see that $\psi(s)$ can only have a pole at a point s_0 for which the denominator vanishes identically in w. For such an s_0, (2.19.4) must hold. However $\alpha(u)$ is clearly non-zero for real u, whence $\psi(w)$ can have at most a single, simple pole for real $w > \frac{1}{2}$, and this is at $w = s_0$. Since it is clear that $\zeta(s)$ does in fact have a singularity at $s = 1$ we see that $s_0 = 1$.

Much of the inelegance of the above analysis arises from the fact that, in the general case where one uses the Eisenstein series rather than the Epstein zeta-function, one has a single function $\rho(s) = \psi(s)/\phi(s)$ rather than two separate ones. Here $\rho(s)$ will indeed have poles to the left of $\mathbf{R}(s) = \frac{1}{2}$. In our special case we can extract the functional equation for $\zeta(s)$ itself, rather than the weaker relation $\rho(s)\rho(1-s) = 1$ (see (2.19.1)), by using (2.18.4) and (2.18.5). We observe that

$$n^{s-1/2}\sigma_{1-2s}(n) = n^{1/2-s}\sigma_{2s-1}(n)$$

and that $K_u(z) = K_{-u}(z)$, whence $\pi^{-s}\Gamma(s)B_0(z, s)$ is invariant under the transformation $s \to 1-s$. It follows that

$$\pi^{-s}\Gamma(s)B(z, s) - \pi^{s-1}\Gamma(1-s)B(z, 1-s)$$

$$= \{A(s) - A(\tfrac{1}{2}-s)\}y^s + \{A(s-\tfrac{1}{2}) - A(1-s)\}y^{1-s},$$

where we have written temporarily $A(s) = 2\pi^{-s}\Gamma(s)\zeta(2s)$. The left-hand side is invariant under the transformation $z \to -1/z$, by (2.18.1), and so, taking $z = iy$ for example, we see that $A(s) = A(\frac{1}{2}-s)$ and $A(s-\frac{1}{2}) = A(1-s)$. These produce the functional equation in the form (2.6.4) and indeed yield

$$\pi^{-s}\Gamma(s)B(z,s) = \pi^{s-1}\Gamma(1-s)B(z,1-s).$$

2.20. An insight into the nature of the zeta-function and its functional equation may be obtained from the work of Tate [1]. He considers an algebraic number field k and a general zeta-function

$$\zeta(f,c) = \int f(a)c(a)\,d^*a,$$

where the integral on the right is over the ideles J of k. Here f is one of a certain class of functions and c is any quasi-character of J, (that is to say, a continuous homomorphism from J to $\mathbb{C}^\times$) which is trivial on $k^\times$. We may write $c(a)$ in the form $c_0(a)|a|^s$, where $c_0(a)$ is a character on J (i.e. $|c_0(a)| = 1$ for $a \in J$). Then $c_0(a)$ corresponds to χ, a 'Hecke character' for k, and $\zeta(f,c)$ differs from

$$\zeta(s,\chi) = \prod_P \{1-\chi(P)(NP)^{-s}\}^{-1}$$

(where P runs over prime ideals of k), in only a finite number of factors. In particular, if $k = \mathbb{Q}$, then $\zeta(f,c)$ is essentially a Dirichlet L-series $L(s,\chi)$. Thus these are essentially the only functions which can be associated to the rational field in this manner.

Tate goes on to prove a Poisson summation formula in this idèlic setting, and deduces the elegant functional equation

$$\zeta(f,c) = \zeta(\hat{f},\hat{c})$$

where $\hat{f}$ is the 'Fourier transform' of f, and $\hat{c}(a) = \overline{c_0(a)}|a|^{1-s}$. The functional equation for $\zeta(s,\chi)$ may be extracted from this. In the case $k = \mathbb{Q}$ we may take c_0 identically equal to 1, and make a particular choice $f = f_0$, such that $\hat{f}_0 = f_0$ and

$$\zeta(f_0, |\cdot|^s) = \pi^{-\frac{1}{2}s}\Gamma(\tfrac{1}{2}s)\zeta(s).$$

The functional equation (2.6.4) is then immediate. Moreover it is now apparent that the factor $\pi^{-\frac{1}{2}s}\Gamma(\frac{1}{2}s)$ should be viewed as the natural term to be included in the Euler product, to correspond to the real valuation of $\mathbb{Q}$.

2.21. It is remarkable that the values of $\zeta(s)$ for $s = 0, -1, -2, \ldots$, are all rational, and this suggests the possibility of a p-adic analogue of $\zeta(s)$, interpolating these numbers. In fact it can be shown that for any prime p and any integer n there is a unique meromorphic function $\zeta_{p,n}(s)$ defined for $s \in \mathbb{Z}_p$, (the p-adic integers) such that

$$\zeta_{p,n}(k) = (1-p^{-k})\zeta(k) \quad \text{for} \quad k \leqslant 0,\ k \equiv n \ (\text{mod}\ p-1).$$

Indeed if $n \not\equiv 1 \ (\text{mod}\ p-1)$ then $\zeta_{p,n}(s)$ will be analytic on $\mathbb{Z}_p$, and if $n \equiv 1 \ (\text{mod}\ p-1)$ then $\zeta_{p,n}(s)$ will be analytic apart from a simple pole at $s = 1$, of residue $1-(1/p)$. These results are due to Leopoldt and Kubota [**1**]. While these p-adic zeta-functions seem to have little interest in the simple case above, their generalizations to Dirichlet L-functions yield important algebraic information about the corresponding cyclotomic fields.

III

THE THEOREM OF HADAMARD AND DE LA VALLÉE POUSSIN, AND ITS CONSEQUENCES

3.1. As we have already observed, it follows from the formula

$$\zeta(s) = \prod_p \left(1 - \frac{1}{p^s}\right)^{-1} \quad (\sigma > 1) \tag{3.1.1}$$

that $\zeta(s)$ has no zeros for $\sigma > 1$. For the purpose of prime-number theory, and indeed to determine the general nature of $\zeta(s)$, it is necessary to extend as far as possible this zero-free region.

It was conjectured by Riemann that *all the complex zeros of* $\zeta(s)$ *lie on the 'critical line'* $\sigma = \frac{1}{2}$. This conjecture, now known as the Riemann hypothesis, has never been either proved or disproved.

The problem of the zero-free region appears to be a question of extending the sphere of influence of the Euler product (3.1.1) beyond its actual region of convergence; for examples are known of functions which are extremely like the zeta-function in their representation by Dirichlet series, functional equation, and so on, but which have no Euler product, and for which the analogue of the Riemann hypothesis is false. In fact the deepest theorems on the distribution of the zeros of $\zeta(s)$ are obtained in the way suggested. But the problem of extending the sphere of influence of (3.1.1) to the left of $\sigma = 1$ in any effective way appears to be of extreme difficulty.

By (1.1.4)
$$\frac{1}{\zeta(s)} = \sum_{n=1}^{\infty} \frac{\mu(n)}{n^s} \quad (\sigma > 1),$$

where $|\mu(n)| \leqslant 1$. Hence for σ near to 1

$$\left|\frac{1}{\zeta(s)}\right| \leqslant \sum_{n=1}^{\infty} \frac{1}{n^\sigma} = \zeta(\sigma) < \frac{A}{\sigma - 1},$$

i.e.
$$|\zeta(s)| > A(\sigma - 1).$$

Hence if $\zeta(s)$ has a zero on $\sigma = 1$ it must be a simple zero. But to prove that there cannot be even simple zeros, a much more subtle argument is required.

It was proved independently by Hadamard and de la Vallée Poussin in 1896 *that* $\zeta(s)$ *has no zeros on the line* $\sigma = 1$. Their methods are similar in principle, and they form the main topic of this chapter.

The main object of both these mathematicians was to prove the prime-number theorem, that as $x \to \infty$

$$\pi(x) \sim \frac{x}{\log x}.$$

This had previously been conjectured on empirical grounds. It was shown by arguments depending on the theory of functions of a complex variable that the prime-number theorem is a consequence of the Hadamard–de la Vallée Poussin theorem. The proof of the prime-number theorem so obtained was therefore not elementary.

An elementary proof of the prime-number theorem, i.e. a proof not depending on the theory of $\zeta(s)$ and complex function theory, has recently been obtained by A. Selberg and Erdős. Since the prime-number theorem implies the Hadamard–de la Vallée Poussin theorem, this leads to a new proof of the latter. However, the Selberg–Erdős method does not lead to such good estimations as the Hadamard–de la Vallée Poussin method, so that the latter is still of great interest.

3.2. Hadamard's argument is, roughly, as follows. We have for $\sigma > 1$

$$\log\zeta(s) = \sum_p \sum_{m=1}^{\infty} \frac{1}{mp^{ms}} = \sum_p \frac{1}{p^s} + f(s), \tag{3.2.1}$$

where $f(s)$ is regular for $\sigma > \frac{1}{2}$. Since $\zeta(s)$ has a simple pole at $s = 1$, it follows in particular that, as $\sigma \to 1$ $(\sigma > 1)$,

$$\sum_p \frac{1}{p^\sigma} \sim \log\frac{1}{\sigma-1}. \tag{3.2.2}$$

Suppose now that $s = 1+it_0$ is a zero of $\zeta(s)$. Then if $s = \sigma+it_0$, as $\sigma \to 1$ $(\sigma > 1)$

$$\sum_p \frac{\cos(t_0\log p)}{p^\sigma} = \log|\zeta(s)| - \mathbf{R}f(s) \sim \log(\sigma-1). \tag{3.2.3}$$

Comparing (3.2.2) and (3.2.3), we see that $\cos(t_0 \log p)$ must, in some sense, be approximately -1 for most values of p. But then $\cos(2t_0 \log p)$ is approximately 1 for most values of p, and

$$\log|\zeta(\sigma+2it_0)| \sim \sum_p \frac{\cos(2t_0\log p)}{p^\sigma} \sim \sum_p \frac{1}{p^\sigma} \sim \log\frac{1}{\sigma-1},$$

so that $1+2it_0$ is a pole of $\zeta(s)$. Since this is false, it follows that $\zeta(1+it_0) \neq 0$.

To put the argument in a rigorous form, let

$$S = \sum_p \frac{1}{p^\sigma}, \qquad P = \sum_p \frac{\cos(t_0\log p)}{p^\sigma}, \qquad Q = \sum_p \frac{\cos(2t_0\log p)}{p^\sigma}.$$

Let S', P', Q' be the parts of these sums for which

$$(2k+1)\pi-\alpha \leqslant t_0 \log p \leqslant (2k+1)\pi+\alpha$$

for any integer k, and α fixed, $0 < \alpha < \frac{1}{4}\pi$. Let S'', etc., be the remainders. Let $\lambda = S'/S$.

If ϵ is any positive number, it follows from (3.2.2) and (3.2.3) that

$$P < -(1-\epsilon)S$$

if $\sigma-1$ is small enough. But

$$P' \geqslant -S' = -\lambda S$$

and
$$P'' \geqslant -S'' \cos\alpha = -(1-\lambda)S\cos\alpha.$$

Hence
$$-\{\lambda+(1-\lambda)\cos\alpha\}S < -(1-\epsilon)S,$$

i.e.
$$(1-\lambda)(1-\cos\alpha) < \epsilon.$$

Hence $\lambda \to 1$ as $\sigma \to 1$.

Also
$$Q' \geqslant S'\cos 2\alpha, \qquad Q'' \geqslant -S'',$$

so that
$$Q \geqslant S(\lambda\cos 2\alpha - 1+\lambda).$$

Since $\lambda \to 1$, $S \to \infty$, it follows that $Q \to \infty$ as $\sigma \to 1$. Hence $1+2it_0$ is a pole, and the result follows as before.

The following form of the argument was suggested by Dr. F. V. Atkinson. We have

$$\left\{\sum_p \frac{\cos(t_0\log p)}{p^\sigma}\right\}^2 = \left\{\sum_p \frac{\cos(t_0\log p)}{p^{\frac{1}{2}\sigma}}\frac{1}{p^{\frac{1}{2}\sigma}}\right\}^2$$

$$\leqslant \sum_p \frac{\cos^2(t_0\log p)}{p^\sigma}\sum_p \frac{1}{p^\sigma}$$

$$= \frac{1}{2}\sum_p \frac{1+\cos(2t_0\log p)}{p^\sigma}\sum_p \frac{1}{p^\sigma},$$

i.e.
$$P^2 \leqslant \tfrac{1}{2}(S+Q)S.$$

Suppose now that, for some t_0, $P \sim \log(\sigma-1)$. Since $S \sim \log\{1/(\sigma-1)\}$, it follows that, for a given ϵ and $\sigma-1$ small enough,

$$(1-\epsilon)^2\log^2\frac{1}{\sigma-1} \leqslant \frac{1}{2}\left\{(1+\epsilon)\log\frac{1}{\sigma-1}+Q\right\}(1+\epsilon)\log\frac{1}{\sigma-1},$$

i.e.
$$Q \geqslant \left\{\frac{2(1-\epsilon)^2}{1+\epsilon}-1-\epsilon\right\}\log\frac{1}{\sigma-1}.$$

Hence $Q \to \infty$, and this involves a contradiction as before.

3.3. In de la Vallée Poussin's argument a relation between $\zeta(\sigma+it)$ and $\zeta(\sigma+2it)$ is also fundamental; but the result is now deduced from the fact that

$$3+4\cos\phi+\cos 2\phi = 2(1+\cos\phi)^2 \geqslant 0 \tag{3.3.1}$$

for all values of ϕ.

We have
$$\zeta(s) = \exp\sum_p \sum_{m=1}^{\infty} \frac{1}{mp^{ms}},$$
and hence
$$|\zeta(s)| = \exp\sum_p \sum_{m=1}^{\infty} \frac{\cos(mt\log p)}{mp^{m\sigma}}.$$
Hence

$$\zeta^3(\sigma)|\zeta(\sigma+it)|^4|\zeta(\sigma+2it)|$$
$$= \exp\left\{\sum_p \sum_{m=1}^{\infty} \frac{3+4\cos(mt\log p)+\cos(2mt\log p)}{mp^{m\sigma}}\right\}. \tag{3.3.2}$$

Since every term in the last sum is positive or zero, it follows that

$$\zeta^3(\sigma)|\zeta(\sigma+it)|^4|\zeta(\sigma+2it)| \geqslant 1 \quad (\sigma > 1). \tag{3.3.3}$$

Now, keeping t fixed, let $\sigma \to 1$. Then

$$\zeta^3(\sigma) = O\{(\sigma-1)^{-3}\},$$

and, if $1+it$ is a zero of $\zeta(s)$, $\zeta(\sigma+it) = O(\sigma-1)$. Also $\zeta(\sigma+2it) = O(1)$, since $\zeta(s)$ is regular at $1+2it$. Hence the left-hand side of (3.3.3) is $O(\sigma-1)$, giving a contradiction. This proves the theorem.

There are other inequalities of the same type as (3.3.1), which can be used for the same purpose; e.g. from

$$5+8\cos\phi+4\cos 2\phi+\cos 3\phi = (1+\cos\phi)(1+2\cos\phi)^2 \geqslant 0 \tag{3.3.4}$$

we deduce that

$$\zeta^5(\sigma)|\zeta(\sigma+it)|^8|\zeta(\sigma+2it)|^4|\zeta(\sigma+3it)| \geqslant 1. \tag{3.3.5}$$

This, however, has no particular advantage over (3.3.3).

3.4. Another alternative proof has been given by Ingham.† This depends on the identity

$$\frac{\zeta^2(s)\zeta(s+ai)\zeta(s-ai)}{\zeta(2s)} = \sum_{n=1}^{\infty} \frac{|\sigma_{ai}(n)|^2}{n^s} \quad (\sigma > 1), \tag{3.4.1}$$

where a is any real number other than zero, and

$$\sigma_{ai}(n) = \sum_{d|n} d^{ai}.$$

† Ingham (3).

This is the particular case of (1.3.3) obtained by putting ai for a and $-ai$ for b.

Let σ_0 be the abscissa of convergence of the series (3.4.1). Then $\sigma_0 \leqslant 1$, and (3.4.1) is valid by analytic continuation for $\sigma > \sigma_0$, the function $f(s)$ on the left-hand side being of necessity regular in this half-plane. Also, since all the coefficients in the Dirichlet series are positive, the real point of the line of convergence, viz. $s = \sigma_0$, is a singularity of the function.

Suppose now that $1+ai$ is a zero of $\zeta(s)$. Then $1-ai$ is also a zero, and these two zeros cancel the double pole of $\zeta^2(s)$ at $s = 1$. Hence $f(s)$ is regular on the real axis as far as $s = -1$, where $\zeta(2s) = 0$; and so $\sigma_0 = -1$. This is easily seen in various ways to be impossible; for example (3.4.1) would then give $f(\frac{1}{2}) \geqslant 1$, whereas in fact $f(\frac{1}{2}) = 0$.

3.5. In the following sections we extend as far as we can the ideas suggested by § 3.1.

Since $\zeta(s)$ has a finite number of zeros in the rectangle $0 \leqslant \sigma \leqslant 1$, $0 \leqslant t \leqslant T$ and none of them lie on $\sigma = 1$, it follows that there is a rectangle $1-\delta \leqslant \sigma \leqslant 1$, $0 \leqslant t \leqslant T$, which is free from zeros. Here $\delta = \delta(T)$ may, for all we can prove, tend to zero as $T \to \infty$; but we can obtain a positive lower bound for $\delta(T)$ for each value of T.

Again, since $1/\zeta(s)$ is regular for $\sigma = 1$, $1 \leqslant t \leqslant T$, it has an upper bound in the interval, which is a function of T. We also investigate the behaviour of this upper bound as $t \to \infty$. There is, of course, a similar problem for $\zeta(s)$, in which the distribution of the zeros is not immediately involved. It is convenient to consider all these problems together, and we begin with $\zeta(s)$.

THEOREM 3.5. *We have*

$$\zeta(s) = O(\log t) \tag{3.5.1}$$

uniformly in the region

$$1-\frac{A}{\log t} \leqslant \sigma \leqslant 2 \quad (t > t_0),$$

where A is any positive constant. In particular

$$\zeta(1+it) = O(\log t). \tag{3.5.2}$$

In (2.1.3), take $\sigma > 1$, $a = N$, and make $b \to \infty$. We obtain

$$\zeta(s) - \sum_{n=1}^{N} \frac{1}{n^s} = s \int_N^\infty \frac{[x]-x+\frac{1}{2}}{x^{s+1}}\, dx + \frac{N^{1-s}}{s-1} - \tfrac{1}{2}N^{-s}, \tag{3.5.3}$$

the result holding by analytic continuation for $\sigma > 0$. Hence for $\sigma > 0$, $t > 1$,

$$\zeta(s) - \sum_{n=1}^{N} \frac{1}{n^s} = O\left(t \int_N^\infty \frac{dx}{x^{\sigma+1}}\right) + O\left(\frac{N^{1-\sigma}}{t}\right) + O(N^{-\sigma})$$

$$= O\left(\frac{t}{\sigma N^\sigma}\right) + O\left(\frac{N^{1-\sigma}}{t}\right) + O(N^{-\sigma}). \tag{3.5.4}$$

In the region considered, if $n \leqslant t$,

$$|n^{-s}| = n^{-\sigma} = e^{-\sigma \log n} \leqslant \exp\left\{-\left(1 - \frac{A}{\log t}\right)\log n\right\} \leqslant n^{-1}e^A.$$

Hence, taking $N = [t]$,

$$\zeta(s) = \sum_{n=1}^{N} O\left(\frac{1}{n}\right) + O\left(\frac{t}{N}\right) + O\left(\frac{1}{t}\right) + O\left(\frac{1}{N}\right)$$

$$= O(\log N) + O(1) = O(\log t).$$

This result will be improved later (Theorems 5.16, 6.11), but at the cost of far more difficult proofs.

It is also easy to see that

$$\zeta'(s) = O(\log^2 t) \tag{3.5.5}$$

in the above region. For, differentiating (3.5.3),

$$\zeta'(s) = -\sum_{n=2}^{N} \frac{\log n}{n^s} + \int_N^\infty \frac{[x] - x + \frac{1}{2}}{x^{s+1}} (1 - s\log x)\, dx -$$

$$- \frac{N^{1-s}\log N}{s-1} - \frac{N^{1-s}}{(s-1)^2} + \tfrac{1}{2}N^{-s}\log N,$$

and a similar argument holds, with an extra factor $\log t$ on the right-hand side. Similarly for higher derivatives of $\zeta(s)$.

We may note in passing that (3.5.3) shows the behaviour of the Dirichlet series (1.1.1) for $\sigma \leqslant 1$. If we take $\sigma = 1$, $t \neq 0$, we obtain

$$\zeta(1+it) - \sum_{1}^{N} \frac{1}{n^{1+it}} = (1+it) \int_N^\infty \frac{[x] - x + \frac{1}{2}}{x^{2+it}}\, dx + \frac{N^{-it}}{it} - \tfrac{1}{2}N^{-1-it},$$

which oscillates finitely as $N \to \infty$. For $\sigma < 1$ the series, of course, diverges (oscillates infinitely).

3.6. Inequalities for $1/\zeta(s)$, $\zeta'(s)/\zeta(s)$, and $\log \zeta(s)$. Inequalities of this type in the neighbourhood of $\sigma = 1$ can now be obtained by a slight elaboration of the argument of § 3.3. We have for $\sigma > 1$

$$\left|\frac{1}{\zeta(\sigma+it)}\right| \leqslant \{\zeta(\sigma)\}^{\frac{3}{4}} |\zeta(\sigma+2it)|^{\frac{1}{4}} = O\left\{\frac{\log^{\frac{1}{4}} t}{(\sigma-1)^{\frac{3}{4}}}\right\}. \tag{3.6.1}$$

Also $$\zeta(1+it)-\zeta(\sigma+it) = -\int_1^\sigma \zeta'(u+it)\,du = O\{(\sigma-1)\log^2 t\} \quad (3.6.2)$$

for $\sigma > 1-A/\log t$. Hence

$$|\zeta(1+it)| > A_1\frac{(\sigma-1)^{\frac{3}{4}}}{\log^{\frac{1}{4}}t}-A_2(\sigma-1)\log^2 t.$$

The two terms on the right are of the same order if $\sigma-1 = \log^{-9}t$. Hence, taking $\sigma-1 = A_3\log^{-9}t$, where A_3 is sufficiently small,

$$|\zeta(1+it)| > A\log^{-7}t. \quad (3.6.3)$$

Next (3.6.2) and (3.6.3) together give, for $1-A\log t < \sigma < 1$,

$$|\zeta(\sigma+it)| > A\log^{-7}t-A(1-\sigma)\log^2 t, \quad (3.6.4)$$

and the right-hand side is positive if $1-\sigma < A\log^{-9}t$. *Hence $\zeta(s)$ has no zeros in the region $\sigma > 1-A\log^{-9}t$*, and in fact, by (3.6.4),

$$\frac{1}{\zeta(s)} = O(\log^7 t) \quad (3.6.5)$$

in this region.

Hence also, by (3.5.5),

$$\frac{\zeta'(s)}{\zeta(s)} = O(\log^9 t), \quad (3.6.6)$$

and $$\log\zeta(s) = \int_2^\sigma \frac{\zeta'(u+it)}{\zeta(u+it)}\,du+\log\zeta(2+it) = O(\log^9 t), \quad (3.6.7)$$

both for $\sigma > 1-A\log^{-9}t$.

We shall see later that all these results can be improved, but they are sufficient for some purposes.

3.7. The Prime-number Theorem. *Let $\pi(x)$ denote the number of primes not exceeding x. Then as $x \to \infty$*

$$\pi(x) \sim \frac{x}{\log x}. \quad (3.7.1)$$

The investigation of $\pi(x)$ was, of course, the original purpose for which $\zeta(s)$ was studied. It is not our purpose to pursue this side of the theory farther than is necessary, but it is convenient to insert here a proof of the main theorem on $\pi(x)$.

We have proved in (1.1.3) that, if $\sigma > 1$,

$$\log\zeta(s) = s\int_2^\infty \frac{\pi(x)}{x(x^s-1)}\,dx.$$

We want an explicit formula for $\pi(x)$, i.e. we want to invert the above integral formula. We can reduce this to a case of Mellin's inversion formula as follows. Let

$$\omega(s) = \int_2^\infty \frac{\pi(x)}{x^{s+1}(x^s-1)}\,dx.$$

Then $$\frac{\log\zeta(s)}{s} - \omega(s) = \int_2^\infty \frac{\pi(x)}{x^{s+1}}\,dx. \tag{3.7.2}$$

This is of the Mellin form, and $\omega(s)$ is a comparatively trivial function; in fact since $\pi(x) \leqslant x$ the integral for $\omega(s)$ converges uniformly for $\sigma \geqslant \frac{1}{2}+\delta$, by comparison with

$$\int_2^\infty \frac{dx}{x^{\frac{1}{2}+\delta}(x^{\frac{1}{2}+\delta}-1)}.$$

Hence $\omega(s)$ is regular and bounded for $\sigma \geqslant \frac{1}{2}+\delta$. Similarly so is $\omega'(s)$, since

$$\omega'(s) = \int_2^\infty \pi(x)\log x\,\frac{1-2x^s}{x^{s+1}(x^s-1)^2}\,dx.$$

We could now use Mellin's inversion formula, but the resulting formula is not easily manageable. We therefore modify (3.7.2) as follows. Differentiating with respect to s,

$$-\frac{\zeta'(s)}{s\zeta(s)} + \frac{\log\zeta(s)}{s^2} + \omega'(s) = \int_2^\infty \frac{\pi(x)\log x}{x^{s+1}}\,dx.$$

Denote the left-hand side by $\phi(s)$, and let

$$g(x) = \int_0^x \frac{\pi(u)\log u}{u}\,du, \qquad h(x) = \int_0^x \frac{g(u)}{u}\,du,$$

$\pi(x)$, $g(x)$, and $h(x)$ being zero for $x < 2$. Then, integrating by parts,

$$\phi(s) = \int_0^\infty g'(x)x^{-s}\,dx = s\int_0^\infty g(x)x^{-s-1}\,dx$$

$$= s\int_0^\infty h'(x)x^{-s}\,dx = s^2\int_0^\infty h(x)x^{-s-1}\,dx \quad (\sigma > 1),$$

or $$\frac{\phi(1-s)}{(1-s)^2} = \int_0^\infty \frac{h(x)}{x}x^{s-1}\,dx.$$

Now $h(x)$ is continuous and of bounded variation in any finite interval; and, since $\pi(x) \leqslant x$, it follows that, for $x > 1$, $g(x) \leqslant x \log x$, and $h(x) \leqslant x \log x$. Hence $h(x)x^{k-2}$ is absolutely integrable over $(0, \infty)$ if $k < 0$. Hence

$$\frac{h(x)}{x} = \frac{1}{2\pi i} \int\limits_{k-i\infty}^{k+i\infty} \frac{\phi(1-s)}{(1-s)^2} x^{-s}\, ds \quad (k < 0),$$

or

$$h(x) = \frac{1}{2\pi i} \int\limits_{c-i\infty}^{c+i\infty} \frac{\phi(s)}{s^2} x^s\, ds \quad (c > 1).$$

The integral on the right is absolutely convergent, since by (3.6.6) and (3.6.7) $\phi(s)$ is bounded for $\sigma \geqslant 1$, except in the neighbourhood of $s = 1$.

In the neighbourhood of $s = 1$

$$\phi(s) = \frac{1}{s-1} + \log \frac{1}{s-1} + \dots,$$

and we may write

$$\phi(s) = \frac{1}{s-1} + \psi(s),$$

where $\psi(s)$ is bounded for $\sigma \geqslant 1$, $|s-1| \geqslant 1$, and $\psi(s)$ has a logarithmic infinity as $s \to 1$. Now

$$h(x) = \frac{1}{2\pi i} \int\limits_{c-i\infty}^{c+i\infty} \frac{x^s}{(s-1)s^2}\, ds + \frac{1}{2\pi i} \int\limits_{c-i\infty}^{c+i\infty} \frac{\psi(s)}{s^2} x^s\, ds.$$

The first term is equal to the sum of the residues on the left of the line $\mathbf{R}(s) = c$, and so is

$$x - \log x - 1.$$

In the other term we may put $c = 1$, i.e. apply Cauchy's theorem to the rectangle $(1 \pm iT,\ c \pm iT)$, with an indentation of radius ϵ round $s = 1$, and make $T \to \infty$, $\epsilon \to 0$. Hence

$$h(x) = x - \log x - 1 + \frac{x}{2\pi} \int\limits_{-\infty}^{\infty} \frac{\psi(1+it)}{(1+it)^2} x^{it}\, dt.$$

The last integral tends to zero as $x \to \infty$, by the extension to Fourier integrals of the Riemann–Lebesgue theorem.† Hence

$$h(x) \sim x. \tag{3.7.3}$$

† See my *Introduction to the Theory of Fourier Integrals*, Theorem 1.

To get back to $\pi(x)$ we now use the following lemma:

Let $f(x)$ *be positive non-decreasing, and as* $x \to \infty$ *let*

$$\int_1^x \frac{f(u)}{u}\,du \sim x.$$

Then
$$f(x) \sim x.$$

If δ is a given positive number,

$$(1-\delta)x < \int_1^x \frac{f(t)}{t}\,dt < (1+\delta)x \quad (x > x_0(\delta)).$$

Hence for any positive ϵ

$$\int_x^{x(1+\epsilon)} \frac{f(u)}{u}\,du = \int_1^{x(1+\epsilon)} \frac{f(u)}{u}\,du - \int_1^x \frac{f(u)}{u}\,du$$
$$< (1+\delta)(1+\epsilon)x-(1-\delta)x$$
$$= (2\delta+\epsilon+\delta\epsilon)x.$$

But, since $f(x)$ is non-decreasing,

$$\int_x^{x(1+\epsilon)} \frac{f(u)}{u}\,du \geqslant f(x)\int_x^{x(1+\epsilon)} \frac{du}{u} > f(x)\int_x^{x(1+\epsilon)} \frac{du}{x(1+\epsilon)} = \frac{\epsilon}{1+\epsilon}f(x).$$

Hence
$$f(x) < x(1+\epsilon)\left(1+\delta+\frac{2\delta}{\epsilon}\right).$$

Taking, for example, $\epsilon = \sqrt{\delta}$, it follows that

$$\overline{\lim}\,\frac{f(x)}{x} \leqslant 1.$$

Similarly, by considering
$$\int_{x(1-\epsilon)}^x \frac{f(u)}{u}\,du,$$

we obtain
$$\underline{\lim}\,\frac{f(x)}{x} \geqslant 1,$$

and the lemma follows.

Applying the lemma twice, we deduce from (3.7.3) that

$$g(x) \sim x,$$

and hence that
$$\pi(x)\log x \sim x.$$

3.8. THEOREM 3.8. *There is a constant* A *such that* $\zeta(s)$ *is not zero for*

$$\sigma \geqslant 1-\frac{A}{\log t} \quad (t > t_0).$$

We have for $\sigma > 1$

$$-\mathbf{R}\left\{\frac{\zeta'(s)}{\zeta(s)}\right\} = \sum_{p,m} \frac{\log p}{p^{m\sigma}} \cos(mt \log p). \tag{3.8.1}$$

Hence, for $\sigma > 1$ and any real γ,

$$-3\frac{\zeta'(\sigma)}{\zeta(\sigma)} - 4\mathbf{R}\frac{\zeta'(\sigma+i\gamma)}{\zeta(\sigma+i\gamma)} - \mathbf{R}\frac{\zeta'(\sigma+2i\gamma)}{\zeta(\sigma+2i\gamma)}$$

$$= \sum_{p,m} \frac{\log p}{p^{m\sigma}}\{3+4\cos(m\gamma\log p)+\cos(2m\gamma\log p)\} \geqslant 0. \tag{3.8.2}$$

Now

$$-\frac{\zeta'(\sigma)}{\zeta(\sigma)} < \frac{1}{\sigma-1} + O(1). \tag{3.8.3}$$

Also, by (2.12.7),

$$-\frac{\zeta'(s)}{\zeta(s)} = O(\log t) - \sum_{\rho} \left(\frac{1}{s-\rho} + \frac{1}{\rho}\right), \tag{3.8.4}$$

where $\rho = \beta + i\gamma$ runs through complex zeros of $\zeta(s)$. Hence

$$-\mathbf{R}\left\{\frac{\zeta'(s)}{\zeta(s)}\right\} = O(\log t) - \sum_{\rho} \left\{\frac{\sigma-\beta}{(\sigma-\beta)^2+(t-\gamma)^2} + \frac{\beta}{\beta^2+\gamma^2}\right\}.$$

Since every term in the last sum is positive, it follows that

$$-\mathbf{R}\left\{\frac{\zeta'(s)}{\zeta(s)}\right\} < O(\log t), \tag{3.8.5}$$

and also, if $\beta+i\gamma$ is a particular zero of $\zeta(s)$, that

$$-\mathbf{R}\left\{\frac{\zeta'(\sigma+i\gamma)}{\zeta(\sigma+i\gamma)}\right\} < O(\log\gamma) - \frac{1}{\sigma-\beta}. \tag{3.8.6}$$

From (3.8.2), (3.8.3), (3.8.5), (3.8.6) we obtain

$$\frac{3}{\sigma-1} - \frac{4}{\sigma-\beta} + O(\log\gamma) \geqslant 0,$$

or say

$$\frac{3}{\sigma-1} - \frac{4}{\sigma-\beta} \geqslant -A_1 \log\gamma.$$

Solving for β, we obtain

$$1-\beta \geqslant \frac{1-(\sigma-1)A_1\log\gamma}{3/(\sigma-1)+A_1\log\gamma}.$$

The right-hand side is positive if $\sigma-1 = \frac{1}{2}A_1/\log\gamma$, and then

$$1-\beta \geqslant \frac{A_2}{\log\gamma},$$

the required result.

3.9. There is an alternative method, due to Landau,† of obtaining results of this kind, in which the analytic character of $\zeta(s)$ for $\sigma \leqslant 0$ need not be known. It depends on the following lemmas.

LEMMA α. *If $f(s)$ is regular, and*

$$\left|\frac{f(s)}{f(s_0)}\right| < e^M \quad (M > 1)$$

in the circle $|s-s_0| \leqslant r$, then

$$\left|\frac{f'(s)}{f(s)} - \sum_\rho \frac{1}{s-\rho}\right| < \frac{AM}{r} \quad (|s-s_0| \leqslant \tfrac{1}{4}r),$$

where ρ runs through the zeros of $f(s)$ such that $|\rho-s_0| \leqslant \frac{1}{2}r$.

The function $g(s) = f(s) \prod_\rho (s-\rho)^{-1}$ is regular for $|s-s_0| \leqslant r$, and not zero for $|s-s_0| \leqslant \frac{1}{2}r$. On $|s-s_0| = r$, $|s-\rho| \geqslant \frac{1}{2}r \geqslant |s_0-\rho|$, so that

$$\left|\frac{g(s)}{g(s_0)}\right| = \left|\frac{f(s)}{f(s_0)} \prod \left(\frac{s_0-\rho}{s-\rho}\right)\right| \leqslant \left|\frac{f(s)}{f(s_0)}\right| < e^M.$$

This inequality therefore holds inside the circle also. Hence the function

$$h(s) = \log\left\{\frac{g(s)}{g(s_0)}\right\},$$

where the logarithm is zero at $s = s_0$, is regular for $|s-s_0| \leqslant \frac{1}{2}r$, and

$$h(s_0) = 0, \qquad \mathbf{R}\{h(s)\} < M.$$

Hence by the Borel–Carathéodory theorem‡

$$|h(s)| < AM \quad (|s-s_0| \leqslant \tfrac{3}{8}r), \tag{3.9.1}$$

and so, for $|s-s_0| \leqslant \frac{1}{4}r$,

$$|h'(s)| = \left|\frac{1}{2\pi i} \int\limits_{|z-s|=\frac{1}{8}r} \frac{h(z)}{(z-s)^2}\,dz\right| < \frac{AM}{r}.$$

This gives the result stated.

LEMMA β. *If $f(s)$ satisfies the conditions of the previous lemma, and has no zeros in the right-hand half of the circle $|s-s_0| \leqslant r$, then*

$$-\mathbf{R}\left\{\frac{f'(s_0)}{f(s_0)}\right\} < \frac{AM}{r};$$

while if $f(s)$ has a zero ρ_0 between $s_0-\frac{1}{2}r$ and s_0, then

$$-\mathbf{R}\left\{\frac{f'(s_0)}{f(s_0)}\right\} < \frac{AM}{r} - \frac{1}{s_0-\rho_0}.$$

† Landau (14). ‡ Titchmarsh, *Theory of Functions*, § 5.5.

Lemma α gives

$$-\mathbf{R}\left\{\frac{f'(s_0)}{f(s_0)}\right\} < \frac{AM}{r} - \sum \mathbf{R}\frac{1}{s_0-\rho},$$

and since $\mathbf{R}\{1/(s_0-\rho)\} \geqslant 0$ for every ρ, both results follow at once.

LEMMA γ. *Let $f(s)$ satisfy the conditions of Lemma α, and let*

$$\left|\frac{f'(s_0)}{f(s_0)}\right| < \frac{M}{r}.$$

Suppose also that $f(s) \neq 0$ in the part $\sigma \geqslant \sigma_0 - 2r'$ of the circle $|s-s_0| \leqslant r$, where $0 < r' < \frac{1}{4}r$. Then

$$\left|\frac{f'(s)}{f(s)}\right| < A\frac{M}{r} \quad (|s-s_0| \leqslant r').$$

Lemma α now gives

$$-\mathbf{R}\left\{\frac{f'(s)}{f(s)}\right\} < A\frac{M}{r} - \sum \mathbf{R}\frac{1}{s-\rho} < A\frac{M}{r}$$

for all s in $|s-s_0| \leqslant \frac{1}{4}r$, $\sigma \geqslant \sigma_0 - 2r'$, each term of the sum being positive in this region. The result then follows on applying the Borel–Carathéodory theorem to the function $-f'(s)/f(s)$ and the circles $|s-s_0| = 2r'$, $|s-s_0| = r'$.

3.10. We can now prove the following general theorem, which we shall apply later with special forms of the functions $\theta(t)$ and $\phi(t)$.

THEOREM 3.10. *Let*

$$\zeta(s) = O(e^{\phi(t)})$$

as $t \to \infty$ in the region

$$1-\theta(t) \leqslant \sigma \leqslant 2 \quad (t \geqslant 0),$$

where $\phi(t)$ and $1/\theta(t)$ are positive non-decreasing functions of t for $t \geqslant 0$, such that $\theta(t) \leqslant 1$, $\phi(t) \to \infty$, and

$$\frac{\phi(t)}{\theta(t)} = o\,(e^{\phi(t)}). \tag{3.10.1}$$

Then there is a constant A_1 such that $\zeta(s)$ has no zeros in the region

$$\sigma \geqslant 1 - A_1\frac{\theta(2t+1)}{\phi(2t+1)}. \tag{3.10.2}$$

Let $\beta+i\gamma$ be a zero of $\zeta(s)$ in the upper half-plane. Let

$$1+e^{-\phi(2\gamma+1)} \leqslant \sigma_0 \leqslant 2,$$

$$s_0 = \sigma_0 + i\gamma, \qquad s_0' = \sigma_0 + 2i\gamma, \qquad r = \theta(2\gamma+1).$$

Then the circles $|s-s_0| \leqslant r$, $|s-s_0'| \leqslant r$ both lie in the region

$$\sigma \geqslant 1-\theta(t).$$

Now
$$\left|\frac{1}{\zeta(s_0)}\right| < \frac{A}{\sigma_0-1} < Ae^{\phi(2\gamma+1)},$$

and similarly for s_0'. Hence there is a constant A_2 such that

$$\left|\frac{\zeta(s)}{\zeta(s_0)}\right| < e^{A_2\phi(2\gamma+1)}, \qquad \left|\frac{\zeta(s)}{\zeta(s_0')}\right| < e^{A_2\phi(2\gamma+1)},$$

in the circles $|s-s_0| \leqslant r$, $|s-s_0'| \leqslant r$ respectively. We can therefore apply Lemma β with $M = A_2\phi(2\gamma+1)$. We obtain

$$-\mathbf{R}\left\{\frac{\zeta'(\sigma_0+2i\gamma)}{\zeta(\sigma_0+2i\gamma)}\right\} < \frac{A_3\phi(2\gamma+1)}{\theta(2\gamma+1)}, \tag{3.10.3}$$

and, if
$$\beta > \sigma_0-\tfrac{1}{2}r, \tag{3.10.4}$$

$$-\mathbf{R}\left\{\frac{\zeta'(\sigma_0+i\gamma)}{\zeta(\sigma_0+i\gamma)}\right\} < \frac{A_3\phi(2\gamma+1)}{\theta(2\gamma+1)}-\frac{1}{\sigma_0-\beta}. \tag{3.10.5}$$

Also as $\sigma_0 \to 1$
$$-\frac{\zeta'(\sigma_0)}{\zeta(\sigma_0)} \sim \frac{1}{\sigma_0-1}.$$

Hence
$$-\frac{\zeta'(\sigma_0)}{\zeta(\sigma_0)} < \frac{a}{\sigma_0-1}, \tag{3.10.6}$$

where a can be made as near 1 as we please by choice of σ_0.

Now (3.8.2), (3.10.3), (3.10.5), and (3.10.6) give

$$\frac{3a}{\sigma_0-1}+\frac{5A_3\phi(2\gamma+1)}{\theta(2\gamma+1)}-\frac{4}{\sigma_0-\beta} \geqslant 0,$$

$$\sigma_0-\beta \geqslant \left\{\frac{3a}{4(\sigma_0-1)}+\frac{5A_3}{4}\frac{\phi(2\gamma+1)}{\theta(2\gamma+1)}\right\}^{-1},$$

$$1-\beta \geqslant \left\{\frac{3a}{4(\sigma_0-1)}+\frac{5A_3}{4}\frac{\phi(2\gamma+1)}{\theta(2\gamma+1)}\right\}^{-1}-(\sigma_0-1)$$

$$= \left\{1-\frac{3a}{4}-\frac{5A_3}{4}\frac{\phi(2\gamma+1)(\sigma_0-1)}{\theta(2\gamma+1)}\right\}\Big/\left\{\frac{3a}{4(\sigma_0-1)}+\frac{5A_3}{4}\frac{\phi(2\gamma+1)}{\theta(2\gamma+1)}\right\}.$$

To make the numerator positive, take $a = \frac{5}{4}$, and

$$\sigma_0-1 = \frac{1}{40A_3}\frac{\theta(2\gamma+1)}{\phi(2\gamma+1)},$$

this being consistent with the previous conditions, by (3.10.1), if γ is large enough. It follows that

$$1-\beta \geqslant \frac{\theta(2\gamma+1)}{1240A_3\phi(2\gamma+1)}$$

as required. If (3.10.4) is not satisfied,

$$\beta \leqslant \sigma_0 - \tfrac{1}{2}r = 1 + \frac{1}{40A_3}\frac{\theta(2\gamma+1)}{\phi(2\gamma+1)} - \tfrac{1}{2}\theta(2\gamma+1),$$

which also leads to (3.10.2). This proves the theorem.

In particular, we can take $\theta(t) = \frac{1}{2}$, $\phi(t) = \log(t+2)$. This gives a new proof of Theorem 3.8.

3.11. THEOREM 3.11. *Under the hypotheses of Theorem* 3.10 *we have*

$$\frac{\zeta'(s)}{\zeta(s)} = O\left\{\frac{\phi(2t+3)}{\theta(2t+3)}\right\}, \qquad \frac{1}{\zeta(s)} = O\left\{\frac{\phi(2t+3)}{\theta(2t+3)}\right\} \tag{3.11.1), (3.11.2}$$

uniformly for

$$\sigma \geqslant 1 - \frac{A_1}{4}\frac{\theta(2t+3)}{\phi(2t+3)}. \tag{3.11.3}$$

In particular

$$\frac{\zeta(1+it)}{\zeta'(1+it)} = O\left\{\frac{\phi(2t+3)}{\theta(2t+3)}\right\}, \qquad \frac{1}{\zeta(1+it)} = O\left\{\frac{\phi(2t+3)}{\theta(2t+3)}\right\}. \tag{3.11.4), (3.11.5}$$

We apply Lemma γ, with

$$s_0 = 1 + \frac{A_1}{2}\frac{\theta(2t_0+3)}{\phi(2t_0+3)} + it_0, \qquad r = \theta(2t_0+3).$$

In the circle $|s-s_0| \leqslant r$

$$\frac{\zeta(s)}{\zeta(s_0)} = O\left\{\frac{e^{\phi(t)}}{\sigma_0-1}\right\} = O\left\{\frac{\phi(2t_0+3)}{\theta(2t_0+3)}e^{\phi(t_0+1)}\right\} = O\{e^{A\phi(2t_0+3)}\},$$

and
$$\frac{\zeta'(s_0)}{\zeta(s_0)} = O\left\{\frac{1}{\sigma_0-1}\right\} = O\left\{\frac{\phi(2t_0+3)}{\theta(2t_0+3)}\right\} = O\left\{\frac{\phi(2t_0+3)}{r}\right\}.$$

We can therefore take $M = A\phi(2t_0+3)$. Also, by the previous theorem, $\zeta(s)$ has no zeros for

$$t \leqslant t_0+1, \qquad \sigma \geqslant 1 - A_1\frac{\theta\{2(t_0+1)+1\}}{\phi\{2(t_0+1)+1\}} = 1 - A_1\frac{\theta(2t_0+3)}{\phi(2t_0+3)}.$$

Hence we can take
$$2r' = \frac{3A_1}{2}\frac{\theta(2t_0+3)}{\phi(2t_0+3)}.$$

Hence
$$\frac{\zeta'(s)}{\zeta(s)} = O\left\{\frac{\phi(2t_0+3)}{\theta(2t_0+3)}\right\}$$

for
$$|s-s_0| \leqslant \frac{3A_1}{4}\frac{\theta(2t_0+3)}{\phi(2t_0+3)},$$

and in particular for

$$t = t_0, \qquad \sigma \geqslant 1 - \frac{A_1}{4}\frac{\theta(2t_0+3)}{\phi(2t_0+3)}.$$

This is (3.11.1), with t_0 instead of t.

Also, if
$$1-\frac{A_1}{4}\frac{\theta(2t+3)}{\phi(2t+3)} \leqslant \sigma \leqslant 1+\frac{\theta(2t+3)}{\phi(2t+3)}, \tag{3.11.6}$$

$$\log\frac{1}{|\zeta(s)|} = -\mathbf{R}\log\zeta(s)$$
$$= -\mathbf{R}\log\zeta\left\{1+\frac{\theta(2t+3)}{\phi(2t+3)}+it\right\}+\int\limits_{\sigma}^{1+\frac{\theta(2t+3)}{\phi(2t+3)}} \mathbf{R}\frac{\zeta'(u+it)}{\zeta(u+it)}\,du$$
$$\leqslant \log\zeta\left\{1+\frac{\theta(2t+3)}{\phi(2t+3)}\right\}+\int\limits_{\sigma}^{1+\frac{\theta(2t+3)}{\phi(2t+3)}} O\left\{\frac{\phi(2t+3)}{\theta(2t+3)}\right\}du$$
$$< \log\frac{A\phi(2t+3)}{\theta(2t+3)}+O(1).$$

Hence (3.11.2) follows if σ is in the range (3.11.6); and for larger σ it is trivial.

Since we may take $\theta(t) = \frac{1}{2}$, $\phi(t) = \log(t+2)$, it follows that
$$\frac{\zeta'(s)}{\zeta(s)} = O(\log t), \qquad \frac{1}{\zeta(s)} = O(\log t) \qquad (3.11.7), (3.11.8)$$
in a region $\sigma \geqslant 1-A/\log t$; and in particular
$$\frac{\zeta'(1+it)}{\zeta(1+it)} = O(\log t), \qquad \frac{1}{\zeta(1+it)} = O(\log t). \qquad (3.11.9), (3.11.10)$$

3.12. For the next theorem we require the following lemma.

Lemma 3.12. *Let* $$f(s) = \sum_{n=1}^{\infty}\frac{a_n}{n^s} \quad (\sigma > 1),$$
where $a_n = O\{\psi(n)\}$, $\psi(n)$ *being non-decreasing, and*
$$\sum_{n=1}^{\infty}\frac{|a_n|}{n^\sigma} = O\left\{\frac{1}{(\sigma-1)^\alpha}\right\}$$
as $\sigma \to 1$. *Then if* $c > 0$, $\sigma+c > 1$, x *is not an integer, and* N *is the integer nearest to* x,
$$\sum_{n<x}\frac{a_n}{n^s} = \frac{1}{2\pi i}\int\limits_{c-iT}^{c+iT} f(s+w)\frac{x^w}{w}\,dw+O\left\{\frac{x^c}{T(\sigma+c-1)^\alpha}\right\}+$$
$$+O\left\{\frac{\psi(2x)x^{1-\sigma}\log x}{T}\right\}+O\left\{\frac{\psi(N)x^{1-\sigma}}{T|x-N|}\right\}. \tag{3.12.1}$$

If x is an integer, the corresponding result is

$$\sum_{n=1}^{x-1}\frac{a_n}{n^s}+\frac{a_x}{2x^s}=\frac{1}{2\pi i}\int\limits_{c-iT}^{c+iT}f(s+w)\frac{x^w}{w}\,dw+O\left\{\frac{x^c}{T(\sigma+c-1)^\alpha}\right\}+$$

$$+O\left\{\frac{\psi(2x)x^{1-\sigma}\log x}{T}\right\}+O\left\{\frac{\psi(x)x^{-\sigma}}{T}\right\}. \quad (3.12.2)$$

Suppose first that x is not an integer. If $n<x$, the calculus of residues gives

$$\frac{1}{2\pi i}\left(\int\limits_{-\infty-iT}^{c-iT}+\int\limits_{c-iT}^{c+iT}+\int\limits_{c+iT}^{-\infty+iT}\right)\left(\frac{x}{n}\right)^w\frac{dw}{w}=1.$$

Now

$$\int\limits_{-\infty+iT}^{c+iT}\left(\frac{x}{n}\right)^w\frac{dw}{w}=\left[\frac{(x/n)^w}{w\log x/n}\right]_{-\infty+iT}^{c+iT}+\frac{1}{\log x/n}\int\limits_{-\infty+iT}^{c+iT}\left(\frac{x}{n}\right)^w\frac{dw}{w^2}$$

$$=O\left\{\frac{(x/n)^c}{T\log x/n}\right\}+O\left\{\frac{(x/n)^c}{\log x/n}\int\limits_{-\infty}^{\infty}\frac{du}{u^2+T^2}\right\}$$

$$=O\left\{\frac{(x/n)^c}{T\log x/n}\right\},$$

and similarly for the integral over $(-\infty-iT, c-iT)$. Hence

$$\frac{1}{2\pi i}\int\limits_{c-iT}^{c+iT}\left(\frac{x}{n}\right)^w\frac{dw}{w}=1+O\left\{\frac{(x/n)^c}{T\log x/n}\right\}.$$

If $n>x$ we argue similarly with $-\infty$ replaced by $+\infty$, and there is no residue term. We therefore obtain a similar result without the term 1.

Multiplying by $a_n n^{-s}$ and summing,

$$\frac{1}{2\pi i}\int\limits_{c-iT}^{c+iT}f(s+w)\frac{x^w}{w}\,dw=\sum_{n<x}\frac{a_n}{n^s}+O\left\{\frac{x^c}{T}\sum_{n=1}^{\infty}\frac{|a_n|}{n^{\sigma+c}|\log x/n|}\right\}.$$

If $n<\frac{1}{2}x$ or $n>2x$, $|\log x/n|>A$, and these parts of the sum are

$$O\left(\sum_{n=1}^{\infty}\frac{|a_n|}{n^{\sigma+c}}\right)=O\left\{\frac{1}{(\sigma+c-1)^\alpha}\right\}.$$

If $N<n\leqslant 2x$, let $n=N+r$. Then

$$\log\frac{n}{x}\geqslant\log\frac{N+r}{N+\frac{1}{2}}>\frac{Ar}{N}>\frac{Ar}{x}.$$

Hence this part of the sum is

$$O\left\{\psi(2x)x^{1-\sigma-c}\sum_{1\leqslant r\leqslant x}\frac{1}{r}\right\} = O\{\psi(2x)x^{1-\sigma-c}\log x\}.$$

A similar argument applies to the terms with $\frac{1}{2}x \leqslant n < N$. Finally

$$\frac{|a_N|}{N^{\sigma+c}|\log x/N|} = O\left\{\frac{\psi(N)}{N^{\sigma+c}\log\{1+(x-N)/N\}}\right\} = O\left\{\frac{\psi(N)x^{1-\sigma-c}}{|x-N|}\right\}.$$

Hence (3.12.1) follows.

If x is an integer, all goes as before except for the term

$$\frac{a_x}{2\pi i x^s}\int_{c-iT}^{c+iT}\frac{dw}{w} = \frac{a_x}{2\pi i x^s}\log\frac{c+iT}{c-iT} = \frac{a_x}{2\pi i x^s}\left\{i\pi+O\left(\frac{1}{T}\right)\right\}.$$

Hence (3.12.2) follows.

3.13. THEOREM 3.13. *We have*

$$\frac{1}{\zeta(s)} = \sum_{n=1}^{\infty}\frac{\mu(n)}{n^s}$$

at all points of the line $\sigma = 1$.

Take $a_n = \mu(n)$, $\alpha = 1$, $\sigma = 1$, in the lemma, and let x be half an odd integer. We obtain

$$\sum_{n<x}\frac{\mu(n)}{n^s} = \frac{1}{2\pi i}\int_{c-iT}^{c+iT}\frac{1}{\zeta(s+w)}\frac{x^w}{w}\,dw+O\left(\frac{x^c}{Tc}\right)+O\left(\frac{\log x}{T}\right).$$

The theorem of residues gives

$$\frac{1}{2\pi i}\int_{c-iT}^{c+iT}\frac{1}{\zeta(s+w)}\frac{x^w}{w}\,dw = \frac{1}{\zeta(s)}+\frac{1}{2\pi i}\left(\int_{c-iT}^{-\delta-iT}+\int_{-\delta-iT}^{-\delta+iT}+\int_{-\delta+iT}^{c+iT}\right)$$

if δ is so small that $\zeta(s+w)$ has no zeros for

$$\mathbf{R}(w) \geqslant -\delta, \quad |\mathbf{I}(s+w)| \leqslant |t|+T.$$

By § 3.6 we can take $\delta = A\log^{-9}T$. Then

$$\int_{-\delta-iT}^{-\delta+iT}\frac{1}{\zeta(s+w)}\frac{x^w}{w}\,dw = O\left(x^{-\delta}\log^7 T\int_{-T}^{T}\frac{dv}{\sqrt{(\delta^2+v^2)}}\right)$$

$$= O\left\{x^{-\delta}\log^7 T\int_{-T/\delta}^{T/\delta}\frac{dv}{\sqrt{(1+v^2)}}\right\} = O(x^{-\delta}\log^8 T),$$

and

$$\int_{-\delta+iT}^{c+iT} \frac{1}{\zeta(s+w)} \frac{x^w}{w} \, dw = O\left(\frac{\log^7 T}{T} \int_{-\delta}^{c} x^u \, du\right) = O\left(\frac{x^c \log^7 T}{T}\right),$$

and similarly for the other integral. Hence

$$\sum_{n<x} \frac{\mu(n)}{n^s} - \frac{1}{\zeta(s)} = O\left(\frac{x^c}{Tc}\right) + O\left(\frac{\log x}{T}\right) + O\left(\frac{x^c \log^7 T}{T}\right) + O\left(\frac{\log^8 T}{x^\delta}\right).$$

Take $c = 1/\log x$, so that $x^c = e$; and take $T = \exp\{(\log x)^{1/10}\}$, so that $\log T = (\log x)^{1/10}$, $\delta = A(\log x)^{-9/10}$, $x^\delta = T^A$. Then the right-hand side tends to zero, and the result follows.

In particular $$\sum_{n=1}^{\infty} \frac{\mu(n)}{n} = 0.$$

3.14. *The series for* $\zeta'(s)/\zeta(s)$ *and* $\log \zeta(s)$ *on* $\sigma = 1$.

Taking† $a_n = \Lambda(n) = O(\log n)$, $\alpha = 1$, $\sigma = 1$, in the lemma, we obtain

$$\sum_{n<x} \frac{\Lambda(n)}{n^s} = -\frac{1}{2\pi i} \int_{c-iT}^{c+iT} \frac{\zeta'(s+w)}{\zeta(s+w)} \frac{x^w}{w} \, dw + O\left(\frac{x^c}{Tc}\right) + O\left(\frac{\log^2 x}{T}\right).$$

In this case there is a pole at $w = 1-s$, giving a residue term

$$\frac{\zeta'(s)}{\zeta(s)} - \frac{x^{1-s}}{1-s} \quad (s \neq 1), \qquad a - \log x \quad (s = 1),$$

where a is a constant. Hence if $s \neq 1$ we obtain

$$\sum_{n<x} \frac{\Lambda(n)}{n^s} + \frac{\zeta'(s)}{\zeta(s)} - \frac{x^{1-s}}{1-s} = O\left(\frac{x^c}{Tc}\right) + O\left(\frac{\log^2 x}{T}\right) + O\left(\frac{\log^{10} T}{x^\delta}\right) + O\left(\frac{x^c \log^9 T}{T}\right).$$

Taking $c = 1/\log x$, $T = \exp\{(\log x)^{1/10}\}$, we obtain as before

$$\sum_{n<x} \frac{\Lambda(n)}{n^s} + \frac{\zeta'(s)}{\zeta(s)} - \frac{x^{1-s}}{1-s} = o(1). \tag{3.14.1}$$

The term $x^{1-s}/(1-s)$ oscillates finitely, so that if $\mathbf{R}(s) = 1$, $s \neq 1$, the series $\sum \Lambda(n) n^{-s}$ is not convergent, but its partial sums are bounded.

If $s = 1$, we obtain

$$\sum_{n<x} \frac{\Lambda(n)}{n} = \log x + O(1), \tag{3.14.2}$$

or, since

$$\sum_{n<x} \frac{\Lambda(n)}{n} = \sum_{p<x} \frac{\log p}{p} + \sum_{m=2}^{\infty} \sum_{p^m<x} \frac{\log p}{p^m} = \sum_{p<x} \frac{\log p}{p} + O(1),$$

$$\sum_{p<x} \frac{\log p}{p} = \log x + O(1). \tag{3.14.3}$$

† See (1.1.8).

Since $\Lambda_1(n) = \Lambda(n)/\log n$, and $1/\log n$ tends steadily to zero, it follows that

$$\sum \frac{\Lambda_1(n)}{n^s}$$

is convergent on $\sigma = 1$, except for $t = 0$. Hence, by the continuity theorem for Dirichlet series, the equation

$$\log \zeta(s) = \sum_{n=2}^{\infty} \frac{\Lambda_1(n)}{n^s}$$

holds for $\sigma = 1$, $t \neq 0$.

To determine the behaviour of this series for $s = 1$ we have, as in the case of $1/\zeta(s)$,

$$\sum_{n<x} \frac{\Lambda_1(n)}{n} = \frac{1}{2\pi i} \int_{c-iT}^{c+iT} \log \zeta(w+1) \frac{x^w}{w}\, dw + O\left(\frac{\log x}{T}\right),$$

where $c = 1/\log x$, and T is chosen as before. Now

$$\frac{1}{2\pi i} \int_{c-iT}^{c+iT} \log \zeta(w+1) \frac{x^w}{w}\, dw = \frac{1}{2\pi i}\left(\int_{c-iT}^{-\delta-iT} + \int_{-\delta-iT}^{-\delta+iT} + \int_{-\delta+iT}^{c+iT} \right) + \frac{1}{2\pi i} \int_C,$$

where C is a loop starting and finishing at $s = -\delta$, and encircling the origin in the positive direction. Defining δ as before, the integral along $\sigma = -\delta$ is $O(x^{-\delta} \log^{10} T)$, and the integrals along the horizontal sides are $O(x^c T^{-1} \log^9 T)$, by (3.6.7). Since

$$\frac{1}{w}\left\{\log \zeta(w+1) - \log \frac{1}{w}\right\}$$

is regular at the origin, the last term is equal to

$$\frac{1}{2\pi i} \int_C \log \frac{1}{w} \frac{x^w}{w}\, dw.$$

Since

$$\frac{1}{2\pi i} \int_C \log \frac{1}{w} \frac{dw}{w} = -\frac{1}{4\pi i} \Delta_C \log^2 w$$

$$= -\frac{1}{4\pi i} \{\log^2(\delta e^{i\pi}) - \log^2(\delta e^{-i\pi})\} = -\log \delta,$$

this term is also equal to

$$\frac{1}{2\pi i} \int_C \log \frac{1}{w} \frac{x^w - 1}{w}\, dw - \log \delta.$$

Take C to be a circle with centre $w = 0$ and radius ρ ($\rho < \delta$), together with the segment $(-\delta, -\rho)$ of the real axis described twice. The integrals along the real segments together give

$$-\frac{1}{2\pi i}\int_{\delta}^{\rho}\log\left(\frac{1}{ue^{-i\pi}}\right)\frac{x^{-u}-1}{-u}\,du-\frac{1}{2\pi i}\int_{\rho}^{\delta}\log\left(\frac{1}{ue^{i\pi}}\right)\frac{x^{-u}-1}{-u}\,du$$

$$= -\int_{\rho}^{\delta}\frac{x^{-u}-1}{u}\,du = -\int_{\rho\log x}^{\delta\log x}\frac{e^{-v}-1}{v}\,dv$$

$$= \int_{\rho\log x}^{1}\frac{1-e^{-v}}{v}\,dv-\int_{1}^{\delta\log x}\frac{e^{-v}}{v}\,dv+\log(\delta\log x)$$

$$= \gamma+\log(\delta\log x)+o(1)$$

if $\rho\log x \to 0$ and $\delta\log x \to \infty$. Also

$$\int_{|w|=\rho}\log\frac{1}{w}\frac{x^{w}-1}{w}\,dw = O\left(\rho\log\frac{1}{\rho}\log x\right).$$

Taking $\rho = 1/\log^2 x$, say, it follows that

$$\sum_{n<x}\frac{\Lambda_1(n)}{n} = \log\log x+\gamma+o(1). \tag{3.14.4}$$

The left-hand side can also be written in the form

$$\sum_{p<x}\frac{1}{p}+\sum_{m\geqslant 2}\sum_{p^m<x}\frac{1}{mp^m}.$$

As $x \to \infty$, the second term clearly tends to the limit

$$\sum_{m=2}^{\infty}\sum_{p}\frac{1}{mp^m}.$$

Hence
$$\sum_{p<x}\frac{1}{p} = \log\log x+\gamma-\sum_{m=2}^{\infty}\sum_{p}\frac{1}{mp^m}+o(1). \tag{3.14.5}$$

3.15. Euler's product on $\sigma = 1$. The above analysis shows that for $\sigma = 1$, $t \neq 0$,

$$\log\zeta(s) = \sum_{p}\frac{1}{p^s}+\sum_{q}\frac{\Lambda_1(q)}{q^s},$$

where p runs through primes and q through powers of primes. In fact the second series is absolutely convergent on $\sigma = 1$, since it is merely a rearrangement of

$$\sum_{p,}\sum_{m=2}^{\infty}\frac{1}{mp^{ms}},$$

which is absolutely convergent by comparison with

$$\sum_p \sum_{m=2}^{\infty} \frac{1}{p^m} = \sum_p \frac{1}{p(p-1)}.$$

Hence also

$$\log \zeta(s) = \sum_p \frac{1}{p^s} + \sum_p \sum_{m=2}^{\infty} \frac{1}{mp^{ms}}$$

$$= \sum_p \sum_{m=1}^{\infty} \frac{1}{mp^{ms}}$$

$$= \sum_p \log \frac{1}{1-p^{-s}} \quad (\sigma = 1,\ t \neq 0).$$

Taking exponentials,

$$\zeta(s) = \prod_p \frac{1}{1-p^{-s}}, \tag{3.15.1}$$

i.e. Euler's product holds on $\sigma = 1$, except at $t = 0$.

At $s = 1$ the product is, of course, not convergent, but we can obtain an asymptotic formula for its partial products, viz.

$$\prod_{p \leqslant x} \left(1 - \frac{1}{p}\right) \sim \frac{e^{-\gamma}}{\log x}. \tag{3.15.2}$$

To prove this, we have to prove that

$$f(x) = -\log \prod_{p \leqslant x} \left(1 - \frac{1}{p}\right) = \log\log x + \gamma + o(1).$$

Now we have proved that

$$g(x) = \sum_{n \leqslant x} \frac{\Lambda_1(n)}{n} = \log\log x + \gamma + o(1).$$

Also

$$f(x) - g(x) = \sum_{p \leqslant x} \sum_{m=1}^{\infty} \frac{1}{mp^m} - \sum_{p^m \leqslant x} \sum \frac{1}{mp^m}$$

$$= \frac{1}{2} \sum_{x^{\frac{1}{2}} \leqslant p \leqslant x} \frac{1}{p^2} + \frac{1}{3} \sum_{x^{\frac{1}{3}} < p \leqslant x} \frac{1}{p^3} + \dots$$

$$< \sum_{\substack{p \\ p^m > x}} \sum_{m=2}^{\infty} \frac{1}{mp^m},$$

which tends to zero as $x \to \infty$, since the double series is absolutely convergent. This proves (3.15.2).

It will also be useful later to note that

$$\prod_{p \leqslant x}\left(1+\frac{1}{p}\right) \sim \frac{6e^{\gamma}\log x}{\pi^2}. \tag{3.15.3}$$

For the left-hand side is

$$\prod_{p \leqslant x} \frac{1-1/p^2}{1-1/p} \sim e^{\gamma}\log x \prod_{p}\left(1-\frac{1}{p^2}\right) = \frac{e^{\gamma}\log x}{\zeta(2)} = \frac{6e^{\gamma}\log x}{\pi^2}.$$

Note also that (3.14.3), (3.14.5) with error term $O(1)$, and (3.15.2) can be proved in an elementary way, i.e. without the theory of the Riemann zeta-function; see Hardy and Wright, *The Theory of Numbers* (5th edn), Theorems 425 and 427–429. Indeed the proof of Theorem 427 yields (3.14.5) with the error term $O\left(\dfrac{1}{\log x}\right)$.

NOTES FOR CHAPTER 3

3.16. The original elementary proofs of the prime number theorem may be found in Selberg [**7**] and Erdős [**1**], and a thorough survey of the ideas involved is given by Diamond [**1**]. The sharpest error term obtained by elementary methods to date is

$$\pi(x) = \mathrm{Li}(x) + O[x\exp\{-(\log x)^{\frac{1}{6}-\varepsilon}\}], \tag{3.16.1}$$

for any $\varepsilon > 0$, due to Lavrik and Sobirov [**1**]. Pintz [**1**] has obtained a very precise relationship between zero-free regions of $\zeta(s)$ and the error term in the prime-number theorem. Specifically, if we define

$$R(x) = \max\{|\pi(t) - \mathrm{Li}(t)|: 2 \leqslant t \leqslant x\},$$

then

$$\log\frac{x}{R(x)} \sim \min_{\rho}\{(1-\beta)\log x + \log|\gamma|\}, \quad (x \to \infty),$$

the minimum being over non-trivial zeros ρ of $\zeta(s)$. Thus (3.16.1) yields

$$(1-\beta)\log x + \log|\gamma| \gg (\log x)^{\frac{1}{6}-\varepsilon}$$

for any ρ and any x. Now, on taking $\log x = (1-\beta)^{-1}\log|\gamma|$ we deduce that

$$1-\beta \gg (\log|\gamma|)^{-5-\varepsilon'},$$

for any $\varepsilon' > 0$. This should be compared with Theorem 3.8.

3.17. It may be observed in the proof of Theorem 3.10 that the bound $\zeta(s) = O(e^{\phi(t)})$ is only required in the immediate vicinity of s_0 and s_0'. It would be nice to eliminate consideration of s_0' and so to have a result of the strength of Theorem 3.10, giving a zero-free region around $1+it$ solely in terms of an estimate for $\zeta(s)$ in a neighbourhood of $1+it$.

Ingham's method in § 3.4 is of special interest because it avoids any reference to the behaviour of $\zeta(s)$ near $1+2i\gamma$. It is possible to get quantitative zero-free regions in this way, by incorporating simple sieve estimates (Balasubramanian and Ramachandra [**1**]). Thus, for example, the analysis of § 3.8 yields

$$\sum_{p,m} \frac{\log p}{p^{m\sigma}}\{1+\cos(m\gamma \log p)\} \leqslant \frac{1}{\sigma-1} - \frac{1}{\sigma-\beta} + O(\log \gamma).$$

However one can show that

$$\sum_{X<p\leqslant 2X} \{1+\cos(\gamma \log p)\} \gg \frac{X}{\log X}$$

for $X \geqslant \gamma^2$, by using a lower bound of Chebychev type for the number of primes $X<p\leqslant 2X$, coupled with an upper bound $O(h/\log h)$ for the number of primes in certain short intervals $X' < p \leqslant X'+h$. One then derives the estimate

$$\sum_{p \geqslant \gamma^2} \frac{\log p}{p^{\sigma}}\{1+\cos(\gamma \log p)\} \gg \frac{\gamma^{2(1-\sigma)}}{\sigma-1},$$

and an appropriate choice of $\sigma = 1+(A/\log \gamma)$ leads to the lower bound $1-\beta \gg (\log \gamma)^{-1}$.

3.18. Another approach to zero-free regions via sieve methods has been given by Motohashi [**1**]. This is distinctly complicated, but has the advantage of applying to the wider regions discussed in §§ 5.17, 6.15 and 6.19.

One may also obtain zero-free regions from a result of Montgomery [**1**; Theorem 11.2] on the proliferation of zeros. Let $n(t, w, h)$ denote the number of zeros $\rho = \beta + i\gamma$ of $\zeta(s)$ in the rectangle $1-w \leqslant \beta \leqslant 1$, $t-\frac{1}{2}h \leqslant \gamma \leqslant t+\frac{1}{2}h$. Suppose ρ is any zero with $\beta > \frac{1}{2}$, $\gamma > 0$, and that δ satisfies $1-\beta \leqslant \delta \leqslant (\log \gamma)^{-\frac{1}{4}}$. Then there is some r with $\delta \leqslant r \leqslant 1$ for which

$$n(\gamma, r, r) + n(2\gamma, r, r) \gg \frac{r^3}{\delta^2(1-\beta)}. \tag{3.18.1}$$

Roughly speaking, this says that if $1-\beta$ is small, there must be many other zeros near either $1+i\gamma$ or $1+2i\gamma$. Montgomery gives a more precise version of this principle, as do Ramachandra [**1**] and Balasubramanian and Ramachandra [**3**]. To obtain a zero-free region one couples hypotheses of the type used in Theorem 3.10 with Jensen's Theorem, to obtain an upper bound for $n(t, r, r)$. For example, the bound

$$\zeta(s) \ll (1+T^{1-\sigma})\log T, \qquad T = |t|+2,$$

which follows from Theorem 4.11, leads to

$$n(t, r, r) \ll r \log T + \log\log T + \log\frac{1}{r}. \tag{3.18.2}$$

On choosing $\delta = (\log\log\gamma)/(\log\gamma)$, a comparison of (3.18.1) and (3.18.2) produces Theorem 3.8 again.

One can also use the Epstein zeta-function of §2.18 and the Maass–Selberg formula (2.18.9) to prove the non-vanishing of $\zeta(s)$ for $\sigma = 1$. For, if $s = \frac{1}{2}+it$ and $\phi(s) = 2\zeta(2s) = 0$, then

$$|\psi(\tfrac{1}{2}+it)|^2 = \psi(s)\psi(1-s) = \phi(s)\phi(1-s) = |\phi(\tfrac{1}{2}+it)|^2 = 0,$$

by the functional equation (2.19.1). Thus (2.18.9) yields

$$\iint_D \tilde{B}(z, s)\tilde{B}(z, w)\frac{dxdy}{y^2} = 0$$

for any $w \neq s, 1-s$. This, of course, may be extended to $w = s$ or $w = 1-s$ by continuity. Taking $w = \frac{1}{2}-it = \bar{s}$ we obtain

$$\iint_D |\tilde{B}(z, s)|^2\frac{dxdy}{y^2} = 0$$

so that $\tilde{B}(z, s)$ must be identically zero. This however is impossible since the Fourier coefficient for $n = 1$ is

$$8\pi^s y^{\frac{1}{2}} K_{s-\frac{1}{2}}(2\pi y)/\Gamma(s),$$

according to (2.18.5), and this does not vanish identically. The above contradiction shows that $\zeta(2s) \neq 0$. One can get quantitative estimates by such methods, but only rather weak ones. It seems that the proof given here has its origins in unpublished work of Selberg.

3.19. Lemma 3.12 is a version of Perron's formula. It is sometimes useful to have a form of this in which the error is bounded as $x \to N$.

LEMMA 3.19. *Under the hypotheses of Lemma* 3.12 *one has*

$$\sum_{n \leqslant x} \frac{a_n}{n^s} = \frac{1}{2\pi i} \int_{c-iT}^{c+iT} f(s+w) \frac{x^w}{w} \, dw + O\left\{\frac{x^c}{T(\sigma+c-1)^\alpha}\right\}$$

$$+ O\left\{\frac{\psi(2x)x^{1-\sigma}\log x}{T}\right\} + O\left\{\psi(N)x^{-\sigma}\min\left(\frac{x}{T|x-N|}, 1\right)\right\}.$$

This follows at once from Lemma 3.12 unless $x - N = O(x/T)$. In the latter case one merely estimates the contribution from the term $n = N$ as

$$\int_{c-iT}^{c+iT} \frac{a_N}{N^s}\left(\frac{x}{N}\right)^w \frac{dw}{w} = \int_{c-iT}^{c+iT} \frac{a_N}{N^s}\left\{1 + O\left(\frac{w}{T}\right)\right\}\frac{dw}{w}$$

$$= \frac{a_N}{N^s}\left\{\log\frac{c+iT}{c-iT} + O(1)\right\}$$

$$= O\{\psi(N)N^{-\sigma}\},$$

and the result follows.

IV

APPROXIMATE FORMULAE

4.1. IN this chapter we shall prove a number of approximate formulae for $\zeta(s)$ and for various sums related to it. We shall begin by proving some general results on integrals and series of a certain type.

4.2. LEMMA 4.2. *Let $F(x)$ be a real differentiable function such that $F'(x)$ is monotonic, and $F'(x) \geqslant m > 0$, or $F'(x) \leqslant -m < 0$, throughout the interval $[a, b]$. Then*

$$\left|\int_a^b e^{iF(x)}\,dx\right| \leqslant \frac{4}{m}. \tag{4.2.1}$$

Suppose, for example, that $F'(x)$ is positive increasing. Then by the second mean-value theorem

$$\int_a^b \cos\{F(x)\}\,dx = \int_a^b \frac{F'(x)\cos\{F(x)\}}{F'(x)}\,dx$$

$$= \frac{1}{F'(a)}\int_a^\xi F'(x)\cos\{F(x)\}\,dx = \frac{\sin\{F(\xi)\}-\sin\{F(a)\}}{F'(a)},$$

and the modulus of this does not exceed $2/m$. A similar argument applies to the imaginary part, and the result follows.

4.3. More generally, we have

LEMMA 4.3. *Let $F(x)$ and $G(x)$ be real functions, $G(x)/F'(x)$ monotonic, and $F'(x)/G(x) \geqslant m > 0$, or $\leqslant -m < 0$. Then*

$$\left|\int_a^b G(x)e^{iF(x)}\,dx\right| \leqslant \frac{4}{m}.$$

The proof is similar to that of the previous lemma.

The values of the constants in these lemmas are usually not of any importance.

4.4. LEMMA 4.4. *Let $F(x)$ be a real function, twice differentiable, and let $F''(x) \geqslant r > 0$, or $F''(x) \leqslant -r < 0$, throughout the interval $[a, b]$. Then*

$$\left|\int_a^b e^{iF(x)}\,dx\right| \leqslant \frac{8}{\sqrt{r}}. \tag{4.4.1}$$

Consider, for example, the first alternative. Then $F'(x)$ is steadily increasing, and so vanishes at most once in the interval (a, b), say at c. Let

$$I = \int_a^b e^{iF(x)}\,dx = \int_a^{c-\delta} + \int_{c-\delta}^{c+\delta} + \int_{c+\delta}^{b} = I_1 + I_2 + I_3,$$

where δ is a positive number to be chosen later, and it is assumed that $a+\delta \leqslant c \leqslant b-\delta$. In I_3

$$F'(x) = \int_c^x F''(t)\,dt \geqslant r(x-c) \geqslant r\delta.$$

Hence, by Lemma 4.2, $$|I_3| \leqslant \frac{4}{r\delta}.$$

I_1 satisfies the same inequality, and $|I_2| \leqslant 2\delta$. Hence

$$|I| \leqslant \frac{8}{r\delta} + 2\delta.$$

Taking $\delta = 2r^{-\frac{1}{2}}$, we obtain the result. If $c < a+\delta$, or $c > b-\delta$, the argument is similar.

4.5. LEMMA 4.5. *Let $F(x)$ satisfy the conditions of the previous lemma, and let $G(x)/F'(x)$ be monotonic, and $|G(x)| \leqslant M$. Then*

$$\left| \int_a^b G(x) e^{iF(x)}\,dx \right| \leqslant \frac{8M}{\sqrt{r}}.$$

The proof is similar to the previous one, but uses Lemma 4.3 instead of Lemma 4.2.

4.6. LEMMA 4.6. *Let $F(x)$ be real, with derivatives up to the third order. Let*

$$0 < \lambda_2 \leqslant F''(x) < A\lambda_2, \tag{4.6.1}$$

or

$$0 < \lambda_2 \leqslant -F''(x) < A\lambda_2, \tag{4.6.2}$$

and

$$|F'''(x)| \leqslant A\lambda_3, \tag{4.6.3}$$

throughout the interval (a, b). Let $F'(c) = 0$, where

$$a \leqslant c \leqslant b. \tag{4.6.4}$$

Then in the case (4.6.1)

$$\int_a^b e^{iF(x)}\,dx = (2\pi)^{\frac{1}{2}} \frac{e^{\frac{1}{4}i\pi + iF(c)}}{|F''(c)|^{\frac{1}{2}}} + O(\lambda_2^{-\frac{4}{5}}\lambda_3^{\frac{1}{5}}) + \\ + O\left\{\min\left(\frac{1}{|F'(a)|}, \lambda_2^{-\frac{1}{2}}\right)\right\} + O\left\{\min\left(\frac{1}{|F'(b)|}, \lambda_2^{-\frac{1}{2}}\right)\right\}. \tag{4.6.5}$$

In the case (4.6.2) *the factor $e^{\frac{1}{4}i\pi}$ is replaced by $e^{-\frac{1}{4}i\pi}$. If $F'(x)$ does not vanish on $[a, b]$ then* (4.6.5) *holds without the leading term.*

If $F'(x)$ does not vanish on $[a, b]$ the result follows from Lemmas 4.2 and 4.4. Otherwise either (4.6.1) or (4.6.2) shows that $F'(x)$ is monotonic, and so vanishes at only one point c. We put

$$\int_a^b e^{iF(x)}\,dx = \int_a^{c-\delta} + \int_{c-\delta}^{c+\delta} + \int_{c+\delta}^{b},$$

assuming that $a+\delta \leqslant c \leqslant b-\delta$. By (4.2.1)

$$\int_{c+\delta}^{b} = O\left\{\frac{1}{|F'(c+\delta)|}\right\} = O\left\{1\bigg/\left|\int_c^{c+\delta} F''(x)\,dx\right|\right\} = O\left(\frac{1}{\delta\lambda_2}\right).$$

Similarly
$$\int_a^{c-\delta} = O\left(\frac{1}{\delta\lambda_2}\right).$$

Also

$$\int_{c-\delta}^{c+\delta} = \int_{c-\delta}^{c+\delta} \exp[i\{F(c)+(x-c)F'(c)+\tfrac{1}{2}(x-c)^2F''(c)+$$
$$+\tfrac{1}{6}(x-c)^3F'''(c+\theta(x-c))\}]\,dx$$
$$= e^{iF(c)}\int_{c-\delta}^{c+\delta} e^{\frac{1}{2}i(x-c)^2F''(c)}[1+O\{(x-c)^3\lambda_3\}]\,dx$$
$$= e^{iF(c)}\int_{c-\delta}^{c+\delta} e^{\frac{1}{2}i(x-c)^2F''(c)}\,dx+O(\delta^4\lambda_3).$$

Supposing $F''(c) > 0$, and putting

$$\tfrac{1}{2}(x-c)^2F''(c) = u,$$

the integral becomes

$$\frac{2^{\frac{1}{2}}}{\{F''(c)\}^{\frac{1}{2}}}\int_0^{\frac{1}{2}\delta^2F''(c)} \frac{e^{iu}}{\sqrt{u}}\,du = \frac{2^{\frac{1}{2}}}{\{F''(c)\}^{\frac{1}{2}}}\left\{\int_0^{\infty} \frac{e^{iu}}{\sqrt{u}}\,du+O\left(\frac{1}{\delta\sqrt{\lambda_2}}\right)\right\}$$
$$= \frac{(2\pi)^{\frac{1}{2}}e^{\frac{1}{4}i\pi}}{\{F''(c)\}^{\frac{1}{2}}}+O\left(\frac{1}{\delta\lambda_2}\right).$$

Taking $\delta = (\lambda_2\lambda_3)^{-\frac{1}{3}}$, the result follows.

If $b-\delta < c \leqslant b$, there is also an error

$$e^{iF(c)}\int_b^{c+\delta} e^{\frac{1}{2}i(x-c)^2F''(c)}\,dx = O\left\{\frac{1}{(b-c)\lambda_2}\right\} = O\left\{\frac{1}{|F'(b)|}\right\} \text{ and also } O(\lambda_2^{-\frac{1}{2}});$$

and similarly if $a \leqslant c \leqslant a+\delta$.

4.7. We now turn to the consideration of exponential sums, i.e. sums of the form

$$\sum e^{2\pi i f(n)},$$

where $f(n)$ is a real function. If the numbers $f(n)$ are the values taken by a function $f(x)$ of a simple kind, we can approximate to such a sum by an integral, or by a sum of integrals.

LEMMA 4.7.† *Let $f(x)$ be a real function with a continuous and steadily decreasing derivative $f'(x)$ in (a,b), and let $f'(b) = \alpha$, $f'(a) = \beta$. Then*

$$\sum_{a<n\leqslant b} e^{2\pi i f(n)} = \sum_{\alpha-\eta<\nu<\beta+\eta} \int\limits_a^b e^{2\pi i\{f(x)-\nu x\}}\,dx + O\{\log(\beta-\alpha+2)\}, \tag{4.7.1}$$

where η is any positive constant less than 1.

We may suppose without loss of generality that $\eta-1<\alpha\leqslant\eta$, so that $\nu\geqslant 0$; for if k is the integer such that $\eta-1<\alpha-k\leqslant\eta$, and

$$h(x) = f(x)-kx,$$

then (4.7.1) is

$$\sum_{a<n\leqslant b} e^{2\pi i h(n)} = \sum_{\alpha'-\eta<\nu-k<\beta'+\eta} \int\limits_a^b e^{2\pi i\{h(x)-(\nu-k)x\}}\,dx + O\{\log(\beta'-\alpha'+2)\},$$

where $\alpha' = \alpha-k$, $\beta' = \beta-k$, i.e. the same formula for $h(x)$.

In (2.1.2), let $\phi(x) = e^{2\pi i f(x)}$. Then

$$\sum_{a<n\leqslant b} e^{2\pi i f(n)} = \int\limits_a^b e^{2\pi i f(x)}\,dx + \int\limits_a^b (x-[x]-\tfrac{1}{2})2\pi i f'(x)e^{2\pi i f(x)}\,dx + O(1).$$

Also
$$x-[x]-\tfrac{1}{2} = -\frac{1}{\pi}\sum_{\nu=1}^{\infty}\frac{\sin 2\nu\pi x}{\nu}$$

if x is not an integer; and the series is boundedly convergent, so that we may multiply by an integrable function and integrate term-by-term. Hence the second term on the right is equal to

$$-2i\sum_{\nu=1}^{\infty}\int\limits_a^b \frac{\sin 2\nu\pi x}{\nu}\, e^{2\pi i f(x)} f'(x)\,dx$$

$$= \sum_{\nu=1}^{\infty}\frac{1}{\nu}\int\limits_a^b (e^{-2\pi i\nu x}-e^{2\pi i\nu x})e^{2\pi i f(x)} f'(x)\,dx.$$

The integral may be written

$$\frac{1}{2\pi i}\int\limits_a^b \frac{f'(x)}{f'(x)-\nu}\,d(e^{2\pi i\{f(x)-\nu x\}}) - \frac{1}{2\pi i}\int\limits_a^b \frac{f'(x)}{f'(x)+\nu}\,d(e^{2\pi i\{f(x)+\nu x\}}).$$

† van der Corput (1).

Since $\dfrac{f'(x)}{f'(x)+\nu}$ is steadily decreasing, the second term is

$$O\left(\frac{\beta}{\beta+\nu}\right),$$

by applying the second mean-value theorem to the real and imaginary parts. Hence this term contributes

$$O\left(\sum_{\nu=1}^{\infty}\frac{\beta}{\nu(\beta+\nu)}\right) = O\left(\sum_{\nu\leqslant\beta}\frac{1}{\nu}\right)+O\left(\sum_{\nu>\beta}\frac{\beta}{\nu^2}\right)$$
$$= O\{\log(\beta+2)\}+O(1).$$

Similarly the first term is $O\{\beta/(\nu-\beta)\}$ for $\nu\geqslant\beta+\eta$, and this contributes

$$O\left(\sum_{\nu\geqslant\beta+\eta}\frac{\beta}{\nu(\nu-\beta)}\right) = O\left(\sum_{\beta+\eta\leqslant\nu<2\beta}\frac{1}{\nu-\beta}\right)+O\left(\sum_{\nu\geqslant 2\beta}\frac{\beta}{\nu^2}\right)$$
$$= O\{\log(\beta+2)\}+O(1).$$

Finally

$$\sum_{\nu=1}^{\beta+\eta}\frac{1}{\nu}\int_a^b e^{2\pi i\{f(x)-\nu x\}}f'(x)\,dx = \sum_{\nu=1}^{\beta+\eta}\left[\frac{e^{2\pi i\{f(x)-\nu x\}}}{2\pi i\nu}\right]_a^b+\sum_{\nu=1}^{\beta+\eta}\int_a^b e^{2\pi i\{f(x)-\nu x\}}\,dx,$$

and the integrated terms are $O\{\log(\beta+2)\}$. The result therefore follows.

4.8. As a particular case, we have

LEMMA 4.8. *Let $f(x)$ be a real differentiable function in the interval $[a,b]$, let $f'(x)$ be monotonic, and let $|f'(x)|\leqslant\theta<1$. Then*

$$\sum_{a<n\leqslant b} e^{2\pi i f(n)} = \int_a^b e^{2\pi i f(x)}\,dx+O(1). \tag{4.8.1}$$

Taking $\eta<1-\theta$, the sum on the right of (4.7.1) either reduces to the single term $\nu=0$, or, if $f'(x)\geqslant\eta$ or $\leqslant-\eta$ throughout $[a,b]$, it is null, and

$$\int_a^b e^{2\pi i f(x)}\,dx = O(1)$$

by Lemma 4.2.

4.9. THEOREM 4.9.† *Let $f(x)$ be a real function with derivatives up to the third order. Let $f'(x)$ be steadily decreasing in $a\leqslant x\leqslant b$, and $f'(b)=\alpha$, $f'(a)=\beta$. Let x_ν be defined by*

$$f'(x_\nu)=\nu \quad (\alpha<\nu\leqslant\beta).$$

† van der Corput (2).

Let $\lambda_2 \leqslant |f''(x)| < A\lambda_2, \qquad |f'''(x)| < A\lambda_3.$

Then

$$\sum_{a<n\leqslant b} e^{2\pi i f(n)} = e^{-\frac{1}{4}\pi i} \sum_{\alpha<\nu\leqslant\beta} \frac{e^{2\pi i\{f(x_\nu)-\nu x_\nu\}}}{|f''(x_\nu)|^{\frac{1}{2}}} + O(\lambda_2^{-\frac{1}{2}}) +$$
$$+O[\log\{2+(b-a)\lambda_2\}] + O\{(b-a)\lambda_2^{\frac{1}{5}}\lambda_3^{\frac{1}{5}}\}.$$

We use Lemma 4.7, where now

$$\beta-\alpha = O\{(b-a)\lambda_2\}.$$

Also we can replace the limits of summation on the right-hand side by $(\alpha+1, \beta-1)$, with error $O(\lambda_2^{-\frac{1}{2}})$. Lemma 4.6. then gives

$$\sum_{\alpha+1<\nu<\beta-1} \int_a^b e^{2\pi i\{f(x)-\nu x\}}\, dx = e^{-\frac{1}{4}i\pi} \sum_{\alpha+1<\nu<\beta-1} \frac{e^{2\pi i\{f(x_\nu)-\nu x_\nu\}}}{|f''(x_\nu)|^{\frac{1}{2}}} +$$
$$+\sum_{\alpha+1<\nu<\beta-1} O(\lambda_2^{-\frac{4}{5}}\lambda_3^{\frac{1}{5}}) + \sum_{\alpha+1<\nu<\beta-1} \left\{O\left(\frac{1}{\nu-\alpha}\right) + O\left(\frac{1}{\beta-\nu}\right)\right\}.$$

The second term on the right is

$$O\{(\beta-\alpha)\lambda_2^{-\frac{4}{5}}\lambda_3^{\frac{1}{5}}\} = O\{(b-a)\lambda_2^{\frac{1}{5}}\lambda_3^{\frac{1}{5}}\},$$

and the last term is

$$O\{\log(2+\beta-\alpha)\} = O[\log\{2+(b-a)\lambda_2\}].$$

Finally we can replace the limits $(\alpha+1, \beta-1)$ by $(\alpha, \beta]$ with error $O(\lambda_2^{-\frac{1}{2}})$.

4.10. LEMMA 4.10. *Let $f(x)$ satisfy the same conditions as in Lemma 4.7, and let $g(x)$ be a real positive decreasing function, with a continuous derivative $g'(x)$, and let $|g'(x)|$ be steadily decreasing. Then*

$$\sum_{a<n\leqslant b} g(n)e^{2\pi i f(n)} = \sum_{\alpha-\eta<\nu<\beta+\eta} \int_a^b g(x)e^{2\pi i\{f(x)-\nu x\}}\, dx +$$
$$+O\{g(a)\log(\beta-\alpha+2)\} + O\{|g'(a)|\}.$$

We proceed as in § 4.7, but with

$$\phi(x) = g(x)e^{2\pi i f(x)}.$$

We encounter terms of the form

$$\int_a^b g(x)\frac{f'(x)}{f'(x)\pm\nu}\, d(e^{2\pi i\{f(x)\pm\nu x\}}),$$

and also

$$\int_a^b \frac{g'(x)}{f'(x)\pm\nu}\, d(e^{2\pi i\{f(x)\pm\nu x\}}).$$

The former lead to $O\{g(a)\log(\beta-\alpha+2)\}$ as before. The latter give, for example,

$$\sum_{\nu=1}^{\infty} \frac{|g'(a)|}{\nu^2} = O(|g'(a)|),$$

and the result follows.

4.11. We now come to the simplest theorem† on the approximation to $\zeta(s)$ in the critical strip by a partial sum of its Dirichlet series.

THEOREM 4.11. *We have*

$$\zeta(s) = \sum_{n\leqslant x} \frac{1}{n^s} - \frac{x^{1-s}}{1-s} + O(x^{-\sigma}) \tag{4.11.1}$$

uniformly for $\sigma \geqslant \sigma_0 > 0$, $|t| < 2\pi x/C$, *when* C *is a given constant greater than* 1.

We have, by (3.5.3),

$$\zeta(s) = \sum_{n=1}^{N} \frac{1}{n^s} - \frac{N^{1-s}}{1-s} + s\int_N^\infty \frac{[u]-u+\frac{1}{2}}{u^{s+1}}\,du - \tfrac{1}{2}N^{-s}$$

$$= \sum_{n=1}^{N} \frac{1}{n^s} - \frac{N^{1-s}}{1-s} + O\left(\frac{|s|}{N^\sigma}\right) + O(N^{-\sigma}). \tag{4.11.2}$$

The sum
$$\sum_{x<n\leqslant N} \frac{1}{n^s} = \sum_{x<n\leqslant N} \frac{n^{-it}}{n^\sigma}$$

is of the form considered in the above lemma, with $g(u) = u^{-\sigma}$, and

$$f(u) = -\frac{t\log u}{2\pi}, \qquad f'(u) = -\frac{t}{2\pi u}.$$

Thus
$$|f'(u)| \leqslant \frac{t}{2\pi x} < \frac{1}{C}.$$

Hence
$$\sum_{x<n\leqslant N} \frac{1}{n^s} = \int_x^N \frac{du}{u^s} + O(x^{-\sigma})$$

$$= \frac{N^{1-s}-x^{1-s}}{1-s} + O(x^{-\sigma}).$$

Hence
$$\zeta(s) = \sum_{n\leqslant x} \frac{1}{n^s} - \frac{x^{1-s}}{1-s} + O(x^{-\sigma}) + O\left(\frac{|s|+1}{N^\sigma}\right).$$

Making $N \to \infty$, the result follows.

† Hardy and Littlewood (3).

4.12. For many purposes the sum involved in Theorem 4.11 contains too many terms (at least $A|t|$) to be of use. We therefore consider the result of taking smaller values of x in the above formulae. The form of the result is given by Theorem 4.9, with an extra factor $g(n)$ in the sum. If we ignore error terms for the moment, this gives

$$\sum_{a<n\leqslant b} g(n)e^{2\pi i f(n)} \sim e^{-\frac{1}{4}\pi i} \sum_{\alpha<\nu\leqslant\beta} \frac{e^{2\pi i\{f(x_\nu)-\nu x_\nu\}}}{|f''(x_\nu)|^{\frac{1}{2}}} g(x_\nu).$$

Taking

$$g(u) = u^{-\sigma}, \qquad f(u) = \frac{t\log u}{2\pi},$$

$$f'(u) = \frac{t}{2\pi u}, \qquad f''(u) = -\frac{t}{2\pi u^2},$$

$$x_\nu = \frac{t}{2\pi\nu}, \qquad f''(x_\nu) = -\frac{2\pi\nu^2}{t},$$

and replacing a, b by x, N, and i by $-i$, we obtain

$$\sum_{x<n\leqslant N} \frac{1}{n^s} \sim e^{\frac{1}{4}\pi i} \sum_{t/2\pi N<\nu\leqslant t/2\pi x} \frac{e^{-2\pi i\{(t/2\pi)\log(t/2\pi\nu)-(t/2\pi)\}}}{(t/2\pi\nu)^\sigma(2\pi\nu^2/t)^{\frac{1}{2}}}$$

$$= \left(\frac{t}{2\pi}\right)^{\frac{1}{2}-\sigma} e^{\frac{1}{4}\pi i-it\log(t/2\pi e)} \sum_{t/2\pi N<\nu\leqslant t/2\pi x} \frac{1}{\nu^{1-s}}.$$

Now the functional equation is

$$\zeta(s) = \chi(s)\zeta(1-s),$$

where

$$\chi(s) = 2^{s-1}\pi^s \sec\tfrac{1}{2}s\pi/\Gamma(s).$$

In any fixed strip $\alpha \leqslant \sigma \leqslant \beta$, as $t \to \infty$

$$\log\Gamma(\sigma+it) = (\sigma+it-\tfrac{1}{2})\log(it)-it+\tfrac{1}{2}\log 2\pi+O\left(\frac{1}{t}\right). \tag{4.12.1}$$

Hence

$$\Gamma(\sigma+it) = t^{\sigma+it-\frac{1}{2}}e^{-\frac{1}{2}\pi t-it+\frac{1}{2}i\pi(\sigma-\frac{1}{2})}(2\pi)^{\frac{1}{2}}\left\{1+O\left(\frac{1}{t}\right)\right\}, \tag{4.12.2}$$

$$\chi(s) = \left(\frac{2\pi}{t}\right)^{\sigma+it-\frac{1}{2}} e^{i(t+\frac{1}{4}\pi)}\left\{1+O\left(\frac{1}{t}\right)\right\}. \tag{4.12.3}$$

Hence the above relation is equivalent to

$$\sum_{x<n\leqslant N} \frac{1}{n^s} \sim \chi(s) \sum_{t/2\pi N<\nu\leqslant t/2\pi x} \frac{1}{\nu^{1-s}}.$$

The formulae therefore suggest that, with some suitable error terms,

$$\zeta(s) \sim \sum_{n\leqslant x} \frac{1}{n^s}+\chi(s) \sum_{\nu\leqslant y} \frac{1}{\nu^{1-s}},$$

where $2\pi xy = |t|$.

Actually the result is that

$$\zeta(s) = \sum_{n \leqslant x} \frac{1}{n^s} + \chi(s) \sum_{n \leqslant y} \frac{1}{n^{1-s}} + O(x^{-\sigma}) + O(|t|^{\frac{1}{2}-\sigma} y^{\sigma-1}) \quad (4.12.4)$$

for $0 < \sigma < 1$. This is known as the *approximate functional equation.*†

4.13. Theorem 4.13. *If h is a positive constant,*

$$0 < \sigma < 1, \qquad 2\pi xy = t, \qquad x > h > 0, \qquad y > h > 0,$$

then

$$\zeta(s) = \sum_{n \leqslant x} \frac{1}{n^s} + \chi(s) \sum_{n \leqslant y} \frac{1}{n^{1-s}} + O(x^{-\sigma} \log|t|) + O(|t|^{\frac{1}{2}-\sigma} y^{\sigma-1}). \quad (4.13.1)$$

This is an imperfect form of the approximate functional equation in which a factor $\log|t|$ appears in one of the O-terms; but for most purposes it is quite sufficient. The proof depends on the same principle as Theorem 4.9, but Theorem 4.9 would not give a sufficiently good O-result, and we have to reconsider the integrals which occur in this problem. Let $t > 0$. By Lemma 4.10

$$\sum_{x < n \leqslant N} \frac{1}{n^s} = \sum_{t/2\pi N - \eta < \nu \leqslant y + \eta} \int_x^N \frac{e^{2\pi i \nu u}}{u^s}\, du + O\left\{x^{-\sigma} \log\left(\frac{t}{x} - \frac{t}{N} + 2\right)\right\},$$

and the last term is $O(x^{-\sigma} \log t)$. If $2\pi N\eta > t$, the first term is $\nu = 0$, i.e.

$$\int_x^N \frac{du}{u^s} = \frac{N^{1-s} - x^{1-s}}{1-s}.$$

Hence by (4.11.2)

$$\zeta(s) = \sum_{n \leqslant x} \frac{1}{n^s} + \sum_{1 \leqslant \nu \leqslant y + \eta} \int_x^N \frac{e^{2\pi i \nu u}}{u^s}\, du + O(x^{-\sigma} \log t) + O(tN^{-\sigma}),$$

since $$x^{1-s}/(1-s) = O(x^{-\sigma}) = O(x^{-\sigma} \log t).$$

Now $$\int_0^\infty \frac{e^{2\pi i \nu u}}{u^s}\, du = \Gamma(1-s)\left(\frac{2\pi\nu}{i}\right)^{s-1},$$

and by Lemma 4.3

$$\int_N^\infty u^{-\sigma} e^{-2\pi i \{(t/2\pi)\log u - \nu u\}}\, du = O\left(\frac{N^{-\sigma}}{\nu - (t/2\pi N)}\right) = O\left(\frac{N^{-\sigma}}{\nu}\right),$$

$$\int_0^x u^{-s} e^{2\pi i \nu u}\, du = \left[\frac{u^{1-s}}{1-s} e^{2\pi i \nu u}\right]_0^x - \frac{2\pi i \nu}{1-s} \int_0^x u^{1-s} e^{2\pi i \nu u}\, du$$

$$= O\left(\frac{x^{1-\sigma}}{t}\right) + O\left(\frac{\nu}{t} \frac{x^{1-\sigma}}{\nu - (t/2\pi x)}\right).$$

† Hardy and Littlewood (3), (4), (6), Siegel (2).

Hence

$$\sum_{1\leqslant\nu\leqslant y-\eta}\int_x^N \frac{e^{2\pi i\nu u}}{u^s}\,du = \left(\frac{2\pi}{i}\right)^{s-1}\Gamma(1-s)\sum_{1\leqslant\nu\leqslant y-\eta}\frac{1}{\nu^{1-s}}+$$

$$+O(N^{-\sigma}\log y)+O\left(\frac{x^{1-\sigma}y}{t}\right)+O\left(\frac{x^{1-\sigma}}{t}\sum_{1\leqslant\nu\leqslant y-\eta}\frac{\nu}{\nu-y}\right)$$

$$= \left(\frac{2\pi}{i}\right)^{s-1}\Gamma(1-s)\sum_{1\leqslant\nu\leqslant y-\eta}\frac{1}{\nu^{1-s}}+O(N^{-\sigma}\log t)+O\left(\frac{x^{1-\sigma}y\log t}{t}\right).$$

There is still a possible term corresponding to $y-\eta < \nu \leqslant y+\eta$; for this, by Lemma 4.5,

$$\int_0^x u^{1-s}e^{2\pi i\nu u}\,du = O\left\{x^{1-\sigma}\left(\frac{t}{x^2}\right)^{-\frac{1}{2}}\right\},$$

giving a term

$$O\left\{\frac{y}{t}x^{1-\sigma}\left(\frac{t}{x^2}\right)^{-\frac{1}{2}}\right\} = O\left\{x^{-\sigma}\left(\frac{t}{x^2}\right)^{-\frac{1}{2}}\right\} = O(x^{1-\sigma}t^{-\frac{1}{2}}) = O(t^{\frac{1}{2}-\sigma}y^{\sigma-1}).$$

Finally we can replace $\nu \leqslant y-\eta$ by $\nu \leqslant y$ with error

$$O\left\{\left|\left(\frac{2\pi}{i}\right)^{s-1}\Gamma(1-s)\right|y^{\sigma-1}\right\} = O(t^{\frac{1}{2}-\sigma}y^{\sigma-1}).$$

Also for $t > 0$

$$\chi(s) = 2^s\pi^{s-1}\sin\tfrac{1}{2}s\pi\,\Gamma(1-s)$$

$$= 2^s\pi^{s-1}\left\{-\frac{e^{-\frac{1}{2}is\pi}}{2i}+O(e^{-\frac{1}{2}\pi t})\right\}\Gamma(1-s)$$

$$= \left(\frac{2\pi}{i}\right)^{s-1}\Gamma(1-s)\{1+O(e^{-\pi t})\}.$$

Hence the result follows on taking N large enough.

It is possible to prove the full result by a refinement of the above methods. We shall not give the details here, since the result will be obtained by another method, depending on contour integration.

4.14. Complex-variable methods. An extremely powerful method of obtaining approximate formulae for $\zeta(s)$ is to express $\zeta(s)$ as a contour integral, and then move the contour into a position where it can be suitably dealt with. The following is a simple example.

Alternative proof of Theorem 4.11. We may suppose without loss of generality that x is half an odd integer, since the last term in the sum, which might be affected by the restriction, is $O(x^{-\sigma})$, and so is the possible variation in $x^{1-s}/(1-s)$.

Suppose first that $\sigma > 1$. Then a simple application of the theorem of residues shows that

$$\zeta(s) - \sum_{n<x} n^{-s} = \sum_{n>x} n^{-s} = -\frac{1}{2i} \int_{x-i\infty}^{x+i\infty} z^{-s} \cot \pi z \, dz$$

$$= -\frac{1}{2i} \int_{x-i\infty}^{x} (\cot \pi z - i) z^{-s} \, dz - \frac{1}{2i} \int_{x}^{x+i\infty} (\cot \pi z + i) z^{-s} \, dz - \frac{x^{1-s}}{1-s}.$$

The final formula holds, by the theory of analytic continuation, for all values of s, since the last two integrals are uniformly convergent in any finite region. In the second integral we put $z = x+ir$, so that

$$|\cot \pi z + i| = \frac{2}{1+e^{2\pi r}} < 2e^{-2\pi r},$$

and $$|z^{-s}| = |z|^{-\sigma} e^{t \arg z} < x^{-\sigma} e^{|t| \arctan(r/x)} < x^{-\sigma} e^{|t| r/x}.$$

Hence the modulus of this term does not exceed

$$x^{-\sigma} \int_0^{\infty} e^{-2\pi r + |t| r/x} \, dr = \frac{x^{-\sigma}}{2\pi - |t|/x}.$$

A similar result holds for the other integral, and the theorem follows.

It is possible to prove the approximate functional equation by an extension of this argument; we may write

$$-\cot \pi z - i = 2i \sum_{\nu=1}^{n} e^{2\nu\pi iz} + \frac{2ie^{2(n+1)\pi iz}}{1-e^{2\pi iz}}.$$

Proceeding as before, this leads to an O-term

$$O\left\{x^{-\sigma} \int_0^{\infty} e^{-2(n+1)\pi r - |t| r/x} \, dr\right\} = O\left(\frac{x^{-\sigma}}{2(n+1)\pi - |t|/x}\right),$$

and this is $O(x^{-\sigma})$ if $2(n+1)\pi - |t|/x > A$, i.e. for comparatively small values of x, if n is large. However, the rest of the argument suggested is not particularly simple, and we prefer another proof, which will be more useful for further developments.

4.15. THEOREM 4.15. *The approximate functional equation* (4.12.4) *holds for* $0 \leqslant \sigma \leqslant 1$, $x > h > 0$, $y > h > 0$.

It is possible to extend the result to any strip $-k < \sigma < k$ by slight changes in the argument.

For $\sigma > 1$ $$\zeta(s) = \sum_{n=1}^{m} \frac{1}{n^s} + \frac{1}{\Gamma(s)} \int_0^{\infty} \frac{x^{s-1} e^{-mx}}{e^x - 1} \, dx.$$

Transforming the integral into a loop-integral as in § 2.4, we obtain

$$\zeta(s) = \sum_{n=1}^{m} \frac{1}{n^s} + \frac{e^{-i\pi s}\Gamma(1-s)}{2\pi i} \int_C \frac{w^{s-1}e^{-mw}}{e^w-1}\, dw,$$

where C excludes the zeros of e^w-1 other than $w = 0$. This holds for all values of s except positive integers.

Let $t > 0$ and $x \leqslant y$, so that $x \leqslant \sqrt{(t/2\pi)}$. Let $\sigma \leqslant 1$,

$$m = [x], \qquad y = t/(2\pi x), \qquad q = [y], \qquad \eta = 2\pi y.$$

We deform the contour C into the straight lines C_1, C_2, C_3, C_4 joining ∞, $c\eta+i\eta(1+c)$, $-c\eta+i\eta(1-c)$, $-c\eta-(2q+1)\pi i$, ∞, where c is an absolute constant, $0 < c \leqslant \frac{1}{2}$. If y is an integer, a small indentation is made above the pole at $w = i\eta$. We have then

$$\zeta(s) = \sum_{n=1}^{m} \frac{1}{n^s} + \chi(s) \sum_{n=1}^{q} \frac{1}{n^{1-s}} + \frac{e^{-i\pi s}\Gamma(1-s)}{2\pi i} \left(\int_{C_1} + \int_{C_2} + \int_{C_3} + \int_{C_4} \right).$$

Let $w = u+iv = \rho e^{i\phi}$ $(0 < \phi < 2\pi)$. Then

$$|w^{s-1}| = \rho^{\sigma-1}e^{-t\phi}.$$

On C_4, $\phi \geqslant \frac{5}{4}\pi$, $\rho > A\eta$, and $|e^w-1| > A$. Hence

$$\left| \int_{C_4} \right| = O\left(\eta^{\sigma-1} e^{-\frac{5}{4}\pi t} \int_{-c\eta}^{\infty} e^{-mu}\, du \right) = O(e^{mc\eta-\frac{5}{4}\pi t}) = O(e^{t(c-\frac{5}{4}\pi)}).$$

On C_3, $\phi \geqslant \frac{1}{2}\pi + \arctan\dfrac{c}{1-c} > \frac{1}{2}\pi + c + A$ where $A > 0$, since

$$\arctan\theta = \int_0^\theta \frac{d\mu}{1+\mu^2} > \int_0^\theta \frac{d\mu}{(1+\mu)^2} = \frac{\theta}{1+\theta}.$$

Hence

$$w^{s-1}e^{-mw} = O(\eta^{\sigma-1}e^{-t(\frac{1}{2}\pi+c+A)+mc\eta}) = O(\eta^{\sigma-1}e^{-t(\frac{1}{2}\pi+A)})$$

and $|e^w-1| > A$. Hence

$$\int_{C_3} = O(\eta^{\sigma}e^{-t(\frac{1}{2}\pi+A)}).$$

On C_1, $|e^w-1| > Ae^u$. Hence

$$\frac{w^{s-1}e^{-mw}}{e^w-1} = O\left[\eta^{\sigma-1}\exp\left\{-t\arctan\frac{(1+c)\eta}{u} - (m+1)u\right\}\right].$$

Since $m+1 \geqslant x = t/\eta$, and

$$\frac{d}{du}\left\{\arctan\frac{(1+c)\eta}{u} + \frac{u}{\eta}\right\} = -\frac{(1+c)\eta}{u^2+(1+c)^2\eta^2} + \frac{1}{\eta} > 0,$$

we have

$$\arctan\frac{(1+c)\eta}{u}+\frac{u}{\eta} \geqslant \arctan\frac{1+c}{c}+c$$

$$= \tfrac{1}{2}\pi+c-\arctan\frac{c}{1+c} = \tfrac{1}{2}\pi+A,$$

since for $0 < \theta < 1$

$$\arctan\theta < \int_0^\theta \frac{d\mu}{(1-\mu)^2} = \frac{\theta}{1-\theta}.$$

Hence

$$\int_{C_1} = O\left(\eta^{\sigma-1}\int_0^{\pi\eta} e^{-(\frac{1}{2}\pi+A)t}\,du\right)+O\left(\eta^{\sigma-1}\int_{\pi\eta}^{\infty} e^{-xu}\,du\right)$$

$$= O(\eta^{\sigma}e^{-(\frac{1}{2}\pi+A)t})+O(\eta^{\sigma-1}e^{-\pi\eta x}) = O(\eta^{\sigma}e^{-(\frac{1}{2}\pi+A)t}).$$

Finally consider C_2. Here $w = i\eta+\lambda e^{\frac{1}{4}i\pi}$, where λ is real, $|\lambda| \leqslant \sqrt{2}c\eta$. Hence

$$w^{s-1} = \exp[(s-1)\{\tfrac{1}{2}i\pi+\log(\eta+\lambda e^{-\frac{1}{4}i\pi})\}]$$

$$= \exp\left[(s-1)\left\{\tfrac{1}{2}i\pi+\log\eta+\frac{\lambda}{\eta}e^{-\frac{1}{4}i\pi}-\frac{1}{2}\frac{\lambda^2}{\eta^2}e^{-\frac{1}{2}i\pi}+O\left(\frac{\lambda^3}{\eta^3}\right)\right\}\right]$$

$$= O\left[\eta^{\sigma-1}\exp\left[\left\{-\tfrac{1}{2}\pi+\frac{\lambda}{\eta\sqrt{2}}-\frac{1}{2}\frac{\lambda^2}{\eta^2}+O\left(\frac{\lambda^3}{\eta^3}\right)\right\}t\right]\right].$$

Also

$$\frac{e^{-mw+xw}}{e^w-1} = O\left(\frac{e^{(x-m-1)u}}{1-e^{-u}}\right) \quad (u \geqslant 0), \quad = O\left(\frac{e^{(x-m)u}}{e^u-1}\right) \quad (u < 0),$$

which is bounded for $u < -\frac{1}{2}\pi$ and $u > \frac{1}{2}\pi$; and

$$|e^{-xw}| = e^{-\lambda t/\eta\sqrt{2}}.$$

Hence the part with $|u| > \frac{1}{2}\pi$ is

$$O\left\{\eta^{\sigma-1}e^{-\frac{1}{2}\pi t}\int_{-c\eta\sqrt{2}}^{c\eta\sqrt{2}} \exp\left[\left\{-\frac{1}{2}\frac{\lambda^2}{\eta^2}+O\left(\frac{\lambda^3}{\eta^3}\right)\right\}t\right]\,d\lambda\right\}$$

$$= O\left\{\eta^{\sigma-1}e^{-\frac{1}{2}\pi t}\int_{-\infty}^{\infty} e^{-A\lambda^2\eta^{-2}t}\,d\lambda\right\} = O(\eta^{\sigma}t^{-\frac{1}{2}}e^{-\frac{1}{2}\pi t}).$$

The argument also applies to the part $|u| \leqslant \frac{1}{2}\pi$ if $|e^w-1| > A$ on this part. If not, suppose, for example, that the contour goes too near to the pole at $w = 2q\pi i$. Take it round an arc of the circle $|w-2q\pi i| = \frac{1}{2}\pi$. On this circle,

$$w = 2q\pi i+\tfrac{1}{2}\pi e^{i\theta}$$

and

$$\log(w^{s-1}e^{-mw}) = -\tfrac{1}{2}m\pi e^{i\theta}+(s-1)\{\tfrac{1}{2}i\pi+\log(2q\pi+\tfrac{1}{2}\pi e^{i\theta}/i)\}$$
$$= -\tfrac{1}{2}m\pi e^{i\theta}-\tfrac{1}{2}\pi t+(s-1)\log(2q\pi)+\frac{te^{i\theta}}{4q}+O(1).$$

Since $$m\pi-\frac{t}{2q} = \frac{2mq\pi-t}{2q} = O(1),$$

this is $$-\tfrac{1}{2}\pi t+(s-1)\log(2q\pi)+O(1).$$

Hence $$|w^{s-1}e^{-mw}| = O(q^{\sigma-1}e^{-\frac{1}{2}\pi t}).$$

The contribution of this part is therefore

$$O(\eta^{\sigma-1}e^{-\frac{1}{2}\pi t}).$$

Since $$e^{-i\pi s}\Gamma(1-s) = O(t^{\frac{1}{2}-\sigma}e^{\frac{1}{2}\pi t})$$

we have now proved that

$$\zeta(s) = \sum_{n=1}^{m}\frac{1}{n^s}+\chi(s)\sum_{n=1}^{q}\frac{1}{n^{1-s}}+O\{t^{\frac{1}{2}-\sigma}(e^{-At}+\eta^{\sigma}t^{-\frac{1}{2}}+\eta^{\sigma-1})\}.$$

The O-terms are

$$O(e^{-At})+O\left\{\left(\frac{t}{x}\right)^{\sigma}t^{-\sigma}\right\}+O\left\{t^{\frac{1}{2}-\sigma}\left(\frac{t}{x}\right)^{\sigma-1}\right\}$$
$$= O(e^{-At})+O(x^{-\sigma})+O(t^{-\frac{1}{2}}x^{1-\sigma}) = O(x^{-\sigma}).$$

This proves the theorem in the case considered.

To deduce the case $x \geqslant y$, change s into $1-s$ in the result already obtained. Then

$$\zeta(1-s) = \sum_{n\leqslant x}\frac{1}{n^{1-s}}+\chi(1-s)\sum_{n\leqslant y}\frac{1}{n^s}+O(x^{\sigma-1}).$$

Multiplying by $\chi(s)$, and using the functional equation and

$$\chi(s)\chi(1-s) = 1,$$

we obtain $$\zeta(s) = \chi(s)\sum_{n\leqslant x}\frac{1}{n^{1-s}}+\sum_{n\leqslant y}\frac{1}{n^s}+O(t^{\frac{1}{2}-\sigma}x^{\sigma-1}).$$

Interchanging x and y, this gives the theorem with $x \geqslant y$.

4.16. Further approximations.† A closer examination of the above analysis, together with a knowledge of the formulae of § 2.10, shows that the O-terms in the approximate functional equation can be replaced by an asymptotic series, each term of which contains trigonometrical functions and powers of t only.

† Siegel (2).

We shall consider only the simplest case in which $x = y = \sqrt{(t/2\pi)}$, $\eta = \sqrt{(2\pi t)}$. In the neighbourhood of $w = i\eta$ we have

$$(s-1)\log\frac{w}{i\eta} = (s-1)\log\left(1+\frac{w-i\eta}{i\eta}\right)$$

$$= (\sigma+it-1)\left\{\frac{w-i\eta}{i\eta}-\frac{1}{2}\left(\frac{w-i\eta}{i\eta}\right)^2+\dots\right\}$$

$$= \frac{\eta}{2\pi}(w-i\eta)+\frac{i}{4\pi}(w-i\eta)^2+\dots.$$

Hence we write

$$e^{(s-1)\log(w/i\eta)} = e^{(\eta/2\pi)(w-i\eta)+(i/4\pi)(w-i\eta)^2}\phi\left(\frac{w-i\eta}{i\sqrt{(2\pi)}}\right),$$

where

$$\phi(z) = \exp\left\{(s-1)\log\left(1+\frac{z}{\sqrt{t}}\right)-iz\sqrt{t}+\tfrac{1}{2}iz^2\right\}$$

$$= \sum_{n=0}^{\infty} a_n z^n,$$

say. Now

$$\frac{d\phi}{dz} = \left(\frac{s-1}{z+\sqrt{t}}-i\sqrt{t}+iz\right)\phi(z) = \frac{\sigma-1+iz^2}{z+\sqrt{t}}\phi(z).$$

Hence
$$(z+\sqrt{t})\sum_{n=1}^{\infty} na_n z^{n-1} = (\sigma-1+iz^2)\sum_{n=0}^{\infty} a_n z^n,$$

and the coefficients a_n are determined in succession by the recurrence formula

$$(n+1)\sqrt{t}.a_{n+1} = (\sigma-n-1)a_n+ia_{n-2} \quad (n = 2, 3,\dots),$$

this being true for $n = 0$, $n = 1$ also if we write $a_{-2} = a_{-1} = 0$. Thus

$$a_0 = 1, \qquad a_1 = \frac{\sigma-1}{\sqrt{t}}, \qquad a_2 = \frac{(\sigma-1)(\sigma-2)}{2t}, \qquad \dots.$$

It follows that
$$a_n = O(t^{-\frac{1}{2}n+[\frac{1}{3}n]}) \tag{4.16.1}$$

(not uniformly in n); for if this is true up to n, then

$$a_{n+1} = O(t^{-\frac{1}{2}n+[\frac{1}{3}n]-\frac{1}{2}})+O(t^{-\frac{1}{2}(n-2)+[\frac{1}{3}(n-2)]-\frac{1}{2}}) = O(t^{-\frac{1}{2}(n+1)+[\frac{1}{3}(n+1)]}).$$

Hence (4.16.1) follows for all n by induction.

Now let
$$\phi(z) = \sum_{n=0}^{N-1} a_n z^n + r_N(z).$$

Then
$$r_N(z) = \frac{1}{2\pi i}\int_{\Gamma} \frac{\phi(w)z^N}{w^N(w-z)}\,dw,$$

where Γ is a contour including the points 0 and z. Now

$$\log\phi(w) = (s-1)\log\left(1+\frac{w}{\sqrt{t}}\right)+\tfrac{1}{2}iw^2-iw\sqrt{t}$$

$$= (\sigma-1)\log\left(1+\frac{w}{\sqrt{t}}\right)+iw^2\sum_{k=1}^{\infty}\frac{(-1)^{k-1}}{k+2}\left(\frac{w}{\sqrt{t}}\right)^k.$$

Hence for $|w| \leqslant \frac{3}{5}\sqrt{t}$ we have

$$\mathbf{R}\{\log\phi(w)\} \leqslant |\sigma-1|\log\frac{8}{5}+|w|^2.\frac{5}{6}\frac{|w|}{\sqrt{t}}.$$

Let $|z| < \frac{4}{7}\sqrt{t}$, and let Γ be a circle with centre $w = 0$, radius ρ_N, where

$$\tfrac{21}{20}|z| \leqslant \rho_N \leqslant \tfrac{3}{5}\sqrt{t}.$$

Then $$r_N(z) = O(|z|^N\rho_N^{-N}e^{5\rho_N^3/6\sqrt{t}}).$$

The function $\rho^{-N}e^{5\rho^3/6\sqrt{t}}$ has the minimum $(5e/2N\sqrt{t})^{\frac{1}{3}N}$ for $\rho = (2N\sqrt{t}/5)^{\frac{1}{3}}$; ρ_N can have this value if

$$\frac{21}{20}|z| \leqslant \left(\frac{2N\sqrt{t}}{5}\right)^{\frac{1}{3}} \leqslant \frac{3}{5}\sqrt{t}.$$

Hence

$$r_N(z) = O\left\{|z|^N\left(\frac{5e}{2N\sqrt{t}}\right)^{\frac{1}{3}N}\right\} \qquad \left\{N \leqslant \frac{27}{50}t,\quad |z| \leqslant \frac{20}{21}\left(\frac{2N\sqrt{t}}{5}\right)^{\frac{1}{3}}\right\}.$$

For $|z| \leqslant \frac{4}{7}\sqrt{t}$ we can also take $\rho_N = \frac{21}{20}|z|$, giving

$$r_N(z) = O\left[\left(\frac{20}{21}\right)^N\left\{\exp\frac{5}{6\sqrt{t}}\left(\frac{21}{20}|z|\right)^3\right\}\right] = O\left\{\exp\left(\frac{14}{29}|z|^2\right)\right\} \quad (|z| \leqslant \tfrac{1}{2}\sqrt{t}).$$

Now consider the integral along C_2, and take $c = 2^{-\frac{3}{2}}$. Then

$$\int_{C_2}\frac{w^{s-1}e^{-mw}}{e^w-1}\,dw = \int_{C_2}(i\eta)^{s-1}\frac{e^{(i/4\pi)(w-i\eta)^2+(\eta/2\pi)(w-i\eta)-mw}}{e^w-1}\sum_{n=0}^{N-1}a_n\left(\frac{w-i\eta}{i\sqrt{(2\pi)}}\right)^n dw+$$

$$+\int_{C_2}(i\eta)^{s-1}\frac{e^{(i/4\pi)(w-i\eta)^2+(\eta/2\pi)(w-i\eta)-mw}}{e^w-1}r_N\left(\frac{w-i\eta}{i\sqrt{(2\pi)}}\right)dw.$$

If $|e^w-1| > A$ on C_2, the last integral is, as in the previous section,

$$O\left[\eta^{\sigma-1}e^{-\frac{1}{2}\pi t}\left\{\int_0^{A(N\sqrt{t})^{\frac{1}{3}}}e^{-\lambda^2/4\pi}\left(\frac{\lambda}{\sqrt{(2\pi)}}\right)^N\left(\frac{5e}{2N\sqrt{t}}\right)^{\frac{1}{3}N}d\lambda+\int_{A(N\sqrt{t})^{\frac{1}{3}}}^{\frac{1}{2}\eta}e^{-(\lambda^2/4\pi)+(7\lambda^2/29\pi)}\,d\lambda\right\}\right]$$

$$= O\left[\eta^{\sigma-1}e^{-\frac{1}{2}\pi t}\left\{\left(\frac{5e}{2N\sqrt{t}}\right)^{\frac{1}{3}N}2^{\frac{1}{2}N}\Gamma(\tfrac{1}{2}N+\tfrac{1}{2})+e^{-A(N\sqrt{t})^{\frac{2}{3}}}\right\}\right]$$

$$= O\left\{\eta^{\sigma-1}e^{-\frac{1}{2}\pi t}\left(\frac{AN}{t}\right)^{\frac{1}{6}N}\right\}$$

for $N < At$. The case where the contour goes near a pole gives a similar result, as in the previous section.

In the first N terms we now replace C_2 by the infinite straight line of which it is a part, C_2' say. The integral multiplying a_n changes by

$$O\left\{\eta^{\sigma-1}e^{-\frac{1}{2}\pi t}\int\limits_{\frac{1}{2}\eta}^{\infty} e^{-(\lambda^2/4\pi)+(\eta\lambda/2\pi\sqrt{2})-(m+1)(\lambda/\sqrt{2})}\left(\frac{\lambda}{\sqrt{(2\pi)}}\right)^n d\lambda\right\}.$$

Since $m+1 \geqslant t/\eta = \eta/(2\pi)$, this is

$$O\left\{\eta^{\sigma-1}e^{-\frac{1}{2}\pi t}\int\limits_{\frac{1}{2}\eta}^{\infty} e^{-\lambda^2/4\pi}\left(\frac{\lambda}{\sqrt{(2\pi)}}\right)^n d\lambda\right\}.$$

We can write the integrand as

$$e^{-\lambda^2/8\pi}\times e^{-\lambda^2/8\pi}\left(\frac{\lambda}{\sqrt{(2\pi)}}\right)^n,$$

and the second factor is steadily decreasing for $\lambda > 2\sqrt{(n\pi)}$, and so throughout the interval of integration if $n < N < At$ with A small enough. The whole term is then

$$O\left\{\eta^{\sigma-1}e^{-\frac{1}{2}\pi t-(\eta^2/32\pi^2)}\left(\frac{\eta}{2\sqrt{(2\pi)}}\right)^n\right\} = O\{\eta^{\sigma-1}e^{-\frac{1}{2}\pi t-(t/16\pi)}(\tfrac{1}{2}\sqrt{t})^n\}.$$

Also $$a_n = (r_n - r_{n+1})z^{-n} = O\left\{\left(\frac{5e}{2n\sqrt{t}}\right)^{\frac{1}{3}n}\right\}.$$

Hence the total error is

$$O\left\{\eta^{\sigma-1}e^{-\frac{1}{2}\pi t-(t/16\pi)}\sum_{n=0}^{N-1}(\tfrac{1}{2}\sqrt{t})^n\left(\frac{5e}{2n\sqrt{t}}\right)^{\frac{1}{3}n}\right\} = O\left\{\eta^{\sigma-1}e^{-\frac{1}{2}\pi t-(t/16\pi)}\sum_{n=0}^{N-1}\left(\frac{5et}{16n}\right)^{\frac{1}{3}n}\right\}.$$

Now $(t/n)^{\frac{1}{3}n}$ increases steadily up to $n = t/e$, and so if $n < At$, where $A < 1/e$, it is

$$O(e^{\frac{1}{3}tA\log 1/A}).$$

Hence if $N < At$, with A small enough, the whole term is

$$O(e^{-(\frac{1}{2}\pi+A)t}).$$

We have finally the sum

$$(i\eta)^{s-1}\sum_{n=0}^{N-1}\frac{a_n}{i^n(2\pi)^{\frac{1}{2}n}}\int\limits_{C_2'}\frac{e^{(i/4\pi)(w-i\eta)^2+(\eta/2\pi)(w-i\eta)-mw}}{e^w-1}(w-i\eta)^n\,dw.$$

The integral may be expressed as

$$-\int\limits_L \exp\left\{\frac{i}{4\pi}(w+2m\pi i-i\eta)^2+\frac{\eta}{2\pi}(w+2m\pi i-i\eta)-mw\right\}\times$$
$$\times\frac{(w+2m\pi i-i\eta)^n}{e^w-1}\,dw,$$

where L is a line in the direction $\arg w = \frac{1}{4}\pi$, passing between 0 and $2\pi i$.

This is $n!$ times the coefficient of ξ^n in

$$-\int_L \exp\left\{\frac{i}{4\pi}(w+2m\pi i-i\eta)^2+\right.$$

$$\left.+\frac{\eta}{2\pi}(w+2m\pi i-i\eta)-mw+\xi(w+2m\pi i-i\eta)\right\}\frac{dw}{e^w-1}$$

$$=-\exp\left\{i(2m\pi-\eta)\left(\frac{3\eta}{4\pi}-\tfrac{1}{2}m+\xi\right)\right\}\int_L \exp\left\{\frac{iw^2}{4\pi}+w\left(\frac{\eta}{\pi}-2m+\xi\right)\right\}\frac{dw}{e^w-1}$$

$$=-2\pi\Psi\left(\frac{\eta}{\pi}-2m+\xi\right)\exp\left\{\frac{i\pi}{2}\left(\frac{\eta}{\pi}-2m+\xi\right)^2-\frac{5i\pi}{8}+\right.$$

$$\left.+i(2m\pi-\eta)\left(\frac{3\eta}{4\pi}-\tfrac{1}{2}m+\xi\right)\right\},$$

where
$$\Psi(a)=\frac{\cos\pi(\frac{1}{2}a^2-a-\frac{1}{8})}{\cos\pi a},$$

$$=2\pi(-1)^{m-1}e^{-\frac{1}{2}it-(5i\pi/8)}\Psi\left(\frac{\eta}{\pi}-2m+\xi\right)e^{\frac{1}{2}i\pi\xi^2}$$

$$=2\pi(-1)^{m-1}e^{-\frac{1}{2}it-(5i\pi/8)}\sum_{\mu=0}^{\infty}\Psi^{(\mu)}\left(\frac{\eta}{\pi}-2m\right)\frac{\xi^\mu}{\mu!}\sum_{\nu=0}^{\infty}\frac{(\frac{1}{2}i\pi\xi^2)^\nu}{\nu!}.$$

Hence we obtain

$$e^{\frac{1}{2}i\pi(s-1)}(2\pi t)^{\frac{1}{2}s-\frac{1}{2}}2\pi(-1)^{m-1}e^{-\frac{1}{2}it-(5i\pi/8)}\sum_{n=0}^{N-1}\sum_{\nu\leqslant\frac{1}{2}n}\frac{n!\,i^{\nu-n}}{\nu!\,(n-2\nu)!\,2^n}\times$$

$$\times\left(\frac{2}{\pi}\right)^{\frac{1}{2}n-\nu}a_n\Psi^{(n-2\nu)}\left(\frac{\eta}{\pi}-2m\right).$$

Denoting the last sum by S_N, we have the following result.

THEOREM 4.16. *If* $0\leqslant\sigma\leqslant 1$, $m=[\sqrt{(t/2\pi)}]$, *and* $N<At$, *where* A *is a sufficiently small constant,*

$$\zeta(s)=\sum_{n=1}^{m}\frac{1}{n^s}+\chi(s)\sum_{n=1}^{m}\frac{1}{n^{1-s}}+$$

$$+(-1)^{m-1}e^{-\frac{1}{2}i\pi(s-1)}(2\pi t)^{\frac{1}{2}s-\frac{1}{2}}e^{-\frac{1}{2}it-(i\pi/8)}\Gamma(1-s)\left\{S_N+O\left\{\left(\frac{AN}{t}\right)^{\frac{1}{6}N}\right\}+O(e^{-At})\right\}.$$

4.17. Special cases

In the approximate functional equation, let $\sigma=\frac{1}{2}$ and

$$x=y=\{t/(2\pi)\}^{\frac{1}{2}}.$$

Then (4.12.4) gives

$$\zeta(\tfrac{1}{2}+it)=\sum_{n\leqslant x}n^{-\frac{1}{2}-it}+\chi(\tfrac{1}{2}+it)\sum_{n\leqslant x}n^{-\frac{1}{2}+it}+O(t^{-\frac{1}{4}}). \qquad (4.17.1)$$

This can also be put into another form which is sometimes useful. We have

$$\chi(\tfrac{1}{2}+it)\chi(\tfrac{1}{2}-it) = 1,$$

so that

$$|\chi(\tfrac{1}{2}+it)| = 1.$$

Let

$$\vartheta = \vartheta(t) = -\tfrac{1}{2}\arg\chi(\tfrac{1}{2}+it),$$

so that

$$\chi(\tfrac{1}{2}+it) = e^{-2i\vartheta}.$$

Let

$$Z(t) = e^{i\vartheta}\zeta(\tfrac{1}{2}+it) = \{\chi(\tfrac{1}{2}+it)\}^{-\frac{1}{2}}\zeta(\tfrac{1}{2}+it). \tag{4.17.2}$$

Since

$$\{\chi(\tfrac{1}{2}+it)\}^{-\frac{1}{2}} = \pi^{-\frac{1}{2}it}\left\{\frac{\Gamma(\frac{1}{4}+\frac{1}{2}it)}{\Gamma(\frac{1}{4}-\frac{1}{2}it)}\right\}^{\frac{1}{2}} = \frac{\pi^{-\frac{1}{2}it}\Gamma(\frac{1}{4}+\frac{1}{2}it)}{|\Gamma(\frac{1}{4}+\frac{1}{2}it)|},$$

we have also

$$Z(t) = -2\pi^{\frac{1}{4}}\frac{\Xi(t)}{(t^2+\frac{1}{4})|\Gamma(\frac{1}{4}+\frac{1}{2}it)|}. \tag{4.17.3}$$

The function $Z(t)$ is thus real for real t, and

$$|Z(t)| = |\zeta(\tfrac{1}{2}+it)|.$$

Multiplying (4.17.1) by $e^{i\vartheta}$, we obtain

$$Z(t) = e^{i\vartheta}\sum_{n\leqslant x} n^{-\frac{1}{2}-it}+e^{-i\vartheta}\sum_{n\leqslant x} n^{-\frac{1}{2}+it}+O(t^{-\frac{1}{4}})$$

$$= 2\sum_{n\leqslant x} n^{-\frac{1}{2}}\cos(\vartheta-t\log n)+O(t^{-\frac{1}{4}}). \tag{4.17.4}$$

Again, in Theorem 4.16, let $N = 3$. Then

$$S_3 = a_0\Psi\left(\frac{\eta}{\pi}-2m\right)+\frac{1}{2i}\left(\frac{2}{\pi}\right)^{\frac{1}{2}}a_1\Psi'\left(\frac{\eta}{\pi}-2m\right)-$$

$$-\frac{a_2}{2\pi}\Psi''\left(\frac{\eta}{\pi}-2m\right)+\frac{a_2}{2i}\Psi\left(\frac{\eta}{\pi}-2m\right)$$

$$= \Psi\left(\frac{\eta}{\pi}-2m\right)+O(t^{-\frac{1}{2}})$$

$$= \frac{\cos\{t-(2m+1)\sqrt{(2\pi t)}-\frac{1}{8}\pi\}}{\cos\sqrt{(2\pi t)}}+O(t^{-\frac{1}{2}}),$$

and the O-term gives, for $\zeta(s)$, a term $O(t^{-\frac{1}{2}\sigma-\frac{1}{2}})$. In the case $\sigma = \frac{1}{2}$ we obtain, on multiplying by $e^{i\vartheta}$ and proceeding as before,

$$Z(t) = 2\sum_{n=1}^{m}\frac{\cos(\vartheta-t\log n)}{n^{\frac{1}{2}}}+$$

$$+(-1)^{m-1}\left(\frac{2\pi}{t}\right)^{\frac{1}{4}}\frac{\cos\{t-(2m+1)\sqrt{(2\pi t)}-\frac{1}{8}\pi\}}{\cos\sqrt{(2\pi t)}}+O(t^{-\frac{3}{4}}). \tag{4.17.5}$$

4.18. A different type of approximate formula has been obtained by Meulenbeld.† Instead of using finite partial sums of the original Dirichlet series, we can approximate to $\zeta(s)$ by sums of the form

$$\sum_{n\leqslant x}\frac{\phi(n/x)}{n^s},$$

where $\phi(u)$ decreases from 1 to 0 as u increases from 0 to 1. This reduces considerably the order of the error terms. The simplest result of this type is

$$\zeta(s) = 2\sum_{n\leqslant x}\frac{1-n/x}{n^s}+\chi(s)\sum_{n\leqslant y}\frac{1}{n^{1-s}}-$$
$$-\chi(s)\sum_{y<n<2y}\frac{1}{n^{1-s}}+\frac{2\chi(s-1)}{x}\sum_{y<n<2y}\frac{1}{n^{2-s}}+O\left(\frac{1}{t^{2\sigma}}+\frac{1}{x^{\sigma}t^{\frac{1}{2}}}+\frac{1}{x^{\sigma-1}t}\right),$$

valid for $2\pi xy = |t|$, $|t| \geqslant (x+1)^{\frac{1}{2}}$, $-2<\sigma<2$.

There is also an approximate functional equation‡ for $\{\zeta(s)\}^2$. This is

$$\{\zeta(s)\}^2 = \sum_{n\leqslant x}\frac{d(n)}{n^s}+\chi^2(s)\sum_{n\leqslant y}\frac{d(n)}{n^{1-s}}+O(x^{\frac{1}{2}-\sigma}\log t), \qquad (4.18.1)$$

where $0\leqslant\sigma\leqslant 1$, $xy = (t/2\pi)^2$, $x\geqslant h>0$, $y\geqslant h>0$. The proofs of this are rather elaborate.

NOTES FOR CHAPTER 4

4.19. Lemmas 4.2 and 4.4 can be generalized by taking F to be k times differentiable, and satisfying $|F^{(k)}(x)|\geqslant\lambda>0$ throughout $[a, b]$. By using induction, in the same way that Lemma 4.4 was deduced from Lemma 4.2, one finds that

$$\int_a^b e^{iF(x)}\,dx \ll_k \lambda^{-1/k}.$$

The error term $O(\lambda_2^{-\frac{4}{5}}\lambda_3^{-\frac{1}{5}})$ in Lemma 4.6 may be replaced by $O(\lambda_2^{-1}\lambda_3^{\frac{1}{3}})$, which is usually sharper in applications. To do this one chooses $\delta = \lambda_3^{-\frac{1}{3}}$ in the proof. It then suffices to show that

$$\int_{-\delta}^{\delta} e^{i\lambda x^2}(e^{if(x)}-1)\,dx \ll (\lambda\delta)^{-1}, \qquad (4.19.1)$$

if f has a continuous first derivative and satisfies $f(x)\ll x^3\delta^{-3}$,

† Meulenbeld (1).
‡ Hardy and Littlewood (6), Titchmarsh (21).

$f'(x) \ll x^2\delta^{-3}$. Here we have written $\lambda = \frac{1}{2}F''(c)$ and

$$f(x) = F(x+c) - F(c) - \tfrac{1}{2}x^2 F''(c).$$

If $\delta \leqslant (\lambda\delta)^{-1}$ then (4.19.1) is immediate. Otherwise we have

$$\int_{-\delta}^{\delta} = \int_{-\delta}^{-(\lambda\delta)^{-1}} + \int_{-(\lambda\delta)^{-1}}^{(\lambda\delta)^{-1}} + \int_{(\lambda\delta)^{-1}}^{\delta}.$$

The second integral on the right is trivially $O\{(\lambda\delta)^{-1}\}$, while the third, for example, is, on integrating by parts,

$$\int_{(\lambda\delta)^{-1}}^{\delta} (2i\lambda x e^{i\lambda x^2}) \frac{e^{if(x)}-1}{2i\lambda x}\, dx =$$

$$\left[e^{i\lambda x^2}\frac{e^{if(x)}-1}{2i\lambda x}\right]_{(\lambda\delta)^{-1}}^{\delta} - \int_{(\lambda\delta)^{-1}}^{\delta} e^{i\lambda x^2}\frac{d}{dx}\left(\frac{e^{if(x)}-1}{2i\lambda x}\right)dx$$

$$\ll \max_{x=(\lambda\delta)^{-1},\delta}\left|\frac{f(x)}{\lambda x}\right| + \int_{(\lambda\delta)^{-1}}^{\delta}\left|\frac{xif'(x)e^{if(x)}-(e^{if(x)}-1)}{2i\lambda x^2}\right|dx$$

$$\ll (\lambda\delta)^{-1} + \int_{(\lambda\delta)^{-1}}^{\delta}\left|\frac{x^3\delta^{-3}}{2i\lambda x^2}\right|dx$$

$$\ll (\lambda\delta)^{-1}$$

as required. Similarly the error term $O\{(b-a)\lambda_2^{\frac{1}{5}}\lambda_3^{\frac{1}{5}}\}$ in Theorem 4.9 may be replaced by $O\{(b-a)\lambda_3^{\frac{1}{3}}\}$.

For further estimates along these lines see Vinogradov [**2**; pp. 86–91] and Heath-Brown [**11**; Lemmas 6 and 10]. These papers show that the error term $O((b-a)\lambda_2^{\frac{1}{5}}\lambda_3^{\frac{1}{5}})$ can be dropped entirely, under suitable conditions.

Lemmas 4.2 and 4.8 have the following corollary, which is sometimes useful.

LEMMA 4.19. *Let $f(x)$ be a real differentiable function on the interval $[a, b]$, let $f'(x)$ be monotonic, and let $0 < \lambda \leqslant |f'(x)| \leqslant \vartheta < 1$. Then*

$$\sum_{a<n\leqslant b} e^{2\pi i f(n)} \ll_{\vartheta} \lambda^{-1}.$$

4.20. Weighted approximate functional equations related to those mentioned in §4.18 have been given by Lavrik [**1**] and Heath-Brown [**3**; Lemma 1], [**4**; Lemma 1]. As a typical example one has

$$\zeta(s)^k = \sum_1^\infty d_k(n) n^{-s} w_s\left(\frac{n}{x}\right) + \chi(s)^k \sum_1^\infty d_k(n) n^{s-1} w_{1-s}\left(\frac{n}{y}\right) + O(x^{1-\sigma}\log^k(2+x)e^{-t^2/4}) \tag{4.20.1}$$

uniformly for $t \geqslant 1$, $|\sigma| \leqslant \frac{1}{2}t$, $xy = (t/2\pi)^k$, $x, y \gg 1$, for any fixed positive integer k. Here

$$w_s(u) = \frac{1}{2\pi i}\int_{c-i\infty}^{c+i\infty}\left((\tfrac{1}{2}t)^{-z/2}\frac{\Gamma\{\frac{1}{2}(s+z)\}}{\Gamma(\frac{1}{2}s)}\right)^k u^{-z}e^{z^2}\frac{dz}{z} \qquad (c > \max(0, -\sigma)).$$

The advantage of (4.20.1) is the very small error term.

Although the weight $w_s(u)$ is a little awkward, it is easy to see, by moving the line of integration to $c = \pm 1$, for example, that

$$w_s(u) = \begin{cases} O(u^{-1}) & (u \geqslant 1), \\ 1 + O(u) + O\left\{u^\sigma\left(\log\frac{2}{u}\right)^k e^{-\frac{1}{2}t^2}\right\} & (0 < u \leqslant 1), \end{cases}$$

uniformly for $0 \leqslant \sigma \leqslant 1$, $t \geqslant 1$. More accurate estimates are however possible.

To prove (4.20.1) one writes

$$\sum_1^\infty d_k(n) n^{-s} w_s\left(\frac{n}{x}\right) = \frac{1}{2\pi i}\int_{c-i\infty}^{c+i\infty}\left((\tfrac{1}{2}t)^{-\frac{1}{2}z}\frac{\Gamma\{\frac{1}{2}(s+z)\}}{\Gamma(\frac{1}{2}s)}\zeta(s+z)\right)^k x^z e^{z^2}\frac{dz}{z} \qquad (c > \max(0, 1-\sigma)),$$

and moves the line of integration to $\mathbf{R}(z) = -d$, $d > \max(0, \sigma)$, giving

$$\frac{1}{2\pi i}\int_{-d-i\infty}^{-d+i\infty}\left((\tfrac{1}{2}t)^{-\frac{1}{2}z}\frac{\Gamma\{\frac{1}{2}(s+z)\}}{\Gamma(\frac{1}{2}s)}\zeta(s+z)\right)^k x^z e^{z^2}\frac{dz}{z} + \zeta(s)^k + \operatorname{Res}(z = 1-s).$$

The residue term is easily seen to be $O\{x^{1-\sigma}\log^k(2+x)e^{-t^2/4}\}$. In the integral we substitute $z = -w$, $x = (t/2\pi)^k y^{-1}$, and we apply the functional equation (2.6.4). This yields

$$\frac{1}{2\pi i}\int_{-d-i\infty}^{-d+i\infty}\left((\tfrac{1}{2}t)^{-z/2}\frac{\Gamma\{\frac{1}{2}(s+z)\}}{\Gamma(\frac{1}{2}s)}\zeta(s+z)\right)^k x^z e^{z^2}\frac{dz}{z}$$

$$= -\frac{\chi(s)^k}{2\pi i}\int_{d-i\infty}^{d+i\infty}\left((\tfrac{1}{2}t)^{-w/2}\frac{\Gamma\{(\frac{1}{2}(1-s+w)\}}{\Gamma\{\frac{1}{2}(1-s)\}}\zeta(1-s+w)\right)^k y^w e^{w^2}\frac{dw}{w}$$

$$= -\chi(s)^k\sum_1^\infty d_k(n)n^{s-1}w_{1-s}\left(\frac{n}{y}\right),$$

as required.

Another result of the same general nature is

$$|\zeta(\tfrac{1}{2}+it)|^{2k} = \sum_{m,n=1}^{\infty} d_k(m)d_k(n)m^{-\frac{1}{2}-it}n^{-\frac{1}{2}+it}W_t(mn)+O(e^{-t^2/2}) \tag{4.20.2}$$

for $t \geqslant 1$, and any fixed positive integer k, where

$$W_t(u) = \frac{1}{\pi i}\int_{1-i\infty}^{1+i\infty}\left(\pi^{-z}\frac{\Gamma\{\frac{1}{2}(\frac{1}{2}+it+z)\}\,\Gamma\{\frac{1}{2}(\frac{1}{2}-it+z)\}}{\Gamma\{\frac{1}{2}(\frac{1}{2}+it)\}\,\Gamma\{\frac{1}{2}(\frac{1}{2}-it)\}}\right)^k u^{-z}e^{z^2}\frac{dz}{z}.$$

This type of formula has the advantage that the cross terms which would arise on multipling (4.20.1) by its complex conjugate are absent. By moving the line of integration to $\mathbf{R}(z) = \pm\frac{1}{2}$ one finds that

$$W_t(u) = 2+O\left\{u^{\frac{1}{2}}\log^k\left(\frac{2}{u}\right)\right\} \quad (0<u\leqslant 1),$$

and $W_t(u) = O(u^{-\frac{1}{2}})$ for $u \geqslant 1$. Again better estimates are possible. The proof of (4.20.2) is similar to that of (4.20.1), and starts from the formula

$$\tfrac{1}{2}\sum_{m,n=1}^{\infty} d_k(m)d_k(n)m^{-\frac{1}{2}-it}n^{-\frac{1}{2}+it}W_t(mn)$$

$$= \frac{1}{2\pi i}\int_{1-i\infty}^{1+i\infty}\left(\pi^{-z}\frac{\Gamma\{\frac{1}{2}(\frac{1}{2}+it+z)\}\,\Gamma\{\frac{1}{2}(\frac{1}{2}-it+z)\}}{\Gamma\{\frac{1}{2}(\frac{1}{2}+it)\}\,\Gamma\{\frac{1}{2}(\frac{1}{2}-it)\}}\right.$$

$$\left.\zeta(\tfrac{1}{2}+it+z)\zeta(\tfrac{1}{2}-it+z)\right)^k e^{z^2}\frac{dz}{z}.$$

4.21. We may write the approximate functional equation (4.18.1) in the form

$$\zeta(s)^2 = S(s, x) + \chi(s)^2 S(1-s, y) + R(s, x).$$

The estimate $R(s, x) \ll x^{\frac{1}{2}-\sigma} \log t$ has been shown by Jutila (see Ivic [**3**; §4.2]) to be best possible for

$$t^{\frac{1}{2}} \ll \left| x - \frac{t}{2\pi} \right| \ll t^{\frac{1}{2}}.$$

Outside this range however, one can do better. Thus Jutila (in work to appear) has proved that

$$R(s, x) \ll t^{\frac{1}{2}} x^{-\sigma} (\log t) \log\left(1 + \frac{x}{t}\right) + t^{-1} x^{1-\sigma} (y^{\varepsilon} + \log t)$$

for $0 \leqslant \sigma \leqslant 1$ and $x \gg t \gg 1$. (The corresponding result for $x \ll t$ may be deduced from this, via the functional equation.) For the special case $x = y = t/2\pi$ one may also improve on (4.18.1). Motohashi [**2**], [**3**], and in work in the course of publication, has established some very precise results in this direction. In particular he has shown that

$$\chi(1-s) R\left(s, \frac{t}{2\pi}\right) = -\left(\frac{4\pi}{t}\right)^{\frac{1}{2}} \Delta\left(\frac{t}{2\pi}\right) + O(t^{-\frac{1}{4}}),$$

where $\Delta(x)$ is the remainder term in the Dirichlet divisor problem (see §12.1). Jutila, in the work to appear, cited above, gives another proof of this. In fact, for the special case $\sigma = \frac{1}{2}$, the problem was considered 40 years earlier by Taylor (**1**).

V

THE ORDER OF $\zeta(s)$ IN THE CRITICAL STRIP

5.1. THE main object of this chapter is to discuss the order of $\zeta(s)$ as $t \to \infty$ in the 'critical strip' $0 \leqslant \sigma \leqslant 1$. We begin with a general discussion of the order problem. It is clear from the original Dirichlet series (1.1.1) that $\zeta(s)$ is bounded in any half-plane $\sigma \geqslant 1+\delta > 1$; and we have proved in (2.12.2) that

$$\zeta(s) = O(|t|) \quad (\sigma \geqslant \tfrac{1}{2}).$$

For $\sigma < \frac{1}{2}$, corresponding results follow from the functional equation

$$\zeta(s) = \chi(s)\zeta(1-s).$$

In any fixed strip $\alpha \leqslant \sigma \leqslant \beta$, as $t \to \infty$

$$|\chi(s)| \sim \left(\frac{t}{2\pi}\right)^{\frac{1}{2}-\sigma}$$

by (4.12.3). Hence

$$\zeta(s) = O(t^{\frac{1}{2}-\sigma}) \quad (\sigma \leqslant -\delta < 0), \tag{5.1.1}$$

and

$$\zeta(s) = O(t^{\frac{3}{2}+\delta}) \quad (\sigma \geqslant -\delta).$$

Thus in any half-plane $\sigma \geqslant \sigma_0$

$$\zeta(s) = O(|t|^k), \qquad k = k(\sigma_0),$$

i.e. $\zeta(s)$ is a function of finite order in the sense of the theory of Dirichlet series.†

For each σ we define a number $\mu(\sigma)$ as the lower bound of numbers ξ such that

$$\zeta(\sigma+it) = O(|t|^{\xi}).$$

It follows from the general theory of Dirichlet series‡ that, as a function of σ, $\mu(\sigma)$ is continuous, non-increasing, and convex downwards in the sense that no arc of the curve $y = \mu(\sigma)$ has any point above its chord; also $\mu(\sigma)$ is never negative.

Since $\zeta(s)$ is bounded for $\sigma \geqslant 1+\delta$ $(\delta > 0)$, it follows that

$$\mu(\sigma) = 0 \quad (\sigma > 1), \tag{5.1.2}$$

and then from the functional equation that

$$\mu(\sigma) = \tfrac{1}{2}-\sigma \quad (\sigma < 0). \tag{5.1.3}$$

These equations also hold by continuity for $\sigma = 1$ and $\sigma = 0$ respectively.

† See Titchmarsh, *Theory of Functions*, §§ 9.4, 9.41.
‡ Ibid., §§ 5.65, 9.41.

The chord joining the points $(0, \frac{1}{2})$ and $(1, 0)$ on the curve $y = \mu(\sigma)$ is $y = \frac{1}{2}-\frac{1}{2}\sigma$. It therefore follows from the convexity property that

$$\mu(\sigma) \leqslant \tfrac{1}{2}-\tfrac{1}{2}\sigma \quad (0 < \sigma < 1). \tag{5.1.4}$$

In particular, $\mu(\frac{1}{2}) \leqslant \frac{1}{4}$, i.e.

$$\zeta(\tfrac{1}{2}+it) = O(t^{\frac{1}{4}+\epsilon}) \tag{5.1.5}$$

for every positive ϵ.

The exact value of $\mu(\sigma)$ is not known for any value of σ between 0 and 1. It will be shown later that $\mu(\frac{1}{2}) < \frac{1}{4}$, and the simplest possible hypothesis is that the graph of $\mu(\sigma)$ consists of two straight lines

$$\mu(\sigma) = \tfrac{1}{2}-\sigma \quad (\sigma \leqslant \tfrac{1}{2}), \quad 0 \quad (\sigma > \tfrac{1}{2}). \tag{5.1.6}$$

This is known as Lindelöf's hypothesis. It is equivalent to the statement that

$$\zeta(\tfrac{1}{2}+it) = O(t^{\epsilon}) \tag{5.1.7}$$

for every positive ϵ.

The approximate functional equation gives a slight refinement on the above results. For example, taking $\sigma = \frac{1}{2}$, $x = y = \sqrt{(t/2\pi)}$ in (4.12.4), we obtain

$$\begin{aligned} \zeta(\tfrac{1}{2}+it) &= \sum_{n \leqslant \sqrt{(t/2\pi)}} \frac{1}{n^{\frac{1}{2}+it}} + O(1) \sum_{n \leqslant \sqrt{(t/2\pi)}} \frac{1}{n^{\frac{1}{2}-it}} + O(t^{-\frac{1}{4}}) \\ &= O\left(\sum_{n \leqslant \sqrt{(t/2\pi)}} \frac{1}{n^{\frac{1}{2}}}\right) + O(t^{-\frac{1}{4}}) \\ &= O(t^{\frac{1}{4}}). \end{aligned} \tag{5.1.8}$$

5.2. To improve upon this we have to show that a certain amount of cancelling occurs between the terms of such a sum. We have

$$\sum_{n=a+1}^{b} n^{-s} = \sum_{n=a+1}^{b} n^{-\sigma} e^{-it\log n}$$

and we apply the familiar lemma of 'partial summation'. *Let*

$$b_1 \geqslant b_2 \geqslant \ldots \geqslant b_n \geqslant 0,$$

and

$$s_m = a_1+a_2+\ldots+a_m$$

where the a's are any real or complex numbers. Then if

$$|s_m| \leqslant M \quad (m = 1, 2, \ldots),$$

$$|a_1 b_1+a_2 b_2+\ldots+a_n b_n| \leqslant M b_1. \tag{5.2.1}$$

For

$$\begin{aligned} a_1 b_1+\ldots+a_n b_n &= b_1 s_1+b_2(s_2-s_1)+\ldots+b_n(s_n-s_{n-1}) \\ &= s_1(b_1-b_2)+s_2(b_2-b_3)+\ldots+s_{n-1}(b_{n-1}-b_n)+s_n b_n. \end{aligned}$$

Hence

$$|a_1 b_1 + \ldots + a_n b_n| \leqslant M(b_1 - b_2 + \ldots + b_{n-1} - b_n + b_n) = Mb_1.$$

If $0 \leqslant b_1 \leqslant b_2 \leqslant \ldots \leqslant b_n$ we obtain similarly

$$|a_1 b_1 + \ldots + a_n b_n| \leqslant 2Mb_n.$$

If $a_n = e^{-it\log n}$, $b_n = n^{-\sigma}$, where $\sigma \geqslant 0$, it follows that

$$\sum_{n=a+1}^{b} n^{-s} = O\Big(a^{-\sigma} \max_{a<c\leqslant b} \Big|\sum_{n=a+1}^{c} e^{-it\log n}\Big|\Big). \tag{5.2.2}$$

This raises the general question of the order of sums of the form

$$\Sigma = \sum_{n=a+1}^{b} e^{2\pi i f(n)}, \tag{5.2.3}$$

when $f(n)$ is a real function of n. In the above case,

$$f(n) = \frac{-t\log n}{2\pi}.$$

The earliest method of dealing with such sums is that of Weyl,† largely developed by Hardy and Littlewood.‡ This is roughly as follows. We can reduce the problem of Σ to that of

$$S = \sum_{n=a+1}^{b} e^{2\pi i g(n)},$$

where $g(n)$ is a polynomial of sufficiently high degree, say of degree k. Now

$$|S|^2 = \sum_m \sum_n e^{2\pi i\{g(m)-g(n)\}} = \sum_\nu \sum_n e^{2\pi i\{g(n+\nu)-g(n)\}}$$

$$\leqslant \sum_\nu \Big|\sum_n e^{2\pi i\{g(n+\nu)-g(n)\}}\Big| \tag{5.2.4}$$

with suitable limits for the sums; and $g(n+\nu)-g(n)$ is of degree $k-1$. By repeating the process we ultimately obtain a sum of the form

$$S_k = \sum_{n=a+1}^{b} e^{2\pi i(\lambda n+\mu)}.$$

We can now actually carry out the summation. We obtain

$$|S_k| = \left|\frac{1-e^{2\pi i(b-a)\lambda}}{1-e^{2\pi i\lambda}}\right| \leqslant \frac{1}{|\sin \pi\lambda|}. \tag{5.2.5}$$

If $|\operatorname{cosec} \pi\lambda|$ is small compared with $b-a$, this is a favourable result, and can be used to give a non-trivial result for the original sum S.

An alternative method is due to van der Corput.§ In this method we approximate to the sum Σ by the corresponding integral

$$\int_a^b e^{2\pi i f(x)}\, dx,$$

† Weyl (1), (2). ‡ Littlewood (2), Landau (15).

§ van der Corput (1)–(7), van der Corput and Koksma (1), Titchmarsh (8)–(12).

and then estimate the integral by the principle of stationary phase, or some such method. Actually the original sum is usually not suitable for this process, and intermediate steps of the form (5.2.4) have to be used.

Still another method has been introduced by Vinogradov. This is in some ways very complicated; but it avoids the k-fold repetition used in the Weyl–Hardy–Littlewood method, which for large k is very 'uneconomical'. An account of this method will be given in the next chapter.

5.3. The Weyl–Hardy–Littlewood method. The relation of the general sum to the sum involving polynomials is as follows:

LEMMA 5.3. *Let k be a positive integer,*

$$t \geqslant 1, \qquad \frac{b-a}{a} \leqslant \tfrac{1}{2}t^{-1/(k+1)},$$

and $$\left|\sum_{m=1}^{\mu} \exp\left\{-it\left(\frac{m}{a}-\frac{1}{2}\frac{m^2}{a^2}+\ldots+\frac{(-1)^{k-1}m^k}{ka^k}\right)\right\}\right| \leqslant M \quad (\mu \leqslant b-a).$$

Then $$\left|\sum_{n=a+1}^{b} e^{-it\log n}\right| < AM.$$

For

$$\left|\sum_{n=a+1}^{b} e^{-it\log n}\right| = \left|\sum_{m=1}^{b-a} e^{-it\log(a+m)}\right|$$

$$= \left|\sum_{m=1}^{b-a} \exp\left\{-it\left(\frac{m}{a}-\ldots+\frac{(-1)^{k-1}m^k}{ka^k}\right)-it\left(\frac{(-1)^k m^{k+1}}{(k+1)a^{k+1}}+\ldots\right)\right\}\right|$$

$$= \left|\sum_{m=1}^{b-a} \exp\left\{-it\left(\frac{m}{a}-\ldots+\frac{(-1)^{k-1}m^k}{ka^k}\right)\right\} \sum_{\nu=0}^{\infty} e_\nu(t)\left(\frac{m}{a}\right)^\nu\right|, \text{ say,}$$

$$= \left|\sum_{\nu=0}^{\infty} \frac{e_\nu(t)}{a^\nu} \sum_{m=1}^{b-a} m^\nu \exp\left\{-it\left(\frac{m}{a}-\ldots+\frac{(-1)^{k-1}m^k}{ka^k}\right)\right\}\right|$$

$$\leqslant 2M \sum_{\nu=0}^{\infty} |e_\nu(t)| \left(\frac{b-a}{a}\right)^\nu$$

$$\leqslant 2M \exp\left[t\left\{\frac{(b-a)^{k+1}}{(k+1)a^{k+1}}+\ldots\right\}\right]$$

$$\leqslant 2M \exp\left\{t\frac{(b-a)^{k+1}}{a^{k+1}}\Big/\left(1-\frac{b-a}{a}\right)\right\} \leqslant 2Me^2.$$

5.4. The simplest case is that of $\zeta(\frac{1}{2}+it)$, and we begin by working this out. We require the case $k = 2$ of the above lemma, and also the following

LEMMA. *Let*
$$S = \sum_{m=1}^{\mu} e^{2\pi i(\alpha m^2+\beta m)}.$$

Then
$$|S|^2 \leqslant \mu+2\sum_{r=1}^{\mu-1} \min(\mu, |\operatorname{cosec} 2\pi\alpha r|).$$

For
$$|S|^2 = \sum_{m=1}^{\mu} \sum_{m'=1}^{\mu} e^{2\pi i(\alpha m^2+\beta m-\alpha m'^2-\beta m')}.$$

Putting $m' = m-r$, this takes the form

$$\sum_{m}\sum_{r} e^{2\pi i(2\alpha mr-\alpha r^2+\beta r)} \leqslant \sum_{r=-\mu+1}^{\mu-1} \left|\sum_{m} e^{4\pi i\alpha mr}\right|,$$

where, corresponding to each value of r, m runs over at most μ consecutive integers. Hence, by (5.2.5),

$$|S|^2 \leqslant \sum_{r=-\mu+1}^{\mu-1} \min(\mu, |\operatorname{cosec} 2\pi\alpha r|)$$
$$= \mu+2\sum_{r=1}^{\mu-1} \min(\mu, |\operatorname{cosec} 2\pi\alpha r|).$$

5.5. THEOREM 5.5. $\zeta(\frac{1}{2}+it) = O(t^{\frac{1}{6}}\log^{\frac{3}{2}}t)$.

Let $2t^{\frac{1}{3}} \leqslant a < At$, $b \leqslant 2a$, and let

$$\mu = [\tfrac{1}{2}at^{-\frac{1}{3}}]. \tag{5.5.1}$$

Then

$$\Sigma = \sum_{n=a+1}^{b} e^{-it\log n} = \sum_{n=a+1}^{a+\mu} + \sum_{a+\mu+1}^{a+2\mu} + \ldots + \sum_{a+N\mu+1}^{b} = \Sigma_1+\Sigma_2+\ldots+\Sigma_{N+1},$$

where
$$N = \left[\frac{b-a}{\mu}\right] = O\left(\frac{a}{\mu}\right) = O(t^{\frac{1}{3}}).$$

By § 5.3, $\Sigma_\nu = O(M)$, where M is the maximum of

$$S_\nu = \sum_{m=1}^{\mu'} \exp\left\{-it\left(\frac{m}{a+\nu\mu} - \frac{1}{2}\frac{m^2}{(a+\nu\mu)^2}\right)\right\}$$

for $\mu' \leqslant \mu$. By § 5.4 this is

$$O\left[\left\{\mu+\sum_{r=1}^{\mu-1}\min\left(\mu, \left|\operatorname{cosec}\frac{tr}{2(a+\nu\mu)^2}\right|\right)\right\}\right]^{\frac{1}{2}}.$$

Hence

$$\Sigma = O\{(N+1)\mu^{\frac{1}{2}}\}+O\left[\left\{\sum_{\nu=1}^{N+1} 1 \sum_{\nu=1}^{N+1}\sum_{r=1}^{\mu-1} \min\left(\mu, \left|\operatorname{cosec}\frac{tr}{2(a+\nu\mu)^2}\right|\right)\right\}^{\frac{1}{2}}\right]$$

$$= O\{(N+1)\mu^{\frac{1}{2}}\}+O\left[(N+1)^{\frac{1}{2}}\left\{\sum_{r=1}^{\mu-1}\sum_{\nu=1}^{N+1} \min\left(\mu, \left|\operatorname{cosec}\frac{tr}{2(a+\nu\mu)^2}\right|\right)\right\}^{\frac{1}{2}}\right].$$

Now

$$\frac{tr}{2(a+\nu\mu)^2}-\frac{tr}{2\{a+(\nu+1)\mu\}^2} = \frac{tr\mu\{2a+(2\nu+1)\mu\}}{2(a+\nu\mu)^2\{a+(\nu+1)\mu\}^2},$$

which, as ν varies, lies between constant multiples of $tr\mu/a^3$, or, by (5.5.1), of r/μ^2. Hence for the values of ν for which $\frac{1}{2}tr/(a+\nu\mu)^2$ lies in a certain interval $\{l\pi, (l\pm\frac{1}{2})\pi\}$, the least value but one of

$$\left|\sin\frac{tr}{2(a+\nu\mu)^2}\right|$$

is greater than Ar/μ^2, the least but two is greater than $2Ar/\mu^2$, the least but three is greater than $3Ar/\mu^2$, and so on to $O(N) = O(t^{\frac{1}{3}})$ terms. Hence these values of ν contribute

$$\mu+O\left(\frac{\mu^2}{r}+\frac{\mu^2}{2r}+\ldots\right) = \mu+O\left(\frac{\mu^2}{r}\log t\right) = O\left(\frac{\mu^2}{r}\log t\right).$$

The number of such intervals $\{l\pi, (l\pm\frac{1}{2})\pi\}$ is

$$O\left\{(N+1)\frac{r}{\mu^2}+1\right\}.$$

Hence the ν-sum is

$$O\{(N+1)\log t\}+O\left(\frac{\mu^2}{r}\log t\right).$$

Hence

$$\Sigma = O\{(N+1)\mu^{\frac{1}{2}}\}+O(N+1)^{\frac{1}{2}}\left[\sum_{r=1}^{\mu-1}\left\{(N+1)\log t+\frac{\mu^2}{r}\log t\right\}\right]^{\frac{1}{2}}$$

$$= O\{(N+1)\mu^{\frac{1}{2}}\}+O\{(N+1)\mu^{\frac{1}{2}}\log^{\frac{1}{2}}t\}+O\{(N+1)^{\frac{1}{2}}\mu\log t\}$$

$$= O(a^{\frac{1}{2}}t^{\frac{1}{6}}\log^{\frac{1}{2}}t)+O(at^{-\frac{1}{6}}\log t).$$

If $a = O(t^{\frac{1}{2}})$, the second term can be omitted. Then by partial summation

$$\sum_{n=a+1}^{b}\frac{1}{n^{\frac{1}{2}+it}} = O(t^{\frac{1}{6}}\log^{\frac{1}{2}}t) \quad (b \leqslant 2a).$$

By adding $O(\log t)$ sums of the above form, we get

$$\sum_{2t^{\frac{1}{3}}\leqslant n\leqslant(t/2\pi)^{\frac{1}{2}}}\frac{1}{n^{\frac{1}{2}+it}} = O(t^{\frac{1}{6}}\log^{\frac{3}{2}}t).$$

Also
$$\sum_{n<2t^{\frac{1}{3}}} \frac{1}{n^{\frac{1}{2}+it}} = O\Big(\sum_{n<2t^{\frac{1}{3}}} \frac{1}{n^{\frac{1}{2}}}\Big) = O(t^{\frac{1}{6}}).$$

The result therefore follows from the approximate functional equation.

5.6. We now proceed to the general case. We require the following lemmas.

LEMMA 5.6. *Let* $$f(x) = \alpha x^k + \ldots$$
be a polynomial of degree k with real coefficients. Let
$$S = \sum e^{2\pi i f(m)}$$
where m ranges over at most μ consecutive integers. Let $K = 2^{k-1}$. Then for $k \geqslant 2$
$$|S|^K \leqslant 2^{2K}\mu^{K-1} + 2^K\mu^{K-k} \sum_{r_1,\ldots,r_{k-1}} \min(\mu, |\operatorname{cosec}(\pi\alpha k!\, r_1 \ldots r_{k-1})|)$$
where each r varies from 1 to $\mu-1$. For $k = 1$ the sum is replaced by the single term $\min(\mu, |\operatorname{cosec} \pi\alpha|)$.

We have
$$|S|^2 = \sum_m \sum_{m'} e^{2\pi i\{f(m)-f(m')\}}$$
$$= \sum_m \sum_r e^{2\pi i\{f(m)-f(m-r_1)\}} \quad (m' = m-r_1)$$
$$\leqslant \sum_{r_1=-\mu+1}^{\mu-1} |S_1|,$$
where
$$S_1 = \sum_m e^{2\pi i\{f(m)-f(m-r_1)\}} = \sum_m e^{2\pi i(\alpha k r_1 m^{k-1}+\ldots)}$$
and, for each r_1, m ranges over at most μ consecutive integers. Hence by Hölder's inequality
$$|S|^2 \leqslant \Big(\sum_{r_1=-\mu+1}^{\mu-1} 1\Big)^{1-2/K} \Big(\sum_{r_1=-\mu+1}^{\mu-1} |S_1|^{\frac{1}{2}K}\Big)^{2/K}$$
$$\leqslant (2\mu)^{1-2/K} \Big(\mu^{\frac{1}{2}K} + \sum_{r_1=-\mu+1}^{\mu-1}{}' |S_1|^{\frac{1}{2}K}\Big)^{2/K},$$
where the dash denotes that the term $r_1 = 0$ is omitted. Hence
$$|S|^K \leqslant (2\mu)^{\frac{1}{2}K-1} \Big(\mu^{\frac{1}{2}K} + \sum_{r_1=-\mu+1}^{\mu-1}{}' |S_1|^{\frac{1}{2}K}\Big).$$

If the theorem is true for $k-1$, then
$$|S_1|^{\frac{1}{2}K} \leqslant 2^K \mu^{\frac{1}{2}K-1} +$$
$$+ 2^{\frac{1}{2}K} \mu^{\frac{1}{2}K-k+1} \sum_{r_2,\ldots,r_{k-1}} \min(\mu, |\operatorname{cosec}\{\pi(\alpha k r_1)(k-1)!\, r_2 \ldots r_{k-1}\}|).$$

Hence

$$|S|^K \leqslant 2^{\frac{1}{2}K-1}\mu^{K-1}+$$
$$+2^{\frac{3}{2}K}\mu^{K-1}+2^K\mu^{K-k}\sum_{r_1,\ldots,r_{k-1}}\min(\mu,|\operatorname{cosec}(\pi\alpha k!\,r_1\ldots r_{k-1})|),$$

and the result for k follows. Since by § 5.4 the result is true for $k = 2$, it holds generally.

5.7. LEMMA 5.7. *For $a < b \leqslant 2a$, $k \geqslant 2$, $K = 2^{k-1}$, $a = O(t)$, $t > t_0$,*

$$\Sigma = \sum_{n=a+1}^{b} n^{-it} = O(a^{1-1/K}t^{1/\{(k+1)K\}}\log^{1/K}t)+O(at^{-1/\{(k+1)K\}}\log^{k/K}t).$$

If $a \leqslant 4t^{1/(k+1)}$, then

$$\Sigma = O(a) = O(a^{1-1/K}t^{1/\{(k+1)K\}})$$

as required. Otherwise, let

$$\mu = [\tfrac{1}{2}at^{-1/(k+1)}],$$

and write

$$\Sigma = \sum_{n=a+1}^{a+\mu}+\sum_{a+\mu+1}^{a+2\mu}+\ldots+\sum_{a+N\mu+1}^{b} = \Sigma_1+\ldots+\Sigma_{N+1}.$$

Then $\Sigma_\nu = O(M)$, where M is the maximum, for $\mu' \leqslant \mu$, of

$$S_\nu = \sum_{m=1}^{\mu'}\exp\left\{-it\left(\frac{m}{a+\nu\mu}-\frac{1}{2}\frac{m^2}{(a+\nu\mu)^2}+\ldots+(-1)^{k-1}\frac{m^k}{k(a+\nu\mu)^k}\right)\right\}.$$

By Lemma 5.6

$$S_\nu = O(\mu^{1-1/K})+O\left[\mu^{1-k/K}\left\{\sum_{r_1,\ldots,r_{k-1}}\min\left(\mu,\left|\operatorname{cosec}\frac{t(k-1)!\,r_1\ldots r_{k-1}}{2(a+\nu\mu)^k}\right|\right)\right\}^{1/K}\right].$$

Hence

$$\Sigma = O\{(N+1)\mu^{1-1/K}\}+$$
$$+O\left[\mu^{1-k/K}\sum_{\nu=1}^{N+1}\left\{\sum_{r_1,\ldots,r_{k-1}}\min\left(\mu,\left|\operatorname{cosec}\frac{t(k-1)!\,r_1\ldots r_{k-1}}{2(a+\nu\mu)^k}\right|\right)\right\}^{1/K}\right]$$
$$= O\{(N+1)\mu^{1-1/K}\}+$$
$$+O\left[\mu^{1-k/K}(N+1)^{1-1/K}\left\{\sum_{\nu=1}^{N+1}\sum_{r_1,\ldots,r_{k-1}}\min\left(\mu,\left|\operatorname{cosec}\frac{t(k-1)!\,r_1\ldots r_{k-1}}{2(a+\nu\mu)^k}\right|\right)\right\}^{1/K}\right]$$

by Hölder's inequality.

Now as ν varies,

$$\frac{t(k-1)!\,r_1...r_{k-1}}{2(a+\nu\mu)^k}-\frac{t(k-1)!\,r_1...r_{k-1}}{2\{a+(\nu-1)\mu\}^k}$$

lies between constant multiples of $t(k-1)!\,r_1...r_{k-1}\mu a^{-k-1}$, i.e. of $(k-1)!\,r_1...r_{k-1}\mu^{-k}$. The number of intervals of the form $\{l\pi, (l\pm\frac{1}{2})\pi\}$ containing values of $\frac{1}{2}t(k-1)!\,r_1...r_{k-1}(a+\nu\mu)^{-k}$ is therefore

$$O\{(N+1)(k-1)!\,r_1...r_{k-1}\mu^{-k}+1\}.$$

The part of the ν-sum corresponding to each of these intervals is, as in the previous case,

$$\mu+O\left(\frac{\mu^k}{(k-1)!\,r_1...r_{k-1}}\right)+O\left(\frac{\mu^k}{2.(k-1)!\,r_1...r_{k-1}}\right)+...$$

$$=\mu+O\left(\frac{\mu^k\log t}{(k-1)!\,r_1...r_{k-1}}\right)=O\left(\frac{\mu^k\log t}{r_1...r_{k-1}}\right).$$

Hence the ν-sum is

$$O\{(N+1)\log t\}+O\left(\frac{\mu^k\log t}{r_1...r_{k-1}}\right).$$

Summing with respect to $r_1,..., r_{k-1}$, we obtain

$$O\{(N+1)\mu^{k-1}\log t\}+O(\mu^k\log^k t).$$

Hence

$$\Sigma = O\{(N+1)\mu^{1-1/K}\}+O\{(N+1)\mu^{1-1/K}\log^{1/K}t\}+O\{(N+1)^{1-1/K}\mu\log^{k/K}t\}.$$

The first term on the right can be omitted, and since

$$N+1 = O\left(\frac{b-a}{\mu}+1\right) = O(t^{1/(k+1)})$$

the result stated follows.

5.8. THEOREM 5.8. *If l is a fixed integer greater than 2, and $L = 2^{l-1}$, then*

$$\zeta(s) = O(t^{1/\{(l+1)L\}}\log^{1+1/L}t) \quad (\sigma = 1-1/L). \tag{5.8.1}$$

The second term in Lemma 5.7 can be omitted if

$$a \leqslant t^{2/(k+1)}\log^{1-k}t.$$

Taking $k = l$ and applying the result $O(\log t)$, times we obtain

$$\sum_{n\leqslant N} n^{-it} = O(N^{1-1/L}t^{1/\{(l+1)L\}}\log^{1/L}t), \tag{5.8.2}$$

for $N \leqslant t^{2/(l+1)}\log^{1-l}t$. Similarly, for $k < l$, we find

$$\sum_{t^{2/(k+2)}\log^{-k}t<n\leqslant N} n^{-it} = O(N^{1-1/K}t^{1/\{(k+1)K\}}\log^{1/K}t)$$

for $t^{2/(k+2)}\log^{-k}t < N \leqslant t^{2/(k+1)}\log^{1-k}t$. The error term here is at most $O(N^{1-1/L}t^{\alpha}\log^{\beta}t)$ with

$$\alpha = \left(\frac{1}{L}-\frac{1}{K}\right)\frac{2}{k+2}+\frac{1}{(k+1)K}, \quad \beta = -\left(\frac{1}{L}-\frac{1}{K}\right)k+\frac{1}{K}.$$

Thus $\beta \leqslant 1/L$. When $k = l-1$ we have

$$\alpha = \left(\frac{1}{L}-\frac{2}{L}\right)\frac{2}{l+1}+\frac{2}{lL} = \frac{2}{l(l+1)L} < \frac{1}{(l+1)L},$$

and for $2 \leqslant k \leqslant l-2$ we have

$$\alpha \leqslant \left(\frac{1}{4K}-\frac{1}{K}\right)\frac{2}{k+2}+\frac{1}{(k+1)K} = -\frac{k-1}{2(k+1)(k+2)K} \leqslant 0 < \frac{1}{(l+1)L}.$$

It therefore follows, on summing over k, that (5.8.2) holds for $N \leqslant t^{\frac{2}{3}}\log^{-1}t$. Hence, by partial summation, we have

$$\sum_{n \leqslant (t/2\pi)^{\frac{1}{2}}} n^{-s} = O(t^{1/\{(l+1)L\}}\log^{l+1/L}t),$$

$$\sum_{n \leqslant (t/2\pi)^{\frac{1}{2}}} n^{s-1} = O(t^{2\sigma-1+1/\{(l+1)L\}}\log^{1/L}t),$$

and the theorem follows from the approximate functional equation.

5.9. van der Corput's method. In this method we approximate to sums by integrals as in Chapter IV.

THEOREM 5.9. *If $f(x)$ is real and twice differentiable, and*

$$0 < \lambda_2 \leqslant f''(x) \leqslant h\lambda_2 \quad (or \quad \lambda_2 \leqslant -f''(x) \leqslant h\lambda_2)$$

throughout the interval $[a, b]$, and $b \geqslant a+1$, then

$$\sum_{a<n\leqslant b} e^{2\pi i f(n)} = O\{h(b-a)\lambda_2^{\frac{1}{2}}\}+O(\lambda_2^{-\frac{1}{2}}).$$

If $\lambda_2 \geqslant 1$ the result is trivial, since the sum is $O(b-a)$. Otherwise Lemmas 4.7 and 4.4 give

$$O\{(\beta-\alpha+1)\lambda_2^{-\frac{1}{2}}\}+O\{\log(\beta-\alpha+2)\},$$

where $$\beta-\alpha = f'(a)-f'(b) = O\{(b-a)h\lambda_2\}.$$

Since $$\log(\beta-\alpha+2) = O(\beta-\alpha+2) = O\{(b-a)h\lambda_2\}+O(1) = O\{(b-a)h\lambda_2^{\frac{1}{2}}\}+O(1),$$

the result follows.

5.10. Lemma 5.10. *Let $f(n)$ be a real function, $a < n \leqslant b$, and q a positive integer not exceeding $b-a$. Then*

$$\Bigl|\sum_{a<n\leqslant b} e^{2\pi i f(n)}\Bigr| < A\frac{b-a}{q^{\frac{1}{2}}}+A\left\{\frac{b-a}{q}\sum_{r=1}^{q-1}\Bigl|\sum_{a<n\leqslant b-r} e^{2\pi i\{f(n+r)-f(n)\}}\Bigr|\right\}^{\frac{1}{2}}.$$

For convenience in the proof, let $e^{2\pi i f(n)}$ denote 0 if $n \leqslant a$ or $n > b$. Then

$$\sum_n e^{2\pi i f(n)} = \frac{1}{q}\sum_n \sum_{m=1}^{q} e^{2\pi i f(m+n)},$$

the inner sum vanishing if $n \leqslant a-q$ or $n > b-1$. Hence

$$\Bigl|\sum_n e^{2\pi i f(n)}\Bigr| \leqslant \frac{1}{q}\sum_n\Bigl|\sum_{m=1}^{q} e^{2\pi i f(m+n)}\Bigr|$$

$$\leqslant \frac{1}{q}\left\{\sum_n 1 \sum_n\Bigl|\sum_{m=1}^{q} e^{2\pi i f(m+n)}\Bigr|^2\right\}^{\frac{1}{2}}.$$

Since there are at most $b-a+q \leqslant 2(b-a)$ values of n for which the inner sum does not vanish, this does not exceed

$$\frac{1}{q}\left\{2(b-a)\sum_n\Bigl|\sum_{m=1}^{q} e^{2\pi i f(m+n)}\Bigr|^2\right\}^{\frac{1}{2}}.$$

Now

$$\Bigl|\sum_{m=1}^{q} e^{2\pi i f(m+n)}\Bigr|^2 = \sum_{m=1}^{q}\sum_{\mu=1}^{q} e^{2\pi i\{f(m+n)-f(\mu+n)\}}$$

$$= q+\sum_{\mu<m}\sum e^{2\pi i\{f(m+n)-f(\mu+n)\}}+\sum_{m<\mu}\sum e^{2\pi i\{f(m+n)-f(\mu+n)\}}.$$

Hence

$$\sum_n\Bigl|\sum_{m=1}^{q} e^{2\pi i f(m+n)}\Bigr|^2 \leqslant 2(b-a)q+2\Bigl|\sum_n\sum_{\mu<m}\sum e^{2\pi i\{f(m+n)-f(\mu+n)\}}\Bigr|.$$

In the last sum, $f(m+n)-f(\mu+n) = f(\nu+r)-f(\nu)$, for given values of ν and r, $1 \leqslant r \leqslant q-1$, just $q-r$ times, namely $\mu = 1$, $m = r+1$, up to $\mu = q-r$, $m = q$, with a consequent value of n in each case. Hence the modulus of this sum is equal to

$$\Bigl|\sum_{r=1}^{q-1}(q-r)\sum_\nu e^{2\pi i\{f(\nu+r)-f(\nu)\}}\Bigr| \leqslant q\sum_{r=1}^{q-1}\Bigl|\sum_\nu e^{2\pi i\{f(\nu+r)-f(\nu)\}}\Bigr|. \qquad (5.10.1)$$

Hence

$$\Bigl|\sum_n e^{2\pi i f(n)}\Bigr| \leqslant \frac{1}{q}\left\{4(b-a)^2q+4(b-a)q\sum_{r=1}^{q-1}\Bigl|\sum_\nu e^{2\pi i\{f(\nu+r)-f(\nu)\}}\Bigr|\right\}^{\frac{1}{2}},$$

and the result stated follows.

5.11. THEOREM 5.11. *Let $f(x)$ be real and have continuous derivatives up to the third order, and let $\lambda_3 \leqslant f'''(x) \leqslant h\lambda_3$, or $\lambda_3 \leqslant -f'''(x) \leqslant h\lambda_3$, and $b-a \geqslant 1$. Then*

$$\sum_{a<n\leqslant b} e^{2\pi i f(n)} = O\{h^{\frac{1}{2}}(b-a)\lambda_3^{\frac{1}{6}}\}+O\{(b-a)^{\frac{1}{2}}\lambda_3^{-\frac{1}{6}}\}.$$

Let
$$g(x) = f(x+r)-f(x).$$
Then
$$g''(x) = f''(x+r)-f''(x) = rf'''(\xi),$$
where $x < \xi < x+r$. Hence
$$r\lambda_3 \leqslant g''(x) \leqslant hr\lambda_3,$$
or the same for $-g''(x)$. Hence by Theorem 5.9
$$\sum_{a<n\leqslant b-r} e^{2\pi i g(n)} = O\{h(b-a)r^{\frac{1}{2}}\lambda_3^{\frac{1}{2}}\}+O(r^{-\frac{1}{2}}\lambda_3^{-\frac{1}{2}}).$$
Hence, by Lemma 5.10,
$$\sum_{a<n\leqslant b} e^{2\pi i f(n)} = O\left(\frac{b-a}{q^{\frac{1}{2}}}\right)+O\left[\frac{b-a}{q}\sum_{r=1}^{q-1}\{h(b-a)r^{\frac{1}{2}}\lambda_3^{\frac{1}{2}}+r^{-\frac{1}{2}}\lambda_3^{-\frac{1}{2}}\}\right]^{\frac{1}{2}}$$
$$= O\left(\frac{b-a}{q^{\frac{1}{2}}}\right)+O\{h(b-a)^2q^{\frac{1}{2}}\lambda_3^{\frac{1}{2}}+(b-a)q^{-\frac{1}{2}}\lambda_3^{-\frac{1}{2}}\}^{\frac{1}{2}}$$
$$= O\{(b-a)q^{-\frac{1}{2}}\}+O\{h^{\frac{1}{2}}(b-a)q^{\frac{1}{4}}\lambda_3^{\frac{1}{4}}\}+O\{(b-a)^{\frac{1}{2}}q^{-\frac{1}{4}}\lambda_3^{-\frac{1}{4}}\}.$$
The first two terms are of the same order in λ_3 if $q = [\lambda_3^{-\frac{1}{3}}]$ provided that $\lambda_3 \leqslant 1$. This gives
$$O\{h^{\frac{1}{2}}(b-a)\lambda_3^{\frac{1}{6}}\}+O\{(b-a)^{\frac{1}{2}}\lambda_3^{-\frac{1}{6}}\}$$
as stated. The theorem is plainly trivial if $\lambda_3 > 1$. The proof also requires that $q \leqslant b-a$. If this is not satisfied, then $b-a = O(\lambda_3^{-\frac{1}{3}})$,
$$b-a = O\{(b-a)^{\frac{1}{2}}\lambda_3^{-\frac{1}{6}}\},$$
and the result again follows.

5.12. THEOREM 5.12.
$$\zeta(\tfrac{1}{2}+it) = O(t^{\frac{1}{6}}\log t).$$

Taking $f(x) = -(2\pi)^{-1}t\log x$, we have
$$f'''(x) = -\frac{t}{\pi x^3}.$$
Hence if $b \leqslant 2a$ the above theorem gives
$$\sum_{a<n\leqslant b} n^{-it} = O\left\{a\left(\frac{t}{a^3}\right)^{\frac{1}{6}}\right\}+O\left\{a^{\frac{1}{2}}\left(\frac{t}{a^3}\right)^{-\frac{1}{6}}\right\}$$
$$= O(a^{\frac{1}{2}}t^{\frac{1}{6}})+O(at^{-\frac{1}{6}}),$$

and the second term can be omitted if $a \leqslant t^{\frac{2}{3}}$. Then by partial summation

$$\sum_{a<n\leqslant b} \frac{1}{n^{\frac{1}{2}+it}} = O(t^{\frac{1}{6}}). \tag{5.12.1}$$

Also, by Theorem 5.9,

$$\sum_{a<n\leqslant b} n^{-it} = O(t^{\frac{1}{2}})+O(at^{-\frac{1}{2}}),$$

and hence by partial summation

$$\sum_{a<n\leqslant b} \frac{1}{n^{\frac{1}{2}+it}} = O\left\{\left(\frac{t}{a}\right)^{\frac{1}{2}}\right\}+O\left\{\left(\frac{a}{t}\right)^{\frac{1}{2}}\right\}.$$

Hence (5.12.1) is also true if $t^{\frac{2}{3}} < a < t$. Hence, applying (5.12.1) $O(\log t)$ times, we obtain

$$\sum_{n<t} \frac{1}{n^{\frac{1}{2}+it}} = O(t^{\frac{1}{6}}\log t),$$

and the result follows.

5.13. THEOREM 5.13. *Let $f(x)$ be real and have continuous derivatives up to the k-th order, where $k \geqslant 4$. Let $\lambda_k \leqslant f^{(k)}(x) \leqslant h\lambda_k$ (or the same for $-f^{(k)}(x)$). Let $b-a \geqslant 1$, $K = 2^{k-1}$. Then*

$$\sum_{a<n\leqslant b} e^{2\pi i f(n)} = O\{h^{2/K}(b-a)\lambda_k^{1/(2K-2)}\}+O\{(b-a)^{1-2/K}\lambda_k^{-1/(2K-2)}\},$$

where the constants implied are independent of k.

If $\lambda_k \geqslant 1$ the theorem is trivial, as before. Otherwise, suppose the theorem true for all integers up to $k-1$. Let

$$g(x) = f(x+r)-f(x).$$

Then
$$g^{(k-1)}(x) = f^{(k-1)}(x+r)-f^{(k)}(x) = rf^{(k)}(\xi),$$

where $x < \xi < x+r$. Hence

$$r\lambda_k \leqslant g^{(k-1)}(x) \leqslant hr\lambda_k.$$

Hence the theorem with $k-1$ for k gives

$$\left|\sum_{a<n\leqslant b-r} e^{2\pi i g(n)}\right| < A_1 h^{4/K}(b-a)(r\lambda_k)^{1/(K-2)}+A_2(b-a)^{1-4/K}(r\lambda_k)^{-1/(K-2)}$$

(writing constants A_1, A_2 instead of the O's). Hence

$$\sum_{r=1}^{q-1}\left|\sum_{a<n\leqslant b-r} e^{2\pi i g(n)}\right| < A_1 h^{4/K}(b-a)q^{1+1/(K-2)}\lambda_k^{1/(K-2)}+$$
$$+2A_2(b-a)^{1-4/K}q^{1-1/(K-2)}\lambda_k^{-1/(K-2)}$$

since
$$\sum_{r=1}^{q-1} r^{-1/(K-2)} < \int_0^q r^{-1/(K-2)}\,dr = \frac{q^{1-1/(K-2)}}{1-1/(K-2)} \leqslant 2q^{1-1/(K-2)}$$

for $K \geqslant 4$. Hence, by Lemma 5.10,

$$\sum_{a<n\leqslant b} e^{2\pi i f(n)} \leqslant A_3(b-a)q^{-\frac{1}{2}}+A_4(b-a)^{\frac{1}{2}}q^{-\frac{1}{2}}\{A_1 h^{4/K}(b-a)q^{1+1/(K-2)}\lambda_k^{1/(K-2)}+$$
$$+2A_2(b-a)^{1-4/K}q^{1-1/(K-2)}\lambda_k^{-1/(K-2)}\}^{\frac{1}{2}}$$
$$\leqslant A_3(b-a)q^{-\frac{1}{2}}+A_4 A_1^{\frac{1}{2}} h^{2/K}(b-a)q^{1/(2K-4)}\lambda_k^{1/(2K-4)}+$$
$$+A_4(2A_2)^{\frac{1}{2}}(b-a)^{1-2/K}q^{-1/(2K-4)}\lambda_k^{-1/(2K-4)}.$$

To make the first two terms of the same order in λ_k, let

$$q = [\lambda_k^{-1/(K-1)}]+1.$$

Then

$$\lambda_k^{-1/(K-1)} \leqslant q \leqslant 2\lambda_k^{-1/(K-1)},$$
$$q^{1/(2K-4)}\lambda_k^{1/(2K-4)} \leqslant 2^{1/(2K-4)}\lambda_k^{1/(2K-4)\{1-1/(K-1)\}} \leqslant 2\lambda_k^{1/(2K-2)},$$
$$q^{-1/(2K-4)}\lambda_k^{-1/(2K-4)} \leqslant \lambda_k^{-1/(2K-2)},$$

and we obtain

$$\left|\sum_{a<n\leqslant b} e^{2\pi i f(n)}\right| \leqslant (A_3+2A_4 A_1^{\frac{1}{2}})h^{2/K}(b-a)\lambda_k^{1/2(K-2)}+$$
$$+A_4(2A_2)^{\frac{1}{2}}(b-a)^{1-2/K}\lambda_k^{-1/(2K-2)}.$$

This gives the result for k; the constants are the same for k as for $k-1$ if

$$A_3+2A_4 A_1^{\frac{1}{2}} \leqslant A_1, \qquad A_4(2A_2)^{\frac{1}{2}} \leqslant A_2,$$

which are satisfied if A_1 and A_2 are large enough.

We have assumed in the proof that $q \leqslant b-a$, which is true if $2\lambda_k^{-1/(K-1)} \leqslant b-a$. Otherwise

$$\left|\sum_{a<n\leqslant b} e^{2\pi i f(n)}\right| \leqslant b-a \leqslant (b-a)^{\frac{1}{2}}(2\lambda_k^{-1/(K-1)})^{\frac{1}{2}} \leqslant 2^{\frac{1}{2}}(b-a)^{1-2/K}\lambda_k^{-1/(2K-2)},$$

and the result again holds.

5.14. THEOREM 5.14. *If $l \geqslant 3$, $L = 2^{l-1}$, $\sigma = 1-l/(2L-2)$,*

$$\zeta(s) = O(t^{1/(2L-2)}\log t). \tag{5.14.1}$$

We apply the above theorem with

$$f(x) = -\frac{t\log x}{2\pi}, \qquad f^{(k)}(x) = \frac{(-1)^k(k-1)!\,t}{2\pi x^k}.$$

If $a < n \leqslant b \leqslant 2a$, then

$$\frac{(k-1)!\,t}{2\pi(2a)^k} \leqslant |f^{(k)}(x)| \leqslant \frac{(k-1)!\,t}{2\pi a^k},$$

and we may apply the theorem with

$$\lambda_k = \frac{(k-1)!\,t}{2\pi(2a)^k}, \qquad h = 2^k.$$

Hence

$$\sum_{a<n\leqslant b} n^{-it} = O\left[2^{2k/K}a\left\{\frac{(k-1)!\,t}{2\pi(2a)^k}\right\}^{1/(2K-2)}\right]+O\left[a^{1-2/K}\left\{\frac{(k-1)!\,t}{2\pi(2a)^k}\right\}^{-1/(2K-2)}\right]$$

$$= O(a^{1-k/(2K-2)}t^{1/(2K-2)})+O(a^{1-2/K+k/(2K-2)}t^{-1/(2K-2)}). \quad (5.14.2)$$

The second term can be omitted if

$$a < At^{K/(kK-2K+2)}. \quad (5.14.3)$$

Hence by partial summation

$$\sum_{a<n\leqslant b} n^{-s} = O(a^{1-\sigma-k/(2K-2)}t^{1/(2K-2)}) \quad (5.14.4)$$

subject to (5.14.3). Taking $\sigma = 1-l/(2L-2)$,

$$\sum_{a<n\leqslant b} n^{-s} = O(a^{l/(2L-2)-k/(2K-2)}t^{1/(2K-2)}). \quad (5.14.5)$$

First take $k = l$. We obtain

$$\sum_{a<n\leqslant b} n^{-s} = O(t^{1/(2L-2)}) \quad (a < At^{L/(lL-2L+2)}).$$

Hence

$$\sum_{n\leqslant t^{L/(lL-2L+2)}} n^{-s} = \sum_{\frac{1}{2}t^{L/(lL-2L+2)}<n\leqslant t^{L/(lL-2L+2)}} +\dots$$

$$= O(t^{1/(2L-2)})+O(t^{1/(2L-2)})+\dots$$

$$= O(t^{1/(2L-2)}\log t). \quad (5.14.6)$$

Next

$$\sum_{t^{L/(lL-2L+2)}<n\leqslant t} \frac{1}{n^s} = \sum_{\frac{1}{2}t<n\leqslant t} + \sum_{\frac{1}{4}t<n\leqslant \frac{1}{2}t} +\dots,$$

and to each term $\sum_{2^{-m}t<n\leqslant 2^{1-m}t}$ corresponds a $k < l$ such that

$$t^{K/\{(k+1)K-2K+1\}} < 2^{-m}t \leqslant t^{K/(kK-2K+2)}.$$

Then

$$\sum_{2^{-m}t<n\leqslant 2^{1-m}t} \frac{1}{n^s} = O\{t^{\{l/(2L-2)-k/(2K-2)\}K/\{(k+1)K-2K+1\}+1/(2K-2)}\}.$$

The index does not exceed that in (5.14.6) if

$$\left(\frac{l}{2L-2}-\frac{k}{2K-2}\right)\frac{K}{(k+1)K-2K+1}+\frac{1}{2K-2} \leqslant \frac{1}{2L-2},$$

which reduces to

$$L-K \geqslant (l-k)K,$$

i.e.

$$2^{l-k}-1 \geqslant l-k$$

which is true. Since there are again $O(\log t)$ terms,

$$\sum_{t^{L/(lL-2L+2)}<n\leqslant t} \frac{1}{n^s} = O(t^{1/(2L-2)}\log t).$$

The result therefore follows. Theorem 5.12 is the particular case $l = 3$, $L = 4$.

5.15. Comparison between the Hardy–Littlewood result and the van der Corput result. The Hardy–Littlewood method shows that the function $\mu(\sigma)$ satisfies

$$\mu\left(1-\frac{1}{2^{k-1}}\right) \leqslant \frac{1}{(k+1)2^{k-1}}, \tag{5.15.1}$$

and the van der Corput method that

$$\mu\left(1-\frac{l}{2^l-2}\right) \leqslant \frac{1}{2^l-2}. \tag{5.15.2}$$

For a given k, determine l so that

$$1-\frac{l-1}{2^{l-1}-2} < 1-\frac{1}{2^{k-1}} \leqslant 1-\frac{l}{2^l-2}.$$

Then (5.15.2) and the convexity of $\mu(\sigma)$ give

$$\mu\left(1-\frac{1}{2^{k-1}}\right) \leqslant \frac{\dfrac{1}{2^{k-1}}-\dfrac{l}{2^l-2}}{\dfrac{l-1}{2^{l-1}-2}-\dfrac{l}{2^l-2}}\frac{1}{2^{l-1}-2}+\frac{\dfrac{l-1}{2^{l-1}-2}-\dfrac{1}{2^{k-1}}}{\dfrac{l-1}{2^{l-1}-2}-\dfrac{l}{2^l-2}}\frac{1}{2^l-2}$$

$$= \frac{2^{l-k}-1}{l2^{l-1}-2^l+2} \leqslant \frac{1}{(k+1)2^{k-1}}$$

if
$$(k+1)(2^{l-1}-2^{k-1}) \leqslant (l-2)2^{l-1}+2.$$

Since $2^{k-1} > (2^{l-1}-2)/(l-1)$, this is true if

$$(k+1)\left(2^{l-1}-\frac{2^{l-1}-2}{l-1}\right) \leqslant (l-2)2^{l-1}+2,$$

i.e. if
$$k+1 \leqslant l-1.$$

Now
$$2^{k-1} \leqslant \frac{2^l-2}{l} < \frac{2^l}{l} \leqslant 2^{l-3}$$

if $l \geqslant 8$. Hence the Hardy–Littlewood result follows from the van der Corput result if $l \geqslant 8$.

For $4 \leqslant l \leqslant 7$ the relevant values of $1-\sigma$ are

H.–L.	$\frac{1}{4}$,	$\frac{1}{8}$,	$\frac{1}{16}$.	
v. d. C.	$\frac{2}{7}$,	$\frac{1}{6}$,	$\frac{3}{31}$,	$\frac{1}{18}$.

The values of k and l in these cases are 3, 4, 5 and 5, 6, 7 respectively. Hence $k \leqslant l-2$ in all cases.

5.16. THEOREM 5.16.

$$\zeta(1+it) = O\left(\frac{\log t}{\log\log t}\right).$$

We have to apply the above results with k variable; in fact it will be seen from the analysis of § 5.13 and § 5.14 that the constants implied in the O's are independent of k. In particular, taking $\sigma = 1$ in (5.14.4), we have

$$\sum_{a<n\leqslant b} \frac{1}{n^{1+it}} = O(a^{-k/(2K-2)}t^{1/(2K-2)}) \quad (a < b \leqslant 2a)$$

uniformly with respect to k, subject to (5.14.3). If

$$t^{K/\{(k+1)K-2K+1\}} < a \leqslant t^{K/(kK-2K+2)}$$

it follows that

$$\sum_{a<n\leqslant b} \frac{1}{n^{1+it}} = O(t^{1/(2K-2)-kK/\{(2K-2)(kK-K+1)\}}) = O(t^{-1/\{2(k-1)K+2\}}).$$

Writing
$$\sum_{t^{R/\{(r-1)R+1\}}<n\leqslant t} \frac{1}{n^{1+it}} = \sum_{\frac{1}{2}t<n\leqslant t} + \sum_{\frac{1}{4}t<n\leqslant\frac{1}{2}t} + \dots,$$

and applying the above result with $k = 2, 3, \dots,$ or r, we obtain, since there are $O(\log t)$ terms,

$$\sum_{t^{R/\{(r-1)R+1\}}<n\leqslant t} \frac{1}{n^{1+it}} = O(t^{-1/\{2(r-1)R+2\}}\log t). \tag{5.16.1}$$

Let $r = [\log\log t]$. Then

$$2R \leqslant 2^{\log\log t} = (\log t)^{\log 2},$$

and

$$t^{1/\{2(r-1)R+2\}} \geqslant \exp\left(\frac{\log t}{(\log t)^{\log 2}\log\log t+2}\right) > \exp\{(\log t)^{A}\} > A\log t.$$

Hence the above sum is bounded. Also

$$\sum_{n\leqslant t^{R/\{(r-1)R+1}} \frac{1}{n^{1+it}} = O(\log t^{R/\{(r-1)R+1\}}) = O\left\{\frac{R\log t}{(r-1)R+1}\right\}$$
$$= O\left(\frac{\log t}{r}\right) = O\left(\frac{\log t}{\log\log t}\right).$$

This proves the theorem.

The same result can also be deduced from the Weyl–Hardy–Littlewood analysis.

5.17. THEOREM 5.17. *For $t > A$*

$$\zeta(s) = O(\log^5 t), \qquad \sigma \geqslant 1-\frac{(\log\log t)^2}{\log t}, \tag{5.17.1}$$

$$\zeta(s) \neq 0, \qquad \sigma \geqslant 1-A_1\frac{\log\log t}{\log t} \tag{5.17.2}$$

(*with some A_1*), *and*

$$\frac{1}{\zeta(1+it)} = O\left(\frac{\log t}{\log\log t}\right), \tag{5.17.3}$$

$$\frac{\zeta'(1+it)}{\zeta(1+it)} = O\left(\frac{\log t}{\log\log t}\right). \tag{5.17.4}$$

We observe that (5.14.1) holds with a constant independent of l, and also, by the Phragmén–Lindelöf theorem, uniformly for

$$\sigma \geqslant 1-l/(2L-2).$$

Let t be given (sufficiently large), and let

$$l = \left[\frac{1}{\log 2}\log\left(\frac{\log t}{\log\log t}\right)\right].$$

Then
$$L \leqslant 2^{(1/\log 2)\log(\log t/\log\log t)-1} = \frac{1}{2}\frac{\log t}{\log\log t},$$

and similarly
$$L \geqslant \frac{1}{4}\frac{\log t}{\log\log t}.$$

Hence

$$\frac{l}{2L-2} \geqslant \frac{l}{2L} \geqslant \frac{\log\log t-\log\log\log t-\log 2}{\log 2}\frac{\log\log t}{\log t} \geqslant \frac{(\log\log t)^2}{\log t}$$

for $t > A$ (large enough). Hence if

$$\sigma \geqslant 1-\frac{(\log\log t)^2}{\log t}$$

then
$$\sigma \geqslant 1-\frac{l}{2L-2}.$$

Hence (5.14.1) is applicable, and gives

$$\zeta(s) = O(t^{1/(2L-2)}\log t) = O(t^{1/L}\log t)$$
$$= O(t^{(4\log\log t/\log t)}\log t) = O(\log^5 t).$$

This proves (5.17.1). The remaining results then follow from Theorems 3.10 and 3.11, taking (for $t > A$)

$$\theta(t) = \frac{(\log\log t)^2}{\log t}, \qquad \phi(t) = 5\log\log t.$$

5.18. In this section we reconsider the problem of the order of $\zeta(\frac{1}{2}+it)$. Small improvements on Theorem 5.12 have been obtained by various different methods. Results of the form

$$\zeta(\tfrac{1}{2}+it) = O(t^{\alpha}\log^{\beta}t)$$

with
$$\alpha = \frac{163}{988},\ \frac{27}{164},\ \frac{229}{1392},\ \frac{19}{116},\ \frac{15}{92}$$

were proved by Walfisz (1), Titchmarsh (9), Phillips (1), Titchmarsh (24), and Min (1) respectively.† We shall give here the argument which leads to the index $\frac{27}{164}$. The main idea of the proof is that we combine Theorem 5.13 with Theorem 4.9, which enables us to transform a given exponential sum into another, which may be easier to deal with.

THEOREM 5.18.
$$\zeta(\tfrac{1}{2}+it) = O(t^{27/164}).$$

Consider the sum
$$\Sigma_1 = \sum_{a<n\leqslant b} n^{-it} = \sum_{a<n\leqslant b} e^{-it\log n},$$
where $a < b \leqslant 2a$, $a < A\surd t$. By § 5.10
$$\Sigma_1 = O\left(\frac{a}{q^{\frac{1}{2}}}\right)+O\left\{\left(\frac{a}{q}\sum_{r=1}^{q-1}|\Sigma_2|\right)^{\frac{1}{2}}\right\}, \tag{5.18.1}$$
where $q \leqslant b-a$, and
$$\Sigma_2 = \sum_{a<n\leqslant b-r} e^{-it\{\log(n+r)-\log n\}}.$$
We now apply Theorem 4.9. to Σ_2. We have
$$f(x) = -\frac{t}{2\pi}\{\log(x+r)-\log x\}, \qquad f'(x) = \frac{tr}{2\pi x(x+r)},$$
$$f''(x) = -\frac{tr}{2\pi}\frac{2x+r}{x^2(x+r)^2}, \qquad f'''(x) = \frac{tr}{\pi}\frac{3x^2+3xr+r^2}{x^3(x+r)^3}.$$
We can therefore apply Theorem 4.9 with $\lambda_2 = tra^{-3}$, $\lambda_3 = tra^{-4}$. Thus
$$\Sigma_2 = e^{-\frac{1}{4}\pi i}\sum_{\alpha<\nu\leqslant\beta}\frac{e^{2\pi i\phi(\nu)}}{|f''(x_\nu)|^{\frac{1}{2}}}+O\left(\frac{a^{\frac{3}{2}}}{t^{\frac{1}{2}}r^{\frac{1}{2}}}\right)+O\left\{\log\left(2+\frac{tr}{a^2}\right)\right\}+O\left(\frac{t^{\frac{2}{5}}r^{\frac{2}{5}}}{a^{\frac{2}{5}}}\right), \tag{5.18.2}$$
where $\phi(\nu) = f(x_\nu)-\nu x_\nu$, $\alpha = f'(b-r)$, $\beta = f'(a)$. Actually the log-term can be omitted, since it is $O(t^{\frac{2}{5}}r^{\frac{2}{5}}a^{-\frac{4}{5}})$.

Consider next the sum
$$\sum_{\alpha<\nu\leqslant\gamma} e^{2\pi i\phi(\nu)} \quad (\alpha < \gamma \leqslant \beta).$$

† Note that the proof of the lemma in Titchmarsh (24) is incorrect. The lemma should be replaced by the corresponding theorem in Titchmarsh (16).

The numbers x_ν are given by

$$\frac{tr}{2\pi x_\nu(x_\nu+r)} = \nu, \quad \text{i.e.} \quad x_\nu = \frac{1}{2}\left(r^2+\frac{2tr}{\pi\nu}\right)^{\frac{1}{2}}-\tfrac{1}{2}r.$$

Hence

$$\phi'(\nu) = \{f'(x_\nu)-\nu\}\frac{dx_\nu}{d\nu}-x_\nu = -x_\nu = \tfrac{1}{2}r-\frac{1}{2}\left(r^2+\frac{2tr}{\pi\nu}\right)^{\frac{1}{2}},$$

$$\phi''(\nu) = \frac{tr}{2\pi\nu^2(r^2+2tr/\pi\nu)^{\frac{1}{2}}} = \frac{1}{2}\left(\frac{tr}{2\pi}\right)^{\frac{1}{2}}\left(\frac{1}{\nu^{\frac{3}{2}}}-\frac{\pi r}{4t}\frac{1}{\nu^{\frac{1}{2}}}+\ldots\right),$$

since

$$r\nu \leqslant rf'(a) = \frac{tr^2}{2\pi a(a+r)} \leqslant \frac{tr^2}{2\pi a^2} \leqslant \frac{t}{2\pi}.$$

It follows that

$$\frac{K_1(tr)^{\frac{1}{2}}}{\nu^{k-\frac{1}{2}}} < |\phi^{(k)}(\nu)| < \frac{K_2(tr)^{\frac{1}{2}}}{\nu^{k-\frac{1}{2}}} \quad (t > t_k),$$

where K_1, K_2,..., and t_k depend on k only. We may therefore apply Theorem 5.13, with $h = O(1)$, and

$$\lambda_k = K_3(tr)^{\frac{1}{2}}(tr/a^2)^{\frac{1}{2}-k} = K_3(tr)^{1-k}a^{2k-1}.$$

Hence

$$\sum_{\alpha<\nu\leqslant\gamma} e^{2\pi i\phi(\nu)} = O\left\{\frac{tr}{a^2}\left(\frac{a^{2k-1}}{t^{k-1}r^{k-1}}\right)^{1/(2K-2)}\right\}+O\left\{\left(\frac{tr}{a^2}\right)^{1-2/K}\left(\frac{a^{2k-1}}{t^{k-1}r^{k-1}}\right)^{-1/(2K-2)}\right\}.$$

Also $|f''(x_\nu)|^{-\frac{1}{2}}$ is monotonic and of the form $O(t^{-\frac{1}{2}}r^{-\frac{1}{2}}a^{\frac{3}{2}})$. Hence by partial summation

$$\sum_{\alpha<\nu\leqslant\beta}\frac{e^{2\pi i\phi(\nu)}}{|f''(x_\nu)|^{\frac{1}{2}}} = O\{(tr)^{\frac{1}{2}-(k-1)/(2K-2)}a^{(2k-1)/(2K-2)-\frac{1}{2}}\}+$$
$$+O\{(tr)^{\frac{1}{2}-2/K+(k-1)/(2K-2)}a^{4/K-\frac{1}{2}-(2k-1)/(2K-2)}\}.$$

Hence

$$\frac{1}{q}\sum_{r=1}^{q-1}|\Sigma_2| = O\{(tq)^{\frac{1}{2}-(k-1)/(2K-2)}a^{(2k-1)/(2K-2)-\frac{1}{2}}\}+$$
$$+O\{(tq)^{\frac{1}{2}-2/K+(k-1)/(2K-2)}a^{4/K-\frac{1}{2}-(2k-1)/(2K-2)}\}+O\{(tq)^{-\frac{1}{2}}a^{\frac{3}{2}}\}+O\{(tq)^{\frac{2}{5}}a^{-\frac{2}{5}}\}.$$

Inserting this in (5.18.1), and using the inequality

$$(X+Y+\ldots)^{\frac{1}{2}} \leqslant X^{\frac{1}{2}}+Y^{\frac{1}{2}}+\ldots$$

we obtain

$$\Sigma_1 = O(aq^{-\frac{1}{2}})+O\{(tq)^{\frac{1}{4}-(k-1)/(4K-4)}a^{(2k-1)/(4K-4)+\frac{1}{4}}\}+$$
$$+O\{(tq)^{\frac{1}{4}-1/K+(k-1)/(4K-4)}a^{2/K+\frac{1}{4}-(2k-1)/(4K-4)}\}+O\{(tq)^{-\frac{1}{4}}a^{\frac{5}{4}}\}+O\{(tq)^{\frac{1}{5}}a^{\frac{3}{10}}\}.$$

The first two terms on the right are of the same order if

$$q = [a^{(3K-2k-2)/(3K-k-2)}t^{-(K-k)/(3K-k-2)}],$$

and they are then of the form

$$O(a^{(3K-2)/\{2(3K-k-2)\}}t^{(K-k)/\{2(3K-k-2)\}}) = O(t^{(5K-2k-2)/\{4(3K-k-2)\}}) \quad (a < A\sqrt{t}).$$

For $k = 2, 3, 4, 5, 6,...$, the index has the values

$$\frac{1}{2}, \quad \frac{3}{7}, \quad \frac{5}{12}, \quad \frac{17}{41}, \quad \frac{73}{176},...$$

and of these $\frac{17}{41}$ is the smallest. We therefore take $k = 5$,

$$q = [a^{\frac{36}{41}} t^{-\frac{11}{41}}] \quad (a > t^{\frac{11}{36}}),$$

and obtain

$$\Sigma_1 = O(a^{\frac{23}{41}} t^{\frac{11}{82}}) + O(a^{\frac{147}{328}} t^{\frac{61}{328}}) + O(a^{\frac{169}{164}} t^{-\frac{15}{82}}) + O(a^{\frac{39}{82}} t^{\frac{6}{41}}).$$

This also holds if $q \geqslant b-a$, since then

$$\Sigma_1 = O(b-a) = O(q) = O(a^{\frac{36}{41}} t^{-\frac{11}{41}}),$$

which is of smaller order than the third term in the above right-hand side.

It is easily seen that the last two terms are negligible compared with the first if $a = O(\sqrt{t})$. Hence by partial summation

$$\sum_{a<n\leqslant b} \frac{1}{n^{\frac{1}{2}+it}} = O(a^{\frac{5}{82}} t^{\frac{11}{82}}) + O(a^{-\frac{17}{328}} t^{\frac{61}{328}}) \quad (a > t^{\frac{11}{36}}).$$

Applying this with $a = N$, $b = 2N-1$; $a = 2N$, $b = 4N-1$,... until $b = [A\sqrt{t}]$, we obtain

$$\sum_{N<n\leqslant A\sqrt{t}} \frac{1}{n^{\frac{1}{2}+it}} = O(t^{\frac{27}{164}}) + O(N^{-\frac{17}{328}} t^{\frac{61}{328}})$$
$$= O(t^{\frac{27}{164}}) \quad (N > t^{\frac{7}{17}}).$$

We require a subsidiary argument for $n \leqslant t^{\frac{7}{17}}$, and in fact (5.14.2) with $k = 4$ gives

$$\sum_{a<n\leqslant b} n^{-it} = O(a^{\frac{5}{7}} t^{\frac{1}{14}}) \qquad (a < At^{\frac{4}{9}}),$$

$$\sum_{a<n\leqslant b} \frac{1}{n^{\frac{1}{2}+it}} = O(a^{\frac{3}{14}} t^{\frac{1}{14}}),$$

and by adding terms of this type as before

$$\sum_{n\leqslant t^{7/17}} \frac{1}{n^{\frac{1}{2}+it}} = O(t^{\frac{3}{14}\frac{7}{17}+\frac{1}{14}}) = O(t^{\frac{19}{119}}) = O(t^{\frac{27}{164}}).$$

The result therefore follows from the approximate functional equation.

NOTES FOR CHAPTER 5

5.19. Two more completely different arguments have been given, leading to the estimate

$$\mu(\tfrac{1}{2}) \leqslant \tfrac{1}{6}. \tag{5.19.1}$$

Firstly Bombieri, in unpublished work, has used a method related to that of §6.12, together with the bound

$$\int_0^1\int_0^1 \left|\sum_{1\leqslant x\leqslant P} \exp\{2\pi i(\alpha x+\beta x^2)\}\right|^6 d\alpha d\beta \ll P^3\log P,$$

to prove (5.19.1). Secondly, (5.19.1) follows from the mean-value bound (7.24.4) of Iwaniec [1]. (This deep result is described in §7.24.)

Heath-Brown [9] has shown that the weaker estimate $\mu(\frac{1}{2})\leqslant\frac{3}{16}$ follows from an argument analogous to Burgess's [1] treatment of character sums. Moreover the bound $\mu(\frac{1}{2})\leqslant\frac{7}{32}$, which is weaker still, but none the less non-trivial, follows from Heath-Brown's [4] fourth-power moment (7.21.1), based on Weil's estimate for the Kloosterman sum. Thus there are some extremely diverse arguments leading to non-trivial bounds for $\mu(\frac{1}{2})$.

5.20. The argument given in §5.18 is generalized by the 'method of exponent pairs' of van der Corput (1), (2) and Phillips (1). Let s, c be positive constants, and let $\mathscr{F}(s,c)$ be the set of quadruples (N, I, f, y) as follows:

(i) N and y are positive and satisfy $yN^{-s}\geqslant 1$,

(ii) I is a subinterval of $(N, 2N]$,

(iii) f is a real valued function on I, with derivatives of all orders, satisfying

$$\left|f^{(n+1)}(x)-\frac{d^n}{dx^n}(yx^{-s})\right|\leqslant c\left|\frac{d^n}{dx^n}(yx^{-s})\right|. \tag{5.20.1}$$

for $n\geqslant 0$.

We then say that (p, q) is an 'exponent pair' if $0\leqslant p\leqslant\frac{1}{2}\leqslant q\leqslant 1$ and if for each $s>0$ there exists a sufficiently small $c=c(p,q,s)>0$ such that

$$\sum_{n\in I}\exp\{2\pi i f(n)\}\ll_{p,q,s}(yN^{-s})^p N^q, \tag{5.20.2}$$

uniformly for $(N, I, f, y)\in\mathscr{F}(s,c)$.

We may observe that yN^{-s} is the order of magnitude of $f'(x)$. It is immediate that (0, 1) is an exponent pair. Moreover Theorems 5.9, 5.11, and 5.13 correspond to the exponent pairs $(\frac{1}{2},\frac{1}{2})$, $(\frac{1}{6},\frac{2}{3})$, and

$$\left(\frac{1}{2^k-2},\ \frac{2^k-k-1}{2^k-2}\right).$$

By using Lemma 5.10 one may prove that

$$A(p, q) = \left(\frac{p}{2p+2}, \frac{p+q+1}{2p+2}\right)$$

is an exponent pair whenever (p, q) is. Similarly from Theorem 4.9, as sharpened in §4.19, one may show that

$$B(p, q) = (q - \tfrac{1}{2}, p + \tfrac{1}{2})$$

is an exponent pair whenever (p, q) is, providing that $p + 2q \geqslant \frac{3}{2}$. Thus one may build up a range of pairs by repeated applications of these A and B processes. In doing this one should note that $B^2(p, q) = (p, q)$. Examples of exponent pairs are:

$$B(0, 1) = (\tfrac{1}{2}, \tfrac{1}{2}), \qquad AB(0, 1) = (\tfrac{1}{6}, \tfrac{2}{3}), \quad A^2B(0, 1) = (\tfrac{1}{14}, \tfrac{11}{14}),$$

$$A^3B(0, 1) = (\tfrac{1}{30}, \tfrac{26}{30}), \qquad BA^2B(0, 1) = (\tfrac{2}{7}, \tfrac{4}{7}), \quad A^4B(0, 1) = (\tfrac{1}{62}, \tfrac{57}{62}),$$

$$BA^3B(0, 1) = (\tfrac{11}{30}, \tfrac{16}{30}), \qquad ABA^2B(1, 0) = (\tfrac{2}{18}, \tfrac{13}{18}),$$

$$BA^4B(0, 1) = (\tfrac{13}{31}, \tfrac{16}{31}), \qquad ABA^3B(0, 1) = (\tfrac{11}{82}, \tfrac{57}{82}),$$

$$A^2BA^2B(0, 1) = (\tfrac{2}{40}, \tfrac{33}{40}), \quad BABA^2B(0, 1) = (\tfrac{4}{18}, \tfrac{11}{18}).$$

To estimate the sum Σ_1 of §5.18 we may take

$$f(x) = \frac{t}{2\pi} \log x, \qquad y = \frac{t}{2\pi}, \qquad s = 1,$$

so that (5.20.1) holds for any $c \geqslant 0$. The exponent pair $(\frac{11}{82}, \frac{57}{82})$ then yields

$$\sum\nolimits_1 \ll t^{\frac{11}{82}} a^{\frac{46}{82}}$$

whence

$$\sum_{a < n \leqslant b} n^{-\frac{1}{2} - it} \ll t^{\frac{11}{82}} a^{\frac{5}{82}} \ll t^{\frac{27}{164}}$$

for $a \ll t^{\frac{1}{2}}$. We therefore recover Theorem 5.18.

The estimate $\mu(\frac{1}{2}) \leqslant \frac{229}{1392}$ of Phillips (1) comes from a better choice of exponent pair. In general we will have

$$\mu(\tfrac{1}{2}) \leqslant \tfrac{1}{2}(p + q - \tfrac{1}{2}),$$

providing that $q \geqslant p + \frac{1}{2}$. Rankin [1] has shown that the infimum of $\frac{1}{2}(p + q - \frac{1}{2})$, over all pairs generated from (0, 1) by the A and B processes, is 0·16451067.... (Graham, in work in the course of publication, gives further details.) Note however that there are exponent pairs better for certain problems than any which can be got in this way, as we shall see in §§6.17–18. These unfortunately do not seem to help in the estimation of $\mu(\frac{1}{2})$.

5.21. The list of bounds for $\mu(\frac{1}{2})$ may be extended as follows.

$\frac{163}{988} = 0{\cdot}164979\ldots$	Walfisz (**1**),
$\frac{27}{164} = 0{\cdot}164634\ldots$	Titchmarsh (**9**),
$\frac{229}{1392} = 0{\cdot}164511\ldots$	Phillips (**1**)
$0{\cdot}164510\ldots$	Rankin [**1**]
$\frac{19}{116} = 0{\cdot}163793\ldots$	Titchmarsh (**24**)
$\frac{15}{92} = 0{\cdot}163043\ldots$	Min (**1**)
$\frac{6}{37} = 0{\cdot}162162\ldots$	Haneke [**1**]
$\frac{173}{1067} = 0{\cdot}162136\ldots$	Kolesnik [**2**]
$\frac{35}{216} = 0{\cdot}162037\ldots$	Kolesnik [**4**]
$\frac{139}{858} = 0{\cdot}162004\ldots$	Kolesnik [**5**].

The value $\frac{6}{37}$ was obtained by Chen [**1**], independently of Haneke, but a little later.

The estimates from Titchmarsh (**24**) onwards depend on bounds for multiple sums. In proving Lemma 5.10 the sum over r on the left of (5.10.1) is estimated trivially. However, there is scope for further savings by considering the sum over r and ν as a two-dimensional sum, and using two dimensional analogues of the A and B processes given by Lemma 5.10 and Theorem 4.9. Indeed since further variables are introduced each time an A process is used, higher-dimensional sums will occur. Srinivasan [**1**] has given a treatment of double sums, but it is not clear whether it is sufficiently flexible to give, for example, new exponent pairs for one-dimensional sums.

VI

VINOGRADOV'S METHOD

6.1. STILL another method of dealing with exponential sums is due to Vinogradov.† This has passed through a number of different forms of which the one given here is the most successful. In the theory of the zeta-function, the method gives new results in the neighbourhood of the line $\sigma = 1$.

Let
$$f(n) = \alpha_k n^k + \ldots + \alpha_1 n + \alpha_0$$
be a polynomial of degree $k \geqslant 2$ with real coefficients, and let a and q be integers,
$$S(q) = \sum_{a<n\leqslant a+q} e^{2\pi i f(n)},$$
$$J(q,l) = \int_0^1 \ldots \int_0^1 |S(q)|^{2l}\, d\alpha_1 \ldots d\alpha_k.$$
The question of the order of $J(q,l)$ as a function of q is important in the method.

Since $S(q) = O(q)$ we have trivially $J(q,l) = O(q^{2l})$. Less trivially, we could argue as follows. We have
$$\{S(q)\}^k = \sum_{n_1,\ldots,n_k} e^{2\pi i \alpha_k (n_1^k + \ldots + n_k^k) + \ldots}$$
$$|S(q)|^{2k} = \sum_{m_1,\ldots,n_k} e^{2\pi i \alpha_k (m_1^k + \ldots + m_k^k - n_1^k - \ldots - n_k^k) + \ldots}.$$
On integrating over the k-dimensional unit cube, we obtain a zero factor if any of the numbers
$$m_1^h + \ldots + m_k^h - n_1^h - \ldots - n_k^h \quad (h = 1,\ldots, k)$$
is different from zero. Hence $J(q,k)$ is equal to the number of solutions of the system of equations
$$m_1^h + \ldots + m_k^h = n_1^h + \ldots + n_k^h \quad (h = 1,\ldots, k),$$
where $a < m_\nu \leqslant a+q$, $a < n_\nu \leqslant a+q$.

But it follows from these equations that the numbers n_ν are equal (in some order) to the numbers m_ν. Hence only the m_ν can be chosen

† Vinogradov (1)–(4), Tchudakoff (1)–(5), Titchmarsh (20), Hua (1).

arbitrarily, and so the total number of solutions is $O(q^k)$. Hence

$$J(q,k) = O(q^k)$$

and
$$J(q,l) = O\{q^{2l-2k}J(q,k)\} = O(q^{2l-k}).$$

This, however, is not sufficient for the application (see Lemma 6.8).

For any integer l, $J(q,l)$ is equal to the number of solutions of the equations

$$m_1^h+\dots+m_l^h = n_1^h+\dots+n_l^h \quad (h = 1, 2,\dots, k),$$

where $a < m_\nu \leqslant a+q$, $a < n_\nu \leqslant a+q$. Actually $J(q,l)$ is independent of a; for putting $M_\nu = m_\nu - a$, $N_\nu = n_\nu - a$, we obtain

$$\sum_{\nu=1}^{l}(M_\nu+a)^h = \sum_{\nu=1}^{l}(N_\nu+a)^h \quad (h = 1,\dots, k),$$

which is equivalent to

$$\sum_{\nu=1}^{l} M_\nu^h = \sum_{\nu=1}^{l} N_\nu^h \quad (h = 1,\dots, k),$$

and $0 < M_\nu \leqslant q$, $0 < N_\nu \leqslant q$.

Clearly $J(q,l)$ is a non-decreasing function of q.

6.2. LEMMA 6.2. *Let $m_1,\dots, m_k$, $n_1,\dots, n_k$ be two sets of integers, let*

$$s_h = \sum_{\nu=1}^{k} m_\nu^h, \qquad s_h' = \sum_{\nu=1}^{k} n_\nu^h,$$

and let σ_h, σ_h' be the h-th elementary symmetric functions of the m_ν and n_ν respectively. If $|m_\nu| \leqslant q$, $|n_\nu| \leqslant q$, and

$$|s_h - s_h'| \leqslant q^{h-1} \quad (h = 1,\dots, k), \tag{6.2.1}$$

then
$$|\sigma_h - \sigma_h'| \leqslant \tfrac{3}{4}(2kq)^{h-1} \quad (h = 2,\dots, k). \tag{6.2.2}$$

Clearly
$$|s_h| \leqslant kq^h, \qquad |s_h'| \leqslant kq^h,$$

and
$$|\sigma_h'| \leqslant \binom{k}{h} q^h \leqslant k^h q^h.$$

Now
$$\sigma_2 = \tfrac{1}{2}(s_1^2 - s_2).$$

Hence
$$\begin{aligned} |\sigma_2 - \sigma_2'| &= \tfrac{1}{2}|(s_1^2 - s_2) - (s_1'^2 - s_2')| \\ &\leqslant \tfrac{1}{2}|(s_1 - s_1')(s_1 + s_1')| + \tfrac{1}{2}|s_2 - s_2'| \\ &\leqslant kq + \tfrac{1}{2}q \leqslant \tfrac{3}{2}kq, \end{aligned}$$

the result stated for $h = 2$.

Now suppose that (6.2.2) holds with $h = 2,\dots, j-1$, where $3 \leqslant j \leqslant k$, so that

$$|\sigma_h - \sigma_h'| \leqslant (2kq)^{h-1} \quad (h = 1,\dots, j-1).$$

By a well-known theorem on symmetric functions

$$s_j - \sigma_1 s_{j-1} + \sigma_2 s_{j-2} - \dots + (-1)^j j\sigma_j = 0.$$

Hence

$$|\sigma_j-\sigma_j'| \leqslant \frac{1}{j}|s_j-s_j'|+\frac{1}{j}\sum_{h=1}^{j-1}|\sigma_h s_{j-h}-\sigma_h' s_{j-h}'|$$

$$= \frac{|s_j-s_j'|}{j}+\frac{1}{j}\sum_{h=1}^{j-1}|(\sigma_h-\sigma_h')s_{j-h}+\sigma_h'(s_{j-h}-s_{j-h}')|$$

$$\leqslant \frac{q^{j-1}}{j}+\frac{1}{j}\sum_{h=1}^{j-1}\{(2kq)^{h-1}kq^{j-h}+(kq)^h q^{j-h-1}\}$$

$$= \frac{q^{j-1}}{j}\left\{1+\sum_{h=1}^{j-1}(2^{h-1}+1)k^h\right\}$$

$$\leqslant \frac{q^{j-1}}{j}\sum_{h=0}^{j-1}2^h k^h = \frac{q^{j-1}}{j}\frac{(2k)^j-1}{2k-1}$$

$$\leqslant (2kq)^{j-1}\frac{2k}{j(2k-1)} \leqslant \tfrac{2}{3}(2kq)^{j-1} \leqslant \tfrac{3}{4}(2kq)^{j-1}$$

since $2k/(2k-1) \leqslant 2$ and $j \geqslant 3$. This proves the lemma.

6.3. LEMMA 6.3. *Let $1 < G < q$, and let $g_1,\ldots, g_k$ be integers satisfying*

$$1 < g_1 < g_2 < \ldots < g_k \leqslant G, \qquad g_\nu - g_{\nu-1} > 1. \tag{6.3.1}$$

For each value of ν $(1 \leqslant \nu \leqslant k)$ let m_ν be an integer lying in the interval

$$-a+(g_\nu-1)q/G < m_\nu \leqslant -a+g_\nu q/G,$$

where $0 \leqslant a \leqslant q$. Then the number of sets of such integers $m_1,\ldots, m_k$ for which the values of s_h $(h = 1,\ldots, k)$ lie in given intervals of lengths not exceeding q^{h-1}, is $\leqslant (4kG)^{\frac{1}{2}k(k-1)}$.

If x is any number such that $|x| \leqslant q$, the above lemma gives

$$|(x-m_1)\ldots(x-m_k)-(x-n_1)\ldots(x-n_k)| \leqslant \sum_{h=1}^{k}|\sigma_h-\sigma_h'||x|^{k-h}$$

$$\leqslant q^{k-1}\left\{1+\tfrac{3}{4}\sum_{h=2}^{k}(2k)^{h-1}\right\}$$

$$= q^{k-1}\left\{1+\frac{3}{4}\frac{(2k)^k-2k}{2k-1}\right\} \leqslant (2kq)^{k-1}$$

since $k \geqslant 2$. If $n_1,\ldots, n_k$ satisfy the same conditions as $m_1,\ldots, m_k$, then $|m_k-n_\nu| \geqslant q/G$ for $\nu = 1, 2,\ldots, k-1$. Hence, putting $x = n_k$,

$$(q/G)^{k-1}|m_k-n_k| \leqslant (2kq)^{k-1},$$

i.e.

$$|m_k-n_k| \leqslant (2kG)^{k-1}.$$

Thus the number of numbers m_k satisfying the requirements of the theorem does not exceed

$$(2kG)^{k-1}+1 \leqslant (4kG)^{k-1}.$$

Next, for a given value of m_k, the numbers $m_1,\dots, m_{k-1}$ satisfy similar conditions with $k-1$ instead of k, and hence the number of values of m_{k-1} is at most $\{4(k-1)G\}^{k-2} < (4kG)^{k-2}$. Proceeding in this way, we find that the total number of sets does not exceed

$$(4kG)^{(k-1)+(k-2)+\dots} = (4kG)^{\frac{1}{2}k(k-1)}.$$

6.4. LEMMA 6.4. *Under the same conditions as in Lemma* 6.3, *the number of sets of integers* $m_1,\dots, m_k$ *for which the numbers* s_h $(h = 1,\dots, k)$ *lie in given intervals of lengths not exceeding* $cq^{h(1-1/k)}$, *where* $c > 1$, *does not exceed*

$$(2c)^k(4kG)^{\frac{1}{2}k(k-1)}q^{\frac{1}{2}(k-1)}.$$

We divide the hth interval into

$$1+\left[\frac{cq^{h(1-1/k)}}{q^{h-1}}\right] \leqslant 2cq^{1-h/k}$$

parts, and apply Lemma 6.3. Since

$$\prod_{h=1}^{k} (2cq^{1-h/k}) = (2c)^k q^{\frac{1}{2}(k-1)}$$

we have at most $(2c)^k q^{\frac{1}{2}(k-1)}$ sets of sub-intervals, each satisfying the conditions of Lemma 6.3. For each set there are at most $(4kG)^{\frac{1}{2}k(k-1)}$ solutions, so that the result follows.

6.5. LEMMA 6.5. *Let* $k < l$, *let* $f(n)$ *be as in* §6.1, *and let*

$$I = \int_0^1 \dots \int_0^1 |Z_{m,g_1} \dots Z_{m,g_k}|^2 |S(q^{1-1/k})|^{2(l-k)}\, d\alpha_1 \dots d\alpha_k,$$

where

$$Z_{m,g_\nu} = \sum_{(g_\nu-1)2^{-m}q<n\leqslant g_\nu 2^{-m}q} e^{2\pi i f(n)}$$

and the g_ν *satisfy* (6.3.1) *with* $1 < G = 2^m < q$. *Then*

$$I \leqslant 2^{3k+(m+2)\frac{1}{2}k(k-1)-mk}(l-k)^k k^{\frac{1}{2}k(k-1)} q^{\frac{3}{2}k-\frac{1}{2}} J(q^{1-1/k}, l-k).$$

We have

$$I = \sum_{N_1,\dots,N_k} \Psi(N_1,\dots, N_k) \int_0^1 \dots \int_0^1 e^{2\pi i(N_k\alpha_k+\dots+N_1\alpha_1)} |S(q^{1-1/k})|^{2(l-k)}\, d\alpha_1 \dots d\alpha_k$$

$$\leqslant \sum_{N_1,\dots,N_k} \Psi(N_1,\dots, N_k) \int_0^1 \dots \int_0^1 |S(q^{1-1/k})|^{2(l-k)}\, d\alpha_1 \dots d\alpha_k,$$

where $\Psi(N_1,\dots, N_k)$ is the number of solutions of the equations

$$m_1^h+\dots+m_k^h-n_1^h-\dots-n_k^h = N_h \quad (h = 1,\dots, k)$$

for m_ν and n_ν in the interval $(g_\nu - 1)\, 2^{-m}q < x \leqslant g_\nu 2^{-m}q$. Moreover N_h runs over those integers for which one can solve

$$N_h = n_1'^h+\dots+n_{l-k}'^h-m_1'^h-\dots-m_{l-k}'^h,$$

where m'_ν and n'_ν lie in an interval $(a, a+q^{1-1/k}]$. As in §6.1 we can shift each range through $-a$, i.e. replace a by 0. Then N_h ranges over at most $2(l-k)q^{h(1-1/k)}$ values. Hence by Lemma 6.4, for given values of $n_1,\ldots, n_k$, the number of sets of $(m_1,\ldots, m_k)$ does not exceed

$$\{4(l-k)\}^k(2^{m+2}k)^{\frac{1}{2}k(k-1)}q^{\frac{1}{2}(k-1)}.$$

Also $(n_1,\ldots, n_k)$ takes not more than $(1+2^{-m}q)^k \leqslant (2^{1-m}q)^k$ values. Hence

$$\sum_{N_1,\ldots,N_k} \Psi(N_1,\ldots, N_k) \leqslant \{4(l-k)\}^k k^{\frac{1}{2}k(k-1)} 2^{(m+2)\frac{1}{2}k(k-1)-mk+k} q^{\frac{3}{2}k-\frac{1}{2}},$$

and the result follows.

6.6. LEMMA 6.6. *The result of Lemma* 6.5 *holds whether the g's satisfy,* (6.3.1) *or not, if m has the value*

$$M = \left[\frac{\log q}{k\log 2}\right]. \tag{6.6.1}$$

Since

$$|Z_{M,g_\nu}| \leqslant 2^{-M}q+1 \leqslant 2^{1-M}q,$$
$$|Z_{M,g_1} \cdots Z_{M,g_k}|^2 \leqslant (2^{1-M}q)^{2k},$$

it is sufficient to prove that

$$(2^{1-M}q)^{2k} \leqslant 2^{3k+(M+2)\frac{1}{2}k(k-1)-Mk}(l-k)^k k^{\frac{1}{2}k(k-1)}q^{\frac{3}{2}k-\frac{1}{2}},$$

or that
$$q^{\frac{1}{2}k+\frac{1}{2}} \leqslant 2^{(M+2)\frac{1}{2}k(k-1)+Mk}k^{\frac{1}{2}k(k-1)},$$

or that
$$(\tfrac{1}{2}k+\tfrac{1}{2})\log q \leqslant \tfrac{1}{2}k(k+1)M\log 2+\tfrac{1}{2}k(k-1)\log 4k,$$

or that
$$\log q \leqslant kM\log 2+\frac{k(k-1)}{k+1}\log 4k.$$

Since
$$M \geqslant \frac{\log q}{k\log 2}-1,$$

this is true if
$$k\log 2 \leqslant \frac{k(k-1)}{k+1}\log 4k,$$

or
$$\log 2 \leqslant \frac{k-1}{k+1}\log 4k,$$

which is true for $k \geqslant 2$.

6.7. LEMMA 6.7. *The set of integers* $(g_1,\ldots, g_l)$, *where* $k < l$, *and each* g_ν *ranges over* $(1, G]$, *is said to be well-spaced if there are at least k of them, say* $g_{j_1},\ldots, g_{j_k}$, *satisfying*

$$g_{j_\nu}-g_{j_{\nu-1}} > 1 \quad (\nu = 2,\ldots, k).$$

The number of sets which are not well-spaced is at most $l!\,3^l G^{k-1}$.

Let $g'_1,\ldots, g'_l$ denote $g_1,\ldots, g_l$ arranged in increasing order, and let $f_\nu = g'_\nu - g'_{\nu-1}$. If the set is not well-spaced, there are at most $k-2$ of the numbers f_ν for which $f_\nu > 1$.

Consider those sets in which exactly h $(0 \leqslant h \leqslant k-2)$ of the numbers f_ν are greater than 1. The number of ways in which these h f_ν's can be chosen from the total $l-1$ is $\binom{l-1}{h}$. Also each of the h f_ν's can take at most G values, and each of the rest at most 2 values. Since g'_1 takes at most G values, the total number of sets of g'_ν arising in this way is at most

$$\binom{l-1}{h} G^{h+1}2^{l-h-1}.$$

The total number of not well-spaced sets g'_ν is therefore

$$\leqslant \sum_{h=0}^{k-2} \binom{l-1}{h} G^{h+1}2^{l-h-1} \leqslant G^{k-1} \sum_{h=0}^{k-2} \binom{l-1}{h} 2^{l-h-1}$$
$$\leqslant G^{k-1}(1+2)^{l-1} < 3^l G^{k-1}.$$

Since the number of sets g_ν corresponding to each set g'_ν is at most $l!$, the result follows.

6.8. LEMMA 6.8. *If* $l \geqslant \frac{1}{4}k^2+\frac{5}{4}k$ *and* M *is defined by* (6.6.1), *then*

$$J(q,l) \leqslant \max(1,M)48^{2l}(l!)^{2l}k^{k\frac{1}{2}k(k-1)}q^{2(l-k)/k+\frac{3}{2}k-\frac{1}{2}}J(q^{1-1/k},l-k).$$

Suppose first that M is not less than 2, i.e. that $q \geqslant 2^{2k}$. Let μ be a positive integer not greater than $M-1$. Then

$$\mu \leqslant \frac{\log q}{k \log 2}-1, \quad \text{i.e.} \quad 2^{\mu+1} \leqslant q^{1/k}.$$

Let
$$S(q) = \sum_{g=1}^{2^\mu} \sum_{(g-1)2^{-\mu}q<n\leqslant g2^{-\mu}q} e^{2\pi i f(n)} = \sum_{g=1}^{2^\mu} Z_{\mu,g},$$

say. Then
$$\{S(q)\}^l = \sum Z_{\mu,g_1} \dots Z_{\mu,g_l},$$

where each g_ν runs from 1 to 2^μ, and the sum contains $2^{\mu l}$ terms.

We denote those products $Z_{\mu,g_1} \dots Z_{\mu,g_l}$ with well-spaced g's by Z'_μ. The number of these, M_μ say, does not exceed $2^{\mu l}$. In the remaining terms we divide each factor into two parts, so that we obtain products of the type $Z_{\mu+1,g_1} \dots Z_{\mu+1,g_l}$, each g lying in $(1, 2^{\mu+1})$. The number of such terms, $M_{\mu+1}$ say, does not exceed $l!\,3^l 2^{\mu(k-1)}2^l = l!\,6^l 2^{\mu(k-1)}$, by Lemma 6.7. The terms of this type with well-spaced g's we denote by $Z'_{\mu+1}$, and the rest we subdivide again. We proceed in this way until finally Z'_M denotes all the products of order M, whether containing well-spaced g's or not. We then have

$$\{S(q)\}^l = \sum_{m=\mu}^{M} \sum Z'_m,$$

$$|S(q)|^{2l} \leqslant M \sum_{m=\mu}^{M} |\sum Z'_m|^2 \leqslant M \sum_{m=\mu}^{M} M_m \sum |Z'_m|^2, \tag{6.8.1}$$

where M_m is the number of terms in the sum $\sum Z'_m$. By Lemma 6.7,

$$M_m \leqslant l!\, 3^l 2^{(m-1)(k-1)} 2^l = l!\, 6^l 2^{(m-1)(k-1)} \quad (m > \mu).$$

Consider, for example, $\sum |Z'_\mu|^2$. The general Z'_μ can be written

$$Z_{\mu,g_1} \dots Z_{\mu,g_k} Z_{\mu,g_{k+1}} \dots Z_{\mu,g_l},$$

where $g_1,\dots, g_k$ satisfy (6.3.1) with $G = 2^\mu$. Now, since the geometric mean does not exceed the arithmetic mean,

$$|Z_{\mu,g_{k+1}} \dots Z_{\mu,g_l}|^2 \leqslant \frac{1}{l-k} \sum_{\nu=k+1}^{l} |Z_{\mu,g_\nu}|^{2(l-k)}.$$

We divide these Z_{μ,g_ν} into parts of length $q^{1-1/k}-1$ (or less). The number of such parts does not exceed

$$\left[\frac{2^{-\mu}q}{q^{1-1/k}-1}\right]+1 \leqslant \frac{2^{-\mu}q}{q^{1-1/k}-1}+2^{-\mu-1}q^{1/k} \leqslant \frac{2^{-\mu}q}{\frac{3}{4}q^{1-1/k}}+2^{-\mu-1}q^{1/k} \leqslant 2^{1-\mu}q^{1/k},$$

since $q^{1-1/k} \geqslant q^{1/k} \geqslant 2^M \geqslant 4$. Each part is of the form $S(q^{1-1/k})$, or with $q^{1-1/k}$ replaced by a smaller number. Hence by Hölder's inequality†

$$|Z_{\mu,g_\nu}|^{2(l-k)} \leqslant (2^{1-\mu}q^{1/k})^{2(l-k)-1} \sum |S(q^{1-1/k})|^{2(l-k)}.$$

Hence

$$\sum |Z'_\mu|^2 \leqslant \frac{(2^{1-\mu}q^{1/k})^{2(l-k)-1}}{l-k} \sum_{g_\nu} |Z_{\mu,g_1} \dots Z_{\mu,g_k}|^2 \sum_{\nu=k+1}^{l} \sum |S(q^{1-1/k})|^{2(l-k)}.$$

Hence by Lemma 6.5, and the non-decreasing property of $J(q, l)$ as a function of q,

$$\int_0^1 \dots \int_0^1 \sum |Z'_\mu|^2\, d\alpha_1 \dots d\alpha_k \leqslant (2^{1-\mu}q^{1/k})^{2(l-k)-1} M_\mu\, 2^{1-\mu}q^{1/k} \times$$
$$\times 2^{3k+(\mu+2)\frac{1}{2}k(k-1)-\mu k}(l-k)^k k^{\frac{1}{2}k(k-1)} q^{\frac{3}{2}k-\frac{1}{2}} J(q^{1-1/k}, l-k)$$
$$= 2^{\mu(\frac{1}{2}k^2+\frac{1}{2}k-2l)+2l+k^2+k} M_\mu (l-k)^k k^{\frac{1}{2}k(k-1)} q^{2(l-k)/k+\frac{3}{2}k-\frac{3}{2}} J(q^{1-1/k}, l-k).$$

A similar argument applies to Z'_m, with μ replaced by m. Hence

$$J(q,l) \leqslant M \sum_{m=\mu}^{M} 2^{m(\frac{1}{2}k^2+\frac{1}{2}k-2l)} M_m^2 \times$$
$$\times 2^{2l+k^2+k}(l-k)^k k^{\frac{1}{2}k(k-1)} q^{2(l-k)/k+\frac{3}{2}k-\frac{1}{2}} J(q^{1-1/k}, l-k).$$

Also

$$\sum_{m=\mu}^{M} 2^{m(\frac{1}{2}k^2+\frac{1}{2}k-2l)} M_m^2$$
$$\leqslant 2^{\mu(\frac{1}{2}k^2+\frac{1}{2}k-2l)+2\mu l} + \sum_{m=\mu+1}^{M} 2^{m(\frac{1}{2}k^2+\frac{1}{2}k-2l)}(l!)^2 6^{2l} 2^{2(m-1)(k-1)}$$
$$= 2^{\frac{1}{2}\mu(k^2+k)} + (l!)^2 6^{2l} \sum_{m=\mu+1}^{M} 2^{m(\frac{1}{2}k^2+\frac{5}{2}k-2l-2)-2(k-1)}$$
$$\leqslant 2^{2\mu l} + (l!)^2 6^{2l} \leqslant 2(l!)^2 6^{2l},$$

† Here $S(q^{1-1/k})$ denotes any sum of the form $S(p)$ with $p \leqslant q^{1-1/k}$.

since we can start with an integer μ such that $2^{\mu l} < l!$. (Indeed we may take $\mu = 1$.) Hence

$$J(q,l) \leqslant M 2^{2l+k^2+k+1}(l!)^2 6^{2l} l^k k^{\frac{1}{2}k(k-1)} q^{2(l-k)/k+\frac{3}{2}k-\frac{1}{2}} J(q^{1-1/k}, l-k),$$

and since
$$2^{2l+k^2+k+1} 6^{2l} \leqslant 2^{6l} 6^{2l} = 48^{2l}$$

the result follows.

If $M < 2$, i.e. $q < 2^{2k}$, divide $S(q)$ into four parts, each of the form $S(q')$, where $q' \leqslant \frac{1}{4}q \leqslant q^{1-1/k}$. By Hölder's inequality

$$|S(q)|^{2l} \leqslant 4^{2l-1} \sum |S(q')|^{2l} \leqslant 4^{2l-1} q^{2k(1-1/k)} \sum |S(q')|^{2(l-k)}.$$

Integrating over the unit hypercube,

$$\begin{aligned} J(q,l) &\leqslant 4^{2l-1} q^{2k(1-1/k)} \sum J(q', l-k) \\ &\leqslant 4^{2l} q^{2k(1-1/k)} J(q^{1-1/k}, l-k), \end{aligned}$$

and the result again follows.

6.9. LEMMA 6.9. *If r is any non-negative integer, and $l \geqslant \frac{1}{4}k^2 + \frac{1}{4}k + kr$, then*

$$J(q,l) \leqslant K^r \log^r q \,.\, q^{2l-\frac{1}{2}k(k+1)+\delta_r}$$

where
$$\delta_r = \tfrac{1}{2}k(k+1)\left(1-\frac{1}{k}\right)^r, \qquad K = 48^{2l}(l!)^2 l^k k^{\frac{1}{2}k(k-1)}.$$

This is obvious if $r = 0$, since then $\delta_0 = \frac{1}{2}k(k+1)$ and $J(q,l) \leqslant q^{2l}$. Assuming then that it is true up to $r-1$, Lemma 6.8 (in which $M \leqslant \log q$) gives

$$\begin{aligned} J(q,l) \leqslant K \log q \,.\, q^{2(l-k)/k+\frac{3}{2}k-\frac{1}{2}} \,.\, K^{r-1} \log^{r-1}(q^{1-1/k}) \times \\ \times q^{(1-1/k)\{2(l-k)-\frac{1}{2}k(k+1)+\delta_{r-1}\}}, \end{aligned}$$

and the index of q reduces to $2l-\frac{1}{2}k(k+1)+\delta_r$.

6.10. LEMMA 6.10. *If $l = [k^2 \log(k^2+k) + \frac{1}{4}k^2 + \frac{5}{4}k] + 1$, $k \geqslant 7$,*

$$J(q,l) \leqslant e^{64lk\log^2 k} \log^{2l} q \,.\, q^{2l-\frac{1}{2}k(k+1)+\frac{1}{2}}.$$

We have $\delta_r \leqslant \frac{1}{2}$ if
$$k(k+1)\left(1-\frac{1}{k}\right)^r \leqslant 1,$$

i.e. if
$$\log\{k(k+1)\} \leqslant r \log \frac{k}{k-1}.$$

This is true if
$$\log\{k(k+1)\} \leqslant r/k,$$

or if
$$r = [k \log(k^2+k)] + 1.$$

Since
$$r < k\log^3 k + 1 < 4k \log k, \qquad l < k^3,$$

and

$$\begin{aligned} \log K &< 2l \log 48 + 2l \log l + k \log l + \tfrac{1}{2}k(k-1)\log k \\ &< 5l \log l + l \log k < 16 l \log k, \end{aligned}$$

the result follows.

6.11. LEMMA 6.11. *Let M and N be integers, $N > 1$, and let $\phi(n)$ be a real function of n, defined for $M \leqslant n \leqslant M+N-1$, such that*

$$\delta \leqslant \phi(n+1)-\phi(n) \leqslant c\delta \quad (M \leqslant n \leqslant M+N-2),$$

where $\delta > 0$, $c \geqslant 1$, $c\delta \leqslant \frac{1}{2}$. Let $W > 0$. Let $\bar{x}$ denote the difference between x and the nearest integer. Then the number of values of n for which $\overline{\phi(n)} \leqslant W\delta$ is less than

$$(Nc\delta+1)(2W+1).$$

Let α be a given real number, and let G be the number of values of n for which

$$\alpha+h < \phi(n) \leqslant \alpha+h+\delta$$

for some integer h. To each h corresponds at most one n, so that $G \leqslant h_2-h_1+1$, where h_1 and h_2 are the least and greatest values of h. But clearly

$$\phi(M) \leqslant \alpha+h_1+\delta, \qquad \alpha+h_2 < \phi(M+N-1),$$

whence $$h_2-h_1-\delta < \phi(M+N-1)-\phi(M) \leqslant (N-1)c\delta,$$

and $$G \leqslant (N-1)c\delta+\delta+1 \leqslant Nc\delta+1.$$

The result of the lemma now follows from the fact that an interval of length $2W\delta$ may be divided into $[2W+1]$ intervals of length less than δ $(< \frac{1}{2})$.

6.12. LEMMA 6.12. *Let k and Q be integers, $k \geqslant 7$, $Q \geqslant 2$, and let $f(x)$ be real and have continuous derivatives up to the $(k+1)$th order in $[P+1, P+Q]$; let $0 < \lambda < 1$ and*

$$\lambda \leqslant \frac{f^{(k+1)}(x)}{(k+1)!} \leqslant 2\lambda \quad (P+1 \leqslant x \leqslant P+Q) \tag{6.12.1}$$

or the same for $-f^{(k+1)}(x)$, and let

$$\lambda^{-\frac{1}{4}} \leqslant Q \leqslant \lambda^{-1}. \tag{6.12.2}$$

Then $$\left|\sum_{n=P+1}^{P+Q} e^{2\pi i f(n)}\right| < Ae^{33k\log^2 k}Q^{1-\rho}\log Q, \tag{6.12.3}$$

where $$\rho = (24k^2\log k)^{-1}.$$

Let $$q = [\lambda^{-1/(k+1)}]+1,$$

so that $$2 \leqslant q \leqslant [Q^{4/(k+1)}]+1 \leqslant Q,$$

and write $$S = \sum_{n=P+1}^{P+Q} e^{2\pi i f(n)},$$

$$T(n) = \sum_{m=1}^{q} e^{2\pi i\{f(m+n)-f(n)\}} \quad (P+1 \leqslant n \leqslant P+Q-q).$$

Then

$$
\begin{aligned}
q|S| &= \left|\sum_{m=1}^{q} \sum_{n=P+1}^{P+Q} e^{2\pi i f(n)}\right| \\
&\leqslant \left|\sum_{m=1}^{q} \sum_{n=P+1+m}^{P+Q-q+m} e^{2\pi i f(n)}\right| + \sum_{m=1}^{q} q \\
&= \left|\sum_{m=1}^{q} \sum_{n=P+1}^{P+Q-q} e^{2\pi i f(m+n)}\right| + q^2 \\
&= \left|\sum_{n=P+1}^{P+Q-q} \sum_{m=1}^{q} e^{2\pi i f(m+n)}\right| + q^2 \\
&\leqslant \sum_{n=P+1}^{P+Q-q} |T(n)| + q^2 \\
&\leqslant Q^{1-1/(2l)}\left\{\sum_{n=P+1}^{P+Q-q} |T(n)|^{2l}\right\}^{1/(2l)} + q^2, \qquad (6.12.4)
\end{aligned}
$$

by Hölder's inequality, where l is any positive integer.

We now write $A_r = A_r(n) = f^{(r)}(n)/r!$ for $1 \leqslant r \leqslant k$, and define the k-dimensional region Ω_n by the inequalities

$$|\alpha_r - A_r| \leqslant \tfrac{1}{2} q^{-r} \quad (r = 1, \ldots, k). \qquad (6.12.5)$$

If we set

$$\delta(m) = f(m+n) - f(n) - (\alpha_k m^k + \ldots + \alpha_1 m),$$

then, by partial summation, we will have

$$T(n) = S(q) e^{2\pi i \delta(q)} - 2\pi i \int_0^q S(p) \delta'(p) e^{2\pi i \delta(p)} dp.$$

However, by Taylor's theorem together with the bound (6.12.1) we obtain

$$
\begin{aligned}
\delta'(p) &= f'(p+n) - \sum_1^k r\alpha_r p^{r-1} \\
&= f'(n) + p f''(n) + \ldots + \frac{p^{k-1}}{(k-1)!} f^{(k)}(n) + \frac{p^k}{k!} f^{(k+1)}(n + \vartheta p) - \sum_1^k r\alpha_r p^{r-1} \\
&= \sum_1^k r(A_r - \alpha_r) p^{r-1} + 2(k+1)\lambda \vartheta' p^k,
\end{aligned}
$$

where $0 < \vartheta$, $\vartheta' \leqslant 1$. If (6.12.5) holds it follows that

$$|\delta'(p)| \leqslant \sum_1^k r \tfrac{1}{2} q^{-r} q^{r-1} + 3k\lambda q^k \leqslant \tfrac{1}{2} k^2 q^{-1} + 3k\lambda q^k \leqslant 2^{k+3} k q^{-1},$$

by our choice of q. We therefore have

$$|T(n)| \leqslant 2^{k+4} k\pi \left(|S(q)| + \frac{1}{q} \int_0^q |S(p)| dp\right) = 2^{k+4} k\pi S_0(q),$$

say. Integrating over the region Ω_n, and dividing by its volume, we obtain

$$|T(n)|^{2l} \leqslant (2^{k+4}k\pi)^{2l} q^{\frac{1}{2}k(k+1)} \int_{\Omega_n} \dots \int |S_0(q)|^{2l} d\alpha_1 \dots d\alpha_k. \qquad (6.12.6)$$

The integral of $|S_0(q)|^{2l}$ over Ω_n is equal to its integral taken over the region obtained by subtracting any integer from each coordinate. We say that such a region is congruent (mod 1) to Ω_n. Now let n, n' be two integers in the interval $[P+1, P+Q-q]$, and let Ω_n, Ω'_n be the corresponding regions defined by (6.12.5). A necessary condition that Ω_n should intersect with any region congruent (mod 1) to Ω'_n is that

$$\overline{A_k(n) - A_k(n')} \leqslant q^{-k} \leqslant \lambda q. \qquad (6.12.7)$$

Let $\phi(n) = A_k(n) - A_k(n')$. Then

$$\phi(n+1) - \phi(n) = \frac{1}{k!}\{f^{(k)}(n+1) - f^{(k)}(n)\} = \frac{f^{(k+1)}(\xi)}{k!},$$

where $n < \xi < n+1$. The conditions of Lemma 6.11 are therefore satisfied, with $c = 2$ and $\delta = \lambda(k+1)$. Taking $W = q/(k+1)$, we see that the number of numbers n in $[P+1, P+Q-q]$ for which (6.12.7) is possible, does not exceed

$$\{2Q\lambda(k+1)+1\}\left(\frac{2q}{k+1}+1\right) \leqslant (2k+3)\left(\frac{2q}{k+1}+1\right) \leqslant 3kq.$$

Since this is independent of n', it follows that

$$\sum_{n=P+1}^{P+Q-q} \int_{\Omega_n} \dots \int |S_0(q)|^{2l} d\alpha_1 \dots d\alpha_k \leqslant 3kq \int_0^1 \dots \int_0^1 |S_0(q)|^{2l} d\alpha_1 \dots d\alpha_k$$
$$\leqslant 3kq 2^{2l} J(q, l), \qquad (6.12.8)$$

since

$$S_0(q)^{2l} \leqslant 2^{2l-1}\left(|S(q)|^{2l} + \frac{1}{q}\int_0^q |S(p)|^{2l} dp\right).$$

Defining l as in Lemma 6.10, we obtain from (6.12.4), (6.12.6), (6.12.8) and Lemma 10

$$|S| \leqslant 2^{k+5}k\pi Q^{1-\frac{1}{2l}} q^{-1}\{q^{\frac{1}{2}k(k+1)} 3kq J(q,l)\}^{\frac{1}{2l}} + q$$

$$\leqslant 2^{k+5}k\pi Q^{1-\frac{1}{2l}}\{3ke^{64lk\log^2 k} q^{\frac{3}{2}}\}^{\frac{1}{2l}} \log q + q.$$

Now $q \leqslant 2\lambda^{-1/(k+1)} \leqslant 2Q^{4/(k+1)}$. Hence

$$|S| \leqslant Ae^{33k\log^2 k} Q^{1-\frac{1}{2l}+3/\{(k+1)l\}} \log Q + 2Q^{4/(k+1)}$$

and the result follows, since $\frac{1}{2l}-3/\{(k+1)l\} \geqslant \frac{1}{8l}$ and $l < 3k^2\log k$.

6.13. LEMMA 6.13. *If $f(x)$ satisfies the conditions of Lemma* 6.12 *in an interval* $[P+1, P+N]$, *where* $N \leqslant Q$, *and*

$$\lambda^{-\frac{1}{3}} \leqslant Q \leqslant \lambda^{-1}, \tag{6.13.1}$$

then
$$\left|\sum_{n=P+1}^{P+N} e^{2\pi i f(n)}\right| < Ae^{33k\log^2 k}Q^{1-\rho}\log Q. \tag{6.13.2}$$

If $\lambda^{-\frac{1}{4}} \leqslant N$, the conditions of the previous theorem are satisfied when Q is replaced by N, and (6.13.2) follows at once from (6.12.3). On the other hand, if $\lambda^{-\frac{1}{4}} > N$, then

$$\left|\sum_{n=P+1}^{P+N} e^{2\pi i f(n)}\right| \leqslant N < \lambda^{-\frac{1}{4}} \leqslant Q^{\frac{3}{4}} \leqslant Q^{1-\rho},$$

and (6.13.1) again follows.

6.14. THEOREM 6.14.

$$\zeta(1+it) = O\{(\log t \log\log t)^{\frac{3}{4}}\}.$$

Let
$$f(x) = -\frac{t\log x}{2\pi}, \qquad f^{(k+1)}(x) = \frac{(-1)^{k+1}k!\,t}{2\pi x^{k+1}}.$$

Let $a < x \leqslant b \leqslant 2a$. Since $(-1)^{k+1}f^{(k+1)}(x)$ is steadily decreasing, we can divide the interval $[a, b]$ into not more than $k+1$ intervals, in each of which inequalities of the form (6.12.1) hold, where λ depends on the particular interval, and satisfies

$$\frac{t}{2\pi(k+1)(2a)^{k+1}} \leqslant \lambda \leqslant \frac{t}{4\pi(k+1)a^{k+1}}. \tag{6.14.1}$$

Let $Q = a \leqslant t$, $\log a > 2\log^{\frac{1}{2}} t$, and

$$k = \left[\frac{\log t}{\log a}\right]+1.$$

Then
$$Q < a^{k+1}t^{-1} \leqslant Q^2.$$

Clearly $\lambda \leqslant Q^{-1}$, while $\lambda \geqslant Q^{-3}$ if $Q \geqslant 2^{k+2}\pi(k+1)$, or if

$$\log a \geqslant \left(\frac{\log t}{\log a}+3\right)\log 2+\log\left(\frac{\log t}{\log a}+2\right)+\log \pi,$$

and this is true if t is large enough. It follows from Lemma 6.13 that

$$\sum_{a<n\leqslant b} e^{-it\log n} = O(ke^{33k\log^2 k}a^{1-\rho}\log a),$$

where ρ is defined as in § 6.12. Hence

$$\sum_{a<n\leqslant b}\frac{1}{n^{1+it}} = O(ke^{33k\log^2 k}a^{-\rho}\log a)$$
$$= O\left\{\log t\exp\left(33k\log^2 k-\frac{\log a}{24k^2\log k}\right)\right\}.$$

Suppose that
$$k \log k < A \log^{\frac{1}{3}} a,$$
with a sufficiently small A, or
$$\log a > A(\log t \log\log t)^{\frac{3}{4}}$$
with a sufficiently large A. Then
$$\sum_{a<n\leqslant b} \frac{1}{n^{1+it}} = O\left\{\log t \exp\left(\frac{-A \log^3 a}{\log^2 t \log\log t}\right)\right\}$$
$$= O[\log t \exp\{-A \log^{\frac{1}{4}} t (\log\log t)^{\frac{5}{4}}\}],$$
and the sum of $O(\log t)$ such terms is bounded.

Since $k \geqslant 7$, we also require that $a \leqslant t^{\frac{1}{7}}$. Using (5.16.1) with $r = 8$, and writing $\beta = t^{128/(7\times 128+1)}$, we obtain
$$\zeta(1+it) = \sum_{n\leqslant\beta} \frac{1}{n^{1+it}} + O(1) = O(\log\alpha) + \sum_{\alpha<n\leqslant\beta} \frac{1}{n^{1+it}} + O(1).$$
The last sum is bounded if
$$\log\alpha = A(\log t \log\log t)^{\frac{3}{4}}$$
with a suitable A, and the theorem follows.

6.15. If $0 < \sigma < 1$, we obtain similarly
$$\sum_{\alpha<n\leqslant\beta} \frac{1}{n^{\sigma+it}} = O[\alpha^{1-\sigma} \exp\{-A \log^{\frac{1}{4}} t (\log\log t)^{\frac{5}{4}}\} \log t],$$
and this is bounded if
$$1-\sigma < \frac{A(\log\log t)^{\frac{1}{2}}}{\log^{\frac{1}{2}} t} = 1-\sigma_0,$$
with a sufficiently small A. Hence in this region
$$\zeta(s) = O\left(\sum_{n\leqslant\alpha} \frac{1}{n^{\sigma_0}}\right) + O(1)$$
$$= O\left(\frac{\alpha^{1-\sigma_0}}{1-\sigma_0}\right) + O(1)$$
$$= O\left[\exp\{A \log^{\frac{1}{4}} t (\log\log t)^{\frac{5}{4}}\} \frac{\log^{\frac{1}{2}} t}{(\log\log t)^{\frac{1}{2}}}\right].$$
We can now apply Theorem 3.10, with
$$\theta(t) = \frac{A(\log\log t)^{\frac{1}{2}}}{\log^{\frac{1}{2}} t}, \qquad \phi(t) = A \log^{\frac{1}{4}} t (\log\log t)^{\frac{5}{4}}.$$
Hence there is a region
$$\sigma \geqslant 1 - \frac{A}{\log^{\frac{3}{4}} t (\log\log t)^{\frac{3}{4}}} \tag{6.15.1}$$

which is free from zeros of $\zeta(s)$; and by Theorem 3.11 we have also

$$\frac{1}{\zeta(1+it)} = O\{\log^{\frac{3}{4}}t(\log\log t)^{\frac{3}{4}}\}, \qquad \frac{\zeta'(1+it)}{\zeta(1+it)} = O\{\log^{\frac{3}{4}}t(\log\log t)^{\frac{3}{4}}\}. \tag{6.15.2), (6.15.3}$$

NOTES FOR CHAPTER 6

6.16. Further improvements have been made in the estimation of $J(q, l)$. The most important of these is due to Karatsuba [**2**] who used a p-adic analogue of the argument given here, thereby producing a considerable simplication of the proof. Moreover, as was shown by Stečkin [**1**], one is then able to sharpen Lemma 6.9 to yield the bound

$$J(q, l) \leqslant C^{k^3 \log k} q^{2l - \frac{1}{2}k(k+1) + \delta_r},$$

for $l \geqslant kr$, where $k \geqslant 2$, r is a positive integer, C is an absolute constant, and $\delta_r = \frac{1}{2}k^2(1-1/k)^r$. Here one has a smaller value for δ_r than formerly, but more significantly, the condition $l \geqslant \frac{1}{4}k^2 + \frac{1}{4}k + kr$ has been relaxed.

6.17. One can use Lemma 6.13 to obtain exponent pairs. To avoid confusion of notation, we take f to be defined on $(a, b]$, with $a < b \leqslant 2a$ and $\lambda^{-\frac{1}{3}} \leqslant a \leqslant \lambda^{-1}$. Then

$$\left| \sum_{a < n \leqslant b} e^{2\pi i f(n)} \right| \leqslant A e^{33k \log^2 k} a^{1-\rho} \log a.$$

Now suppose that (N, I, f, y) is in the set $\mathscr{F}(s, \frac{1}{4})$ of §5.20, whence

$$\tfrac{3}{4}\alpha_k x^{-s-k} \leqslant \frac{|f^{(k+1)}(x)|}{(k+1)!} \leqslant \tfrac{5}{4}\alpha_k x^{-s-k}$$

with

$$\alpha_k = y \frac{s(s+1)\ldots(s+k-1)}{(k+1)!}.$$

We may therefore break up I into $O(s+k)$ subintervals $(a, b]$ with $b \leqslant (\frac{6}{5})^{1/(s+k)} a$, on each of which one has

$$\lambda \leqslant \frac{|f^{(k+1)}(x)|}{(k+1)!} \leqslant 2\lambda,$$

with $\lambda = \frac{5}{8}\alpha_k a^{-s-k}$. We now choose k so that $\lambda^{-\frac{1}{3}} \leqslant N \leqslant 2N \leqslant \lambda^{-1}$ for all a in the range $N \leqslant a \leqslant 2N$. To do this we take $k \geqslant 7$ such that

$$\frac{N^{k-1}}{\alpha_{k-1}} < \tfrac{5}{4}N^{1-s} \leqslant \frac{N^k}{\alpha_k}. \tag{6.17.1}$$

Note that N^k/α_k tends to infinity with k, if $N \geqslant 2$, so this is always possible, providing that

$$\frac{N^6}{\alpha_6} < \tfrac{5}{4}N^{1-s}. \tag{6.17.2}$$

The estimate (6.17.1) ensures that $2N \leqslant \lambda^{-1}$, and hence, incidentally, that $\lambda < 1$. Moreover we also have

$$N^k < \tfrac{5}{4}\alpha_{k-1}N^{2-s} \leqslant \tfrac{5}{8}\alpha_k 2^{-s-k}N^{3-s}$$

if $N \geqslant 2^{s+k+2}$, and so $\lambda^{-\frac{1}{3}} \leqslant N$. It follows that

$$\sum_{n\in I} e^{2\pi i f(n)} \ll_s k e^{33k^2\log k} N^{1-\rho}\log N \tag{6.17.3}$$

for $N \geqslant 2^{s+k+2}$, subject to (6.17.2).

We shall now show that

$$(p, q) = \left(\frac{1}{25(m-2)m^2\log m}, 1 - \frac{1}{25m^2\log m}\right) \tag{6.17.4}$$

is an exponent pair whenever $m \geqslant 3$. If $yN^{2-s-m} \geqslant 1$ then $(yN^{-s})^pN^q \geqslant N$, and the required bound (5.20.2) is trivial. If (6.17.2) fails, then $yN^{-s} \ll_s N^5$ and, using the exponent pair $(\frac{1}{126}, \frac{120}{126}) = A^5B(0, 1)$ (in the notation of §5.20) we have

$$\sum_{n\in I} e^{2\pi i f(n)} \ll_s (yN^{-s})^{\frac{1}{126}} N^{\frac{120}{126}} \ll_s N^{\frac{125}{126}} \ll N^q \ll (yN^{-s})^pN^q$$

as required. We may therefore assume that $yN^{2-s-m} < 1$, and that (6.17.2) holds. Let us suppose that $N \geqslant \max(2^{s+m+2}, 2(\frac{1}{2}s+1)^m)$. Then (6.17.1) yields

$$N^{k-1} < \frac{5}{4}\cdot\frac{s}{2}\cdot\frac{s+1}{3}\cdot\frac{s+2}{4}\cdot\ldots\cdot\frac{s+k-2}{k}yN^{1-s}$$

$$\leqslant \tfrac{5}{4}\left(\max\left(\frac{s}{2}, 1\right)\right)^{k-1} yN^{1-s} < 2(\tfrac{1}{2}s+1)^{k-1}N^{m-1},$$

whence

$$\left(\frac{N}{\frac{1}{2}s+1}\right)^{k-m} < 2(\tfrac{1}{2}s+1)^{m-1}.$$

Since $N \geqslant 2(\frac{1}{2}s+1)^m$ we deduce that $k \leqslant m$. Moreover we then have $N \geqslant 2^{s+m+2} \geqslant 2^{s+k+2}$, so that (6.17.3) applies. Since k is bounded in terms of p, q and s, it follows that

$$\sum_{n \in I} e^{2\pi i f(n)} \ll_{p,q,s} N^{1-\rho} \log N \ll_{p,q,s} N^q$$

if $N \gg_{p,q,s} 1$, and the required estimate (5.20.2) follows.

6.18. We now show that the exponent pair (6.17.4) is better than any pair (α, β) obtainable by the A and B processes from $(0, 1)$, if $m \geqslant 10^6$. By this we mean that there is no pair (α, β) with both $p \geqslant \alpha$ and $q \geqslant \beta$. To do this we shall show that

$$\beta + 5\alpha^{\frac{3}{4}} \geqslant 1. \tag{6.18.1}$$

Then, since $5^4 25 m^2 \log m < (m-2)^3$ for $m \geqslant 10^6$, we have $q + 5p^{\frac{3}{4}} < 1$, and the result will follow. Certainly (6.18.1) holds for $(0, 1)$. Thus it suffices to prove (6.18.1) by induction on the number of A and B processes needed to obtain (α, β). Since $B^2(\alpha, \beta) = (\alpha, \beta)$ and $A(0, 1) = (0, 1)$, we may suppose that either $(\alpha, \beta) = A(\gamma, \delta)$ or $(\alpha, \beta) = BA(\gamma, \delta)$, where (γ, δ) satisfies (6.18.1). In the former case we have

$$\beta + 5\alpha^{\frac{3}{4}} = \frac{\gamma + \delta + 1}{2\gamma + 2} + 5\left(\frac{\gamma}{2\gamma + 2}\right)^{\frac{3}{4}} \geqslant \frac{\gamma + 2 - 5\gamma^{\frac{3}{4}}}{2\gamma + 2} + 5\left(\frac{\gamma}{2\gamma + 2}\right)^{\frac{3}{4}} \geqslant 1$$

for $0 \leqslant \gamma \leqslant \frac{1}{2}$, and in the latter case

$$\beta + 5\alpha^{\frac{3}{4}} = \frac{2\gamma + 1}{2\gamma + 2} + 5\left(\frac{\delta}{2\gamma + 2}\right)^{\frac{3}{4}} \geqslant \frac{2\gamma + 1}{2\gamma + 2} + 5\left(\frac{\frac{1}{2}}{2\gamma + 2}\right)^{\frac{3}{4}} \geqslant 1$$

for $0 \leqslant \gamma \leqslant \frac{1}{2}$. This completes the proof of our assertion.

The exponent pairs (6.17.4) are not likely to be useful in practice. The purpose of the above analysis is to show that Lemma 6.13 is sufficiently general to apply to any function for which the exponent pairs method can be used, and that there do exist exponent pairs not obtainable by the A and B processes.

6.19. Different ways of using $J(q, l)$ to estimate exponential sums have been given by Korobov [**1**] and Vinogradov [**1**] (see Walfisz [**1**; Chapter 2] for an alternative exposition). These methods require more information about f than a bound (6.12.1) for a single derivative, and so we shall give the result for partial sums of the zeta-function only. The two methods give qualitatively similar estimates, but Vinogradov's is slightly simpler, and is quantitatively better. Vinogradov's result, as given by Walfisz [**1**], is

$$\sum_{a < n \leqslant b} n^{-it} \ll a^{1-\rho} \tag{6.19.1}$$

for $a < b \leqslant 2a$, $t \geqslant 1$, where

$$t^{1/k} \leqslant a \leqslant t^{1/(k-1)},$$

$k \geqslant 19$, and

$$\rho = \frac{1}{60000k^2}.$$

The implied constant is absolute. Richert [3] has used this to show that

$$\zeta(\sigma + it) \ll (1 + t^{100(1-\sigma)^{\frac{3}{2}}})(\log t)^{\frac{2}{3}}, \tag{6.19.2}$$

uniformly for $0 \leqslant \sigma \leqslant 2$, $t \geqslant 2$. The choices

$$\theta(t) = \left(\frac{\operatorname{loglog} t}{100 \log t}\right)^{\frac{2}{3}}, \quad \phi(t) = \operatorname{loglog} t$$

in Theorems 3.10 and 3.11 therefore give a region

$$\sigma \geqslant 1 - A(\log t)^{-\frac{2}{3}}(\operatorname{loglog} t)^{-\frac{1}{3}}$$

free of zeros, and in which

$$\frac{\zeta'(s)}{\zeta(s)} \ll (\log t)^{\frac{2}{3}}(\operatorname{loglog} t)^{\frac{1}{3}},$$

$$\frac{1}{\zeta(s)} \ll (\log t)^{\frac{2}{3}}(\operatorname{loglog} t)^{\frac{1}{3}}.$$

The superiority of (6.19.1) over Lemma 6.13 lies mainly in the elimination of the term $\exp(33k^2 \log k)$, rather than in the improvement in the exponent ρ.

Various authors have reduced the constant 100 in (6.19.2), and the best result to date appears to be one in which 100 is replaced by 18.8 (Heath-Brown, unpublished).

6.20. We shall sketch the proof of Vinogradov's bound. The starting point is an estimate of the form (6.12.4), but with

$$\sum_{u,\,v=1}^{q} e^{2\pi i\{f(uv+n)-f(n)\}} \tag{6.20.1}$$

in place of $T(n)$. One replaces $f(uv+n) - f(n)$ by a polynomial

$$F(uv) = A_1 uv + \ldots + A_k u^k v^k$$

as in §6.12, and then uses Hölder's inequality to obtain

$$\left|\sum e^{2\pi iF(uv)}\right|^l \leqslant q^{l-1}\sum_v\left|\sum_u e^{2\pi iF(uv)}\right|^l$$

$$= q^{l-1}\sum_v \eta(v)\left(\sum_u e^{2\pi iF(uv)}\right)^l$$

$$= q^{l-1}\sum_{\sigma_1,\ldots,\sigma_k} n(\sigma_1,\ldots,\sigma_k)\sum_v \eta(v)e^{2\pi iG(\sigma_1,\ldots,\sigma_k;v)},$$

where $|\eta(v)| = 1$, $n(\sigma_1,\ldots,\sigma_k)$ denotes the number of solutions of

$$u_1{}^h + \ldots + u_l{}^h = \sigma_h \quad (1 \leqslant h \leqslant k),$$

and

$$G(\sigma_1,\ldots,\sigma_k;v) = A_1\sigma_1 v + \ldots + A_k\sigma_k v^k.$$

Now, by Hölder's inequality again, one has

$$\left|\sum e^{2\pi iF(uv)}\right|^{2l^2} \leqslant q^{2l(l-1)}\left(\sum n(\sigma_1,\ldots,\sigma_k)\right)^{2l-2}\times\left(\sum n(\sigma_1,\ldots,\sigma_k)^2\right)$$

$$\times\left(\sum_{\sigma_1,\ldots,\sigma_k}\left|\sum_v \eta(v)e^{2\pi iG(\sigma_1,\ldots,\sigma_k;v)}\right|^{2l}\right).$$

Here

$$\sum_{\sigma_1,\ldots,\sigma_k} n(\sigma_1,\ldots,\sigma_k) = q^l,$$

and

$$\sum_{\sigma_1,\ldots,\sigma_k} n(\sigma_1,\ldots,\sigma_k)^2 = J(q,l).$$

Moreover

$$\sum_{\sigma_1,\ldots,\sigma_k}\left|\sum_v \eta(v)e^{2\pi iG}\right|^{2l} = \sum_{\tau_1,\ldots,\tau_k} n^*(\tau_1,\ldots,\tau_k)\sum_{\sigma_1,\ldots,\sigma_k} e^{2\pi iH(\sigma_1,\ldots,\sigma_k;\tau_1,\ldots,\tau_k)},$$

where

$$H(\sigma_1,\ldots,\sigma_k;\tau_1,\ldots,\tau_k) = A_1\sigma_1\tau_1 + \ldots + A_k\sigma_k\tau_k,$$

and $n^*(\tau_1,\ldots,\tau_k)$ is the sum of $\eta(v_1)\ldots\eta(v_{2l})$ subject to

$$v_1{}^h + \ldots + v_l{}^h - v_{l+1}{}^h - \ldots - v_{2l}{}^h = \tau_h \quad (1 \leqslant h \leqslant k).$$

Since $|n^*(\tau_1,\ldots,\tau_k)| \leqslant J(q,l)$, it follows that

$$\left|\sum e^{2\pi i F(uv)}\right|^{2l^2} \leqslant q^{4l^2-4l} J(q,l)^2 \prod_{h=1}^{k} \left(\sum_{\tau_h} \left| \sum_{\sigma_h} \exp\left(2\pi i A_h \sigma_h \tau_h\right) \right| \right)$$

$$\leqslant q^{4l^2-4l} J(q,l)^2 \prod_{h=1}^{k} \left(\sum_{\tau_h} \min\left(lq^h, |\csc \pi A_h \tau_n|\right) \right).$$

At this point one estimates the sum over τ_h, getting a non-trivial bound whenever $q^{-2h} \ll |A_h| \ll 1$. This leads to an appropriate result for the original sum (6.20.1), on taking $l = [ck^2]$ with a suitable constant c. If we use Lemma 6.9, for example, to estimate $J(q, l)$, then

$$(K^{2r})^{(2l^2)^{-1}} \ll 1.$$

One therefore sees that the implied constant in (6.19.1) is indeed independent of k.

VII

MEAN-VALUE THEOREMS

7.1. The problem of the order of $\zeta(s)$ in the critical strip is, as we have seen, unsolved. The problem of the average order, or mean-value, is much easier, and, in its simplest form, has been solved completely. The form which it takes is that of determining the behaviour of

$$\frac{1}{T}\int_1^T |\zeta(\sigma+it)|^2\,dt$$

as $T \to \infty$, for any given value of σ. We also consider mean values of other powers of $\zeta(s)$.

Results of this kind have applications in the problem of the zeros, and also in problems in the theory of numbers. They could also be used to prove O-results if we could push them far enough; and they are closely connected with the Ω-results which are the subject of the next chapter.

We begin by recalling a general mean-value theorem for Dirichlet series.

THEOREM 7.1. *Let*

$$f(s) = \sum_{n=1}^{\infty} \frac{a_n}{n^s}, \qquad g(s) = \sum_{n=1}^{\infty} \frac{b_n}{n^s}$$

be absolutely convergent for $\sigma > \sigma_1$, $\sigma > \sigma_2$ *respectively. Then for* $\alpha > \sigma_1$, $\beta > \sigma_2$,

$$\lim_{T\to\infty} \frac{1}{2T}\int_{-T}^{T} f(\alpha+it)g(\beta-it)\,dt = \sum_{n=1}^{\infty} \frac{a_n b_n}{n^{\alpha+\beta}}. \tag{7.1.1}$$

For

$$f(\alpha+it)g(\beta-it) = \sum_{m=1}^{\infty} \frac{a_m}{m^{\alpha+it}} \sum_{n=1}^{\infty} \frac{b_n}{n^{\beta-it}} = \sum_{n=1}^{\infty} \frac{a_n b_n}{n^{\alpha+\beta}} + \sum\sum_{m\neq n} \frac{a_m b_n}{m^{\alpha} n^{\beta}}\left(\frac{n}{m}\right)^{it},$$

the series being absolutely convergent, and uniformly convergent in any finite t-range. Hence we may integrate term-by-term, and obtain

$$\frac{1}{2T}\int_{-T}^{T} f(\alpha+it)g(\beta-it)\,dt = \sum_{n=1}^{\infty} \frac{a_n b_n}{n^{\alpha+\beta}} + \sum\sum_{m\neq n} \frac{a_m b_n}{m^{\alpha} n^{\beta}} \frac{2\sin(T\log n/m)}{2T\log n/m}.$$

The factor involving T is bounded for all T, m, and n, so that the double series converges uniformly with respect to T; and each term tends to zero as $T \to \infty$. Hence the sum also tends to zero, and the result follows.

In particular, taking $b_n = \bar{a}_n$ *and* $\alpha = \beta = \sigma$, *we obtain*

$$\lim_{T\to\infty}\frac{1}{2T}\int_{-T}^{T}|f(\sigma+it)|^2\,dt = \sum_{n=1}^{\infty}\frac{|a_n|^2}{n^{2\sigma}} \quad (\sigma > \sigma_1). \tag{7.1.2}$$

These theorems have immediate applications to $\zeta(s)$ in the half-plane $\sigma > 1$. We deduce at once

$$\lim_{T\to\infty}\frac{1}{2T}\int_{-T}^{T}|\zeta(\sigma+it)|^2\,dt = \zeta(2\sigma) \quad (\sigma > 1), \tag{7.1.3}$$

and generally

$$\lim_{T\to\infty}\frac{1}{2T}\int_{-T}^{T}\zeta^{(\mu)}(\alpha+it)\zeta^{(\nu)}(\beta-it)\,dt = \zeta^{(\mu+\nu)}(\alpha+\beta) \quad (\alpha > 1, \beta > 1). \tag{7.1.4}$$

Taking $a_n = d_k(n)$, we obtain

$$\lim_{T\to\infty}\frac{1}{2T}\int_{-T}^{T}|\zeta(\sigma+it)|^{2k}\,dt = \sum_{n=1}^{\infty}\frac{d_k^2(n)}{n^{2\sigma}} \quad (\sigma > 1). \tag{7.1.5}$$

By (1.2.10), the case $k = 2$ is

$$\lim_{T\to\infty}\frac{1}{2T}\int_{-T}^{T}|\zeta(\sigma+it)|^4\,dt = \frac{\zeta^4(2\sigma)}{\zeta(4\sigma)} \quad (\sigma > 1). \tag{7.1.6}$$

The following sections are mainly concerned with the attempt to extend these formulae to values of σ less than or equal to 1. The attempt is successful for $k \leqslant 2$, only partially successful for $k > 2$.

7.2. We require the following lemmas.

LEMMA. *We have*

$$\sum_{0<m<n<T}\sum\frac{1}{m^\sigma n^\sigma\log n/m} = O(T^{2-2\sigma}\log T) \tag{7.2.1}$$

for $\frac{1}{2} \leqslant \sigma < 1$, *and uniformly for* $\frac{1}{2} \leqslant \sigma \leqslant \sigma_0 < 1$.

Let Σ_1 denote the sum of the terms for which $m < \frac{1}{2}n$, Σ_2 the remainder. In Σ_1, $\log n/m > A$, so that

$$\Sigma_1 < A\sum_{m<n<T}\sum m^{-\sigma}n^{-\sigma} < A\Big(\sum_{n<T}n^{-\sigma}\Big)^2 < AT^{2-2\sigma}.$$

In Σ_2 we write $m = n-r$, where $1 \leqslant r \leqslant \frac{1}{2}n$, and then

$$\log n/m = -\log(1-r/n) > r/n.$$

Hence

$$\Sigma_2 < A\sum_{n<T}\sum_{r\leqslant\frac{1}{2}n}\frac{(n-r)^{-\sigma}n^{-\sigma}}{r/n} < A\sum_{n<T}n^{1-2\sigma}\sum_{r\leqslant\frac{1}{2}n}\frac{1}{r} < AT^{2-2\sigma}\log T.$$

LEMMA. $$\sum_{0<m<n<\infty}\sum \frac{e^{-(m+n)\delta}}{m^{\sigma}n^{\sigma}\log n/m} = O\left(\delta^{2\sigma-2}\log\frac{1}{\delta}\right). \tag{7.2.2}$$

Dividing up as before, we obtain

$$\Sigma_1 = O\left[\left(\sum_1^{\infty} n^{-\sigma}e^{-\delta n}\right)^2\right] = O(\delta^{2\sigma-2}),$$

and $$\Sigma_2 = O\left(\sum_{n=1}^{\infty} n^{1-2\sigma}e^{-\delta n}\sum_{r=1}^{\frac{1}{2}n}\frac{1}{r}\right) = O\left(\delta^{2\sigma-2}\log\frac{1}{\delta}\right).$$

THEOREM 7.2.

$$\lim_{T\to\infty}\frac{1}{T}\int_1^T |\zeta(\sigma+it)|^2\,dt = \zeta(2\sigma) \quad (\sigma > \tfrac{1}{2}).$$

We have already accounted for the case $\sigma > 1$, so that we now suppose that $\frac{1}{2} < \sigma \leqslant 1$. Since $t \geqslant 1$, Theorem 4.11, with $x = t$, gives

$$\zeta(s) = \sum_{n<t} n^{-s} + O(t^{-\sigma}) = Z + O(t^{-\sigma}),$$

say. Now

$$\begin{aligned}\int_1^T |Z|^2\,dt &= \int_1^T \Big[\sum_{m<t} m^{-\sigma-it}\sum_{n<t} n^{-\sigma+it}\Big]\,dt \\ &= \sum_{m<T}\sum_{n<T} m^{-\sigma}n^{-\sigma}\int_{T_1}^T \left(\frac{n}{m}\right)^{it}dt \quad (T_1 = \max(m,n)) \\ &= \sum_{n<T} n^{-2\sigma}(T-n) + \sum_{m\neq n}\sum m^{-\sigma}n^{-\sigma}\frac{(n/m)^{iT}-(n/m)^{iT_1}}{i\log n/m} \\ &= T\sum_{n<T} n^{-2\sigma} - \sum_{n<T} n^{1-2\sigma} + O\left(\sum_{0<m<n<T}\sum \frac{1}{m^{\sigma}n^{\sigma}\log n/m}\right) \\ &= T\{\zeta(2\sigma)+O(T^{1-2\sigma})\} + O(T^{2-2\sigma}) + O(T^{2-2\sigma}\log T),\end{aligned}$$

provided that $\sigma < 1$. If $\sigma = 1$, we can replace the σ of the last two terms by $\frac{3}{4}$, say. In either case

$$\int_1^T |Z|^2\,dt \sim T\,\zeta(2\sigma).$$

Hence

$$\begin{aligned}\int_1^T |\zeta(s)|^2\,dt &= \int_1^T |Z|^2\,dt + O\left(\int_1^T |Z|t^{-\sigma}\,dt\right) + O\left(\int_1^T t^{-2\sigma}\,dt\right) \\ &= \int_1^T |Z|^2\,dt + O\left(\int_1^T |Z|^2\,dt\int_1^T t^{-2\sigma}\,dt\right)^{\frac{1}{2}} + O(\log T) \\ &= \int_1^T |Z|^2\,dt + O\{(T\log T)^{\frac{1}{2}}\} + O(\log T),\end{aligned}$$

and the result follows.

It will be useful later to have a result of this type which holds uniformly in the strip. It is†

THEOREM 7.2 (A).

$$\int_1^T |\zeta(\sigma+it)|^2\,dt < AT\min\left(\log T, \frac{1}{\sigma-\frac{1}{2}}\right)$$

uniformly for $\frac{1}{2} \leqslant \sigma \leqslant 2$.

Suppose first that $\frac{1}{2} \leqslant \sigma \leqslant \frac{3}{4}$. Then we have, as before,

$$\int_1^T |Z|^2\,dt < T\sum_{n<T} n^{-2\sigma}+O(T^{2-2\sigma}\log T)$$

uniformly in σ. Now

$$\sum_{n<T} n^{-2\sigma} \leqslant \sum_{n<T} n^{-1} < A\log T$$

and also
$$\leqslant 1+\int_1^\infty u^{-2\sigma}\,du < \frac{A}{\sigma-\frac{1}{2}}.$$

Similarly
$$T^{2-2\sigma}\log T \leqslant T\log T,$$

and also, putting $x = (2\sigma-1)\log T$,

$$T^{2-2\sigma}\log T = \tfrac{1}{2}Txe^{-x}/(\sigma-\tfrac{1}{2}) \leqslant \tfrac{1}{2}T/(\sigma-\tfrac{1}{2}).$$

This gives the result for $\sigma \leqslant \frac{3}{4}$, the term $O(t^{-\sigma})$ being dealt with as before.

If $\frac{3}{4} \leqslant \sigma \leqslant 2$, we obtain

$$\int_1^T |Z|^2\,dt < T\sum_{n<T} n^{-\frac{3}{2}}+O(T^{\frac{1}{2}}\log T),$$

and the result follows at once.

7.3. The particular case $\sigma = \frac{1}{2}$ of the above theorem is

$$\int_1^T |\zeta(\tfrac{1}{2}+it)|^2\,dt = O(T\log T).$$

We can improve this O-result to an asymptotic equality.‡ But Theorem 4.11 is not sufficient for this purpose, and we have to use the approximate functional equation.

THEOREM 7.3. *As* $T\to\infty$

$$\int_0^T |\zeta(\tfrac{1}{2}+it)|^2\,dt \sim T\log T.$$

† Littlewood (4). ‡ Hardy and Littlewood (2), (4).

In the approximate functional equation (4.12.4), take $\sigma = \frac{1}{2}$, $t > 2$, and $x = t/(2\pi\sqrt{\log t})$, $y = \sqrt{\log t}$. Then, since $\chi(\frac{1}{2}+it) = O(1)$,

$$\zeta(\tfrac{1}{2}+it) = \sum_{n<x} n^{-\frac{1}{2}-it} + O\Big(\sum_{n<y} n^{-\frac{1}{2}}\Big) + O(t^{-\frac{1}{2}}\log^{\frac{1}{4}} t) + O(\log^{-\frac{1}{4}} t)$$

$$= \sum_{n<x} n^{-\frac{1}{2}-it} + O(\log^{\frac{1}{4}} t)$$

$$= Z + O(\log^{\frac{1}{4}} t),$$

say. Since $$\int_2^T (\log^{\frac{1}{4}} t)^2\, dt = O(T\log^{\frac{1}{2}} T) = o(T\log T),$$

it is, as in the proof of Theorem 7.2, sufficient to prove that

$$\int_0^T |Z|^2\, dt \sim T\log T.$$

Now $$\int_0^T |Z|^2\, dt = \int_0^T \sum_{m<x} m^{-\frac{1}{2}-it} \sum_{n<x} n^{-\frac{1}{2}+it}\, dt.$$

In inverting the order of integration and summation, it must be remembered that x is a function of t. The term in (m, n) occurs if

$$x > \max(m, n) = T_1/(2\pi\sqrt{\log T_1})$$

say, where $T_1 = T_1(m, n)$. Hence, writing $X = T/(2\pi\sqrt{\log T})$,

$$\int_0^T |Z|^2\, dt = \sum_{m,n<X}\sum \int_{T_1}^T m^{-\frac{1}{2}-it} n^{-\frac{1}{2}+it}\, dt$$

$$= \sum_{n<X} \frac{T - T_1(n, n)}{n} + \sum_{m \neq n}\sum \frac{1}{\sqrt{(mn)}} \int_{T_1}^T \left(\frac{n}{m}\right)^{it} dt$$

$$= T\sum_{n<X} \frac{1}{n} + O\bigg(\sum_{n<X} \frac{T_1(n, n)}{n}\bigg) + O\bigg(\sum_{m<n<X}\sum \frac{1}{\sqrt{(mn)}\log n/m}\bigg).$$

The first term is

$$T\log X + O(T) = T\log T + o(T\log T).$$

The second term is

$$O\Big(\sum_{n<X} \sqrt{\log n}\Big) = O(X\sqrt{\log X}) = O(T),$$

and, by the first lemma of § 7.2, the last term is

$$O(X\log X) = O(T\sqrt{\log T}).$$

This proves the theorem.

7.4. We shall next obtain a more precise form of the above mean-value formula.†

THEOREM 7.4.

$$\int_0^T |\zeta(\tfrac{1}{2}+it)|^2\,dt = T\log T+(2\gamma-1-\log 2\pi)T+O(T^{\frac{1}{2}+\epsilon}). \quad (7.4.1)$$

We first prove the following lemma.

LEMMA. *If* $n < T/2\pi$,

$$\frac{1}{2\pi i}\int_{\frac{1}{2}-iT}^{\frac{1}{2}+iT} \chi(1-s)n^{-s}\,ds = 2+O\left(\frac{1}{n^{\frac{1}{2}}\log(T/2\pi n)}\right)+O\left(\frac{\log T}{n^{\frac{1}{2}}}\right). \quad (7.4.2)$$

If $n > T/2\pi$, $c > \frac{1}{2}$,

$$\frac{1}{2\pi i}\int_{c-iT}^{c+iT} \chi(1-s)n^{-s}\,ds = O\left(\frac{T^{c-\frac{1}{2}}}{n^c\log(2\pi n/T)}\right)+O\left(\frac{T^{c-\frac{1}{2}}}{n^c}\right). \quad (7.4.3)$$

We have

$$\chi(1-s) = 2^{1-s}\pi^{-s}\cos\tfrac{1}{2}s\pi\Gamma(s) = \frac{2^{1-s}\pi^{1-s}}{2\sin\frac{1}{2}s\pi\Gamma(1-s)}.$$

This has poles at $s = -2\nu$ ($\nu = 0, 1,...$) with residues

$$\frac{(-1)^\nu 2^{1+2\nu}\pi^{2\nu}}{(2\nu)!}.$$

Also, by Stirling's formula, for $-\pi+\delta < \arg(-s) < \pi-\delta$

$$\chi(1-s) = \left(\frac{2\pi}{-s}\right)^{\frac{1}{2}-s}\frac{e^{-s}}{2\sin\frac{1}{2}s\pi}\left\{1+O\left(\frac{1}{|s|}\right)\right\}.$$

The calculus of residues therefore gives

$$\frac{1}{2\pi i}\left(\int_{-\infty-iT_1}^{\frac{1}{2}-iT_1}+\int_{\frac{1}{2}-iT_1}^{\frac{1}{2}+iT_1}+\int_{\frac{1}{2}+iT_1}^{-\infty+iT_1}\right)\chi(1-s)n^{-s}\,ds$$
$$= \sum_{\nu=0}^{\infty}\frac{(-1)^\nu 2^{1+2\nu}\pi^{2\nu}n^{2\nu}}{(2\nu)!}$$
$$= 2\cos 2\pi n = 2.$$

Also, since $$e^{is\arg(-s)} = O(e^{\frac{1}{2}\pi t}),$$

† Ingham (1) obtained the error term $O(T^{\frac{1}{2}}\log T)$; the method given here is due to Atkinson (1).

$$\int_{\frac{1}{2}+iT_1}^{-\infty+iT_1} \chi(1-s)n^{-s}\,ds = O\left\{\int_{-\infty}^{\frac{1}{2}} \left(\frac{2\pi}{|\sigma+iT_1|}\right)^{\frac{1}{2}-\sigma} e^{-\sigma}n^{-\sigma}\,d\sigma\right\}$$

$$= O\left\{n^{-\frac{1}{2}} \int_{-\infty}^{\frac{1}{2}} \left(\frac{T_1}{2\pi en}\right)^{\sigma-\frac{1}{2}} d\sigma\right\} = O\left(\frac{1}{n^{\frac{1}{2}}\log(T_1/2\pi en)}\right),$$

and similarly for the integral over $(-\infty-iT_1, \frac{1}{2}-iT_1)$.

Again, for a fixed σ,

$$\chi(1-s) = \left(\frac{2\pi}{t}\right)^{\frac{1}{2}-\sigma-it} e^{-it-\frac{1}{4}i\pi}\left\{1+O\left(\frac{1}{t}\right)\right\} \quad (t \geqslant 1).$$

Hence

$$\int_{\frac{1}{2}+iT}^{\frac{1}{2}+iT_1} \chi(1-s)n^{-s}\,ds = n^{-\frac{1}{2}}e^{-\frac{1}{4}i\pi} \int_{T}^{T_1} e^{iF(t)}\,dt+O(n^{-\frac{1}{2}}\log T_1),$$

where

$$F(t) = t\log t-t(\log 2\pi+1+\log n),$$

$$F'(t) = \log t-\log 2\pi n.$$

Hence by Lemma 4.2, the last integral is of the form

$$O\left(\frac{1}{\log(T/2\pi n)}\right)$$

uniformly with respect to T_1. Taking, for example, $T_1 = 2eT > 4\pi en$, we obtain (7.4.2). Again

$$\int_{c+i}^{c+iT} \chi(1-s)n^{-s}\,ds = n^{-c}e^{-\frac{1}{4}i\pi} \int_{1}^{T} \left(\frac{t}{2\pi}\right)^{c-\frac{1}{2}} e^{iF(t)}\,dt+O\left(n^{-c}\int_{1}^{T} t^{c-\frac{3}{2}}\,dt\right),$$

and (7.4.3) follows from Lemma 4.3.

In proving (7.4.1) we may suppose that $T/2\pi$ is half an odd integer; for a change of $O(1)$ in T alters the left-hand side by $O(T^{\frac{1}{2}})$, since $\zeta(\frac{1}{2}+it) = O(t^{\frac{1}{4}})$, and the leading terms on the right-hand side by $O(\log T)$. Now the left-hand side is

$$\frac{1}{2}\int_{-T}^{T} |\zeta(\tfrac{1}{2}+it)|^2\,dt = \frac{1}{2}\int_{-T}^{T} \zeta(\tfrac{1}{2}+it)\zeta(\tfrac{1}{2}-it)\,dt$$

$$= \frac{1}{2i}\int_{\frac{1}{2}-iT}^{\frac{1}{2}+iT} \zeta(s)\zeta(1-s)\,ds = \frac{1}{2i}\int_{\frac{1}{2}-iT}^{\frac{1}{2}+iT} \chi(1-s)\zeta^2(s)\,ds$$

$$= \frac{1}{2i}\int_{\frac{1}{2}-iT}^{\frac{1}{2}+iT} \chi(1-s)\sum_{n\leqslant T/2\pi} \frac{d(n)}{n^s}\,ds+\frac{1}{2i}\int_{\frac{1}{2}-iT}^{\frac{1}{2}+iT} \chi(1-s)\left(\zeta^2(s)-\sum_{n\leqslant T/2\pi}\frac{d(n)}{n^s}\right)ds$$

$$= I_1+I_2, \text{ say.}$$

By (7.4.2),

$$I_1 = 2\pi \sum_{n \leqslant T/2\pi} d(n) + O\left(\sum_{n \leqslant T/2\pi} \frac{d(n)}{n^{\frac{1}{2}} \log(T/2\pi n)}\right) + O\left(\log T \sum_{n \leqslant T/2\pi} \frac{d(n)}{n^{\frac{1}{2}}}\right).$$

The first term is†

$$2\pi\left\{\frac{T}{2\pi}\log\frac{T}{2\pi} + (2\gamma-1)\frac{T}{2\pi} + O(T^{\frac{1}{2}})\right\}$$
$$= T\log T + (2\gamma-1-\log 2\pi)T + O(T^{\frac{1}{2}}).$$

Since‡ $d(n) = O(n^{\epsilon})$, the second term is

$$O\left(\sum_{n \leqslant T/4\pi} \frac{1}{n^{\frac{1}{2}-\epsilon}}\right) + O\left\{T^{\frac{1}{2}+\epsilon} \sum_{T/4\pi < n \leqslant T/2\pi} \frac{1}{(T/2\pi)-n}\right\} = O(T^{\frac{1}{2}+\epsilon}).$$

The last term is also clearly of this form. Hence

$$I_1 = T\log T + (2\gamma-1-\log 2\pi)T + O(T^{\frac{1}{2}+\epsilon}).$$

Next, if $c > 1$,

$$I_2 = \frac{1}{2i}\left(\int_{\frac{1}{2}-iT}^{c-iT} + \int_{c+iT}^{\frac{1}{2}+iT}\right)\chi(1-s)\left(\zeta^2(s) - \sum_{n \leqslant T/2\pi} \frac{d(n)}{n^s}\right) ds +$$
$$+ \frac{1}{2i} \sum_{n > T/2\pi} d(n) \int_{c-iT}^{c+iT} \chi(1-s) n^{-s}\, ds - A,$$

A being the residue of $\pi\chi(1-s)\zeta^2(s)$ at $s = 1$.

Since $\chi(1-s) = O(t^{\sigma-\frac{1}{2}})$, and $\zeta^2(\sigma+iT)$ and $\sum_{n \leqslant T/2\pi} d(n)n^{-s}$ are both of the form

$$O(T^{1-\sigma+\epsilon}) \quad (\sigma \leqslant 1), \qquad O(T^{\epsilon}) \quad (\sigma > 1),$$

the first term is

$$O(T^{\frac{1}{2}+\epsilon}) + O(T^{c-\frac{1}{2}+\epsilon}).$$

By (7.4.3), the second term is

$$O\left\{T^{c-\frac{1}{2}} \sum_{n > T/2\pi} \frac{d(n)}{n^c}\left(\frac{1}{\log(2\pi n/T)} + 1\right)\right\}$$
$$= O\left(T^{\frac{1}{2}+\epsilon} \sum_{T/2\pi < n \leqslant T/\pi} \frac{1}{n-(T/2\pi)}\right) + O\left(T^{c-\frac{1}{2}} \sum_{n > T/\pi} \frac{1}{n^{c-\epsilon}}\right)$$
$$= O(T^{\frac{1}{2}+\epsilon}).$$

Since c may be as near to 1 as we please, this proves the theorem.

A more precise form of the above argument shows that the error-term in (7.4.1) is $O(T^{\frac{1}{2}}\log^2 T)$. But a more complicated argument,§

† See § 12.1, or Hardy and Wright, *An Introduction to the Theory of Numbers*, Theorem 320.

‡ Ibid. Theorem 315.

§ Titchmarsh (12).

depending on van der Corput's method, shows that it is $O(T^{\frac{5}{12}}\log^2 T)$; and presumably further slight improvements could be made by the methods of the later sections of Chapter V.

7.5. We now pass to the more difficult, but still manageable, case of $|\zeta(s)|^4$. We first prove†

THEOREM 7.5.

$$\lim_{T\to\infty}\frac{1}{T}\int_1^T|\zeta(\sigma+it)|^4\,dt = \frac{\zeta^4(2\sigma)}{\zeta(4\sigma)} \quad (\sigma > \tfrac{1}{2}).$$

Take $x = y = \sqrt{(t/2\pi)}$ and $\sigma > \frac{1}{2}$ in the approximate functional equation. We obtain

$$\zeta(s) = \sum_{n<\sqrt{(t/2\pi)}}\frac{1}{n^s}+\chi(s)\sum_{n<\sqrt{(t/2\pi)}}\frac{1}{n^{1-s}}+O(t^{-\frac{1}{4}}) = Z_1+Z_2+O(t^{-\frac{1}{4}}), \tag{7.5.1}$$

say. Now

$$|Z_1|^4 = \sum\frac{1}{m^{\sigma+it}}\sum\frac{1}{n^{\sigma+it}}\sum\frac{1}{\mu^{\sigma-it}}\sum\frac{1}{\nu^{\sigma-it}}$$
$$= \sum\frac{1}{(mn\mu\nu)^\sigma}\left(\frac{\mu\nu}{mn}\right)^{it},$$

where each variable runs over $\{1, \sqrt{(t/2\pi)}\}$. Hence

$$\int_1^T|Z_1|^4\,dt = \int_1^T\sum\frac{1}{(mn\mu\nu)^\sigma}\left(\frac{\mu\nu}{mn}\right)^{it}dt$$
$$= \sum_{m,n,\mu,\nu<\sqrt{(T/2\pi)}}\frac{1}{(mn\mu\nu)^\sigma}\int_{T_1}^T\left(\frac{\mu\nu}{mn}\right)^{it}dt,$$

where $T_1 = 2\pi\max(m^2, n^2, \mu^2, \nu^2)$

$$= \sum_{mn=\mu\nu}\frac{T-T_1}{(mn)^{2\sigma}}+\sum_{mn\neq\mu\nu}O\left(\frac{1}{(mn\mu\nu)^\sigma}\frac{1}{|\log(\mu\nu/mn)|}\right).$$

The number of solutions of the equations $mn = \mu\nu = r$ is $\{d(r)\}^2$ if $r < \sqrt{(T/2\pi)}$, and in any case does not exceed $\{d(r)\}^2$. Hence

$$T\sum_{mn=\mu\nu}\frac{1}{(mn)^{2\sigma}} = T\sum_{r<\sqrt{(T/2\pi)}}\frac{\{d(r)\}^2}{r^{2\sigma}}+O\left(T\sum_{\sqrt{(T/2\pi)}\leqslant r<T/2\pi}\frac{\{d(r)\}^2}{r^{2\sigma}}\right)$$
$$\sim T\sum_{r=1}^{\infty}\frac{\{d(r)\}^2}{r^{2\sigma}} = T\frac{\zeta^4(2\sigma)}{\zeta(4\sigma)}. \tag{7.5.2}$$

† Hardy and Littlewood (4).

Next $$\sum_{mn=\mu\nu} \frac{T_1}{(mn)^{2\sigma}} < \sum_{mn=\mu\nu} \frac{2\pi(m^2+n^2+\mu^2+\nu^2)}{(mn\mu\nu)^\sigma},$$
and the right-hand side, by considerations of symmetry, is
$$8\pi \sum_{mn=\mu\nu} \frac{m^2}{(mn\mu\nu)^\sigma} \leqslant 8\pi \sum \frac{m^2\, d(mn)}{(mn)^{2\sigma}} = O(T^\epsilon \sum m^{2-2\sigma} \sum n^{-2\sigma})$$
$$= O\{T^\epsilon(T^{\frac{1}{2}(3-2\sigma)}+1)\log T\} = O(T^{\frac{3}{2}-\sigma+\epsilon})+O(T^\epsilon).$$
The remaining sum is
$$O\left(\sum_{0<q<r<T/2\pi} \frac{d(q)d(r)}{(qr)^\sigma \log(r/q)}\right) = O\left(T^\epsilon \sum \frac{1}{(qr)^\sigma \log(r/q)}\right) = O(T^{2-2\sigma+\epsilon}),$$
by the lemma of § 7.2. Hence
$$\int_1^T |Z_1|^4\, dt \sim T\frac{\zeta^4(2\sigma)}{\zeta(4\sigma)}.$$
Now let $$j(T) = \int_1^T \Big|\sum_{n<\surd(t/2\pi)} n^{s-1}\Big|^4\, dt.$$
The calculations go as before, but with σ replaced by $1-\sigma$. The term corresponding to (7.5.2) is
$$T\sum_{r<AT} \frac{O(r^\epsilon)}{r^{2-2\sigma}} = O(T^{2\sigma+\epsilon}),$$
and the other two terms are $O(T^{\frac{1}{2}+\sigma+\epsilon})$ and $O(T^{2\sigma+\epsilon})$ respectively. Hence
$$j(T) = O(T^{2\sigma+\epsilon}),$$
and, since $\chi(s) = O(t^{\frac{1}{2}-\sigma})$,
$$\int_1^T |Z_2|^4\, dt < A\int_1^T t^{2-4\sigma} j'(t)\, dt$$
$$= A[t^{2-4\sigma}j(t)]_1^T + A(4\sigma-2)\int_1^T t^{1-4\sigma}j(t)\, dt$$
$$= O(T^{2-2\sigma+\epsilon}) + O\left(\int_1^T t^{1-2\sigma+\epsilon}\, dt\right) = O(T^{2-2\sigma+\epsilon}).$$
The theorem now follows as in previous cases.

7.6. The problem of the mean value of $|\zeta(\frac{1}{2}+it)|^4$ is a little more difficult. If we follow out the above argument, with $\sigma = \frac{1}{2}$, as accurately as possible, we obtain
$$\int_1^T |\zeta(\tfrac{1}{2}+it)|^4\, dt = O(T\log^4 T), \tag{7.6.1}$$

but fail to obtain an asymptotic equality. It was proved by Ingham† by means of the functional equation for $\{\zeta(s)\}^2$ that

$$\int_1^T |\zeta(\tfrac{1}{2}+it)|^4\,dt = \frac{T\log^4 T}{2\pi^2}+O(T\log^3 T). \tag{7.6.2}$$

The relation

$$\int_1^T |\zeta(\tfrac{1}{2}+it)|^4\,dt \sim \frac{T\log^4 T}{2\pi^2} \tag{7.6.3}$$

is a consequence of a result obtained later in this chapter (Theorem 7.16).

7.7. We now pass to still higher powers of $\zeta(s)$. In the general case our knowledge is very incomplete, and we can state a mean-value formula in a certain restricted range of values of σ only.

THEOREM 7.7. *For every positive integer $k > 2$*

$$\lim_{T\to\infty}\frac{1}{T}\int_1^T |\zeta(\sigma+it)|^{2k}\,dt = \sum_{n=1}^{\infty}\frac{d_k^2(n)}{n^{2\sigma}} \quad \left(\sigma > 1-\frac{1}{k}\right). \tag{7.7.1}$$

This can be proved by a straightforward extension of the argument of § 7.5. Starting again from (7.5.1), we have

$$|Z_1|^{2k} = \sum \frac{1}{(m_1 \dots m_k n_1 \dots n_k)^\sigma}\left(\frac{n_1 \dots n_k}{m_1 \dots m_k}\right)^{it},$$

where each variable runs over $\{1, \sqrt{(t/2\pi)}\}$. The leading term goes in the same way as before, $d(r)$ being replaced by $d_k(r)$. The main O-term is of the form

$$O\left(T^\epsilon \sum_{0<q<r<AT^{\frac{1}{2}k}}\sum \frac{1}{(qr)^\sigma \log r/q}\right) = O(T^{k(1-\sigma)+\epsilon}).$$

The corresponding term in

$$j(T) = \int_1^T \left|\sum_{n<\sqrt{(t/2\pi)}} n^{s-1}\right|^{2k} dt$$

is

$$O(T^{k\sigma+\epsilon}),$$

and since $|\chi|^{2k} = O(t^{k-2k\sigma})$, we obtain $O(T^{k(1-\sigma)+\epsilon})$ again. These terms are $o(T)$ if $\sigma > 1-1/k$, and the theorem follows as before.

7.8. It is convenient to introduce at this point the following notation. For each positive integer k and each σ, let $\mu_k(\sigma)$ be the lower bound of positive numbers ξ such that

$$\frac{1}{T}\int_1^T |\zeta(\sigma+it)|^{2k}\,dt = O(T^\xi).$$

† Ingham (1).

Each $\mu_k(\sigma)$ has the same general properties as the function $\mu(\sigma)$ defined in § 5.1. By (7.1.5), $\mu_k(\sigma) = 0$ for $\sigma > 1$. Further, as a function of σ, $\mu_k(\sigma)$ is continuous, non-increasing, and convex downwards. We shall deduce this from a general theorem on mean-values of analytic functions.†

Let $f(s)$ be an analytic function of s, real for real s, regular for $\sigma \geqslant \alpha$ except possibly for a pole at $s = s_0$, and $O(e^{\epsilon|t|})$ as $|t| \to \infty$ for every positive ϵ and $\sigma \geqslant \alpha$. Let $\alpha < \beta$, and suppose that for all $T > 0$

$$\int_0^T |f(\alpha+it)|^2\,dt \leqslant C(T^a+1), \tag{7.8.1}$$

$$\int_0^T |f(\beta+it)|^2\,dt \leqslant C'(T^b+1), \tag{7.8.2}$$

where $a \geqslant 0$, $b \geqslant 0$, and C, C' depend on $f(s)$. Then for $\alpha < \sigma < \beta$, $T \geqslant 2$,

$$\int_{\frac{1}{2}T}^T |f(\sigma+it)|^2\,dt \leqslant K(CT^a)^{(\beta-\sigma)/(\beta-\alpha)}(C'T^b)^{(\sigma-\alpha)/(\beta-\alpha)}, \tag{7.8.3}$$

where K depends on a, b, α, β only, and is bounded if these are bounded.

We may suppose in the proof that $\alpha \geqslant \frac{1}{2}$, since otherwise we could apply the argument to $f(s+\frac{1}{2}-\alpha)$. Suppose first that $f(s)$ is regular for $\sigma \geqslant \alpha$. Let

$$\frac{1}{2\pi i}\int_{\sigma-i\infty}^{\sigma+i\infty} \Gamma(s)f(s)z^{-s}\,ds = \phi(z) \quad (\sigma \geqslant \alpha,\ |\arg z| < \tfrac{1}{2}\pi).$$

Putting $z = ixe^{-i\delta}$ $(0 < \delta < \frac{1}{2}\pi)$, we find that

$$\Gamma(\sigma+it)f(\sigma+it)e^{-i(\sigma+it)(\frac{1}{2}\pi-\delta)}, \qquad \phi(ixe^{-i\delta})$$

are Mellin transforms. Let

$$I(\sigma) = \int_{-\infty}^{\infty} |\Gamma(\sigma+it)f(\sigma+it)|^2 e^{(\pi-2\delta)t}\,dt.$$

Then, using Parseval's formula and Hölder's inequality, we obtain

$$\begin{aligned} I(\sigma) &= 2\pi\int_0^\infty |\phi(ixe^{-i\delta})|^2 x^{2\sigma-1}\,dx \\ &\leqslant 2\pi\left(\int_0^\infty |\phi|^2 x^{2\alpha-1}\,dx\right)^{(\beta-\sigma)/(\beta-\alpha)}\left(\int_0^\infty |\phi|^2 x^{2\beta-1}\,dx\right)^{(\sigma-\alpha)/(\beta-\alpha)} \\ &= \{I(\alpha)\}^{(\beta-\sigma)/(\beta-\alpha)}\{I(\beta)\}^{(\sigma-\alpha)/(\beta-\alpha)}. \end{aligned}$$

† Hardy, Ingham, and Pólya (1), Titchmarsh (23).

Writing $$F(T) = \int_0^T |f(\alpha+it)|^2\,dt \leqslant C(T^a+1)$$

we have by Stirling's theorem (with various values of K)

$$I(\alpha) < K\int_0^\infty (t^{2\alpha-1}+1)|f(\alpha+it)|^2 e^{-2\delta t}\,dt$$

$$= K\int_0^\infty F(t)\{2\delta(t^{2\alpha-1}+1)-(2\alpha-1)t^{2\alpha-2}\}e^{-2\delta t}\,dt$$

$$< KC\int_0^\infty (t^a+1)2\delta(t^{2\alpha-1}+1)e^{-2\delta t}\,dt$$

$$< KC\int_0^\infty (t^{a+2\alpha-1}+1)\delta e^{-2\delta t}\,dt$$

$$= KC\int_0^\infty \left\{\left(\frac{u}{\delta}\right)^{a+2\alpha-1}+1\right\}e^{-2u}\,du$$

$$< KC(\delta^{-a-2\alpha+1}+1) < KC\delta^{-a-2\alpha+1}.$$

Similarly for $I(\beta)$. Hence

$$I(\sigma) < K(C\delta^{-a-2\alpha+1})^{(\beta-\sigma)/(\beta-\alpha)}(C'\delta^{-b-2\beta+1})^{(\sigma-\alpha)/(\beta-\alpha)}$$

$$= K\delta^{-2\sigma+1}(C\delta^{-a})^{(\beta-\sigma)/(\beta-\alpha)}(C'\delta^{-b})^{(\sigma-\alpha)/(\beta-\alpha)}.$$

Also

$$I(\sigma) > K\int_{1/2\delta}^{1/\delta} |f(\sigma+it)|^2 t^{2\sigma-1}\,dt > K\delta^{-2\sigma+1}\int_{1/2\delta}^{1/\delta} |f(\sigma+it)|^2\,dt.$$

Putting $\delta = 1/T$, the result follows.

If $f(s)$ has a pole of order k at s_0, we argue similarly with $(s-s_0)^k f(s)$; this merely introduces a factor T^{2k} on each side of the result, so that (7.8.3) again follows.

Replacing T in (7.8.3) by $\frac{1}{2}T$, $\frac{1}{4}T$,..., and adding, we obtain the result:

If $$\int_0^T |f(\alpha+it)|^2\,dt = O(T^a), \qquad \int_0^T |f(\beta+it)|^2\,dt = O(T^b),$$

then $$\int_0^T |f(\sigma+it)|^2\,dt = O\{T^{\{a(\beta-\sigma)+b(\sigma-\alpha)\}/(\beta-\alpha)}\}.$$

Taking $f(s) = \zeta^k(s)$, the convexity of $\mu_k(\sigma)$ follows.

7.9. An alternative method of dealing with these problems is due to Carlson.† His main result is

THEOREM 7.9. *Let σ_k be the lower bound of numbers σ such that*

$$\frac{1}{T}\int_1^T |\zeta(\sigma+it)|^{2k}\,dt = O(1). \tag{7.9.1}$$

Then
$$\sigma_k \leqslant \max\left(1-\frac{1-\alpha}{1+\mu_k(\alpha)},\ \tfrac{1}{2},\ \alpha\right)$$
for $0 < \alpha < 1$.

We first prove the following lemma.

LEMMA. *Let* $f(s) = \sum_{n=1}^{\infty} a_n n^{-s}$ *be absolutely convergent for* $\sigma > 1$. *Then*

$$\sum_{n=1}^{\infty} \frac{a_n}{n^s} e^{-\delta n} = \frac{1}{2\pi i}\int_{c-i\infty}^{c+i\infty} \Gamma(w-s)f(w)\delta^{s-w}\,dw$$

for $\delta > 0,\ c > 1,\ c > \sigma$.

For the right-hand side is

$$\frac{1}{2\pi i}\int_{c-i\infty}^{c+i\infty} \Gamma(w-s)\sum_{n=1}^{\infty}\frac{a_n}{n^w}\delta^{s-w}\,dw = \sum_{n=1}^{\infty}\frac{a_n}{n^s}\frac{1}{2\pi i}\int_{c-i\infty}^{c+i\infty}\Gamma(w-s)(\delta n)^{s-w}\,dw$$

$$= \sum_{n=1}^{\infty}\frac{a_n}{n^s}\frac{1}{2\pi i}\int_{c-\sigma-i\infty}^{c-\sigma+i\infty}\Gamma(w')(\delta n)^{-w'}\,dw'$$

$$= \sum_{n=1}^{\infty}\frac{a_n}{n^s}e^{-\delta n}.$$

The inversion is justified by the convergence of

$$\int_{-\infty}^{\infty} |\Gamma\{c-\sigma+i(v-t)\}|\sum_{n=1}^{\infty}\frac{|a_n|}{n^c}\delta^{\sigma-c}\,dv.$$

Taking $a_n = d_k(n)$, $f(s) = \zeta^k(s)$, $c = 2$, we obtain

$$\sum_{n=1}^{\infty}\frac{d_k(n)}{n^s}e^{-\delta n} = \frac{1}{2\pi i}\int_{2-i\infty}^{2+i\infty}\Gamma(w-s)\zeta^k(w)\delta^{s-w}\,dw \quad (\sigma < 2).$$

Moving the contour to $\mathbf{R}(w) = \alpha$, where $\sigma-1 < \alpha < \sigma$, we pass the pole of $\Gamma(w-s)$ at $w = s$, with residue $\zeta^k(s)$, and the pole of $\zeta^k(w)$ at $w = 1$, where the residue is a finite sum of terms of the form

$$K_{m,n}\Gamma^{(m)}(1-s)\log^n\delta\,.\,\delta^{s-1}.$$

† Carlson (2), (3).

This residue is therefore of the form $O(\delta^{\sigma-1+\epsilon}e^{-A|t|})$, and, if $\delta > |t|^{-A}$, it is of the form $O(e^{-A|t|})$. Hence

$$\zeta^k(s) = \sum_{n=1}^{\infty} \frac{d_k(n)}{n^s} e^{-\delta n} - \frac{1}{2\pi i} \int_{\alpha-i\infty}^{\alpha+i\infty} \Gamma(w-s)\zeta^k(w)\delta^{s-w}\,dw + O(e^{-A|t|}).$$

Let us call the first two terms on the right Z_1 and Z_2. Then, as in previous proofs, if $\sigma > \frac{1}{2}$,

$$\int_{\frac{1}{2}T}^{T} |Z_1|^2\,dt = O\left(T \sum_{n=1}^{\infty} \frac{d_k^2(n)}{n^{2\sigma}} e^{-2\delta n}\right) + O\left(\sum_{m \neq n}\sum \frac{d_k(m)d_k(n)e^{-(m+n)\delta}}{m^\sigma n^\sigma |\log m/n|}\right)$$

$$= O(T) + O\left(\sum_{m \neq n}\sum \frac{e^{-(m+n)\delta}}{m^{\sigma-\epsilon} n^{\sigma-\epsilon} |\log m/n|}\right)$$

$$= O(T) + O(\delta^{2\sigma-2-\epsilon})$$

by (7.2.2). Also, putting $w = \alpha + iv$,

$$|Z_2| \leqslant \frac{\delta^{\sigma-\alpha}}{2\pi} \int_{-\infty}^{\infty} |\Gamma(w-s)\zeta^k(w)|\,dv$$

$$\leqslant \frac{\delta^{\sigma-\alpha}}{2\pi} \left\{ \int_{-\infty}^{\infty} |\Gamma(w-s)|\,dv \int_{-\infty}^{\infty} |\Gamma(w-s)\zeta^{2k}(w)|\,dv \right\}^{\frac{1}{2}}.$$

The first integral is $O(1)$, while for $|t| \leqslant T$

$$\left(\int_{-\infty}^{-2T} + \int_{2T}^{\infty}\right)|\Gamma(w-s)\zeta^{2k}(w)|\,dv = \left(\int_{-\infty}^{-2T} + \int_{2T}^{\infty}\right) e^{-A|v-t|}|v|^{Ak}\,dv = O(e^{-AT}).$$

Hence

$$\int_{\frac{1}{2}T}^{T} |Z_2|^2\,dt = O\left\{\delta^{2\sigma-2\alpha} \int_{-2T}^{2T} |\zeta(w)|^{2k}\,dv \int_{\frac{1}{2}T}^{T} |\Gamma(w-s)|\,dt\right\} + O(\delta^{2\sigma-2\alpha})$$

$$= O\left\{\delta^{2\sigma-2\alpha} \int_{-2T}^{2T} |\zeta(\alpha+iv)|^{2k}\,dv\right\} + O(\delta^{2\sigma-2\alpha})$$

$$= O(\delta^{2\sigma-2\alpha}T^{1+\mu_k(\alpha)+\epsilon}).$$

Hence

$$\int_{\frac{1}{2}T}^{T} |\zeta(s)|^{2k}\,dt = O(T) + O(\delta^{2\sigma-2-\epsilon}) + O(\delta^{2\sigma-2\alpha}T^{1+\mu_k(\alpha)+\epsilon}).$$

Let $\delta = T^{-\frac{1}{2}\{1+\mu_k(\alpha)\}/(1-\alpha)}$, so that the last two terms are of the same order, apart from ϵ's. These terms are then $O(T)$ if

$$\sigma > 1 - \frac{1-\alpha}{1+\mu_k(\alpha)}.$$

For such values of σ, replacing T by $\frac{1}{2}T$, $\frac{1}{4}T$,..., and adding, it follows that (7.9.1) holds. Hence σ_k is less than any such σ, and the theorem follows.

A similar argument shows that, if we define σ_k' to be the lower bound of numbers σ such that

$$\frac{1}{T}\int_1^T |\zeta(\sigma+it)|^{2k}\,dt = O(T^\epsilon), \tag{7.9.2}$$

then actually $\sigma_k' = \sigma_k$. For clearly $\sigma_k' \leqslant \sigma_k$; and the above argument shows that, if $\alpha > \sigma_k'$, and $\sigma < \alpha$, then

$$\int_{\frac{1}{2}T}^T |\zeta(\sigma+it)|^{2k}\,dt = O(T)+O(\delta^{2\sigma-2-\epsilon})+O(\delta^{2\sigma-2\alpha}T^{1+\epsilon}).$$

Taking $\delta = T^{-\lambda}$, where $0 < \lambda < 1/(2-2\sigma)$, the right-hand side is $O(T)$. Hence $\sigma_k \leqslant \alpha$, and so $\sigma_k \leqslant \sigma_k'$.

It is also easily seen that

$$\frac{1}{T}\int_1^T |\zeta(\sigma+it)|^{2k}\,dt \sim \sum_{n=1}^{\infty} \frac{d_k^2(n)}{n^{2\sigma}} \quad (\sigma > \sigma_k).$$

For the term $O(T)$ of the above argument is actually

$$\tfrac{1}{2}T\sum_{n=1}^{\infty} \frac{d_k^2(n)}{n^{2\sigma}}e^{-2\delta n} = \tfrac{1}{2}T\sum_{n=1}^{\infty} \frac{d_k^2(n)}{n^{2\sigma}}+o(T),$$

and the result follows by obvious modifications of the argument. This is a case of a general theorem on Dirichlet series.†

THEOREM 7.9 (A). *If $\mu(\sigma)$ is the μ-function defined in § 5.1,*

$$1-\sigma_k \geqslant \frac{1-\sigma_{k-1}}{1+2\mu(\sigma_{k-1})}$$

for $k = 1, 2,\ldots$.

Since $\zeta(\sigma+it) = O(t^{\mu(\sigma)+\epsilon})$,

$$\int_1^T |\zeta(\sigma+it)|^{2k}\,dt = O\left\{T^{2\mu(\sigma)+\epsilon}\int_1^T |\zeta(\sigma+it)|^{2k-2}\,dt\right\},$$

and hence $$\mu_k(\sigma) \leqslant 2\mu(\sigma)+\mu_{k-1}(\sigma).$$

Since $\mu_{k-1}(\sigma_{k-1}) = 0$, this gives $\mu_k(\sigma_{k-1}) \leqslant 2\mu(\sigma_{k-1})$, and the result follows on taking $\alpha = \sigma_{k-1}$ in the previous theorem.

† See E. C. Titchmarsh, *Theory of Functions*, § 9.51.

These formulae may be used to give alternative proofs of Theorems 7.2, 7.5, and 7.7. It follows from the functional equation that

$$\mu_k(1-\sigma) = \mu_k(\sigma)+2k(\sigma-\tfrac{1}{2}).$$

Since $\mu_k(\sigma_k) = 0$, $\mu_k(1-\sigma_k) \geqslant 0$, it follows that $\sigma_k \geqslant \frac{1}{2}$. Hence, putting $\alpha = 1-\sigma_k$ in Theorem 7.9, we obtain either $\sigma_k = \frac{1}{2}$ or

$$\sigma_k \leqslant 1-\frac{\sigma_k}{1+2k(\sigma_k-\frac{1}{2})},$$

i.e.
$$2\sigma_k-1 \leqslant 2k(\sigma_k-\tfrac{1}{2})(1-\sigma_k).$$

Hence $\sigma_k = \frac{1}{2}$, or

$$1 \leqslant k(1-\sigma_k), \qquad \sigma_k \leqslant 1-\frac{1}{k}. \tag{7.9.3}$$

For $k = 2$ we obtain $\sigma_2 = \frac{1}{2}$, but for $k > 2$ we must take the weaker alternative (7.9.2).

7.10. The following refinement† on the above results uses the theorems of Chapter V on $\mu(\sigma)$.

THEOREM 7.10. *Let k be an integer greater than* 1, *and let ν be determined by*

$$(\nu-1)2^{\nu-2}+1 < k \leqslant \nu 2^{\nu-1}+1. \tag{7.10.1}$$

Then
$$\sigma_k \leqslant 1-\frac{\nu+1}{2k+2^{\nu}-2}. \tag{7.10.2}$$

The theorem is true for $k = 2$ ($\nu = 1$). We then suppose it true for all l with $1 < l < k$, and deduce it for k.

Take $l = (\nu-1)2^{\nu-2}+1$, where ν is determined by (7.10.1). Then $\mu_l(\alpha) = 0$, provided that

$$\alpha > 1-\frac{\nu}{2l+2^{\nu-1}-2} = 1-\frac{1}{2^{\nu-1}}.$$

Taking $\alpha = 1-2^{-\nu+1}+\epsilon$, we have, since

$$\frac{1}{T}\int\limits_1^T |\zeta(\alpha+it)|^{2k}\,dt \leqslant \max_{1\leqslant t\leqslant T} |\zeta(\alpha+it)|^{2k-2l}\frac{1}{T}\int\limits_1^T |\zeta(\alpha+it)|^{2l}\,dt,$$

$$\begin{aligned}\mu_k(x) &\leqslant 2(k-l)\mu(\alpha)+\mu_i(\alpha)\\ &= 2(k-l)\mu(\alpha)\\ &\leqslant \frac{2\{k-(\nu-1)2^{\nu-2}-1\}}{(\nu+1)2^{\nu-1}}\end{aligned}$$

† Davenport (1), Haselgrove (1).

by Theorem 5.8. Hence, by Theorem 7.9,

$$\sigma_k \leqslant 1-2^{-\nu+1}\left(\frac{2k+2^\nu-2}{(\nu+1)2^{\nu-1}}\right)^{-1} = 1-\frac{\nu+1}{2k+2^\nu-2}.$$

The theorem therefore follows by induction.

For example, if $k = 3$, then $\nu = 2$, and we obtain

$$\sigma_3 \leqslant \tfrac{5}{8}$$

instead of the result $\sigma_3 \leqslant \frac{2}{3}$ given by Theorem 7.7.

7.11. For integral k, $d_k(n)$ denotes the number of decompositions of n into k factors. If k is not an integer, we can define $d_k(n)$ as the coefficient of n^{-s} in the Dirichlet series for $\zeta^k(s)$, which converges for $\sigma > 1$.

We can now extend Theorem 7.7 to certain non-integral values of k.

THEOREM† 7.11. *For* $0 < k \leqslant 2$

$$\lim_{T\to\infty}\frac{1}{T}\int_1^T |\zeta(\sigma+it)|^{2k}\,dt = \sum_{n=1}^{\infty}\frac{d_k^2(n)}{n^{2\sigma}} \quad (\sigma > \tfrac{1}{2}). \qquad (7.11.1)$$

This is the formula already proved for $k = 1$, $k = 2$; we now take $0 < k < 2$. Let

$$\zeta_N(s) = \prod_{p<N}\frac{1}{1-p^{-s}}, \qquad \eta_N(s) = \zeta(s)/\zeta_N(s).$$

The proof depends on showing (i) that the formula corresponding to (7.11.1) with ζ_N instead of ζ is true; and (ii) that $\zeta_N(s)$, though it does not converge to $\zeta(s)$ for $\sigma \leqslant 1$, still approximates to it in a certain average sense in this strip.

We have, if $\lambda > 0$,

$$\{\zeta_N(s)\}^\lambda = \prod_{p<N}(1-p^{-s})^{-\lambda} = \sum_{n=1}^{\infty}\frac{d'_\lambda(n)}{n^s},$$

say, where the series on the right converges absolutely for $\sigma > 0$, and $d'_\lambda(n) = d_\lambda(n)$ if $n < N$, and $0 \leqslant d'_\lambda(n) \leqslant d_\lambda(n)$ for all n. Hence

$$\lim_{T\to\infty}\frac{1}{T}\int_1^T |\zeta_N(\sigma+it)|^{2\lambda}\,dt = \sum_{n=1}^{\infty}\frac{\{d'_\lambda(n)\}^2}{n^{2\sigma}} \quad (\sigma > 0), \qquad (7.11.2)$$

and

$$\lim_{N\to\infty}\lim_{T\to\infty}\frac{1}{T}\int_1^T |\zeta_N(\sigma+it)|^{2\lambda}\,dt = \sum_{n=1}^{\infty}\frac{\{d_\lambda(n)\}^2}{n^{2\sigma}} \quad (\sigma > \tfrac{1}{2}). \qquad (7.11.3)$$

† Ingham (4); proof by Davenport (1).

We shall next prove that

$$\lim_{N\to\infty}\lim_{T\to\infty}\frac{1}{T}\int_1^T|\zeta(\sigma+it)-\zeta_N(\sigma+it)|^{2k}\,dt = 0 \quad (\sigma > \tfrac{1}{2}). \tag{7.11.4}$$

By Hölder's inequality

$$\frac{1}{T}\int_1^T|\zeta-\zeta_N|^{2k}\,dt \leqslant \left\{\frac{1}{T}\int_1^T|\eta_N-1|^4\,dt\right\}^{\frac{1}{2}k}\left\{\frac{1}{T}\int_1^T|\zeta_N|^{4k/(2-k)}\,dt\right\}^{\frac{1}{2}(2-k)}. \tag{7.11.5}$$

Now $\{\eta_N(s)-1\}^2$ is regular everywhere except for a pole at $s = 1$, and is of finite order in t. Also, for $\sigma > \frac{1}{2}$,

$$\int_1^T|\eta_N(\sigma+it)-1|^4\,dt \leqslant \int_1^T\{1+2^N|\zeta(\sigma+it)|\}^4\,dt = O(T).$$

Hence, by a theorem of Carlson,†

$$\lim_{T\to\infty}\frac{1}{T}\int_1^T|\eta_N(\sigma+it)-1|^4\,dt = \sum_{n=1}^{\infty}\frac{\rho_N^2(n)}{n^{2\sigma}}$$

for $\sigma > \frac{1}{2}$, where ρ_N is the coefficient of n^{-s} in the Dirichlet series of $\{\eta_N(s)-1\}^2$. Now $\rho_N(n) = 0$ for $n < N$, and $0 \leqslant \rho_N(n) \leqslant d(n)$ for all n. Since $\sum d^2(n)n^{-2\sigma}$ converges, it follows that

$$\lim_{N\to\infty}\lim_{T\to\infty}\frac{1}{T}\int_1^T|\eta_N(\sigma+it)-1|^4\,dt = 0; \tag{7.11.6}$$

(7.11.4) now follows from (7.11.5), (7.11.6), and (7.11.3).

We can now deduce (7.11.1) from (7.11.3) and (7.11.4). We have‡

$$\left\{\int_1^T|\zeta|^{2k}\,dt\right\}^R = \left\{\int_1^T|\zeta_N+\zeta-\zeta_N|^{2k}\,dt\right\}^R$$

$$\leqslant \left\{\int_1^T|\zeta_N|^{2k}\,dt\right\}^R + \left\{\int_1^T|\zeta-\zeta_N|^{2k}\,dt\right\}^R,$$

where $R = 1$ if $0 < 2k \leqslant 1$, $R = 1/2k$ if $2k > 1$. Similarly

$$\left\{\int_1^T|\zeta_N|^{2k}\,dt\right\}^R \leqslant \left\{\int_1^T|\zeta|^{2k}\,dt\right\}^R + \left\{\int_1^T|\zeta-\zeta_N|^{2k}\,dt\right\}^R,$$

and (7.11.1) clearly follows.

† See Titchmarsh, *Theory of Functions*, § 9.51.

‡ Hardy, Littlewood, and Pólya, *Inequalities*, Theorem 28.

7.12. An alternative set of mean-value theorems.† Instead of considering integrals of the form

$$I(T) = \int_0^T |\zeta(\sigma+it)|^{2k}\,dt$$

where T is large, we shall now consider integrals of the form

$$J(\delta) = \int_0^\infty |\zeta(\sigma+it)|^{2k}e^{-\delta t}\,dt$$

where δ is small.

The behaviour of these two integrals is very similar. If $J(\delta) = O(1/\delta)$, then

$$I(T) < e\int_0^T |\zeta(\sigma+it)|^{2k}e^{-t/T}\,dt < eJ(1/T) = O(T).$$

Conversely, if $I(T) = O(T)$, then

$$J(\delta) = \int_0^\infty I'(t)e^{-\delta t}\,dt = \left[I(t)e^{-\delta t}\right]_0^\infty + \delta\int_0^\infty I(t)e^{-\delta t}\,dt$$

$$= O\left(\delta\int_0^\infty te^{-\delta t}\,dt\right) = O(1/\delta).$$

Similar results plainly hold with other powers of T, and with other functions, such as powers of T multiplied by powers of $\log T$.

We have also more precise results; for example, *if* $I(T) \sim CT$, *then* $J(\delta) \sim C/\delta$, *and conversely.*

If $I(T) \sim CT$, let $|I(T)-CT| \leqslant \epsilon T$ for $T \geqslant T_0$. Then

$$J(\delta) = \delta\int_0^{T_0} I(t)e^{-\delta t}\,dt + \delta\int_{T_0}^\infty \{I(t)-Ct\}e^{-\delta t}\,dt + C\delta\int_{T_0}^\infty te^{-\delta t}\,dt.$$

The last term is $Ce^{-\delta T_0}(T_0+1/\delta)$, and the modulus of the previous term does not exceed $\epsilon(T_0+1/\delta)$. That $J(\delta) \sim C/\delta$ plainly follows on choosing first T_0 and then δ.

The converse deduction is the analogue for integrals of the well-known Tauberian theorem of Hardy and Littlewood,‡ viz. that *if* $a_n \geqslant 0$, *and*

$$\sum_{n=0}^\infty a_n x^n \sim \frac{1}{1-x} \quad (x \to 1)$$

then

$$\sum_{n=0}^N a_n \sim N.$$

† Titchmarsh (1), (19).

‡ See Titchmarsh, *Theory of Functions*, §§ 7.51–7.53.

The theorem for integrals is as follows:

If $f(t) \geqslant 0$ for all t, and

$$\int_0^\infty f(t)e^{-\delta t}\,dt \sim \frac{1}{\delta} \tag{7.12.1}$$

as $\delta \to 0$, then

$$\int_0^T f(t)\,dt \sim T \tag{7.12.2}$$

as $T \to \infty$.

We first show that, if $P(x)$ is any polynomial,

$$\int_0^\infty f(t)e^{-\delta t}P(e^{-\delta t})\,dt \sim \frac{1}{\delta}\int_0^1 P(x)\,dx.$$

It is sufficient to prove this for $P(x) = x^k$. In this case the left-hand side is

$$\int_0^\infty f(t)e^{-(k+1)\delta t}\,dt \sim \frac{1}{(k+1)\delta} = \frac{1}{\delta}\int_0^1 x^k\,dx.$$

Next, we deduce that

$$\int_0^\infty f(t)e^{-\delta t}g(e^{-\delta t})\,dt \sim \frac{1}{\delta}\int_0^1 g(x)\,dx \tag{7.12.3}$$

if $g(x)$ is continuous, or has a discontinuity of the first kind. For, given ϵ, we can† construct polynomials $p(x)$, $P(x)$, such that

$$p(x) \leqslant g(x) \leqslant P(x)$$

and

$$\int_0^1 \{g(x)-p(x)\}\,dx \leqslant \epsilon, \qquad \int_0^1 \{P(x)-g(x)\}\,dx \leqslant \epsilon.$$

Then

$$\overline{\lim_{\delta\to 0}}\,\delta \int_0^\infty f(t)e^{-\delta t}g(e^{-\delta t})\,dt \leqslant \overline{\lim_{\delta\to 0}}\,\delta \int_0^\infty f(t)e^{-\delta t}P(e^{-\delta t})\,dt$$

$$= \int_0^1 P(x)\,dx < \int_0^1 g(x)\,dx+\epsilon,$$

and making $\epsilon \to 0$ we obtain

$$\overline{\lim_{\delta\to 0}}\,\delta \int_0^\infty f(t)e^{-\delta t}g(e^{-\delta t})\,dt \leqslant \int_0^1 g(x)\,dx.$$

Similarly, arguing with $p(x)$, we obtain

$$\underline{\lim_{\delta\to 0}}\,\delta \int_0^\infty f(t)e^{-\delta t}g(e^{-\delta t})\,dt \geqslant \int_0^1 g(x)\,dx,$$

and (7.12.3) follows.

† See Titchmarsh, *Theory of Functions*, § 7.53.

Now let

$$g(x) = 0 \quad (0 \leqslant x < e^{-1}), \qquad = 1/x \quad (e^{-1} \leqslant x \leqslant 1).$$

Then
$$\int_0^\infty f(t)e^{-\delta t}g(e^{-\delta t})\,dt = \int_0^{1/\delta} f(t)\,dt$$

and
$$\int_0^1 g(x)\,dx = \int_{1/e}^1 \frac{dx}{x} = 1.$$

Hence
$$\int_0^{1/\delta} f(t)\,dt \sim \frac{1}{\delta},$$

which is equivalent to (7.12.2).

If $f(t) \geqslant 0$ *for all* t, *and, for a given positive* m,

$$\int_0^\infty f(t)e^{-\delta t}\,dt \sim \frac{1}{\delta}\log^m\frac{1}{\delta}, \tag{7.12.4}$$

then
$$\int_0^T f(t)\,dt \sim T\log^m T. \tag{7.12.5}$$

The proof is substantially the same. We have

$$\int_0^\infty f(t)e^{-(k+1)\delta t}\,dt \sim \frac{1}{(k+1)\delta}\log^m\left\{\frac{1}{(k+1)\delta}\right\} \sim \frac{1}{(k+1)\delta}\log^m\frac{1}{\delta},$$

and the argument runs as before, with $\frac{1}{\delta}$ replaced by $\frac{1}{\delta}\log^m\frac{1}{\delta}$.

We shall also use the following theorem:

If
$$\int_1^\infty f(t)e^{-\delta t}\,dt \sim C\delta^{-\alpha} \quad (\alpha > 0), \tag{7.12.6}$$

then
$$\int_1^\infty t^{-\beta}f(t)e^{-\delta t}\,dt \sim C\frac{\Gamma(\alpha-\beta)}{\Gamma(\alpha)}\delta^{\beta-\alpha} \quad (0 < \beta < \alpha). \tag{7.12.7}$$

Multiplying (7.12.6) by $(\delta-\eta)^{\beta-1}$ and integrating over (η, ∞), we obtain

$$\int_1^\infty f(t)\,dt \int_\eta^\infty e^{-\delta t}(\delta-\eta)^{\beta-1}\,d\delta = C\int_\eta^\infty \{\delta^{-\alpha}+o(\delta^{-\alpha})\}(\delta-\eta)^{\beta-1}\,d\delta.$$

Now

$$\int_{\eta}^{\infty} e^{-\delta t}(\delta-\eta)^{\beta-1}\,d\delta = e^{-\eta t}\int_{0}^{\infty} e^{-xt}x^{\beta-1}\,dx = e^{-\eta t}t^{-\beta}\Gamma(\beta),$$

$$\int_{\eta}^{\infty} \delta^{-\alpha}(\delta-\eta)^{\beta-1}\,d\delta = \int_{0}^{\infty}\frac{x^{\beta-1}}{(\eta+x)^{\alpha}}\,dx = \eta^{\beta-\alpha}\frac{\Gamma(\beta)\Gamma(\alpha-\beta)}{\Gamma(\alpha)},$$

and the remaining term is plainly $o(\eta^{\beta-\alpha})$ as $\eta \to 0$. Hence the result.

7.13. We can approximate to integrals of the form $J(\delta)$ by means of Parseval's formula. If $\mathbf{R}(z) > 0$, we have

$$\frac{1}{2\pi i}\int_{2-i\infty}^{2+i\infty}\Gamma(s)\zeta^{k}(s)z^{-s}\,ds = \sum_{n=1}^{\infty}\frac{d_k(n)}{2\pi i}\int_{2-i\infty}^{2+i\infty}\Gamma(s)(nz)^{-s}\,ds = \sum_{n=1}^{\infty}d_k(n)e^{-nz},$$

the inversion being justified by absolute convergence. Now move the contour to $\sigma = \alpha$ $(0 < \alpha < 1)$. Let $R_k(z)$ be the residue at $s = 1$, so that $R_k(z)$ is of the form

$$\frac{1}{z}(a_0^{(k)}+a_1^{(k)}\log z+\ldots+a_{k-1}^{(k)}\log^{k-1}z).$$

Let
$$\phi_k(z) = \sum_{n=1}^{\infty}d_k(n)e^{-nz}-R_k(z).$$

Then
$$\frac{1}{2\pi i}\int_{\alpha-i\infty}^{\alpha+i\infty}\Gamma(s)\zeta^{k}(s)z^{-s}\,ds = \phi_k(z). \tag{7.13.1}$$

Putting $z = ixe^{-i\delta}$, where $0 < \delta < \frac{1}{2}\pi$, we see that

$$\phi_k(ixe^{-i\delta}), \qquad \Gamma(s)\zeta^{k}(s)e^{-i(\frac{1}{2}\pi-\delta)s} \tag{7.13.2}$$

are Mellin transforms. Hence the Parseval formula gives

$$\frac{1}{2\pi}\int_{-\infty}^{\infty}|\Gamma(\sigma+it)\zeta^{k}(\sigma+it)|^{2}e^{(\pi-2\delta)t}\,dt = \int_{0}^{\infty}|\phi_k(ixe^{-i\delta})|^{2}x^{2\sigma-1}\,dx. \tag{7.13.3}$$

Now as $|t| \to \infty$

$$|\Gamma(\sigma+it)| = e^{-\frac{1}{2}\pi|t|}|t|^{\sigma-\frac{1}{2}}\sqrt{(2\pi)}\{1+O(t^{-1})\}.$$

Hence the part of the t-integral over $(-\infty, 0)$ is bounded as $\delta \to 0$, and we obtain, for $\frac{1}{2} < \sigma < 1$,

$$\int_{0}^{\infty} t^{2\sigma-1}\{1+O(t^{-1})\}|\zeta(\sigma+it)|^{2k}e^{-2\delta t}\,dt = \int_{0}^{\infty}|\phi_k(ixe^{-i\delta})|^{2}x^{2\sigma-1}\,dx+O(1). \tag{7.13.4}$$

In the case $\sigma = \frac{1}{2}$, we have

$$|\Gamma(\tfrac{1}{2}+it)|^2 = \pi \operatorname{sech} \pi t = 2\pi e^{-\pi|t|}+O(e^{-3\pi|t|}).$$

The integral over $(-\infty, 0)$, and the contribution of the O-term to the whole integral, are now bounded, and in fact are analytic functions of δ, regular for sufficiently small $|\delta|$. Hence we have

$$\int_0^\infty |\zeta(\tfrac{1}{2}+it)|^{2k}e^{-2\delta t}\,dt = \int_0^\infty |\phi_k(ixe^{-i\delta})|^2\,dx+O(1). \tag{7.13.5}$$

7.14. We now apply the above formulae to prove

THEOREM 7.14. *As* $\delta \to 0$

$$\int_0^\infty |\zeta(\tfrac{1}{2}+it)|^2 e^{-\delta t}\,dt \sim \frac{1}{\delta}\log\frac{1}{\delta}. \tag{7.14.1}$$

In this case $R_1(z) = 1/z$, and

$$\phi_1(z) = \sum_{n=1}^\infty e^{-nz}-\frac{1}{z} = \frac{1}{e^z-1}-\frac{1}{z}.$$

Hence (7.13.5) gives

$$\int_0^\infty |\zeta(\tfrac{1}{2}+it)|^2 e^{-2\delta t}\,dt = \int_0^\infty \left|\frac{1}{\exp(ixe^{-i\delta})-1}-\frac{1}{ixe^{-i\delta}}\right|^2 dx+O(1). \tag{7.14.2}$$

The x-integrand is bounded uniformly in δ over $(0,\pi)$, so that this part of the integral is $O(1)$. The remainder is

$$\int_\pi^\infty \left\{\frac{1}{\exp(ixe^{-i\delta})-1}-\frac{1}{ixe^{-i\delta}}\right\}\left\{\frac{1}{\exp(-ixe^{i\delta})-1}+\frac{1}{ixe^{i\delta}}\right\} dx$$

$$= \int_\pi^\infty \frac{dx}{\{\exp(ixe^{-i\delta})-1\}\{\exp(-ixe^{i\delta})-1\}}+ie^{i\delta}\int_\pi^\infty \frac{1}{\exp(-ixe^{i\delta})-1}\frac{dx}{x}-$$

$$-ie^{-i\delta}\int_\pi^\infty \frac{1}{\exp(ixe^{-i\delta})-1}\frac{dx}{x}+\int_\pi^\infty \frac{dx}{x^2}. \tag{7.14.3}$$

The last term is a constant. In the second term, turn the line of integration round to $(\pi, \pi+i\infty)$. The integrand is then regular on the contour for sufficiently small $|\delta|$, and is $O\{x^{-1}\exp(-x\cos\delta)\}$ as $x \to \infty$. This integral is therefore bounded; and similarly so is the third term.

The first term is

$$\int_{\pi}^{\infty} \sum_{m=1}^{\infty} \sum_{n=1}^{\infty} \exp(-imxe^{-i\delta}+inxe^{i\delta})\,dx$$

$$= \sum_{m=1}^{\infty} \sum_{n=1}^{\infty} \frac{\exp\{-(m+n)\pi\sin\delta - i(m-n)\pi\cos\delta\}}{(m+n)\sin\delta + i(m-n)\cos\delta}$$

$$= \sum_{n=1}^{\infty} \frac{e^{-2n\pi\sin\delta}}{2n\sin\delta} + 2\sum_{m=2}^{\infty} \sum_{n=1}^{m-1} \frac{(m+n)\sin\delta\cos\{(m-n)\pi\cos\delta\}}{(m+n)^2\sin^2\delta+(m-n)^2\cos^2\delta} e^{-(m+n)\pi\sin\delta} -$$

$$-2\sum_{m=2}^{\infty} \sum_{n=1}^{m-1} \frac{(m-n)\cos\delta\sin\{(m-n)\pi\cos\delta\}}{(m+n)^2\sin^2\delta+(m-n)^2\cos^2\delta} e^{-(m+n)\pi\sin\delta}$$

$$= \Sigma_1+\Sigma_2+\Sigma_3,$$

the series of imaginary parts vanishing identically. Now

$$\Sigma_1 = \frac{1}{2\sin\delta}\log\frac{1}{1-e^{-2\pi\sin\delta}} \sim \frac{1}{2\delta}\log\frac{1}{\delta},$$

$$|\Sigma_2| < 2\sum_{m=2}^{\infty}\sum_{n=1}^{m-1} \frac{2m\sin\delta}{(m-n)^2\cos^2\delta} e^{-m\pi\sin\delta} = O\left(\delta \sum_{m=2}^{\infty} me^{-m\pi\sin\delta}\right) = O\left(\frac{1}{\delta}\right),$$

and, since $|\sin\{(m-n)\pi\cos\delta\}| = |\sin\{2(m-n)\pi\sin^2\tfrac{1}{2}\delta\}| = O\{(m-n)\delta^2\}$,

$$\Sigma_3 = O\left(\delta^2 \sum_{m=2}^{\infty}\sum_{n=1}^{m-1} e^{-m\pi\sin\delta}\right) = O\left(\delta^2 \sum_{m=2}^{\infty} me^{-m\pi\sin\delta}\right) = O(1).$$

This proves the theorem.

The case $\frac{1}{2} < \sigma < 1$ can be dealt with in a similar way. The leading term is

$$\int_{\pi}^{\infty} \sum_{n=1}^{\infty} e^{-2nx\sin\delta} x^{2\sigma-1}\,dx = \int_{\pi}^{\infty} \frac{x^{2\sigma-1}}{e^{2x\sin\delta}-1}\,dx$$

$$= \frac{1}{(2\sin\delta)^{2\sigma}} \int_{2\pi\sin\delta}^{\infty} \frac{y^{2\sigma-1}}{e^y-1}\,dy \sim \frac{1}{(2\delta)^{2\sigma}} \int_{0}^{\infty} \frac{y^{2\sigma-1}}{e^y-1}\,dy$$

$$= \frac{1}{(2\delta)^{2\sigma}}\Gamma(2\sigma)\zeta(2\sigma).$$

Also (turning the line of integration through $-\frac{1}{2}\pi$)

$$\int_{\pi}^{\infty} e^{-\{(m+n)\sin\delta+i(m-n)\cos\delta\}x} x^{2\sigma-1}\,dx$$

$$= O\left\{e^{-(m+n)\pi\sin\delta} \int_{0}^{\infty} e^{-(m-n)y\cos\delta}(\pi^{2\sigma-1}+y^{2\sigma-1})\,dy\right\}$$

$$= O\left(\frac{e^{-(m+n)\pi\sin\delta}}{m-n}\right), \tag{7.14.4}$$

and the terms with $m \neq n$ give

$$O\left(\sum_{m=2}^{\infty} e^{-m\pi\sin\delta}\sum_{n=1}^{m-1}\frac{1}{m-n}\right) = O\left(\frac{1}{\delta}\log\frac{1}{\delta}\right).$$

Hence
$$\int_0^{\infty} t^{2\sigma-1}|\zeta(\sigma+it)|^2 e^{-2\delta t}\,dt \sim \frac{\Gamma(2\sigma)\zeta(2\sigma)}{2^{2\sigma}\delta^{2\sigma}}. \tag{7.14.5}$$

Hence by (7.12.6), (7.12.7)

$$\int_0^{\infty} |\zeta(\sigma+it)|^2 e^{-\delta t}\,dt \sim \frac{\zeta(2\sigma)}{\delta}. \tag{7.14.6}$$

7.15. We shall now show that we can approximate to the integral (7.14.1) by an asymptotic series in positive powers of δ.

We first require†

THEOREM 7.15. *As $z \to 0$ in any angle $|\arg z| \leqslant \lambda$, where $\lambda < \frac{1}{2}\pi$,*

$$\sum_{n=1}^{\infty} d(n)e^{-nz} = \frac{\gamma-\log z}{z}+\frac{1}{4}+\sum_{n=0}^{N-1} b_n z^{2n+1}+O(|z|^{2N}), \tag{7.15.1}$$

where the b_n are constants.

Near $s = 1$

$$\Gamma(s)\zeta^2(s)z^{-s} = \{1-\gamma(s-1)+\ldots\}\left(\frac{1}{s-1}+\gamma+\ldots\right)^2\frac{1}{z}\{1-(s-1)\log z+\ldots\}$$

$$= \frac{1}{z(s-1)^2}+\frac{\gamma-\log z}{z}\frac{1}{s-1}+\ldots.$$

Hence by (7.13.1), with $k = 2$,

$$\sum_{n=1}^{\infty} d(n)e^{-nz} = \frac{\gamma-\log z}{z}+\frac{1}{2\pi i}\int_{\alpha-i\infty}^{\alpha+i\infty} \Gamma(s)\zeta^2(s)z^{-s}\,ds \quad (0 < \alpha < 1).$$

Here we can move the line of integration to $\sigma = -2N$, since $\Gamma(s) = O(|t|^K e^{-\frac{1}{2}\pi|t|})$, $\zeta^2(s) = O(|t|^K)$ and $z^{-s} = O(r^{-\sigma}e^{\lambda t})$. The residue at $s = 0$ is $\zeta^2(0) = \frac{1}{4}$. The poles of $\Gamma(s)$ at $s = -2n$ are cancelled by zeros of $\zeta^2(s)$. The poles of $\Gamma(s)$ at $s = -2n-1$ give residues

$$\frac{-1}{(2n+1)!}\zeta^2(-2n-1)z^{2n+1} = -\frac{B_{n+1}^2}{(2n+1)!(2n+2)^2}z^{2n+1}.$$

The remaining integral is $O(|z|^{2N})$, and the result follows.

The constant implied in the O, of course, depends on N, and the series taken to infinity is divergent, since the function $\sum d(n)\,e^{-nz}$ cannot be continued analytically across the imaginary axis.

† Wigert (1).

We can now prove†

THEOREM 7.15 (A). *As* $\delta \to 0$, *for every positive* N,

$$\int_0^\infty |\zeta(\tfrac{1}{2}+it)|^2 e^{-2\delta t}\,dt = \frac{\gamma-\log 4\pi\delta}{2\sin\delta}+\sum_{n=0}^{N} c_n\,\delta^n+O(\delta^{N+1})$$

the constant of the O *depending on* N, *and the* c_n *being constants.*

We observe that the term $O(1)$ in (7.14.2) is

$$\tfrac{1}{2}\int_0^\infty |\zeta(\tfrac{1}{2}+it)|^2 e^{-(\pi+2\delta)t}\operatorname{sech}\pi t\,dt-\tfrac{1}{2}\int_{-\infty}^0 |\zeta(\tfrac{1}{2}+it)|^2 e^{(\pi-2\delta)t}\operatorname{sech}\pi t\,dt,$$

and is thus an analytic function of δ, regular for $|\delta| < \pi$. Also

$$\int_0^\pi \left\{\frac{1}{\exp(ixe^{-i\delta})-1}-\frac{1}{ixe^{-i\delta}}\right\}\left\{\frac{1}{\exp(-ixe^{i\delta})-1}+\frac{1}{ixe^{i\delta}}\right\}dx$$

is analytic for sufficiently small $|\delta|$. We dissect the remainder of the integral on the right of (7.14.2) as in (7.14.3). As before

$$\int_\pi^\infty \frac{1}{\exp(-ixe^{i\delta})-1}\,\frac{dx}{x}=\int_\pi^{\pi+i\infty}\frac{1}{\exp(-ize^{i\delta})-1}\,\frac{dz}{z},$$

and the integrand is regular on the new line of integration for sufficiently small $|\delta|$, and, if $\delta = \xi+i\eta$, $z = \pi+iy$, it is $O\{y^{-1}\exp(-y\cos\xi e^{-\eta})\}$ as $y \to \infty$. The integral is therefore regular for sufficiently small $|\delta|$. Similarly for the third term on the right of (7.14.3); and the fourth term is a constant.

By the calculus of residues, the first term is equal to

$$2i\pi\sum_{n=1}^\infty \frac{1}{ie^{-i\delta}}\,\frac{1}{\exp(-2in\pi e^{2i\delta})-1}+$$
$$+\int_0^\infty \frac{dy}{[\exp\{(i\pi-y)e^{-i\delta}\}-1][\exp\{(-i\pi+y)e^{i\delta}\}-1]}.$$

As before, the y-integral is an analytic function of δ, regular for $|\delta|$ small enough. Expressing the series as a power series in $\exp(2i\pi e^{2i\delta})$, we therefore obtain

$$\int_0^\infty |\zeta(\tfrac{1}{2}+it)|^2 e^{-2\delta t}\,dt = 2\pi e^{i\delta}\sum_{n=1}^\infty d(n)\exp(2in\pi e^{2i\delta})+\sum_{n=0}^\infty a_n\,\delta^n \qquad (7.15.2)$$

for $|\delta|$ small enough and $\mathbf{R}(\delta) > 0$.

† Kober (4), Atkinson (1).

Let $z = 2i\pi(1-e^{2i\delta})$ in (7.15.1). Multiplying by $2\pi e^{i\delta}$, we obtain

$$2\pi e^{i\delta}\sum_{n=1}^{\infty} d(n)\exp(2in\pi e^{2i\delta}) = \frac{\gamma-\log(4\pi e^{i\delta}\sin\delta)}{2\sin\delta}+\frac{1}{4}+$$
$$+\sum_{n=0}^{N-1} b_n\{2i\pi(1-e^{2i\delta})\}^{2n+1}+O(\delta^{2N}),$$

and the result now easily follows.

7.16. The next case is that of $|\zeta(\frac{1}{2}+it)|^4$.

In (7.14.2) the contribution of the x-integral for small x was negligible. We now take (7.13.5) with $k = 2$, and

$$\phi_2(z) = \sum_{n=1}^{\infty} d(n)e^{-nz}-\frac{\gamma-\log z}{z}. \tag{7.16.1}$$

In this case the contribution of small x is not negligible, but is substantially the same as that of the other part. We have

$$\phi_2\left(\frac{1}{z}\right) = \frac{1}{2\pi i}\int_{\alpha-i\infty}^{\alpha+i\infty} \Gamma(s)\zeta^2(s)z^s\,ds \quad (0<\alpha<1)$$
$$= \frac{1}{2\pi i}\int_{1-\alpha-i\infty}^{1-\alpha+i\infty} \Gamma(1-s)\zeta^2(1-s)z^{1-s}\,ds$$
$$= \frac{z}{2\pi i}\int_{1-\alpha-i\infty}^{1-\alpha+i\infty} \frac{\Gamma(1-s)}{\chi^2(s)}\,\zeta^2(s)z^{-s}\,ds.$$

Now

$$\Gamma(1-s)/\chi^2(s) = 2^{2-2s}\pi^{-2s}\cos^2\tfrac{1}{2}s\pi\,\Gamma^2(s)\Gamma(1-s)$$
$$= 2^{1-2s}\pi^{1-2s}\cot\tfrac{1}{2}s\pi\,\Gamma(s)$$
$$= 2^{1-2s}\pi^{1-2s}\left\{-i+O\left(\frac{e^{-\frac{1}{2}\pi t}}{|\sin\frac{1}{2}s\pi|}\right)\right\}\Gamma(s)\quad (t\to\pm\infty).$$

If $z = ixe^{-i\delta}$ ($x > x_0$, $0 < \delta < \frac{1}{2}\pi$), the O term is

$$O\left\{x\int_{1-\alpha-i\infty}^{1-\alpha+i\infty} \frac{e^{-\frac{1}{2}\pi t}}{|\sin\frac{1}{2}s\pi|}\,|\Gamma(s)|e^{(\frac{1}{2}\pi-\delta)t}(1+|t|)x^{\alpha-1}\,dt\right\} = O(x^{\alpha}),$$

uniformly for small δ. Hence

$$\phi_2\left(\frac{1}{z}\right) = \frac{-iz}{2\pi i}\int_{1-\alpha-i\infty}^{1-\alpha+i\infty} 2^{1-2s}\pi^{1-2s}\Gamma(s)\zeta^2(s)z^{-s}\,ds+O(x^{\alpha})$$
$$= -2\pi iz\phi_2(4\pi^2 z)+O(x^{\alpha}), \tag{7.16.2}$$

where α may be as near zero as we please.

We also use the results

$$\sum_{n=1}^{\infty} d^2(n)e^{-n\eta} = O\left(\frac{1}{\eta}\log^3\frac{1}{\eta}\right), \tag{7.16.3}$$

$$\sum_{n=1}^{\infty} n^2 d^2(n)e^{-n\eta} = O\left(\frac{1}{\eta^3}\log^3\frac{1}{\eta}\right) \tag{7.16.4}$$

as $\eta \to 0$. By (1.2.10)

$$\frac{1}{2\pi i}\int_{2-i\infty}^{2+i\infty} \Gamma(s)\frac{\zeta^4(s)}{\zeta(2s)}\eta^{-s}\,ds = \frac{1}{2\pi i}\sum_{n=1}^{\infty} d^2(n)\int_{2-i\infty}^{2+i\infty} \Gamma(s)(n\eta)^{-s}\,ds$$

$$= \sum_{n=1}^{\infty} d^2(n)e^{-n\eta}. \tag{7.16.5}$$

Hence

$$\sum_{n=1}^{\infty} d^2(n)e^{-n\eta} = R+\frac{1}{2\pi i}\int_{c-i\infty}^{c+i\infty} \Gamma(s)\frac{\zeta^4(s)}{\zeta(2s)}\eta^{-s}\,ds \quad (\tfrac{1}{2}<c<1)$$

$$= R+O(\eta^{-c}),$$

where R is the residue at $s = 1$; and

$$R = \frac{1}{\eta}\left(a\log^3\frac{1}{\eta}+b\log^2\frac{1}{\eta}+c\log\frac{1}{\eta}+d\right),$$

where a, b, c, d are constants, and in fact

$$a = \frac{1}{3!\,\zeta(2)} = \frac{1}{\pi^2}.$$

This proves (7.16.3); and (7.16.4) can be proved similarly by first differentiating (7.16.5) twice with respect to η.

We can now prove†

THEOREM 7.16. *As* $\delta \to 0$

$$\int_0^{\infty} |\zeta(\tfrac{1}{2}+it)|^4 e^{-\delta t}\,dt \sim \frac{1}{2\pi^2}\frac{1}{\delta}\log^4\frac{1}{\delta}.$$

Using (7.13.5), we have

$$\int_0^{\infty} |\zeta(\tfrac{1}{2}+it)|^4 e^{-2\delta t}\,dt = \int_0^{\infty} |\phi_2(ixe^{-i\delta})|^2\,dx+O(1),$$

and it is sufficient to prove that

$$\int_{2\pi}^{\infty} |\phi_2(ixe^{-i\delta})|^2\,dx \sim \frac{1}{8\pi^2}\frac{1}{\delta}\log^4\frac{1}{\delta}.$$

† Titchmarsh (1).

For then, by (7.16.2),

$$\int_0^{2\pi} |\phi_2(ixe^{-i\delta})|^2\, dx = \int_{1/2\pi}^{\infty} \left|\phi_2\left(\frac{ie^{-i\delta}}{x}\right)\right|^2 \frac{dx}{x^2} = \int_{1/2\pi}^{\infty} \left|\phi_2\left(\frac{1}{ixe^{-i\delta}}\right)\right|^2 \frac{dx}{x^2}$$

$$= \int_{1/2\pi}^{\infty} |2\pi xe^{-i\delta}\phi_2(4\pi^2 ixe^{-i\delta})+O(x^\alpha)|^2 \frac{dx}{x^2} \quad (0 < \alpha < \tfrac{1}{2})$$

$$= \int_{2\pi}^{\infty} |\phi_2(ixe^{-i\delta})+O(x^{\alpha-1})|^2\, dx$$

$$= \int_{2\pi}^{\infty} |\phi_2(ixe^{-i\delta})|^2\, dx+O\left(\int_{2\pi}^{\infty} |\phi_2(ixe^{-i\delta})|^2\, dx \int_{2\pi}^{\infty} x^{2\alpha-2}\, dx\right)^{\frac{1}{2}}+ +O\left(\int_{2\pi}^{\infty} x^{2\alpha-2}\, dx\right)$$

$$= \frac{1}{8\pi^2}\frac{1}{\delta}\log^4\frac{1}{\delta}+O\left(\frac{1}{\sqrt{\delta}}\log^2\frac{1}{\delta}\right)+O(1),$$

and the result clearly follows.

It is then sufficient to prove that

$$\int_{2\pi}^{\infty} \left|\sum_{n=1}^{\infty} d(n)\exp(-inxe^{-i\delta})\right|^2 dx \sim \frac{1}{8\pi^2}\frac{1}{\delta}\log^4\frac{1}{\delta},$$

for the remainder of (7.16.1) will then contribute $O(\delta^{-\frac{1}{2}}\log^2 1/\delta)$.

As in the previous proof, the left-hand side is equal to

$$\sum_{n=1}^{\infty} d^2(n)\frac{e^{-4n\pi\sin\delta}}{2n\sin\delta}+$$

$$+2\sum_{m=2}^{\infty}\sum_{n=1}^{m-1} d(m)d(n)\frac{(m+n)\sin\delta\cos\{2(m-n)\pi\cos\delta\}}{(m+n)^2\sin^2\delta+(m-n)^2\cos^2\delta}e^{-2(m+n)\pi\sin\delta}-$$

$$-2\sum_{m=2}^{\infty}\sum_{n=1}^{m-1} d(m)d(n)\frac{(m-n)\cos\delta\sin\{2(m-n)\pi\cos\delta\}}{(m+n)^2\sin^2\delta+(m-n)^2\cos^2\delta}e^{-2(m+n)\pi\sin\delta}$$

$$= \Sigma_1+\Sigma_2+\Sigma_3.$$

Now

$$\Sigma_1 = \frac{1}{2\sin\delta}(1-e^{-4\pi\sin\delta})\sum_{n=1}^{\infty} e^{-4\pi n\sin\delta}\sum_{\nu=1}^{n}\frac{d^2(\nu)}{\nu}$$

$$\sim 2\pi\sum_{n=1}^{\infty} e^{-4\pi n\sin\delta}\frac{\log^4 n}{4\pi^2} \sim \frac{1}{2\pi}\int_0^{\infty} e^{-4\pi x\sin\delta}\log^4 x\, dx$$

$$= \frac{1}{8\pi^2\sin\delta}\int_0^{\infty} e^{-y}\log^4\left(\frac{y}{4\pi\sin\delta}\right) dy \sim \frac{1}{8\pi^2\delta}\log^4\frac{1}{\delta},$$

$$|\Sigma_2| \leqslant 2\sum_{m=2}^{\infty}\sum_{n=1}^{m-1} d(m)d(n)\frac{2m\sin\delta}{(m-n)^2\cos^2\delta}e^{-2\pi m\sin\delta}$$

$$= \frac{4\sin\delta}{\cos^2\delta}\sum_{m=2}^{\infty} m\,d(m)\,e^{-2\pi m\sin\delta}\sum_{r=1}^{m-1}\frac{d(m-r)}{r^2}$$

$$= \frac{4\sin\delta}{\cos^2\delta}\sum_{r=1}^{\infty}\frac{1}{r^2}\sum_{m=r+1}^{\infty} m\,d(m)d(m-r)e^{-2\pi m\sin\delta}.$$

The square of the inner sum does not exceed

$$\sum_{m=r+1}^{\infty} m^2 d^2(m)e^{-2\pi m\sin\delta}\sum_{m=r+1}^{\infty} d^2(m-r)e^{-2\pi m\sin\delta}$$

$$\leqslant \sum_{m=1}^{\infty} m^2 d^2(m)e^{-2\pi m\sin\delta}\sum_{m=1}^{\infty} d^2(m)e^{-2\pi m\sin\delta}$$

$$= O\left(\frac{1}{\delta^3}\log^3\frac{1}{\delta}\right)O\left(\frac{1}{\delta}\log^3\frac{1}{\delta}\right) = O\left(\frac{1}{\delta^4}\log^6\frac{1}{\delta}\right)$$

by (7.16.3) and (7.16.4). Hence

$$\Sigma_2 = O\left(\frac{1}{\delta}\log^3\frac{1}{\delta}\right).$$

Finally (as in the previous proof)

$$\Sigma_3 = O\left(\delta^2\sum_{m=2}^{\infty} m^{1+\epsilon}e^{-2m\pi\sin\delta}\right) = O(\delta^{-\epsilon}).$$

This proves the theorem.

It has been proved by Atkinson (2) that

$$\int_0^{\infty} |\zeta(\tfrac{1}{2}+it)|^4 e^{-\delta t}\,dt$$

$$= \frac{1}{\delta}\left(A\log^4\frac{1}{\delta}+B\log^3\frac{1}{\delta}+C\log^2\frac{1}{\delta}+D\log\frac{1}{\delta}+E\right)+O\left\{\left(\frac{1}{\delta}\right)^{\frac{13}{14}+\epsilon}\right\},$$

where $A = \dfrac{1}{2\pi^2}$, $\qquad B = -\dfrac{1}{\pi^2}\left(2\log 2\pi - 6\gamma + \dfrac{24\zeta'(2)}{\pi^2}\right).$

A method is also indicated by which the index $\frac{13}{14}$ could be reduced to $\frac{8}{9}$.

7.17. The method of residues used in § 7.15 for $|\zeta(\frac{1}{2}+it)|^2$ suggests still another method of dealing with $|\zeta(\frac{1}{2}+it)|^4$. This is primarily a question of approximating to

$$\int_{2\pi}^{\infty}\left|\sum_{n=1}^{\infty} d(n)\exp(-inxe^{-i\delta})\right|^2 dx = \int_{2\pi}^{\infty}\left|\sum_{n=1}^{\infty}\frac{1}{\exp(inxe^{-i\delta})-1}\right|^2 dx$$

$$= \sum_{m=1}^{\infty}\sum_{n=1}^{\infty}\int_{2\pi}^{\infty}\frac{dx}{\{\exp(imxe^{-i\delta})-1\}\{\exp(-inxe^{i\delta})-1\}}.$$

In the terms with $n \geqslant m$, put $x = \xi/m$. We get

$$\sum_{m=1}^{\infty} \frac{1}{m} \sum_{n=m}^{\infty} \int\limits_{2\pi m}^{\infty} \frac{d\xi}{\{\exp(i\xi e^{-i\delta})-1\}\{\exp(-inm^{-1}\xi e^{i\delta})-1\}}.$$

Approximating to the integral by a sum obtained from the residues of the first factor, as in § 7.15, we obtain as an approximation to this

$$2\pi e^{i\delta} \sum_{m=1}^{\infty} \frac{1}{m} \sum_{n=m}^{\infty} \sum_{r=m}^{\infty} \frac{1}{\exp\{-2i(nr/m)\pi e^{2i\delta}\}-1}$$

$$= 2\pi e^{i\delta} \sum_{m=1}^{\infty} \frac{1}{m} \sum_{n=m}^{\infty} \sum_{r=m}^{\infty} \sum_{q=1}^{\infty} \exp\left(2i\,\frac{nqr}{m}\,\pi e^{2i\delta}\right)$$

$$= 2\pi e^{i\delta} \sum_{m=1}^{\infty} \frac{1}{m} \sum_{r=m}^{\infty} \sum_{q=1}^{\infty} \frac{\exp(2iqr\pi e^{2i\delta})}{1-\exp\{2i(qr/m)\pi e^{2i\delta}\}}$$

$$= O\left(\sum_{m=1}^{\infty} \frac{1}{m} \sum_{r=m}^{\infty} \sum_{q=1}^{\infty} \frac{e^{-2qr\pi \sin 2\delta}}{|1-\exp\{2i(qr/m)\pi e^{2i\delta}\}|}\right)$$

$$= O\left(\sum_{m=1}^{\infty} \frac{1}{m} \sum_{\nu=m}^{\infty} \frac{d(\nu)e^{-2\nu\pi \sin 2\delta}}{|1-\exp(2i\nu m^{-1}\pi e^{2i\delta})|}\right).$$

The terms with $m \mid \nu$ are

$$O\left(\sum_{m=1}^{\infty} \frac{1}{m} \sum_{\nu=m}^{\infty} \frac{d(\nu)e^{-2\nu\pi \sin 2\delta}}{\nu m^{-1}\delta}\right) = O\left(\frac{1}{\delta} \sum \sum_{m|\nu} \frac{d(\nu)}{\nu} e^{-2\nu\pi \sin 2\delta}\right)$$

$$= O\left(\frac{1}{\delta} \sum_{\nu=1}^{\infty} \frac{d^2(\nu)}{\nu} e^{-2\nu\pi \sin 2\delta}\right) = O\left(\frac{1}{\delta} \log^4 \frac{1}{\delta}\right).$$

The remaining terms are

$$O\left(\sum_{m=1}^{\infty} \frac{1}{m} \sum_{k=1}^{\infty} \sum_{l=1}^{m-1} \frac{d(km+l)e^{-2(km+l)\pi \sin 2\delta}}{l/m}\right)$$

$$= O\left(\sum_{m=1}^{\infty} \sum_{k=1}^{\infty} e^{-2km\pi \sin 2\delta} \sum_{l=1}^{m-1} \frac{d(km+l)}{l}\right)$$

$$= O\left(\sum_{m=1}^{\infty} \sum_{k=1}^{\infty} e^{-2km\pi \sin 2\delta} \sum_{l=1}^{km} \frac{d(km+l)}{l}\right)$$

$$= O\left(\sum_{\nu=1}^{\infty} d(\nu)e^{-2\nu\pi \sin 2\delta} \sum_{l=1}^{\nu} \frac{d(\nu+l)}{l}\right)$$

$$= O\left(\sum_{l=1}^{\infty} \frac{1}{l} \sum_{\nu=l}^{\infty} d(\nu)d(\nu+l)e^{-2\nu\pi \sin 2\delta}\right)$$

$$= O\left(\sum_{l=1}^{\infty} \frac{e^{-l\pi \sin 2\delta}}{l} \sum_{\nu=l}^{\infty} d(\nu)d(\nu+l)e^{-(2\nu-l)\pi \sin 2\delta}\right).$$

Using Schwarz's inequality and (7.16.3) we obtain

$$O\left(\sum_{l=1}^{\infty} \frac{e^{-l\pi \sin 2\delta}}{l} \frac{1}{\delta} \log^3 \frac{1}{\delta}\right) = O\left(\frac{1}{\delta} \log^4 \frac{1}{\delta}\right).$$

Actually it follows from a theorem of Ingham (1) that this term is

$$O\left(\frac{1}{\delta} \log^3 \frac{1}{\delta}\right).$$

7.18. There are formulae similar to those of § 7.16 for larger values of k, though in the higher cases they fail to give the desired mean-value formula.†

We have

$$\phi_k\left(\frac{1}{z}\right) = \frac{1}{2\pi i} \int_{\alpha-i\infty}^{\alpha+i\infty} \Gamma(s)\zeta^k(s) z^s \, ds$$

$$= \frac{1}{2\pi i} \int_{1-\alpha-i\infty}^{1-\alpha+i\infty} \Gamma(1-s)\zeta^k(1-s) z^{1-s} \, ds$$

$$= \frac{z}{2\pi i} \int_{1-\alpha-i\infty}^{1-\alpha+i\infty} \frac{\Gamma(1-s)}{\chi^k(s)} \zeta^k(s) z^{-s} \, ds.$$

Now

$$\Gamma(1-s)\chi^{-k}(s) = 2^{k-ks}\pi^{-ks} \cos^k \tfrac{1}{2}s\pi \Gamma^k(s)\Gamma(1-s)$$

$$= 2^{k-ks}\pi^{1-ks} \cos^k \tfrac{1}{2}s\pi \operatorname{cosec} \pi s \Gamma^{k-1}(s).$$

For large s

$$\Gamma^{k-1}(s) \sim s^{(k-1)(s-\frac{1}{2})} e^{-(k-1)s} (2\pi)^{\frac{1}{2}(k-1)}.$$

Now

$$\Gamma\{(k-1)s-\tfrac{1}{2}k+1\} \sim \{(k-1)s\}^{(k-1)s-\frac{1}{2}k+\frac{1}{2}} e^{-(k-1)s} (2\pi)^{\frac{1}{2}}.$$

Hence we may expect to be able to replace $\Gamma^{k-1}(s)$ by

$$(k-1)^{-(k-1)s+\frac{1}{2}k-\frac{1}{2}} (2\pi)^{\frac{1}{2}(k-2)} \Gamma\{(k-1)s-\tfrac{1}{2}k+1\}.$$

Also, in the upper half-plane,

$$\cos^k \tfrac{1}{2}s\pi \operatorname{cosec} s\pi \sim (\tfrac{1}{2}e^{-\frac{1}{2}is\pi})^k \frac{-2i}{e^{-is\pi}} = -2^{1-k} i e^{-is\pi(\frac{1}{2}k-1)}.$$

We should thus replace $\Gamma(1-s)\chi^{-k}(s)$ by

$$-i \, . \, 2^{1-ks}\pi^{1-ks} e^{-is\pi(\frac{1}{2}k-1)} (k-1)^{-(k-1)s+\frac{1}{2}k-\frac{1}{2}} (2\pi)^{\frac{1}{2}(k-2)} \Gamma\{(k-1)s-\tfrac{1}{2}k+1\}.$$

Hence an approximation to $\phi_k(1/z)$ should be

$$\psi_k\left(\frac{1}{z}\right) = -i(2\pi)^{\frac{1}{2}k} \sum_{n=1}^{\infty} d_k(n) \frac{z}{2\pi i} \times$$

$$\times \int_{1-\alpha-i\infty}^{1-\alpha+i\infty} \Gamma\{(k-1)s-\tfrac{1}{2}k+1\}(k-1)^{-(k-1)s+\frac{1}{2}k-\frac{1}{2}} e^{-is\pi(\frac{1}{2}k-1)} (2^k\pi^k nz)^{-s} \, ds.$$

† See also Bellman (3).

Putting $s = (w+\frac{1}{2}k-1)/(k-1)$, the integral is

$$-i(2\pi)^{\frac{1}{2}k}\frac{z}{2\pi i}\int \Gamma(w)(k-1)^{-w-\frac{1}{2}}e^{-i\pi(\frac{1}{2}k-1)(w+\frac{1}{2}k-1)/(k-1)}(2^k\pi^k nz)^{-(w+\frac{1}{2}k-1)/(k-1)}\,dw$$

$$= -i(2\pi)^{\frac{1}{2}k}z(k-1)^{-\frac{1}{2}}e^{-i\pi(\frac{1}{2}k-1)^2/(k-1)}(2^k\pi^k nz)^{-(\frac{1}{2}k-1)/(k-1)}\times$$
$$\times\exp\{-(k-1)e^{i\pi(\frac{1}{2}k-1)/(k-1)}2^{k/(k-1)}\pi^{k/(k-1)}(nz)^{1/(k-1)}\}.$$

Putting $z = ixe^{-i\delta}$, we obtain

$$(2\pi)^{k/(2k-2)}(k-1)^{-\frac{1}{2}}x^{k/\{2(k-1)\}}n^{-(\frac{1}{2}k-1)/(k-1)}\times$$
$$\times C_k\exp\{-(k-1)e^{\frac{1}{2}i\pi}2^{k/(k-1)}\pi^{k/(k-1)}(nx)^{1/(k-1)}e^{-i\delta/(k-1)}\},$$

where $|C_k| = 1$.

We have, by (7.13.5),

$$\int_0^\infty |\zeta(\tfrac{1}{2}+it)|^{2k}e^{-2\delta t}\,dt = \int_0^\infty |\phi_k(ixe^{-i\delta})|^2\,dx+O(1)$$
$$= \int_0^\lambda |\phi_k(ixe^{-i\delta})|^2\,dx+\int_\lambda^\infty |\phi_k(ixe^{-i\delta})|^2\,dx+O(1).$$

As in the above cases, the integral over (λ,∞) is

$$\sum_{n=1}^\infty d_k^2(n)\frac{e^{-2\lambda n\sin\delta}}{2n\sin\delta}+$$
$$+2\sum_{m=2}^\infty\sum_{n=1}^{m-1} d_k(m)d_k(n)\frac{(m+n)\sin\delta\cos\{\lambda(m-n)\cos\delta\}}{(m+n)^2\sin^2\delta+(m-n)^2\cos^2\delta}e^{-\lambda(m+n)\sin\delta}-$$
$$-2\sum_{m=2}^\infty\sum_{n=1}^{m-1} d_k(m)d_k(n)\frac{(m-n)\cos\delta\sin\{\lambda(m-n)\cos\delta\}}{(m+n)^2\sin^2\delta+(m-n)^2\cos^2\delta}e^{-\lambda(m+n)\sin\delta}$$
$$= \Sigma_1+\Sigma_2+\Sigma_3. \quad (7.18.1)$$

Using the relation $d_k(n) = O(n^\epsilon)$, we obtain

$$\Sigma_1 = O\left(\frac{1}{\delta}\frac{1}{(\lambda\delta)^\epsilon}\right),$$

and, since $(m+n)^2\sin^2\delta+(m-n)^2\cos^2\delta > A\delta(m+n)(m-n)$,

$$\Sigma_2 = O\left(\sum_{m=2}^\infty m^\epsilon e^{-\lambda m\sin\delta}\sum_{n=1}^{m-1}\frac{1}{m-n}\right) = O\left(\sum_{m=2}^\infty m^\epsilon e^{-\lambda m\sin\delta}\right) = O\left(\frac{1}{(\lambda\delta)^{1+\epsilon}}\right),$$

$$\Sigma_3 = O\left(\sum_{m=2}^\infty m^\epsilon e^{-\lambda m\sin\delta}\sum_{n=1}^{m-1}\frac{1}{m-n}\right) = O\left(\frac{1}{(\lambda\delta)^{1+\epsilon}}\right).$$

Hence, for $\lambda < A$,

$$\int_\lambda^\infty |\phi_k(ixe^{-i\delta})|^2\,dx = O\left(\frac{1}{(\lambda\delta)^{1+\epsilon}}\right).$$

Also
$$\int_0^\lambda |\phi_k(ixe^{-i\delta})|^2\,dx = \int_{1/\lambda}^\infty \left|\phi_k\left(\frac{1}{ixe^{-i\delta}}\right)\right|^2 \frac{dx}{x^2},$$

and by the above formula this should be approximately

$$\frac{(2\pi)^{k/(k-1)}}{k-1}\times$$
$$\times \int_{1/\lambda}^\infty \left|\sum_{n=1}^\infty \frac{d_k(n)}{n^{(\frac{1}{2}k-1)/(k-1)}} \exp\{-(k-1)i(2\pi)^{k/(k-1)}(nx)^{1/(k-1)}e^{-i\delta/(k-1)}\}\right|^2 \frac{dx}{x^{2-k/(k-1)}}.$$

Putting $x = \xi^{k-1}$, this is

$$(2\pi)^{k/(k-1)}\times$$
$$\times \int_{\lambda^{-1/(k-1)}}^\infty \left|\sum_{n=1}^\infty \frac{d_k(n)}{n^{(\frac{1}{2}k-1)/(k-1)}} \exp\{-(k-1)i(2\pi)^{k/(k-1)}n^{1/(k-1)}\xi e^{-i\delta/(k-1)}\}\right|^2 d\xi,$$

and we can integrate as before. We obtain

$$K\sum_{m=1}^\infty\sum_{n=1}^\infty \frac{d_k(m)d_k(n)}{(mn)^{(\frac{1}{2}k-1)/(k-1)}}\times$$
$$\times\frac{\exp[(k-1)(2\pi)^{k/(k-1)}\{(n^{1/(k-1)}-m^{1/(k-1)})i\cos\delta/(k-1)-(m^{1/(k-1)}+n^{1/(k-1)})\sin\delta/(k-1)\}\lambda^{-1/(k-1)}]}{(n^{1/(k-1)}-m^{1/(k-1)})i\cos\delta/(k-1)-(m^{1/(k-1)}+n^{1/(k-1)})\sin\delta/(k-1)},$$

where K depends on k only.

The terms with $m = n$ are

$$O\left\{\frac{1}{\delta}\sum_{n=1}^\infty \frac{d_k^2(n)}{n}\exp(-K\delta n^{1/(k-1)}\lambda^{-1/(k-1)})\right\} = O\left\{\frac{1}{\delta}\frac{1}{(\lambda\delta)^\epsilon}\right\}.$$

The rest are

$$O\left\{\sum_{m>n}\sum \frac{1}{(mn)^{(\frac{1}{2}k-1)/(k-1)}}\frac{\exp(-K\delta m^{1/(k-1)}\lambda^{-1/(k-1)})}{m^{1/(k-1)}-n^{1/(k-1)}}\right\}.$$

Now

$$\sum_{n=1}^{m-1}\frac{1}{n^{(\frac{1}{2}k-1)/(k-1)}(m^{1/(k-1)}-n^{1/(k-1)})}$$
$$= O\left\{\sum_{n=1}^{\frac{1}{2}m}\frac{1}{n^{(\frac{1}{2}k-1)/(k-1)}m^{1/(k-1)}} + \sum_{\frac{1}{2}m}^{m-1}\frac{1}{m^{(\frac{1}{2}k-1)/(k-1)+1/(k-1)-1}(m-n)}\right\}$$
$$= O(m^{1-(\frac{1}{2}k-1)/(k-1)-1/(k-1)+\epsilon}) = O(m^{(\frac{1}{2}k-1)/(k-1)+\epsilon}).$$

Hence we obtain

$$O\left\{\sum_{m=2}^{\infty} m^{\epsilon} \exp(-K\delta m^{1/(k-1)}\lambda^{-1/(k-1)})\right\} = O\left\{\int_0^{\infty} x^{\epsilon} \exp(-K\delta x^{1/(k-1)}\lambda^{-1/(k-1)})\, dx\right\}$$
$$= O\left\{\left(\frac{\lambda}{\delta^{k-1}}\right)^{1+\epsilon}\right\}.$$

Altogether

$$\int_0^{\infty} |\zeta(\tfrac{1}{2}+it)|^{2k} e^{-2\delta t}\, dt = O\left\{\frac{1}{(\lambda\delta)^{1+\epsilon}}\right\} + O\left\{\left(\frac{\lambda}{\delta^{k-1}}\right)^{1+\epsilon}\right\},$$

and taking $\lambda = \delta^{\frac{1}{2}k-1}$, we obtain

$$\int_0^{\infty} |\zeta(\tfrac{1}{2}+it)|^{2k} e^{-2\delta t}\, dt = O(\delta^{-\frac{1}{2}k-\epsilon}) \quad (k \geqslant 2).$$

This index is what we should obtain from the approximate functional equation.

7.19. The attempt to obtain a non-trivial upper bound for

$$\int_0^{\infty} |\zeta(\tfrac{1}{2}+it)|^{2k} e^{-\delta t}\, dt$$

for $k > 2$ fails. But we can obtain a lower bound† for it which may be somewhere near the truth; for in this problem we can ignore $\phi_k(ixe^{-i\delta})$ for small x, since by (7.13.5)

$$\int_0^{\infty} |\zeta(\tfrac{1}{2}+it)|^{2k} e^{-2\delta t}\, dt > \int_1^{\infty} |\phi_k(ixe^{-i\delta})|^2\, dx + O(1), \tag{7.19.1}$$

and we can approximate to the right-hand side by the method already used.

If k is any positive integer, and $\sigma > 1$,

$$\zeta^k(s) = \prod_p \left(1-\frac{1}{p^s}\right)^{-k} = \prod_p \sum_{m=0}^{\infty} \frac{(k+m-1)!}{(k-1)!\,m!} \frac{1}{p^{ms}} = \sum_{n=1}^{\infty} \frac{d_k(n)}{n^s}.$$

If we replace the coefficient of each term p^{-ms} by its square, the coefficient of each n^{-s} is replaced by its square. Hence if

$$F_k(s) = \sum_{n=1}^{\infty} \frac{d_k^2(n)}{n^s},$$

then
$$F_k(s) = \prod_p \sum_{m=0}^{\infty} \left\{\frac{(k+m-1)!}{(k-1)!\,m!}\right\}^2 \frac{1}{p^{ms}} = \prod_p f_k(p^{-s}),$$

† Titchmarsh (4).

say. Thus
$$f_k\left(\frac{1}{p^s}\right) = 1+\frac{k^2}{p^s}+\dots,$$
and
$$\left(1-\frac{1}{p^s}\right)^{k^2} f_k\left(\frac{1}{p^s}\right) = \left(1-\frac{k^2}{p^s}+\dots\right)\left(1+\frac{k^2}{p^s}+\dots\right) = 1+O\left(\frac{1}{p^{2\sigma}}\right).$$

Hence the product
$$\prod_p \left(1-\frac{1}{p^s}\right)^{k^2} f_k\left(\frac{1}{p^s}\right)$$
is absolutely convergent for $\sigma > \frac{1}{2}$, and so represents an analytic function, $g(s)$ say, regular for $\sigma > \frac{1}{2}$, and bounded in any half-plane $\sigma \geqslant \frac{1}{2}+\delta$; and
$$F_k(s) = \zeta^{k^2}(s)g(s).$$

Now
$$\sum_{n=1}^{\infty} d_k^2(n)e^{-2n\sin\delta} = \frac{1}{2\pi i}\int_{2-i\infty}^{2+i\infty} \Gamma(s)F_k(s)(2\sin\delta)^{-s}\,ds.$$

Moving the line of integration just to the left of $\sigma = 1$, and evaluating the residue at $s = 1$, we obtain in the usual way
$$\sum_{n=1}^{\infty} d_k^2(n)e^{-2n\sin\delta} \sim \frac{C_k'}{\delta}\log^{k^2-1}\frac{1}{\delta}.$$

Similarly
$$\sum_{n=1}^{\infty} \frac{d_k^2(n)}{n}e^{-2n\sin\delta} = \frac{1}{2\pi i}\int_{2-i\infty}^{2+i\infty} \Gamma(s)F_k(s+1)(2\sin\delta)^{-s}\,ds \sim C_k\log^{k^2}\frac{1}{\delta},$$

since here there is a pole of order k^2+1 at $s = 0$.

We can now prove

THEOREM 7.19. *For any fixed integer k, and $0 < \delta \leqslant \delta_0 = \delta_0(k)$,*
$$\int_0^{\infty} |\zeta(\tfrac{1}{2}+it)|^{2k}e^{-\delta t}\,dt \geqslant \frac{C_k}{\delta}\log^{k^2}\frac{1}{\delta}.$$

The integral on the right of (7.19.1) is equal to (7.18.1) with $\lambda = 1$; and
$$\Sigma_1 \sim \frac{C_k}{2\delta}\log^{k^2}\frac{1}{\delta},$$
while
$$\Sigma_2+\Sigma_3 = O\left(\frac{1}{\delta}\log^{k^2-1}\frac{1}{\delta}\right).$$

The result therefore follows.

NOTES FOR CHAPTER 7

7.20. When applied (with care) to a general Dirichlet polynomial, the proof of the first lemma of §7.2 leads to

$$\int_0^T \left|\sum_1^N a_n n^{-it}\right|^2 dt = \sum_1^N |a_n|^2 \{T + O(n \log 2n)\}.$$

However Montgomery and Vaughan [1] have given a superior result, namely

$$\int_0^T \left|\sum_1^N a_n n^{-it}\right|^2 dt = \sum_1^N |a_n|^2 \{T + O(n)\}. \tag{7.20.1}$$

Ramachandra [2] has given an alternative proof of this result. Both proofs are more complicated than the argument leading to (7.2.1). However (7.20.1) has the advantage of dealing with the mean value of $\zeta(s)$ uniformly for $\sigma \geqslant \frac{1}{2}$. Suppose for example that $\sigma = \frac{1}{2}$. One takes $x = 2T$ in Theorem 4.11, whence

$$\zeta(\tfrac{1}{2}+it) = \sum_{n \leqslant 2T} n^{-\frac{1}{2}-it} + O(T^{-\frac{1}{2}}) = Z + O(T^{-\frac{1}{2}}),$$

say, for $T \leqslant t \leqslant 2T$. Then

$$\int_T^{2T} |Z|^2 dt = \sum_{n \leqslant 2T} n^{-1}\{T + O(n)\} = T \log T + O(T).$$

Moreover $Z \ll T^{\frac{1}{2}}$, whence

$$\int_T^{2T} |Z| T^{-\frac{1}{2}} dt \ll T.$$

Then, since

$$\int_T^{2T} O(T^{-\frac{1}{2}})^2 dt = O(1),$$

we conclude that

$$\int_T^{2T} |\zeta(\tfrac{1}{2}+it)|^2 dt = T \log T + O(T),$$

and Theorem 7.3 follows (with error term $O(T)$) on summing over $\frac{1}{2}T, \frac{1}{4}T, \ldots$. In particular we see that Theorem 4.11 is sufficient for this purpose, contrary to Titchmarsh's remark at the beginning of §7.3.

We now write

$$\int_0^T |\zeta(\tfrac{1}{2}+it)|^2\,dt = T\log\left(\frac{T}{2\pi}\right)+(2\gamma-1)\,T+E(T).$$

Much further work has been done concerning the error term $E(T)$. It has been shown by Balasubramanian [1] that $E(T) \ll T^{\frac{1}{3}+\varepsilon}$. A different proof was given by Heath-Brown [4]. The estimate may be improved slightly by using exponential sums, and Ivić [3; Corollary 15.4] has sketched the argument leading to the exponent $\frac{35}{108}+\varepsilon$, using a lemma due to Kolesnik [4]. It is no coincidence that this is twice the exponent occurring in Kolesnik's estimate for $\mu(\frac{1}{2})$, since one has the following result.

LEMMA 7.20. *Let k be a fixed positive integer and let $t \geqslant 2$. Then*

$$\zeta(\tfrac{1}{2}+it)^k \ll (\log t)\left(1+\int_{-\log^2 t}^{\log^2 t} |\zeta(\tfrac{1}{2}+it+iu)|^k e^{-|u|}\,du\right). \tag{7.20.2}$$

This is a trivial generalization of Lemma 3 of Heath-Brown [2], which is the case $k = 2$. It follows that

$$\zeta(\tfrac{1}{2}+it)^2 \ll (\log t)^4+(\log t)\max E\{t\pm(\log t)^2\}. \tag{7.20.3}$$

Thus, if μ is the infimum of those α for which $E(T) \ll T^{\alpha}$, then $\mu(\frac{1}{2}) \leqslant \frac{1}{2}\mu$. On the other hand, an examination of the initial stages of the process for estimating $\zeta(\frac{1}{2}+it)$ by van der Corput's method shows that one is, in effect, bounding the mean square of $\zeta(\frac{1}{2}+it)$ over a short range $(t-\Delta, t+\Delta)$. Thus it appears that one can hope for nothing better for $\mu(\frac{1}{2})$, by this method, than is given by (7.20.3).

The connection between estimates for $\zeta(\frac{1}{2}+it)$ and those for $E(T)$ should not be pushed too far however, for Good [1] has shown that $E(T) = \Omega(T^{\frac{1}{4}})$. Indeed Heath-Brown [1] later gave the asymptotic formula

$$\int_0^T E(t)^2\,dt = \tfrac{2}{3}(2\pi)^{-\frac{1}{2}}\frac{\zeta(\frac{3}{2})^4}{\zeta(3)}T^{\frac{3}{2}}+O(T^{\frac{5}{4}}\log^2 T) \tag{7.20.4}$$

from which the above Ω-result is immediate. It is perhaps of interest to note that the error term of (7.20.4) is $\Omega(T^{\frac{3}{4}}(\log T)^{-\frac{1}{4}})$. This has been proved by Meurman in unpublished work, using an argument analogous to that employed in the proof of Lemma α in §14.13. He has also reduced the error term in (7.20.4) to $O(T(\log T)^5)$, so as to include Balasubramanian's bound $E(T) \ll T^{\frac{1}{3}+\varepsilon}$.

Higher mean-values of $E(T)$ have been investigated by Ivić [**1**] who showed, for example, that

$$\int_0^T E(t)^8\,dt \ll T^{3+\varepsilon}. \tag{7.20.5}$$

This readily implies the estimate $E(T) \ll T^{\frac{1}{3}+\varepsilon}$.

The mean-value theorems of Heath-Brown and Ivić depend on a remarkable formula for $E(T)$ due to Atkinson [**1**]. Let $0 < A < A'$ be constants and suppose $AT \leqslant N \leqslant A'T$. Put

$$N' = N'(T) = \frac{T}{2\pi} + \frac{N}{2} - \left(\frac{NT}{2\pi} + \frac{N^2}{4}\right)^{\frac{1}{2}}.$$

Then $E(T) = \Sigma_1 + \Sigma_2 + O(\log^2 T)$, where

$$\Sigma_1 = 2^{-\frac{1}{2}} \sum_{n \leqslant N} (-1)^n d(n) \left(\frac{nT}{2\pi} + \frac{n^2}{4}\right)^{-\frac{1}{4}} \left\{\sinh^{-1}\left(\frac{\pi n}{2T}\right)^{\frac{1}{2}}\right\}^{-1} \sin f(n) \tag{7.20.6}$$

with

$$f(n) = \tfrac{1}{4}\pi + 2T \sinh^{-1}\left(\frac{\pi n}{2T}\right)^{\frac{1}{2}} + (\pi^2 n^2 + 2\pi n T)^{\frac{1}{2}}, \tag{7.20.7}$$

and

$$\Sigma_2 = 2 \sum_{n \leqslant N'} d(n) n^{-\frac{1}{2}} \left(\log \frac{T}{2\pi n}\right)^{-1} \sin g(n)$$

where

$$g(n) = T \log \frac{T}{2\pi n} - T - \tfrac{1}{4}\pi.$$

Atkinson loses a minus sign on [**1**; p 375]. This is corrected above. In applications of the above formula one can usually show that Σ_2 may be ignored. On the Lindelöf hypothesis, for example, one has

$$\sum_{n \leqslant x} d(n) n^{-\frac{1}{2}-iT} \ll T^{\varepsilon}$$

for $x \ll T$, so that $\Sigma_2 \ll T^{\varepsilon}$ by partial summation; and in general one finds $\Sigma_2 \ll T^{2\mu(\frac{1}{2})+\varepsilon}$. The sum Σ_1 is closely analogous to that occuring in the explicit formula (12.4.4) for $\Delta(x)$ in Dirichlet's divisor problem. Indeed, if $n = o(T^{\frac{1}{3}})$ then the summands of (7.20.6) are

$$(-1)^n \left(\frac{2T}{\pi}\right)^{\frac{1}{4}} \frac{d(n)}{n^{\frac{3}{4}}} \cos\left(2\sqrt{(2\pi n T)} - \frac{\pi}{4}\right) + o\left(T^{\frac{1}{4}} \frac{d(n)}{n^{\frac{3}{4}}}\right).$$

7.21. Ingham's result has been improved by Heath-Brown [**4**] to give

$$\int_0^T |\zeta(\tfrac{1}{2}+it)|^4\,dt = \sum_{n=0}^{4} c_n T(\log T)^n + O(T^{\frac{7}{8}+\varepsilon}) \tag{7.21.1}$$

where $c_4 = (2\pi^2)^{-1}$ and

$$c_3 = 2\{4\gamma - 1 - \log(2\pi) - 12\zeta'(2)\pi^{-2}\}\pi^{-2}.$$

The proof requires an asymptotic formula for

$$\sum_{n \leqslant N} d(n)\,d(n+r)$$

with a good error term, uniform in r. Such estimates are obtained in Heath-Brown [**4**] by applying Weil's bound for the Kloosterman sum (see §7.24).

7.22. Better estimates for σ_k are now available. In particular we have $\sigma_3 \leqslant \frac{7}{12}$ and $\sigma_4 \leqslant \frac{5}{8}$. The result on σ_4 is due to Heath-Brown [**8**]. To deduce the estimate for σ_3 one merely uses Gabriel's convexity theorem (see § 9.19), taking $\alpha = \frac{1}{2}$, $\beta = \frac{5}{8}$, $\lambda = \frac{1}{4}$, $\mu = \frac{1}{8}$, and $\sigma = \frac{7}{12}$.

The key ingredient required to obtain $\sigma_4 \leqslant \frac{5}{8}$ is the estimate

$$\int_0^T |\zeta(\tfrac{1}{2}+it)|^{12}\,dt \ll T^2(\log T)^{17} \tag{7.22.1}$$

of Heath-Brown [**2**]. According to (7.20.2) this implies the bound $\mu(\frac{1}{2}) \leqslant \frac{1}{6}$. In fact, in establishing (7.22.1) it is shown that, if $|\zeta(\frac{1}{2}+it_r)| \geqslant V(>0)$ for $1 \leqslant r \leqslant R$, where $0 < t_r \leqslant T$ and $t_{r+1} - t_r \geqslant 1$, then

$$R \ll T^2 V^{-12}(\log T)^{16},$$

and, if $V \geqslant T^{\frac{2}{13}}(\log T)^2$, then

$$R \ll TV^{-6}(\log T)^8.$$

Thus one sees not only that $\zeta(\frac{1}{2}+it) \ll t^{\frac{1}{6}}(\log t)^{\frac{4}{3}}$, but also that the number of points at which this bound is close to being attained is very small. Moreover, for $V \geqslant T^{\frac{2}{13}}(\log T)^2$, the behaviour corresponds to the, as yet unproven, estimate

$$\int_0^T |\zeta(\tfrac{1}{2}+it)|^6\,dt \ll T^{1+\varepsilon}.$$

To prove (7.22.1) one uses Atkinson's formula for $E(T)$ (see §7.20) to show that

$$\int_{T-G}^{T+G} |\zeta(\tfrac{1}{2}+it)|^2\,dt \ll G\log T +$$

$$G\sum_K (TK)^{-\frac{1}{4}}\left(|S(K)| + K^{-1}\int_0^K |S(x)|\,dx\right)e^{-G^2K/T}, \tag{7.22.2}$$

where K runs over powers of 2 in the range $T^{\frac{1}{3}} \leqslant K \leqslant TG^{-2}\log^3 T$, and

$$S(x) = S(x, K, T) = \sum_{K<n\leqslant K+x} (-1)^n d(n) e^{if(n)}$$

with $f(n)$ as in (7.20.7). The bound (7.22.2) holds uniformly for $\log^2 T \leqslant G \leqslant T^{\frac{5}{12}}$. In order to obtain the estimate (7.22.1) one proceeds to estimate how often the sum $S(x, K, T)$ can be large, for varying T. This is done by using a variant of Halász's method, as described in §9.28.

By following similar ideas, Graham, in work in the process of publication, has obtained

$$\int_0^T |\zeta(\tfrac{5}{7}+it)|^{196}\,dt \ll T^{14}(\log T)^{425}. \tag{7.22.3}$$

Of course there is no analogue of Atkinson's formula available here, and so the proof is considerably more involved. The result (7.22.3) contains the estimate $\mu(\frac{5}{7}) \leqslant \frac{1}{14}$ (which is the case $l = 4$ of Theorem 5.14) in the same way that (7.22.1) implies $\mu(\frac{1}{2}) \leqslant \frac{1}{6}$.

7.23. As in §7.9, one may define σ_k, for all positive real k, as the infimum of those σ for which (7.9.1) holds, and σ'_k similarly, for (7.9.2).

Then it is still true that $\sigma_k = \sigma'_k$, and that

$$\int_1^T |\zeta(\sigma+it)|^{2k}\,dt = T\sum_1^\infty d_k(n)^2 n^{-2\sigma} + O(T^{1-\delta})$$

for $\sigma > \sigma_k$, where $\delta = (\sigma, k) > 0$ may be explicitly determined. This may be proved by the method of Haselgrove [**1**]; see also Turganaliev [**1**]. In particular one may take $\delta(\sigma, \frac{1}{2}) = \frac{1}{2}(\sigma - \frac{1}{2})$ for $\frac{1}{2} < \sigma < 1$ (Ivić [**3**; (8.111)] or Turganaliev [**1**]). For some quite general approaches to these fractional moments the reader should consult Ingham (4) and Bohr and Jessen (4).

Mean values for $\sigma = \frac{1}{2}$ are far more difficult, and in no case other than $k = 1$ or 2 is an asymptotic formula for

$$\int_0^T |\zeta(\tfrac{1}{2}+it)|^{2k}\,dt = I_k(T),$$

say, known, even assuming the Riemann hypothesis. However Heath-Brown [**7**] has shown that

$$T(\log T)^{k^2} \ll I_k(T) \ll T(\log T)^{k^2} \quad \left(k = \frac{1}{n}\right),$$

Ramachandra [**3**], [**4**] having previously dealt with the case $k = \frac{1}{2}$. Jutila [**4**] observed that the implied constants may be taken to be independent of k. We also have

$$I_k(T) \gg T(\log T)^{k^2}$$

for any positive rational k. This is due to Ramachandra [**4**] when k is half an integer, and to Heath-Brown [**7**] in the remaining cases. (Titchmarsh [**1**; Theorem 29] states such a result for positive integral k, but the reference given there seems to yield only Theorem 7.19, which is weaker.) When k is irrational the best result known is Ramachandra's estimate [**5**]

$$I_k(T) \gg T(\log T)^{k^2}(\log\log T)^{-k^2}.$$

If one assumes the Riemann hypothesis one can obtain the better results

$$I_k(T) \ll T(\log T)^{k^2} \quad (0 \leqslant k \leqslant 2)$$

and

$$I_k(T) \gg T(\log T)^{k^2} \quad (k \geqslant 0), \tag{7.23.1}$$

for which see Ramachandra [4] or Heath-Brown [7]. Conrey and Ghosh [1] have given a particularly simple proof of (7.23.1) in the form

$$I_k(T) \geqslant \{C_k + o(1)\}\, T(\log T)^{k^2},$$

with

$$C_k = \{\Gamma(k^2+1)\}^{-1} \prod_p \left\{\left(1-\frac{1}{p}\right)^{k^2} \sum_{m=0}^{\infty} \left(\frac{\Gamma(k+m)}{m!\,\Gamma(k)}\right)^2 p^{-m}\right\}.$$

They suggest that this relation may even hold with equality (as it does when $k = 1$ or 2).

7.24. The work of Atkinson (2) alluded to at the end of §7.16 is of special historical interest, since it contains the first occurence of Kloosterman sums in the subject. These sums are defined by

$$S(q;\, a,\, b) = \sum_{\substack{n=1\\(n,\, q)=1}}^{q} \exp\left(\frac{2\pi i}{q}(an + b\bar{n})\right), \tag{7.24.1}$$

where $n\bar{n} \equiv 1 \pmod q$. Such sums have been of great importance in recent work, notably that of Heath-Brown [4] mentioned in §7.21, and of Iwaniec [1] and Deshouillers and Iwaniec [2], [3] referred to later in this section. The key fact about these sums is the estimate

$$|S(q;\, a,\, b)| \leqslant d(q)\, q^{\frac{1}{2}} (q,\, a,\, b)^{\frac{1}{2}}, \tag{7.24.2}$$

which indicates a very considerable amount of cancellation in (7.24.1). This result is due to Weil [1] when q is prime (the most important case) and to Estermann [2] in general. Weil's proof uses deep methods from algebraic geometry. It is possible to obtain further cancellations by averaging $S(q;\, a,\, b)$ over q, a and b. In order to do this one employs the theory of non-holomorphic modular forms, as in the work of Deshouillers and Iwaniec [1]. This is perhaps the most profound area of current research in the subject.

One way to see how Kloosterman sums arise is to use (7.15.2). Suppose for example one considers

$$\int_0^{\infty} |\zeta(\tfrac{1}{2}+it)|^2 \left|\sum_{u \leqslant U} u^{-it}\right|^2 e^{-t/T} dt. \tag{7.24.3}$$

Applying (7.15.2) with $2\delta = 1/T + i\log(v/u)$ one is led to examine

$$\sum_{n=1}^{\infty} d(n) \exp\left(\frac{2\pi i n u}{v} e^{i/T}\right).$$

One may now replace $e^{i/T}$ by $1+(i/T)$ with negligible error, producing

$$\sum_{n=1}^{\infty} d(n)\exp\left(\frac{2\pi inu}{v}\right)\exp\left(-\frac{2\pi nu}{vT}\right) = \frac{1}{2\pi i}\int_{2-i\infty}^{2+i\infty} \Gamma(s)\left(\frac{Tv}{2\pi u}\right)^{s} D\left(s, \frac{u}{v}\right) ds$$

where

$$D\left(s, \frac{u}{v}\right) = \sum_{n=1}^{\infty} d(n)\exp\left(\frac{2\pi inu}{v}\right) n^{-s}.$$

This Dirichlet series was investigated by Estermann [**1**], using the function $\zeta(s, a)$ of §2.17. It has an analytic continuation to the whole complex plane, and satisfies the functional equation

$$D\left(s, \frac{u}{v}\right) = 2v^{1-2s}\frac{\Gamma(1-s)^2}{(2\pi)^{2-2s}}\left\{D\left(1-s, \frac{\bar{u}}{v}\right) - \cos(\pi s)\, D\left(1-s, -\frac{\bar{u}}{v}\right)\right\}$$

providing that $(u, v) = 1$. To evaluate our original integral (7.24.3) it is necessary to average over u and v, so that one is led to consider

$$\sum_{\substack{u,v\leqslant U\\(u,v)=1}} D\left(1-s, \frac{\bar{u}}{v}\right) = \sum_{v\leqslant U}\sum_{n=1}^{\infty} d(n)n^{s-1} \sum_{\substack{u\leqslant U\\(u,v)=1}} \exp\left(\frac{2\pi in\bar{u}}{v}\right),$$

for example. In order to get a sharp bound for the innermost sum on the right one introduces the Kloosterman sum:

$$\begin{aligned}\sum_{\substack{u\leqslant U\\(u,v)=1}} \exp\left(\frac{2\pi in\bar{u}}{v}\right) &= \sum_{\substack{m=1\\(m,v)=1}}^{v} \exp\left(\frac{2\pi in\bar{m}}{v}\right) \sum_{\substack{u\leqslant U\\u\equiv m(\text{mod } v)}} 1\\ &= \sum_{\substack{m=1\\(m,v)=1}}^{v} \exp\left(\frac{2\pi in\bar{m}}{v}\right) \sum_{u\leqslant U}\left\{\frac{1}{v}\sum_{a=1}^{v} \exp\left(\frac{2\pi ia(m-u)}{v}\right)\right\}\\ &= \frac{1}{v}\sum_{a=1}^{v} S(v; a, n) \sum_{u\leqslant U} \exp\left(-\frac{2\pi iau}{v}\right),\end{aligned}$$

and one can now get a significant saving by using (7.24.2). Notice also that $S(v; a, n)$ is averaged over v, a and n, so that estimates for averages of Kloosterman sums are potentially applicable.

By pursuing such ideas and exploiting the connection with non-holomorphic modular forms, Iwaniec [**1**] showed that

$$\sum_{1}^{R}\int_{t_r}^{t_r+\Delta} |\zeta(\tfrac{1}{2}+it)|^4\, dt \ll (R\Delta + TR^{\frac{1}{2}}\Delta^{-\frac{1}{2}})\, T^{\varepsilon}$$

for $0 \leqslant t_r \leqslant T$, $t_{r+1} - t_r \geqslant \Delta \geqslant T^{\frac{1}{2}}$. In particular, taking $R = 1$, one has

$$\int_T^{T+T^{\frac{2}{3}}} |\zeta(\tfrac{1}{2}+it)|^4 \, dt \ll T^{\frac{2}{3}+\varepsilon} \tag{7.24.4}$$

which again implies $\mu(\frac{1}{2}) \leqslant \frac{1}{6}$, by (7.20.2). Moreover, by a suitable choice of the points t_r one can deduce (7.22.1), with $T^{2+\varepsilon}$ on the right.

Mean-value theorems involving general Dirichlet polynomials and partial sums of the zeta function are of interest, particularly in connection with the problems considered in Chapters 9 and 10. Such results may be proved by the methods of this chapter, but sharper estimates can be obtained by using Kloosterman sums and their connection with modular forms. Thus Deshouillers and Iwaniec [**2**], [**3**] established the bounds

$$\int_0^T |\zeta(\tfrac{1}{2}+it)|^4 \left| \sum_{n \leqslant N} a_n n^{it} \right|^2 dt \ll T^{\varepsilon}(T + T^{\frac{1}{2}}N^2 + T^{\frac{3}{4}}N^{\frac{5}{4}}) \sum_{n \leqslant N} |a_n|^2 \tag{7.24.5}$$

and

$$\int_0^T |\zeta(\tfrac{1}{2}+it)|^2 \left| \sum_{m \leqslant M} a_m m^{it} \right|^2 \left| \sum_{n \leqslant N} b_n n^{it} \right|^2 dt$$

$$\ll T^{\varepsilon}(T + T^{\frac{1}{2}}M^{\frac{3}{4}}N + T^{\frac{1}{2}}MN^{\frac{1}{2}} + M^{\frac{7}{4}}N^{\frac{3}{2}}) \left(\sum_{m \leqslant M} |a_m|^2 \right) \left(\sum_{n \leqslant N} |b_n|^2 \right) \tag{7.24.6}$$

for $N \leqslant M$. In a similar vein Balasubramanian, Conrey, and Heath-Brown [**1**] showed that

$$\int_0^T |\zeta(\tfrac{1}{2}+it)|^2 \left| \sum_{m \leqslant M} \mu(m) F(m) m^{-\frac{1}{2}-it} \right|^2 dt = CT + O_A\{T(\log T)^{-A}\}, \tag{7.24.7}$$

$$C = \sum_{m,n \leqslant M} \frac{\mu(m)\mu(n)}{mn} F(m)\overline{F(n)}\,(m,n) \left(\log \frac{T(m,n)^2}{2\pi mn} + 2\gamma - 1 \right)$$

for $M \leqslant T^{\frac{9}{17}-\varepsilon}$, where A is any positive constant, and the function F satisfies $F(x) \ll 1$, $F'(x) \ll x^{-1}$. The proof requires Weil's estimate for the Kloosterman sum, if $T^{\frac{1}{2}} \leqslant M \leqslant T^{\frac{9}{17}-\varepsilon}$.

VIII

Ω-THEOREMS

8.1. Introduction. The previous chapters have been largely concerned with what we may call O-theorems, i.e. results of the form

$$\zeta(s) = O\{f(t)\}, \qquad 1/\zeta(s) = O\{g(t)\}$$

for certain values of σ.

In this chapter we prove a corresponding set of Ω-theorems, i.e. results of the form

$$\zeta(s) = \Omega\{\phi(t)\}, \qquad 1/\zeta(s) = \Omega\{\psi(t)\},$$

the Ω symbol being defined as the negation of o, so that $F(t) = \Omega\{\phi(t)\}$ means that the inequality $|F(t)| > A\phi(t)$ is satisfied for some arbitrarily large values of t.

If, for a given function $F(t)$, we have both

$$F(t) = O\{f(t)\}, \qquad F(t) = \Omega\{f(t)\},$$

we may say that the order of $F(t)$ is determined, and the only remaining question is that of the actual constants involved.

For $\sigma > 1$, the problems of $\zeta(\sigma+it)$ and $1/\zeta(\sigma+it)$ are both solved. For $\frac{1}{2} \leqslant \sigma \leqslant 1$ there remains a considerable gap between the O-results of Chapters V–VI and the Ω-results of the present chapter. We shall see later that, on the Riemann hypothesis, it is the Ω-results which represent the real truth, and the O-results which fall short of it. We are always more successful with Ω-theorems. This is perhaps not surprising, since an O-result is a statement about all large values of t, an Ω-result about some indefinitely large values only.

8.2. The first Ω results were obtained by means of Diophantine approximation, i.e. the approximate solution in integers of given equations. The following two theorems are used.

Dirichlet's Theorem. *Given N real numbers $a_1, a_2,..., a_N$, a positive integer q, and a positive number t_0, we can find a number t in the range*

$$t_0 \leqslant t \leqslant t_0 q^N, \tag{8.2.1}$$

and integers $x_1, x_2,..., x_N$, such that

$$|ta_n - x_n| \leqslant 1/q \quad (n = 1, 2,..., N). \tag{8.2.2}$$

The proof is based on an argument which was introduced and employed extensively by Dirichlet. This argument, in its simplest form, is that, if there are $m+1$ points in m regions, there must be at least one region which contains at least two points.

Consider the N-dimensional unit cube with a vertex at the origin and edges along the coordinate axes. Divide each edge into q equal parts, and thus the cube into q^N equal compartments. Consider the q^N+1 points, in the cube, congruent (mod 1) to the points $(ua_1, ua_2,..., ua_N)$, where $u = 0, t_0, 2t_0,..., q^N t_0$. At least two of these points must lie in the same compartment. If these two points correspond to $u = u_1$, $u = u_2$ $(u_1 < u_2)$, then $t = u_2 - u_1$ clearly satisfies the requirements of the theorem.

The theorem may be extended as follows. Suppose that we give u the values $0, t_0, 2t_0,..., mq^N t_0$. We obtain mq^N+1 points, of which one compartment must contain at least $m+1$. Let these points correspond to $u = u_1,..., u_{m+1}$. Then $t = u_2-u_1,..., u_m-u_1$, all satisfy the requirements of the theorem.

We conclude that the interval $(t_0, mq^N t_0)$ contains at least m solutions of the inequalities (8.2.2), any two solutions differing by at least t_0.

8.3. Kronecker's Theorem. *Let $a_1, a_2,..., a_N$ be linearly independent real numbers, i.e. numbers such that there is no linear relation*

$$\lambda_1 a_1 + ... + \lambda_N a_N = 0$$

in which the coefficients $\lambda_1,...$ are integers not all zero. Let $b_1,..., b_N$ be any real numbers, and q a given positive number. Then we can find a number t and integers $x_1,..., x_N$, such that

$$|ta_n - b_n - x_n| \leqslant 1/q \quad (n = 1, 2,..., N). \tag{8.3.1}$$

If all the numbers b_n are 0, the result is included in Dirichlet's theorem. In the general case, we have to suppose the a_n linearly independent; for example, if the a_n are all zero, and the b_n are not all integers, there is in general no t satisfying (8.3.1). Also the theorem assigns no upper bound for the number t such as the q^N of Dirichlet's theorem. This makes a considerable difference to the results which can be deduced from the two theorems.

Many proofs of Kronecker's theorem are known.† The following is due to Bohr (15).

We require the following lemma

Lemma. *If $\phi(x)$ is positive and continuous for $a \leqslant x \leqslant b$, then*

$$\lim_{n\to\infty} \left\{ \int_a^b \{\phi(x)\}^n \, dx \right\}^{1/n} = \max_{a \leqslant x \leqslant b} \phi(x).$$

A similar result holds for an integral in any number of dimensions.

† Bohr (15), (16), Bohr and Jessen (3), Estermann (3), Lettenmeyer (1).

Let $M = \max \phi(x)$. Then

$$\left\{\int_a^b \{\phi(x)\}^n\,dx\right\}^{1/n} \leqslant \{(b-a)M^n\}^{1/n} = (b-a)^{1/n}M.$$

Also, given ϵ, there is an interval, (α, β) say, throughout which

$$\phi(x) \geqslant M-\epsilon.$$

Hence

$$\left\{\int_a^b \{\phi(x)\}^n\,dx\right\}^{1/n} \geqslant \{(\beta-\alpha)(M-\epsilon)^n\}^{1/n} = (\beta-\alpha)^{1/n}(M-\epsilon),$$

and the result is clear. A similar proof holds in the general case.

Proof of Kronecker's theorem. It is sufficient to prove that we can find a number t such that each of the numbers

$$e^{2\pi i(a_n t-b_n)} \quad (n = 1, 2,..., N)$$

differs from 1 by less than a given ϵ; or, if

$$F(t) = 1+\sum_{n=1}^{N} e^{2\pi i(a_n t-b_n)},$$

that the upper bound of $|F(t)|$ for real values of t is $N+1$. Let us denote this upper bound by L. Clearly $L \leqslant N+1$.

Let $$G(\phi_1, \phi_2,..., \phi_N) = 1+\sum_{n=1}^{N} e^{2\pi i\phi_n},$$

where the numbers ϕ_1, ϕ_2,..., ϕ_N are independent real variables, each lying in the interval $(0, 1)$. Then the upper bound of $|G|$ is $N+1$, this being the value of $|G|$ when $\phi_1 = \phi_2 = ... = \phi_N = 0$.

We consider the polynomial expansions of $\{F(t)\}^k$ and $\{G(\phi_1,..., \phi_N)\}^k$, where k is an arbitrary positive integer; and we observe that each of these expansions contains the same number of terms. For, the numbers a_1, a_2,..., a_N being linearly independent, no two terms in the expansion of $\{F(t)\}^k$ fall together. Also the moduli of corresponding terms are equal. Thus if

$$\{G(\phi_1,..., \phi_N)\}^k = 1+\sum C_q\, e^{2\pi i(\lambda_{q,1}\phi_1+...+\lambda_{q,N}\phi_N)},$$

then
$$\{F(t)\}^k = 1+\sum C_q\, e^{2\pi i\{\lambda_{q,1}(a_1 t-b_1)+...+\lambda_{q,N}(a_N t-b_N)\}}$$
$$= 1+\sum C_q\, e^{2\pi i(\alpha_q t-\beta_q)},$$

say. Now the mean values

$$F_k = \lim_{T\to\infty} \frac{1}{2T}\int_{-T}^{T} |F(t)|^{2k}\,dt$$

and
$$G_k = \int_0^1 \int_0^1 \dots \int_0^1 |G(\phi_1,\dots,\phi_N)|^{2k}\, d\phi_1 \dots d\phi_N$$
are equal, each being equal to
$$1+\sum C_q^2.$$
This is easily seen in each case on expressing the squared modulus as a product of conjugates and integrating term by term.

Since $N+1$ is the upper bound of $|G|$, the lemma gives
$$\lim_{k\to\infty} G_k^{1/2k} = N+1.$$
Hence also
$$\lim_{k\to\infty} F_k^{1/2k} = N+1.$$
But plainly
$$F_k^{1/2k} \leqslant L$$
for all values of k. Hence $L \geqslant N+1$, and so in fact $L = N+1$. This proves the theorem.

8.4. THEOREM 8.4. *If $\sigma > 1$, then*
$$|\zeta(s)| \leqslant \zeta(\sigma) \tag{8.4.1}$$
for all values of t, while
$$|\zeta(s)| \geqslant (1-\epsilon)\zeta(\sigma) \tag{8.4.2}$$
for some indefinitely large values of t.

We have
$$|\zeta(s)| = \left|\sum_{n=1}^{\infty} n^{-s}\right| \leqslant \sum_{n=1}^{\infty} n^{-\sigma} = \zeta(\sigma),$$
so that the whole difficulty lies in the second part. To prove this we use Dirichlet's theorem. For all values of N
$$\zeta(s) = \sum_{n=1}^{N} n^{-\sigma} e^{-it\log n} + \sum_{n=N+1}^{\infty} n^{-\sigma-it},$$
and hence (the modulus of the first sum being not less than its real part)
$$|\zeta(s)| \geqslant \sum_{n=1}^{N} n^{-\sigma}\cos(t\log n) - \sum_{n=N+1}^{\infty} n^{-\sigma}. \tag{8.4.3}$$
By Dirichlet's theorem there is a number t ($t_0 \leqslant t \leqslant t_0 q^N$) and integers $x_1,\dots, x_N$, such that, for given N and q ($q \geqslant 4$),
$$\left|\frac{t\log n}{2\pi} - x_n\right| \leqslant \frac{1}{q} \quad (n = 1, 2,\dots, N).$$
Hence $\cos(t\log n) \geqslant \cos(2\pi/q)$ for these values of n, and so
$$\sum_{n=1}^{N} n^{-\sigma}\cos(t\log n) \geqslant \cos(2\pi/q)\sum_{n=1}^{N} n^{-\sigma} > \cos(2\pi/q)\zeta(\sigma) - \sum_{n=N+1}^{\infty} n^{-\sigma}.$$

Hence by (8.4.3)

$$|\zeta(s)| \geqslant \cos(2\pi/q)\zeta(\sigma) - 2\sum_{N+1}^{\infty} n^{-\sigma}.$$

Now
$$\zeta(\sigma) = \sum_{n=1}^{\infty} n^{-\sigma} > \int_1^{\infty} u^{-\sigma}\,du = \frac{1}{\sigma-1},$$

and
$$\sum_{N+1}^{\infty} n^{-\sigma} < \int_N^{\infty} u^{-\sigma}\,du = \frac{N^{1-\sigma}}{\sigma-1}.$$

Hence
$$|\zeta(s)| \geqslant \{\cos(2\pi/q) - 2N^{1-\sigma}\}\zeta(\sigma), \tag{8.4.4}$$

and the result follows if q and N are large enough.

THEOREM 8.4 (A). *The function $\zeta(s)$ is unbounded in the open region*

$$\sigma > 1,\ t > \delta > 0.$$

This follows at once from the previous theorem, since the upper bound $\zeta(\sigma)$ of $\zeta(s)$ itself tends to infinity as $\sigma \to 1$.

THEOREM 8.4 (B). *The function $\zeta(1+it)$ is unbounded as $t \to \infty$.*

This follows from the previous theorem and the theorem of Phragmén and Lindelöf. Since $\zeta(2+it)$ is bounded, if $\zeta(1+it)$ were also bounded $\zeta(s)$ would be bounded throughout the half-strip $1 \leqslant \sigma \leqslant 2$, $t > \delta$; and this is false, by the previous theorem.

8.5. Dirichlet's theorem also gives the following more precise result.†

THEOREM 8.5. *However large t_1 may be, there are values of s in the region $\sigma > 1$, $t > t_1$, for which*

$$|\zeta(s)| > A \log\log t. \tag{8.5.1}$$

Also
$$\zeta(1+it) = \Omega(\log\log t). \tag{8.5.2}$$

Take $t_0 = 1$ and $q = 6$ in the proof of Theorem 8.4. Then (8.4.4) gives

$$|\zeta(s)| \geqslant (\tfrac{1}{2} - 2N^{1-\sigma})/(\sigma-1) \tag{8.5.3}$$

for a value of t between 1 and 6^N. We choose N to be the integer next above $8^{1/(\sigma-1)}$. Then

$$|\zeta(s)| \geqslant \frac{1}{4(\sigma-1)} \geqslant \frac{\log(N-1)}{4\log 8} > A\log N \tag{8.5.4}$$

for a value of t such that $N > A\log t$. The required inequality (8.5.1) then follows from (8.5.4). It remains only to observe that the value of t in question must be greater than any assigned t_1, if $\sigma-1$ is sufficiently small; otherwise it would follow from (8.5.3) that $\zeta(s)$ was unbounded

† Bohr and Landau (1).

in the region $\sigma > 1$, $1 < t \leqslant t_1$; and we know that $\zeta(s)$ is bounded in any such region.

The second part of the theorem now follows from the first by the Phragmén–Lindelöf method. Consider the function

$$f(s) = \frac{\zeta(s)}{\log\log s},$$

the branch of $\log\log s$ which is real for $s > 1$, and is restricted to $|s| > 1$, $\sigma > 0$, $t > 0$ being taken. Then $f(s)$ is regular for $1 \leqslant \sigma \leqslant 2$, $t > \delta$. Also $|\log\log s| \sim \log\log t$ as $t \to \infty$, uniformly with respect to σ in the strip. Hence $f(2+it) \to 0$ as $t \to \infty$, and so, if $f(1+it) \to 0$, $f(s) \to 0$ uniformly in the strip.† This contradicts (8.5.1), and so (8.5.2) follows.

It is plain that arguments similar to the above may be applied to all Dirichlet series, with coefficients of fixed sign, which are not absolutely convergent on their line of convergence. For example, the series for $\log\zeta(s)$ and its differential coefficients are of this type. The result for $\log\zeta(s)$ is, however, a corollary of that for $\zeta(s)$, which gives at once

$$|\log\zeta(s)| > \log\log\log t - A$$

for some indefinitely large values of t in $\sigma > 1$. For the nth differential coefficient of $\log\zeta(s)$ the result is that

$$\left|\left(\frac{d}{ds}\right)^n \log\zeta(s)\right| > A_n(\log\log t)^n$$

for some indefinitely large values of t in $\sigma > 1$.

8.6. We now turn to the corresponding problems‡ for $1/\zeta(s)$. We cannot apply the argument depending on Dirichlet's theorem to this function, since the coefficients in the series

$$\frac{1}{\zeta(s)} = \sum_{n=1}^{\infty} \frac{\mu(n)}{n^s}$$

are not all of the same sign; nor can we argue similarly with Kronecker's theorem, since the numbers $(\log n)/2\pi$ are not linearly independent. Actually we consider $\log\zeta(s)$, which depends on the series $\sum p^{-s}$, to which Kronecker's theorem can be applied.

THEOREM 8.6. *The function* $1/\zeta(s)$ *is unbounded in the open region* $\sigma > 1$, $t > \delta > 0$.

We have for $\sigma \geqslant 1$

$$\left|\log\zeta(s) - \sum_p \frac{1}{p^s}\right| = \left|\sum_p \sum_{m=2}^{\infty} \frac{1}{mp^{ms}}\right| \leqslant \sum_p \sum_{m=2}^{\infty} \frac{1}{p^m} = \sum_p \frac{1}{p(p-1)} = A.$$

† See e.g. my *Theory of Functions*, § 5.63, with the angle transformed into a strip.
‡ Bohr and Landau (7).

Now

$$\mathbf{R}\left(\sum_p \frac{1}{p^s}\right) = \sum_{n=1}^{\infty} \frac{\cos(t\log p_n)}{p_n^\sigma} \leqslant \sum_{n=1}^{N} \frac{\cos(t\log p_n)}{p_n^\sigma} + \sum_{n=N+1}^{\infty} \frac{1}{p_n^\sigma}.$$

Also *the numbers* $\log p_n$ *are linearly independent.* For it follows from the theorem that an integer can be expressed as a product of prime factors in one way only, that there can be no relation of the form

$$p_1^{\lambda_1} p_2^{\lambda_2} \dots p_N^{\lambda_N} = 1,$$

where the λ's are integers, and therefore no relation of the form

$$\lambda_1 \log p_1 + \dots + \lambda_N \log p_N = 0.$$

Hence also the numbers $(\log p_n)/2\pi$ are linearly independent. It follows therefore from Kronecker's theorem that we can find a number t and integers $x_1, \dots, x_N$ such that

$$\left| t\frac{\log p_n}{2\pi} - \tfrac{1}{2} - x_n \right| \leqslant \tfrac{1}{6} \quad (n = 1, 2, \dots, N),$$

or

$$|t\log p_n - \pi - 2\pi x_n| \leqslant \tfrac{1}{3}\pi \quad (n = 1, 2, \dots, N).$$

Hence for these values of n

$$\cos(t\log p_n) = -\cos(t\log p_n - \pi - 2\pi x_n) \leqslant -\cos\tfrac{1}{3}\pi = -\tfrac{1}{2},$$

and hence

$$\mathbf{R}\left(\sum_p \frac{1}{p^s}\right) \leqslant -\frac{1}{2}\sum_{n=1}^{N} \frac{1}{p_n^\sigma} + \sum_{n=N+1}^{\infty} \frac{1}{p_n^\sigma}.$$

Since $\sum p_n^{-1}$ is divergent, we can, if H is any assigned positive number, choose σ so near to 1 that $\sum p_n^{-\sigma} > H$. Having fixed σ, we can choose N so large that

$$\sum_{n=1}^{N} p_n^{-\sigma} > \tfrac{3}{4}H, \qquad \sum_{n=N+1}^{\infty} p_n^{-\sigma} < \tfrac{1}{4}H.$$

Then

$$\mathbf{R}\left(\sum_p p^{-s}\right) < -\tfrac{3}{8}H + \tfrac{1}{4}H = -\tfrac{1}{8}H.$$

Since H may be as large as we please, it follows that $\mathbf{R}(\sum p^{-s})$, and so $\log|\zeta(s)|$, takes arbitrarily large negative values. This proves the theorem.

THEOREM 8.6 (A). *The function* $1/\zeta(1+it)$ *is unbounded as* $t \to \infty$.

This follows from the previous theorem in the same way as Theorem 8.4 (B) from Theorem 8.4 (A).

We cannot, however, proceed to deduce an analogue of Theorem 8.5 for $1/\zeta(s)$. In proving Theorem 8.5, each of the numbers $\cos(t\log n)$ has to be made as near as possible to 1, and this can be done by Dirichlet's theorem. In Theorem 8.6, each of the numbers $\cos(t\log p_n)$ has to be made as near as possible to -1, and this requires Kronecker's theorem.

Now Theorem 8.5 depends on the fact that we can assign an upper limit to the number t which satisfies the conditions of Dirichlet's theorem. Since there is no such upper limit in Kronecker's theorem, the corresponding argument for $1/\zeta(s)$ fails. We shall see later that the analogue of Theorem 8.5 is in fact true, but it requires a much more elaborate proof.

8.7. Before proceeding to these deeper theorems, we shall give another method of proving some of the above results.† This method deals directly with integrals of high powers of the functions in question, and so might be described as a short cut which avoids explicit use of Diophantine approximation.

We write
$$M\{|f(s)|^2\} = \lim_{T\to\infty} \frac{1}{2T} \int_{-T}^{T} |f(\sigma+it)|^2\,dt,$$
and prove the following lemma.

LEMMA. *Let*
$$g(s) = \sum_{m=1}^{\infty} \frac{b_m}{m^s}, \qquad h(s) = \sum_{n=1}^{\infty} \frac{c_n}{n^s}$$
be absolutely convergent for a given value of σ, *and let every* m *with* $b_m \neq 0$ *be prime to every* n *with* $c_n \neq 0$. *Then for such* σ
$$M\{|g(s)h(s)|^2\} = M\{|g(s)|^2\}M\{|h(s)|^2\}.$$

By Theorem 7.1
$$M\{|g(s)|^2\} = \sum_{m=1}^{\infty} \frac{|b_m|^2}{m^{2\sigma}}, \qquad M\{|h(s)|^2\} = \sum_{n=1}^{\infty} \frac{|c_n|^2}{n^{2\sigma}}.$$

Now
$$g(s)h(s) = \sum_{r=1}^{\infty} \frac{d_r}{r^s},$$
where each term $d_r r^{-s}$ is the product of two terms $b_m m^{-s}$ and $c_n n^{-s}$. Hence
$$M\{|g(s)h(s)|^2\} = \sum_{r=1}^{\infty} \frac{|d_r|^2}{r^{2\sigma}} = \sum\sum \frac{|b_m c_n|^2}{(mn)^{2\sigma}} = M\{|g(s)|^2\}M\{|h(s)|^2\}.$$

We can now prove the analogue for $1/\zeta(s)$ of Theorem 8.4.

THEOREM 8.7. *If* $\sigma > 1$, *then*
$$\left|\frac{1}{\zeta(s)}\right| \leqslant \frac{\zeta(\sigma)}{\zeta(2\sigma)} \tag{8.7.1}$$
for all values of t, *while*
$$\left|\frac{1}{\zeta(s)}\right| \geqslant (1-\epsilon)\frac{\zeta(\sigma)}{\zeta(2\sigma)} \tag{8.7.2}$$
for some indefinitely large values of t.

† Bohr and Landau (7).

We have, for $\sigma > 1$,

$$\left|\frac{1}{\zeta(s)}\right| = \left|\sum_{n=1}^{\infty}\frac{\mu(n)}{n^s}\right| \leqslant \sum_{n=1}^{\infty}\frac{|\mu(n)|}{n^\sigma}.$$

Since

$$\sum_{n=1}^{\infty}\frac{\mu(n)}{n^s} = \prod_p\left(1-\frac{1}{p^s}\right)$$

we have also

$$\sum_{n=1}^{\infty}\frac{|\mu(n)|}{n^\sigma} = \prod_p\left(1+\frac{1}{p^\sigma}\right) = \prod_p\left(\frac{1-p^{-2\sigma}}{1-p^{-\sigma}}\right) = \frac{\zeta(\sigma)}{\zeta(2\sigma)},$$

and the first part follows.

To prove the second part, write

$$\frac{1}{\zeta(s)} = \prod_{n=1}^{N}\left(1-\frac{1}{p_n^s}\right)\eta_N(s),$$

$$\frac{1}{\{\zeta(s)\}^k} = \prod_{n=1}^{N}\left(1-\frac{1}{p_n^s}\right)^k\{\eta_N(s)\}^k.$$

By repeated application of the lemma it follows that

$$M\left\{\frac{1}{|\zeta(s)|^{2k}}\right\} = \prod_{n=1}^{N}M\left\{\left|\left(1-\frac{1}{p_n^s}\right)\right|^{2k}\right\}M\{|\eta_N(s)|^{2k}\}.$$

Now, for every p,

$$M\left\{\left|1-\frac{1}{p^s}\right|^{2k}\right\} = \frac{\log p}{2\pi}\int_0^{2\pi/\log p}\left|1-\frac{1}{p^s}\right|^{2k}dt,$$

since the integrand is periodic with period $2\pi/\log p$; and

$$M\{|\eta_N(s)|^{2k}\} \geqslant 1,$$

since the Dirichlet series for $\{\eta_N(s)\}^k$ begins with $1+\ldots$. Hence

$$M\left\{\frac{1}{|\zeta(s)|^{2k}}\right\} \geqslant \prod_{n=1}^{N}\frac{\log p_n}{2\pi}\int_0^{2\pi/\log p_n}\left|1-\frac{1}{p_n^s}\right|^{2k}dt.$$

Now

$$\lim_{k\to\infty}\left\{\int_0^{2\pi/\log p}\left|1-\frac{1}{p^s}\right|^{2k}dt\right\}^{1/2k} = \max_{0\leqslant t\leqslant 2\pi/\log p}\left|1-\frac{1}{p^s}\right| = 1+\frac{1}{p^\sigma}.$$

Hence

$$\varliminf_{k\to\infty}\left[M\left\{\frac{1}{|\zeta(s)|^{2k}}\right\}\right]^{1/2k} \geqslant \prod_{n=1}^{N}\left(1+\frac{1}{p_n^\sigma}\right).$$

Since the left-hand side is independent of N, we can make $N\to\infty$ on the right, and obtain

$$\varliminf_{k\to\infty}\left[M\left\{\frac{1}{|\zeta(s)|^{2k}}\right\}\right]^{1/2k} \geqslant \frac{\zeta(\sigma)}{\zeta(2\sigma)}.$$

Hence to any ϵ corresponds a k such that

$$\left[M\left\{\frac{1}{|\zeta(s)|^{2k}}\right\}\right]^{1/2k} > (1-\epsilon)\frac{\zeta(\sigma)}{\zeta(2\sigma)},$$

and (8.7.2) now follows.

Since $\zeta(\sigma)/\zeta(2\sigma) \to \infty$ as $\sigma \to 1$, this also gives an alternative proof of Theorem 8.6

It is easy to see that a similar method can be used to prove Theorem 8.4 (A). It is also possible to prove Theorems 8.4 (B) and 8.6 (A) directly by this method without using the Phragmén–Lindelöf theorem. This, however, requires an extension of the general mean-value theorem for Dirichlet series.

8.8. THEOREM 8.8.† *However large t_0 may be, there are values of s in the region $\sigma > 1$, $t > t_0$ for which*

$$\left|\frac{1}{\zeta(s)}\right| > A \log\log t.$$

Also

$$\frac{1}{\zeta(1+it)} = \Omega(\log\log t).$$

As in the case of Theorem 8.5, it is enough to prove the first part. We first prove some lemmas. The object of these lemmas is to supply, for the particular case in hand, what Kronecker's theorem lacks in the general case, viz. an upper bound for the number t which satisfies the conditions (8.3.1).

LEMMA α. *If m and n are different positive integers,*

$$\left|\log\frac{m}{n}\right| > \frac{1}{\max(m,n)}.$$

For if $m < n$

$$\log\frac{n}{m} \geqslant \log\frac{n}{n-1} = \frac{1}{n}+\frac{1}{2}\frac{1}{n^2}+\ldots > \frac{1}{n}.$$

LEMMA β. *If $p_1,\ldots, p_N$ are the first N primes, and $\mu_1,\ldots, \mu_N$ are integers, not all 0 (not necessarily positive), then*

$$\left|\log\prod_{n=1}^{N} p_n^{\mu_n}\right| > p_N^{-\mu N} \quad (\mu = \max|\mu_n|).$$

For $\prod_{n=1}^{N} p_n^{\mu_n} = u/v$, where

$$u = \prod_{\mu_n>0} p_n^{\mu_n}, \qquad v = \prod_{\mu_n<0} p_n^{\mu_n},$$

† Bohr and Landau (7).

and u and v, being mutually prime, are different. Also

$$\max(u,v) \leqslant \prod_{n=1}^{N} p_n^{\mu} \leqslant p_N^{N\mu},$$

and the result follows from Lemma α.

LEMMA γ. *The number of solutions in positive or zero integers of the equation*

$$\nu_0+\nu_1+...+\nu_N = k$$

does not exceed $(k+1)^N$.

For $N = 1$ the number of solutions is $k+1$, so that the theorem holds. Suppose that it holds for any given N. Then for given ν_{N+1} the number of solutions of

$$\nu_0+\nu_1+...+\nu_N = k-\nu_{N+1}$$

does not exceed $(k-\nu_{N+1}+1)^N \leqslant (k+1)^N$; and ν_{N+1} can take $k+1$ values. Hence the total number of solutions is $\leqslant (k+1)^{N+1}$, whence the result.

LEMMA δ. *For $N > A$, there exits a t satisfying $0 \leqslant t \leqslant \exp(N^6)$ for which*

$$\cos(t\log p_n) < -1+\frac{1}{N} \quad (n \leqslant N).$$

Let $N > 1$, $k > 1$. Then

$$\Big(\sum_{n=0}^{N} x_n\Big)^k = \sum c(\nu_0,..., \nu_N)x_0^{\nu_0}... x_N^{\nu_N},$$

where
$$c(\nu_0,..., \nu_N) = \frac{k!}{\nu_0!...\nu_N!}, \qquad \sum \nu_n = k.$$

The number of distinct terms in the expansion is at most $(k+1)^N < k^{2N}$, by Lemma γ. Hence

$$(\sum c)^2 \leqslant \sum c^2 \sum 1 < k^{2N} \sum c^2,$$

so that
$$\sum c^2 > k^{-2N}(\sum c)^2 = k^{-2N}(N+1)^{2k}.$$

Let
$$F(t) = 1 - \sum_{n=1}^{N} e^{it\log p_n},$$

so that

$$\{F(t)\}^k = \sum c(\nu_0,..., \nu_N)(-1)^{\nu_1+...+\nu_N}\exp\Big(it\sum_1^N \nu_n \log p_n\Big),$$

$$|F(t)|^{2k} = \sum_{\nu,\nu'}\sum cc'(-1)^{\Sigma\nu_n+\Sigma\nu'_n}\exp\Big\{it\sum_n (\nu_n-\nu'_n)\log p_n\Big\}$$

$$= \Sigma_1 + \Sigma_2,$$

where Σ_1 is taken over values of (ν,ν') for which $\nu_1 = \nu'_1$, $\nu_2 = \nu'_2$,..., and

Σ_2 over the remainder. Now

$$\frac{1}{T}\int_0^T e^{i\alpha t}\,dt = 1 \quad (\alpha = 0),$$

$$\left|\frac{1}{T}\int_0^T e^{i\alpha t}\,dt\right| = \left|\frac{e^{i\alpha T}-1}{i\alpha T}\right| \leqslant \frac{2}{|\alpha|T} \quad (\alpha \neq 0).$$

Hence
$$\frac{1}{T}\int_0^T |F(t)|^{2k}\,dt \geqslant \Sigma_1\, c^2 - \Sigma_2 \frac{2cc'}{|\sum(\nu_n-\nu_n')\log p_n|T}.$$

By Lemma β, since the numbers $\nu_n-\nu_n'$ are not all 0,

$$|\sum(\nu_n-\nu_n')\log p_n| = \left|\log\prod_1^N p_n^{(\nu_n-\nu_n')}\right| > p_N^{-N\max|\nu_n-\nu_n'|} \geqslant p_N^{-kN}.$$

Hence

$$\frac{1}{T}\int_0^T |F(t)|^{2k}\,dt \geqslant \sum c^2 - \frac{2p_N^{kN}}{T}\sum\sum cc'$$

$$\geqslant k^{-2N}(\sum c)^2 - \frac{2p_N^{kN}}{T}(\sum c)^2$$

$$= \left(k^{-2N} - \frac{2p_N^{kN}}{T}\right)(N+1)^{2k}.$$

In this we take $k = N^4$, $T = e^{N^6}$, and obtain, for $N > A$,

$$k^{-2N} - \frac{2p_N^{kN}}{T} = N^{-8N} - 2\left(\frac{p_N}{e^N}\right)^{kN} > e^{-N^3/(N+1)}.$$

Hence

$$\left\{\frac{1}{T}\int_0^T |F(t)|^{2k}\,dt\right\}^{1/2k} \geqslant (N+1)e^{-1/\{2N(N+1)\}} > N+1-\frac{1}{2N}.$$

Hence there is a t in $(0, e^{N^6})$ such that

$$|F(t)| > N+1-\frac{1}{2N}.$$

Suppose, however, that $\cos(t\log p_n) \geqslant -1+1/N$ for some value of n. Then

$$|F(t)| \leqslant N-1+|1-e^{it\log p_n}| = N-1+\surd 2(1-\cos t\log p_n)^{\frac{1}{2}}$$

$$\leqslant N-1+\surd 2\left(2-\frac{1}{N}\right)^{\frac{1}{2}} \leqslant N+1-\frac{1}{2N},$$

a contradiction. Hence the result.

We can now prove Theorem 8.8. As in § 8.6, for $\sigma > 1$

$$\log\frac{1}{|\zeta(s)|} = -\sum\frac{\cos(t\log p_n)}{p_n^\sigma}+O(1).$$

Let now N be large, $t = t(N)$ the number of Lemma δ, $\delta = 1/\log N$, and $\sigma = 1+\delta$. Then

$$\log\frac{A}{|\zeta(s)|} \geqslant -\sum\frac{\cos(t\log p_n)}{p_n^\sigma} \geqslant \left(1-\frac{1}{N}\right)\sum_1^N\frac{1}{p_n^\sigma}-\sum_{N+1}^\infty\frac{1}{p_n^\sigma}$$

$$> \left(1-\frac{1}{N}\right)\sum\frac{1}{p^\sigma}-2\sum_{N+1}^\infty\frac{1}{p_n^\sigma} > \left(1-\frac{1}{N}\right)\{\log\zeta(\sigma)-A\}-2\sum_{N+1}^\infty\frac{1}{(An\log n)^\sigma}$$

$$> \left(1-\frac{1}{N}\right)\left(\log\frac{1}{\delta}-A\right)-\frac{A}{\log N}\sum_{N+1}^\infty\frac{1}{n^\sigma},$$

$$\log\frac{A\delta}{|\zeta(s)|} > -A-\frac{1}{N}\log\frac{1}{\delta}-\frac{A}{\log N}\frac{N^{1-\sigma}}{\sigma-1} > -A,$$

$$\frac{1}{|\zeta(s)|} > \frac{A}{\delta} = A\log N > A\log\log t.$$

The number $t = t(N)$ evidently tends to infinity with N, since $1/\zeta(s)$ is bounded in $|t| \leqslant A$, $\sigma \geqslant 1$, and the proof is completed.

8.9. In Theorems 8.5 and 8.8 we have proved that each of the inequalities

$$|\zeta(1+it)| > A\log\log t, \qquad 1/|\zeta(1+it)| > A\log\log t$$

is satisfied for some arbitrarily large values of t, if A is a suitable constant. We now consider the question how large the constant can be in the two cases.

Since neither $|\zeta(1+it)|/\log\log t$ nor $|\zeta(1+it)|^{-1}/\log\log t$ is known to be bounded, the question of the constants might not seem to be of much interest. But we shall see later that on the Riemann hypothesis they are both bounded; in fact if

$$\lambda = \overline{\lim_{t\to\infty}}\frac{|\zeta(1+it)|}{\log\log t}, \qquad \mu = \overline{\lim_{t\to\infty}}\frac{1/|\zeta(1+it)|}{\log\log t}, \tag{8.9.1}$$

then, on the Riemann hypothesis,

$$\lambda \leqslant 2e^\gamma, \qquad \mu \leqslant \frac{12}{\pi^2}e^\gamma, \tag{8.9.2}$$

where γ is Euler's constant.

There is therefore a certain interest in proving the following results.†

THEOREM 8.9 (A). $$\overline{\lim_{t\to\infty}}\frac{|\zeta(1+it)|}{\log\log t}\geqslant e^{\gamma}.$$

THEOREM 8.9 (B). $$\overline{\lim_{t\to\infty}}\frac{1/|\zeta(1+it)|}{\log\log t}\geqslant \frac{6}{\pi^2}e^{\gamma}.$$

Thus on the Riemann hypothesis it is only a factor 2 which remains in doubt in each case.

We first prove some identities and inequalities. As in § 7.19, if

$$F_k(s)=\sum_{n=1}^{\infty}\frac{d_k^2(n)}{n^s}\quad(\sigma>1) \tag{8.9.3}$$

and

$$f_k(x)=\sum_{m=0}^{\infty}\left\{\frac{(k+m-1)!}{(k-1)!m!}\right\}^2 x^m, \tag{8.9.4}$$

then

$$F_k(s)=\prod_p f_k(p^{-s}). \tag{8.9.5}$$

Now for real x

$$f_k(x)=\frac{1}{2\pi}\int_{-\pi}^{\pi}\left|\sum_{m=0}^{\infty}\frac{(k+m-1)!}{(k-1)!m!}x^{\frac{1}{2}m}e^{im\phi}\right|^2 d\phi$$

$$=\frac{1}{\pi}\int_0^{\pi}\frac{d\phi}{|1-x^{\frac{1}{2}}e^{i\phi}|^{2k}}=\frac{1}{\pi}\int_0^{\pi}\frac{d\phi}{(1-2\sqrt{x}\cos\phi+x)^k}. \tag{8.9.6}$$

Using the familiar formula

$$P_n(z)=\frac{1}{\pi}\int_0^{\pi}\frac{d\phi}{\{z-\sqrt{(z^2-1)}\cos\phi\}^{n+1}} \tag{8.9.7}$$

for the Legendre polynomial of degree n, we see that

$$f_k(x)=\frac{1}{(1-x)^k}P_{k-1}\left(\frac{1+x}{1-x}\right). \tag{8.9.8}$$

Naturally this identity holds also for complex x; it gives

$$F_k(s)=\prod_p\frac{1}{(1-p^{-s})^k}P_{k-1}\left(\frac{1+p^{-s}}{1-p^{-s}}\right)=\zeta^k(s)\prod_p P_{k-1}\left(\frac{1+p^{-s}}{1-p^{-s}}\right). \tag{8.9.9}$$

A similar set of formulae holds for $1/\zeta(s)$. We have

$$\frac{1}{\{\zeta(s)\}^k}=\prod_p\left(1-\frac{1}{p^s}\right)^k=\prod_p\left(1-\frac{k}{p^s}+\frac{k(k-1)}{1.2}\frac{1}{p^{2s}}-\ldots+\frac{(-1)^k}{p^{ks}}\right).$$

† Littlewood (5), (6), Titchmarsh (4), (14).

Hence
$$\frac{1}{\zeta^k(s)} = \sum_{n=1}^{\infty} \frac{b_k(n)}{n^s}, \tag{8.9.10}$$
where the coefficients $b_k(n)$ are determined in an obvious way from the above product. They are integers, but are not all positive.

The form of these coefficients shows that
$$\sum_{n=1}^{\infty} \frac{|b_k(n)|}{n^s} = \prod_p \left(1+\frac{k}{p^s}+\ldots+\frac{1}{p^{ks}}\right) = \prod_p \left(1+\frac{1}{p^s}\right)^k$$
$$= \prod_p \left(1-\frac{1}{p^{2s}}\right)^k \left(1-\frac{1}{p^s}\right)^{-k} = \left\{\frac{\zeta(s)}{\zeta(2s)}\right\}^k. \tag{8.9.11}$$

Again, let
$$G_k(s) = \sum_{n=1}^{\infty} \frac{b_k^2(n)}{n^s}. \tag{8.9.12}$$
As in the case of $F_k(s)$,
$$G_k(s) = \prod_p \left(1+\frac{k^2}{p^s}+\frac{k^2(k-1)^2}{1^2.2^2}\frac{1}{p^{2s}}+\ldots+\frac{1}{p^{ks}}\right) = \prod_p g_k(p^{-s}), \tag{8.9.13}$$
say. Now, for real x,
$$g_k(x) = \frac{1}{2\pi}\int_{-\pi}^{\pi} \left|\sum_{m=0}^{k} \frac{k!}{m!(k-m)!} x^{\frac{1}{2}m} e^{im\phi}\right|^2 d\phi$$
$$= \frac{1}{\pi}\int_0^{\pi} |1+x^{\frac{1}{2}}e^{i\phi}|^{2k}\, d\phi = \frac{1}{\pi}\int_0^{\pi} (1+2x^{\frac{1}{2}}\cos\phi+x)^k\, d\phi.$$

Comparing this with the formula
$$P_n(z) = \frac{1}{\pi}\int_0^{\pi} \{z+\sqrt{(z^2-1)}\cos\phi\}^n\, d\phi$$
we see that†
$$g_k(x) = (1-x)^k P_k\left(\frac{1+x}{1-x}\right). \tag{8.9.14}$$
Hence
$$G_k(s) = \prod_p (1-p^{-s})^k P_k\left(\frac{1+p^{-s}}{1-p^{-s}}\right) = \frac{1}{\zeta^k(s)} \prod_p P_k\left(\frac{1+p^{-s}}{1-p^{-s}}\right).$$
We have also the identity
$$F_{k+1}(s) = \zeta^{2k+1}(s) G_k(s). \tag{8.9.15}$$

† This formula is, essentially, Murphy's well-known formula
$$P_k(\cos\theta) = \cos^{2k}\tfrac{1}{2}\theta\, F(-k, -k; 1; -\tan^2\tfrac{1}{2}\theta)$$
with $x = -\tan^2\frac{1}{2}\theta$; cf. Hobson, *Spherical and Ellipsoidal Harmonics*, pp. 22, 31.

Again for $0 < x < \frac{1}{2}$

$$f_k(x) > \frac{1}{\pi}\int\limits_0^{\pi/k} \frac{d\phi}{(1-2\sqrt{x}\cos\phi+x)^k}$$

$$= \frac{1}{\pi(1-\sqrt{x})^{2k}}\int\limits_0^{\pi/k}\left\{1-\frac{2\sqrt{x}(1-\cos\phi)}{1-2\sqrt{x}\cos\phi+x}\right\}^k d\phi$$

$$= \frac{1}{\pi(1-\sqrt{x})^{2k}}\int\limits_0^{\pi/k}\left\{1+O\left(\frac{1}{k^2}\right)\right\}^k d\phi > \frac{1}{2k(1-\sqrt{x})^{2k}} \tag{8.9.16}$$

if k is large enough. Hence also

$$g_k(x) = (1-x)^{2k+1}f_{k+1}(x) > \frac{(1+\sqrt{x})^{2k+1}}{2k+2} > \frac{(1+\sqrt{x})^{2k}}{3k} \tag{8.9.17}$$

for k large enough; and

$$g_k(x) \leqslant \frac{1}{\pi}\int\limits_0^{\pi}(1+\sqrt{x})^{2k}\,d\phi = (1+\sqrt{x})^{2k} \tag{8.9.18}$$

for all values of x and k.

8.10. Proof of Theorem 8.9 (A). Let $\sigma > 1$. Then

$$\int\limits_{-T}^{T}\left(1-\frac{|t|}{T}\right)|\zeta(\sigma+it)|^{2k}\,dt = \int\limits_{-T}^{T}\left(1-\frac{|t|}{T}\right)\sum_{m=1}^{\infty}\frac{d_k(m)}{m^{\sigma+it}}\sum_{n=1}^{\infty}\frac{d_k(n)}{n^{\sigma-it}}\,dt$$

$$= \sum_{n=1}^{\infty}\frac{d_k^2(n)}{n^{2\sigma}}\int\limits_{-T}^{T}\left(1-\frac{|t|}{T}\right)dt + \sum_{m\neq n}\sum\frac{d_k(m)d_k(n)}{(mn)^\sigma}\int\limits_{-T}^{T}\left(1-\frac{|t|}{T}\right)\left(\frac{n}{m}\right)^{it}dt$$

$$= T\sum_{n=1}^{\infty}\frac{d_k^2(n)}{n^{2\sigma}} + \sum_{m\neq n}\sum\frac{d_k(m)d_k(n)}{(mn)^\sigma}\,\frac{4\sin^2\{\frac{1}{2}T\log(n/m)\}}{T\log^2(n/m)}$$

$$\geqslant T\sum_{n=1}^{\infty}\frac{d_k^2(n)}{n^{2\sigma}} = TF_k(2\sigma). \tag{8.10.1}$$

Since (from its original definition) $f_k(p^{-2\sigma}) \geqslant 1$ for all values of p,

$$F_k(2\sigma) \geqslant \prod_{p\leqslant x} f_k(p^{-2\sigma}) \geqslant \prod_{p\leqslant x}\left\{\frac{1}{2k}\left(1-\frac{1}{p^\sigma}\right)^{-2k}\right\} \tag{8.10.2}$$

for any positive x and k large enough. Here the number of factors is $\pi(x) < Ax/\log x$. Hence if $x > \sqrt{k}$

$$\prod_{p\leqslant x}\frac{1}{2k} \geqslant \left(\frac{1}{2k}\right)^{Ax/\log x} = \exp\left(-\frac{Ax\log 2k}{\log x}\right) > e^{-Ax}. \tag{8.10.3}$$

Also

$$\log \prod_{p \leqslant x} \frac{1-p^{-\sigma}}{1-p^{-1}} = \sum_{p \leqslant x} \log \frac{1-p^{-\sigma}}{1-p^{-1}} = \sum_{p \leqslant x} O\left(\frac{1}{p^{\sigma}} - \frac{1}{p}\right)$$

$$= \sum_{p \leqslant x} O\left(\log p \int_1^{\sigma} \frac{du}{p^u}\right) = O\left\{(\sigma-1) \sum_{p \leqslant x} \frac{\log p}{p}\right\} = O\{(\sigma-1)\log x\}. \tag{8.10.4}$$

Hence
$$F_k(2\sigma) > e^{-Ax - Ak(\sigma-1)\log x} \prod_{p \leqslant x} \left(1 - \frac{1}{p}\right)^{-2k},$$

and

$$\left\{\frac{1}{T} \int_{-T}^{T} \left(1 - \frac{|t|}{T}\right) |\zeta(\sigma+it)|^{2k}\, dt\right\}^{1/2k} > e^{-Ax/k - A(\sigma-1)\log x} \prod_{p \leqslant x} \left(1 - \frac{1}{p}\right)^{-1}$$

$$> \{e^{\gamma} + o\,(1)\} e^{-Ax/k - A(\sigma-1)\log x} \log x$$

as $x \to \infty$, by (3.15.2).

Let $x = \delta k$, where $k^{-\frac{1}{2}} < \delta < 1$, and $\sigma = 1 + \eta/\log k$, where $0 < \eta < 1$. Then the right-hand side is greater than

$$\{e^{\gamma} + o\,(1)\} e^{-A\delta - A\eta} \left(\log k - \log \frac{1}{\delta}\right).$$

Also, if $m_{\sigma,T} = \max_{1 \leqslant |t| \leqslant T} |\zeta(\sigma+it)|$, the left-hand side does not exceed

$$\left\{\frac{2}{T} \int_0^1 \left(1 - \frac{|t|}{T}\right) \left(\frac{2}{\sigma-1}\right)^{2k} dt\right\}^{1/2k} + \left\{\frac{2}{T} \int_1^T \left(1 - \frac{|t|}{T}\right) m_{\sigma,T}^{2k}\, dt\right\}^{1/2k}$$

$$< \left(\frac{2}{T}\right)^{1/2k} \frac{2}{\sigma-1} + 2^{1/2k} m_{\sigma,T}.$$

Hence

$$m_{\sigma,T} > 2^{-1/2k} \{e^{\gamma} + o\,(1)\} e^{-A\delta - A\eta} \left(\log k - \log \frac{1}{\delta}\right) - \frac{2\log k}{T^{1/2k} \eta}.$$

Let $T = \eta^{-4k}$, so that

$$\log\log T = \log k + \log\left(4 \log \frac{1}{\eta}\right).$$

Then

$$m_{\sigma,T} > 2^{-1/2k} \{e^{\gamma} + o\,(1)\} e^{-A\delta - A\eta} \left\{\log\log T - \log\left(4 \log \frac{1}{\eta}\right) - \log \frac{1}{\delta}\right\} -$$

$$-2\eta \left\{\log\log T - \log\left(4 \log \frac{1}{\eta}\right)\right\}.$$

Giving δ and η arbitrarily small values, and then making $k \to \infty$, i.e. $T \to \infty$, we obtain

$$\overline{\lim} \frac{m_{\sigma,T}}{\log\log T} \geqslant e^{\gamma},$$

where, of course, σ is a function of T.

The result now follows by the Phragmén–Lindelöf method. Let

$$f(s) = \frac{\zeta(s)}{\log\log(s+hi)}$$

where $h > 4$, and let $$\lambda = \overline{\lim} \frac{|\zeta(1+it)|}{\log\log t}.$$

We may suppose λ finite, or there is nothing to prove. On $\sigma = 1$, $t \geqslant 0$, we have

$$|f(s)| \leqslant \frac{|\zeta(s)|}{\log\log t} < \lambda+\epsilon \quad (t > t_0).$$

Also, on $\sigma = 2$, $$|f(s)| = o(1) < \lambda+\epsilon \quad (t > t_1).$$

We can choose h so that $|f(s)| < \lambda+\epsilon$ also on the remainder of the boundary of the strip bounded by $\sigma = 1$, $\sigma = 2$, and $t = 1$. Then, by the Phragmén–Lindelöf theorem, $|f(s)| < \lambda+\epsilon$ in the interior, and so

$$\overline{\lim} \frac{|\zeta(s)|}{\log\log t} = \overline{\lim} \frac{|\zeta(s)|}{\log\log(t+h)} \leqslant \lambda.$$

Hence $\lambda \geqslant e^{\gamma}$, the required result.

8.11. Proof of Theorem 8.9 (B). The above method depends on the fact that $d_k(n)$ is positive. Since $b_k(n)$ is not always positive, a different method is required in this case.

Let $\sigma > 1$, and let N be any positive number. Then

$$\frac{1}{T}\int_0^T \left|\sum_{n \leqslant N} \frac{b_k(n)}{n^s}\right|^2 dt = \frac{1}{T}\int_0^T \sum_{m \leqslant N} \frac{b_k(m)}{m^{\sigma+it}} \sum_{n \leqslant N} \frac{b_k(n)}{n^{\sigma-it}}\, dt$$

$$= \sum_{n \leqslant N} \frac{b_k^2(n)}{n^{2\sigma}} + \frac{1}{T} \sum_{m \neq N}\sum \frac{b_k(m)b_k(n)}{m^{\sigma}n^{\sigma}} \int_0^T \left(\frac{n}{m}\right)^{it} dt$$

$$\geqslant \sum_{n \leqslant N} \frac{b_k^2(n)}{n^{2\sigma}} - \frac{1}{T} \sum_{m \neq n}\sum \frac{|b_k(m)b_k(n)|}{m^{\sigma}n^{\sigma}} \frac{2}{|\log n/m|}.$$

Now $$\left|\log\frac{n}{m}\right| \geqslant \log\frac{n+1}{n} \geqslant \frac{1}{2n} \geqslant \frac{1}{2N},$$

so that the last sum does not exceed

$$\frac{4N}{T} \sum_{m \neq n}\sum \frac{|b_k(m)b_k(n)|}{m^{\sigma}n^{\sigma}} < \frac{4N}{T}\left(\sum_{n=1}^{\infty} \frac{|b_k(n)|}{n^{\sigma}}\right)^2 = \frac{4N}{T}\left\{\frac{\zeta(\sigma)}{\zeta(2\sigma)}\right\}^{2k}.$$

Since $\zeta(\sigma) \sim 1/(\sigma-1)$ as $\sigma \to 1$, and $\zeta(2) > 1$, we have, if σ is sufficiently near to 1,

$$\frac{\zeta(\sigma)}{\zeta(2\sigma)} < \frac{1}{\sigma-1}.$$

Hence the above last sum is less than

$$\frac{4N}{T(\sigma-1)^{2k}}.$$

Also

$$\left|\frac{1}{\zeta^k(s)}-\sum_{n\leqslant N}\frac{b_k(n)}{n^s}\right| \leqslant \sum_{n>N}\frac{|b_k(n)|}{n^\sigma} < \frac{1}{N^{\frac{1}{2}\sigma-\frac{1}{2}}}\sum_{n>N}\frac{|b_k(n)|}{n^{\frac{1}{2}\sigma+\frac{1}{2}}}$$

$$< \frac{1}{N^{\frac{1}{2}\sigma-\frac{1}{2}}}\left\{\frac{\zeta(\frac{1}{2}\sigma+\frac{1}{2})}{\zeta(\sigma+1)}\right\}^k < \frac{1}{N^{\frac{1}{2}\sigma-\frac{1}{2}}}\left(\frac{2}{\sigma-1}\right)^k$$

for σ sufficiently near to 1. Since for $\sigma > 2$

$$G_k(\sigma) \leqslant \prod_p \left(1+\frac{1}{p^{\frac{1}{2}\sigma}}\right)^{2k} = \prod_p \left(\frac{1-p^{-\sigma}}{1-p^{-\frac{1}{2}\sigma}}\right)^{2k} = \left\{\frac{\zeta(\frac{1}{2}\sigma)}{\zeta(\sigma)}\right\}^{2k},$$

we have similarly

$$G_k(2\sigma)-\sum_{n\leqslant N}\frac{b_k^2(n)}{n^{2\sigma}} = \sum_{n>N}\frac{b_k^2(n)}{n^{2\sigma}} < \frac{1}{N^{\sigma-1}}\sum_{n>N}\frac{b_k^2(n)}{n^{\sigma+1}}$$

$$< \frac{G_k(\sigma+1)}{N^{\sigma-1}} < \frac{1}{N^{\sigma-1}}\left\{\frac{\zeta(\frac{1}{2}\sigma+\frac{1}{2})}{\zeta(\sigma+1)}\right\}^{2k} < \frac{1}{N^{\sigma-1}}\left(\frac{2}{\sigma-1}\right)^{2k}.$$

These two differences are therefore both bounded if

$$N = \left(\frac{2}{\sigma-1}\right)^{2k/(\sigma-1)}$$

With this value of N we have

$$\frac{1}{T}\int_0^T \left|\frac{1}{\zeta^k(s)}+O(1)\right|^2 dt = \frac{1}{T}\int_0^T \left|\sum_{n\leqslant N}\frac{b_k(n)}{n^s}\right|^2 dt$$

$$> G_k(2\sigma)-\frac{4N}{T(\sigma-1)^{2k}}+O(1)$$

$$> \prod_{p\leqslant x}\left\{\frac{1}{3k}\left(1+\frac{1}{p^\sigma}\right)^{2k}\right\}-\frac{4N}{T(\sigma-1)^{2k}}+O(1)$$

by (8.9.17). Now

$$\log\prod_{p\leqslant x}\frac{1+p^{-1}}{1+p^{-\sigma}} = O\{(\sigma-1)\log x\}$$

as in (8.10.4). Hence, as in (8.10.3) and (3.15.3),

$$\prod_{p\leqslant x}\left\{\frac{1}{3k}\left(1+\frac{1}{p^\sigma}\right)^{2k}\right\} > e^{-Ax-Ak(\sigma-1)\log x}\{b+o(1)\}^{2k}\log^{2k}x$$

where $b = 6e^\gamma/\pi^2$.

Choosing x and σ as in the last proof,

$$\frac{N}{(\sigma-1)^{2k}} < \left(\frac{2\log k}{\eta}\right)^{2k\log k/\eta+2k},$$

and we obtain

$$\frac{1}{T}\int_0^T \left|\frac{1}{\zeta^k(s)}+O(1)\right|^2 dt > e^{-A\delta k-A\eta k}\{b+o(1)\}^{2k}\log^{2k}\delta k-$$

$$-\frac{4}{T}\left(\frac{2\log k}{\eta}\right)^{2k\log k/\eta+2k}+O(1).$$

Finally, let

$$T = \left(\frac{2\log k}{\eta}\right)^{2k\log k/\eta+2k}.$$

Then

$$\log\log T = \log k+\log\left(\frac{2\log k}{\eta}+2\right)+\log\log\frac{2\log k}{\eta} < (1+\epsilon)\log k$$

for $k > k_1 = k_1(\epsilon, \eta)$. Hence

$$\frac{1}{T}\int_0^T \left|\frac{1}{\zeta^k(s)}+O(1)\right|^2 dt > e^{-A\delta k-A\eta k}\{b+o(1)\}^{2k}\left(\frac{\log\log T}{1+\epsilon}-\log\frac{1}{\delta}\right)^{2k}+O(1).$$

Let

$$M_{\sigma,T} = \max_{0\leqslant t\leqslant T}\frac{1}{|\zeta(\sigma+it)|}.$$

Since the first term on the right of the above inequality tends to infinity with k (for fixed δ, η, and ϵ) it is clear that $M_{\sigma,T}^k$ tends to infinity. Hence

$$\left|\frac{1}{\zeta^k(s)}+O(1)\right| < 2M_{\sigma,T}^k$$

if k is large enough, and we deduce that

$$4M_{\sigma,T}^{2k} > \tfrac{1}{2}e^{-A\delta k-A\eta k}\{b+o(1)\}^{2k}\left(\frac{\log\log T}{1+\epsilon}-\log\frac{1}{\delta}\right)^{2k}$$

for k large enough. Hence

$$M_{\sigma,T} > \frac{1}{8^{1/2k}}e^{-A\delta-A\eta}\{b+o(1)\}\left(\frac{\log\log T}{1+\epsilon}-\log\frac{1}{\delta}\right).$$

Giving δ, ϵ, and η arbitrarily small values, and then varying T, we obtain

$$\overline{\lim}\,\frac{M_{\sigma,T}}{\log\log T} \geqslant b.$$

The theorem now follows as in the previous case.

8.12. The above theorems are mainly concerned with the neighbourhood of the line $\sigma = 1$. We now penetrate further into the critical strip, and prove†

THEOREM 8.12. *Let σ be a fixed number in the range $\frac{1}{2} \leqslant \sigma < 1$. Then the inequality*

$$|\zeta(\sigma+it)| > \exp(\log^{\alpha} t)$$

is satisfied for some indefinitely large values of t, provided that

$$\alpha < 1-\sigma.$$

Throughout the proof k is supposed large enough, and δ small enough, for any purpose that may be required. We take $\frac{1}{2} < \sigma < 1$, and the constants C_1, C_2,..., and those implied by the symbol O, are independent of k and δ, but may depend on σ, and on ϵ when it occurs. The case $\sigma = \frac{1}{2}$ is deduced finally from the case $\sigma > \frac{1}{2}$.

We first prove some lemmas.

LEMMA α. *Let*

$$\Gamma(s)\zeta^k(s) = \sum_{m=0}^{k-1} \frac{(-1)^m m!\, a_m^{(k)}}{(s-1)^{m+1}} + \ldots$$

in the neighbourhood of $s = 1$. Then

$$|a_m^{(k)}| < e^{C_1 k} \quad (1 \leqslant m \leqslant k).$$

The $a_m^{(k)}$ are the same as those of § 7.13. We have

$$\Gamma(s) = \sum_{n=0}^{\infty} c_n (s-1)^n, \qquad \zeta^k(s) = (s-1)^{-k} \sum_{n=0}^{\infty} e_n^{(k)} (s-1)^n,$$

where $\quad |c_n| \leqslant C_2^n, \qquad |e_n^{(1)}| \leqslant C_3^n \quad (C_2 > 1,\ C_3 > 1).$

Hence $e_n^{(k)}$ is less than the coefficient of $(s-1)^n$ in

$$\Big\{\sum_{n=0}^{\infty} C_3^n (s-1)^n\Big\}^k = \{1-C_3(s-1)\}^{-k} = \sum_{n=0}^{\infty} \frac{(k+n-1)!}{(k-1)!\, n!} C_3^n (s-1)^n.$$

Hence

$$m!\,|a_m^{(k)}| = \Big|\sum_{n=0}^{k-m-1} c_{k-m-n-1}\, e_n^{(k)}\Big| < \sum_{n=0}^{k-m-1} C_2^{k-m-n-1} \frac{(k+n-1)!}{(k-1)!\, n!} C_3^n$$

$$< k C_2^k C_3^k \frac{(2k-2)!}{\{(k-1)!\}^2} < e^{C_1 k}.$$

LEMMA β.

$$\frac{1}{\pi} \int_{-\infty}^{\infty} |\Gamma(\sigma+it)\zeta^k(\sigma+it) e^{(\frac{1}{2}\pi-\delta)t}|^2\, dt$$

$$> \int_{1}^{\infty} \Big|\sum_{n=1}^{\infty} d_k(n) \exp(-inxe^{-i\delta})\Big|^2 x^{2\sigma-1}\, dx - \exp(C_4 k \log k).$$

† Titchmarsh (4).

By (7.13.3) the left-hand side is greater than

$$2\int_1^\infty |\phi_k(ixe^{-i\delta})|^2 x^{2\sigma-1}\,dx \geqslant \int_1^\infty \Big|\sum_{n=1}^\infty d_k(n)\exp(-inxe^{-i\delta})\Big|^2 x^{2\sigma-1}\,dx - \\ -2\int_1^\infty |R_k(ixe^{-i\delta})|^2 x^{2\sigma-1}\,dx.$$

Since $|\log(ixe^{-i\delta})| \leqslant \log x+\frac{1}{2}\pi$,

$$|R_k(ixe^{-i\delta})| \leqslant \frac{1}{x}\{|a_0^{(k)}|+|a_1^{(k)}|(\log x+\tfrac{1}{2}\pi)+...+a_{k-1}^{(k)}(\log x+\tfrac{1}{2}\pi)^{k-1}\} \\ \leqslant \frac{ke^{C_1 k}(\log x+\frac{1}{2}\pi)^{k-1}}{x},$$

and

$$\int_1^\infty (\log x+\tfrac{1}{2}\pi)^{2k-2}x^{2\sigma-3}\,dx < \int_1^\infty (2\log x)^{2k-2}x^{2\sigma-3}\,dx+\int_1^\infty \pi^{2k-2}x^{2\sigma-3}\,dx \\ = \frac{\Gamma(2k-1)}{2(1-\sigma)^{2k-1}}+\frac{\pi^{2k-2}}{2-2\sigma}.$$

The result now clearly follows.

LEMMA γ.

$$\int_1^\infty \Big|\sum_{n=1}^\infty d_k(n)\exp(-inxe^{-i\delta})\Big|^2 x^{2\sigma-1}\,dx \\ > \frac{C_5}{\delta^{2\sigma}}\sum_{n=1}^\infty \frac{d_k^2(n)}{n^{2\sigma}}e^{-2n\sin\delta}-C_6\log\frac{1}{\delta}\sum_{n=1}^\infty d_k^2(n)e^{-n\sin\delta}.$$

The left-hand side is equal to

$$\sum_{m=1}^\infty\sum_{n=1}^\infty d_k(m)d_k(n)\int_1^\infty \exp(imxe^{i\delta}-inxe^{-i\delta})x^{2\sigma-1}\,dx \\ = \sum_{m=n}+\sum_{m\neq n} = \Sigma_1+\Sigma_2.$$

Now
$$\int_1^\infty e^{-2nx\sin\delta}x^{2\sigma-1}\,dx = (2n\sin\delta)^{-2\sigma}\int_{2n\sin\delta}^\infty e^{-y}y^{2\sigma-1}\,dy,$$

and for $2n\sin\delta\leqslant 1$

$$\int_{2n\sin\delta}^\infty e^{-y}y^{2\sigma-1}\,dy \geqslant \int_1^\infty e^{-y}y^{2\sigma-1}\,dy = C_7 > C_7 e^{-2n\sin\delta},$$

while for $2n\sin\delta>1$

$$\int_{2n\sin\delta}^\infty e^{-y}y^{2\sigma-1}\,dy > \int_{2n\sin\delta}^\infty e^{-y}\,dy = e^{-2n\sin\delta}.$$

Hence

$$\Sigma_1 = \sum_{n=1}^{\infty} d_k^2(n) \int_1^{\infty} e^{-2nx\sin\delta} x^{2\sigma-1}\,dx > \frac{C_5}{\delta^{2\sigma}} \sum_{n=1}^{\infty} \frac{d_k^2(n)}{n^{2\sigma}} e^{-2n\sin\delta}.$$

Also, using (7.14.4),

$$|\Sigma_2| < C_8 \sum_{m=2}^{\infty} \sum_{n=1}^{m-1} d_k(m) d_k(n) \frac{e^{-m\sin\delta}}{m-n}$$

$$= C_8 \sum_{r=1}^{\infty} \sum_{m=r+1}^{\infty} d_k(m) d_k(m-r) \frac{e^{-m\sin\delta}}{r}$$

$$= C_8 \sum_{r=1}^{\infty} \frac{e^{-\frac{1}{2}r\sin\delta}}{r} \sum_{m=r+1}^{\infty} d_k(m) e^{-\frac{1}{2}m\sin\delta}\, d_k(m-r) e^{-\frac{1}{2}(m-r)\sin\delta}$$

$$\leqslant C_8 \sum_{r=1}^{\infty} \frac{e^{-\frac{1}{2}r\sin\delta}}{r} \left\{ \sum_{m=r+1}^{\infty} d_k^2(m) e^{-m\sin\delta} \sum_{m=r+1}^{\infty} d_k^2(m-r) e^{-(m-r)\sin\delta} \right\}^{\frac{1}{2}}$$

$$< C_8 \sum_{r=1}^{\infty} \frac{e^{-\frac{1}{2}r\sin\delta}}{r} \sum_{m=1}^{\infty} d_k^2(m) e^{-m\sin\delta} < C_6 \log\frac{1}{\delta} \sum_{m=1}^{\infty} d_k^2(m) e^{-m\sin\delta}.$$

This proves the lemma.

Lemma δ. *For* $\sigma > 1$

$$\exp\left\{C_9\left(\frac{k}{\log k}\right)^{2/\sigma}\right\} < F_k(\sigma) < \exp(C_{10}\, k^{2/\sigma}).$$

It is clear from (8.9.6) that

$$f_k(x) \leqslant (1-\sqrt{x})^{-2k} \quad (0 < x < 1).$$

Also it is easily verified that

$$\left\{\frac{(k+m-1)!}{(k-1)!\,m!}\right\}^2 \leqslant \frac{(k^2+m-1)!}{(k^2-1)!\,m!}.$$

Hence, for $0 < x < 1$,

$$f_k(x) \leqslant \sum_{m=0}^{\infty} \frac{(k^2+m-1)!}{(k^2-1)!\,m!} x^m = (1-x)^{-k^2}.$$

Hence

$$\log F_k(\sigma) = \sum_{p^\sigma \leqslant k^2} \log f_k(p^{-\sigma}) + \sum_{p^\sigma > k^2} \log f_k(p^{-\sigma})$$

$$\leqslant 2k \sum_{p^\sigma \leqslant k^2} \log(1-p^{-\frac{1}{2}\sigma})^{-1} + k^2 \sum_{p^\sigma > k^2} \log(1-p^{-\sigma})^{-1}$$

$$= O\Big(k \sum_{p^\sigma \leqslant k^2} p^{-\frac{1}{2}\sigma}\Big) + O\Big(k^2 \sum_{p^\sigma > k^2} p^{-\sigma}\Big)$$

$$= O\{k(k^{2/\sigma})^{1-\frac{1}{2}\sigma}\} + O\{k^2(k^{2/\sigma})^{1-\sigma}\} = O(k^{2/\sigma}).$$

On the other hand, (8.10.2) gives

$$\log F_k(\sigma) > 2k \sum_{p<x} \log(1-p^{-\frac{1}{2}\sigma})^{-1} - \sum_{p<x} \log 2k$$

$$> 2k \sum_{p<x} p^{-\frac{1}{2}\sigma} - C_{11} \frac{x}{\log x} \log 2k$$

$$> C_{12}\, k \frac{x^{1-\frac{1}{2}\sigma}}{\log x} - C_{11} \frac{x}{\log x} \log 2k.$$

Taking

$$x = \left(\frac{C_{12}}{2C_{11}} \frac{k}{\log k}\right)^{2/\sigma}$$

the other result follows.

Proof of Theorem 8.12 *for* $\frac{1}{2} < \sigma < 1$. It follows from Lemmas β and γ and Stirling's theorem that

$$\int_0^\infty |\zeta(\sigma+it)|^{2k} e^{-2\delta t} t^{2\sigma-1}\, dt > \frac{C_{13}}{\delta^{2\sigma}} \sum_{n=1}^\infty \frac{d_k^2(n)}{n^{2\sigma}} e^{-2n\sin\delta} -$$

$$- C_{14} \log\frac{1}{\delta} \sum_{n=1}^\infty d_k^2(n) e^{-n\sin\delta} - C_{15}\, e^{C_4 k \log k}.$$

Now, if $0 < \epsilon < 2\sigma - 1$,

$$\sum_{n=1}^\infty \frac{d_k^2(n)}{n^{2\sigma}} e^{-2n\sin\delta} = F_k(2\sigma) - \sum_{n=1}^\infty \frac{d_k^2(n)}{n^{2\sigma}} (1-e^{-2n\sin\delta})$$

$$> F_k(2\sigma) - C_{16} \sum_{n=1}^\infty \frac{d_k^2(n)}{n^{2\sigma}} (n\delta)^\epsilon$$

$$= F_k(2\sigma) - C_{16}\, \delta^\epsilon F_k(2\sigma-\epsilon)$$

$$> \exp\left\{C_9 \left(\frac{k}{\log k}\right)^{1/\sigma}\right\} - C_{16}\, \delta^\epsilon \exp\{C_{10}\, k^{2/(2\sigma-\epsilon)}\},$$

and

$$\sum_{n=1}^\infty d_k^2(n) e^{-n\sin\delta} < C_{17} \sum_{n=1}^\infty d_k^2(n)(n\delta)^{\epsilon-2\sigma} = C_{17}\, \delta^{\epsilon-2\sigma} F_k(2\sigma-\epsilon)$$

$$< C_{17}\, \delta^{\epsilon-2\sigma} \exp\{C_{10}\, k^{2/(2\sigma-\epsilon)}\}.$$

Let

$$\delta = \exp\left\{-\frac{C_{10}}{\epsilon} k^{2/(2\sigma-\epsilon)}\right\}.$$

Then

$$\int_0^\infty |\zeta(\sigma+it)|^{2k} e^{-2\delta t} t^{2\sigma-1}\, dt > \frac{1}{\delta^{2\sigma}} \left[C_{13} \exp\left\{C_9 \left(\frac{k}{\log k}\right)^{1/\sigma}\right\} - C_{13}\, C_{16} -\right.$$

$$\left. - C_{14}\, C_{17} \frac{C_{10}}{\epsilon} k^{2/(2\sigma-\epsilon)}\right] - C_{15}\, e^{C_4 k \log k}$$

$$> \frac{C_{18}}{\delta^{2\sigma}} \exp\left\{C_9 \left(\frac{k}{\log k}\right)^{1/\sigma}\right\}.$$

Suppose now that

$$|\zeta(\sigma+it)| \leqslant \exp(\log^\alpha t) \quad (t \geqslant t_0)$$

where $0 < \alpha < 1$. Then

$$\int_0^\infty |\zeta(\sigma+it)|^{2k} e^{-2\delta t} t^{2\sigma-1}\, dt \leqslant C_{19}^{2k} + \int_1^\infty e^{2k\log^\alpha t - 2\delta t} t^{2\sigma-1}\, dt.$$

If $t > k^2/\delta^2$, $k > k_0$, then

$$\frac{k}{\delta} < \sqrt{t} < \frac{1}{2}\frac{t}{\log^\alpha t}.$$

Hence

$$\int_1^\infty e^{2k\log^\alpha t - 2\delta t} t^{2\sigma-1}\, dt \leqslant e^{2k\log^\alpha(k^2/\delta^2)} \int_1^{k^2/\delta^2} e^{-2\delta t} t^{2\sigma-1}\, dt + \int_{k^2/\delta^2}^\infty e^{-\delta t} t^{2\sigma-1}\, dt$$

$$< e^{2k\log^\alpha(k^2/\delta^2)} \frac{C_{20}}{\delta^{2\sigma}}.$$

Hence $$\left(\frac{k}{\log k}\right)^{1/\sigma} = O\left(k \log^\alpha \frac{k}{\delta}\right) = O(k^{1+(2\alpha)/(2\sigma-\epsilon)}).$$

Hence $$\frac{1}{\sigma} \leqslant 1 + \frac{2\alpha}{2\sigma-\epsilon},$$

and since ϵ may be as small as we please

$$\frac{1}{\sigma} \leqslant 1 + \frac{\alpha}{\sigma}, \qquad \alpha \geqslant 1-\sigma.$$

The case $\sigma = \frac{1}{2}$. Suppose that

$$\zeta(\tfrac{1}{2}+it) = O\{\exp(\log^\beta t)\} \quad (0 < \beta < \tfrac{1}{2}).$$

Then the function $$f(s) = \zeta(s)\exp(-\log^\beta s)$$

is bounded on the lines $\sigma = \frac{1}{2}$, $\sigma = 2$, $t > t_0$, and it is $O(t)$ uniformly in this strip. Hence by the Phragmén–Lindelöf theorem $f(s)$ is bounded in the strip, i.e.

$$\zeta(\sigma+it) = O\{\exp(\log^\beta t)\}$$

for $\frac{1}{2} < \sigma < 2$. Since this is not true for $\frac{1}{2} < \sigma < 1-\beta$, it follows that $\beta \geqslant \frac{1}{2}$.

NOTES FOR CHAPTER 8

8.13. Levinson [**1**] has sharpened Theorems 8.9(A) and 8.9(B) to show that the inequalities

$$|\zeta(1+it)| \geqslant e^\gamma \log\log t + O(1)$$

and

$$\frac{1}{|\zeta(1+it)|} \geqslant \frac{6e^{\gamma}}{\pi^2}(\log\log t - \log\log\log t) + O(1)$$

each hold for arbitrarily large t. Theorem 8.12 has also been improved, by Montgomery [**3**]. He showed that for any σ in the range $\frac{1}{2} < \sigma < 1$, and for any real ϑ, there are arbitrarily large t such that

$$\mathbf{R}\{e^{i\vartheta}\log\zeta(\sigma+it)\} \geqslant \tfrac{1}{20}(\sigma-\tfrac{1}{2})^{-1}(\log t)^{1-\sigma}(\log\log t)^{-\sigma}.$$

Here $\log\zeta(s)$ is, as usual, defined by continuous variation along lines parallel to the real axis, using the Dirichlet series (1.1.9) for $\sigma > 1$. It follows in particular that

$$\zeta(\sigma+it) = \Omega\left\{\exp\left(\frac{\frac{1}{20}}{\sigma-\frac{1}{2}}\frac{(\log t)^{1-\sigma}}{(\log\log t)^{\sigma}}\right)\right\} \quad (\tfrac{1}{2} < \sigma < 1),$$

and the same for $\zeta(\sigma+it)^{-1}$. For $\sigma = \frac{1}{2}$ the best result is due to Balasubramanian and Ramachandra [**2**], who showed that

$$\max_{T\leqslant t\leqslant T+H} |\zeta(\tfrac{1}{2}+it)| \geqslant \exp\left(\tfrac{3}{4}\frac{(\log H)^{\frac{1}{2}}}{(\log\log H)^{\frac{1}{2}}}\right)$$

if $(\log T)^{\delta} \leqslant H \leqslant T$ and $T \geqslant T(\delta)$, where δ is any positive constant. Their method is akin to that of §8.12, in that it depends on a lower bound for a mean value of $|\zeta(\frac{1}{2}+it)|^{2k}$, uniform in k. By contrast the method of Montgomery [**3**] uses the formula

$$\frac{4}{\pi}\int_{-(\log t)^2}^{(\log t)^2} e^{-i\vartheta}\log\zeta(\sigma+it+iy)\left(\frac{\sin\frac{1}{2}y}{y}\right)^2\{1+\cos(\vartheta+y\log x)\}dy$$

$$= \sum_{|\log n/x| \leqslant \frac{1}{2}} \frac{\Lambda(n)}{\log n} n^{-\sigma-it}\left(\tfrac{1}{2} - |\log\frac{n}{x}|\right) + O\{x(\log t)^{-2}\}. \quad (8.13.1)$$

This is valid for any real x and ϑ, providing that $\zeta(s) \neq 0$ for $\mathbf{R}(s) \geqslant \sigma$ and $|\mathbf{I}(s)-t| \leqslant 2(\log t)^2$. After choosing x suitably one may use the extended version of Dirichlet's theorem given in §8.2 to show that the real part of the sum on the right of (8.13.1) is large at points $t_1 < \ldots < t_N \leqslant T$, spaced at least $4(\log T)^2$ apart. One can arrange that N exceeds $N(\sigma, T)$, whence at least one such t_n will satisfy the condition that $\zeta(s) \neq 0$ in the corresponding rectangle.

IX

THE GENERAL DISTRIBUTION OF THE ZEROS

9.1. In § 2.12 we deduced from the general theory of integral functions that $\zeta(s)$ *has an infinity of complex zeros*. This may be proved directly as follows.

We have

$$\frac{1}{2^2}+\frac{1}{3^2}+\ldots < \frac{1}{2^2}+\frac{1}{2.3}+\frac{1}{3.4}+\ldots = \frac{1}{4}+\left(\frac{1}{2}-\frac{1}{3}\right)+\left(\frac{1}{3}-\frac{1}{4}\right)+\ldots = \frac{3}{4}.$$

Hence for $\sigma \geqslant 2$

$$|\zeta(s)| \leqslant 1+\frac{1}{2^\sigma}+\frac{1}{3^\sigma}+\ldots \leqslant 1+\frac{1}{2^2}+\ldots < \frac{7}{4}, \tag{9.1.1}$$

and

$$|\zeta(s)| \geqslant 1-\frac{1}{2^\sigma}-\ldots \geqslant 1-\frac{1}{2^2}-\ldots > \frac{1}{4}. \tag{9.1.2}$$

Also

$$\mathbf{R}\{\zeta(s)\} = 1+\frac{\cos(t\log 2)}{2^\sigma}+\ldots \geqslant 1-\frac{1}{2^2}-\ldots > \frac{1}{4}. \tag{9.1.3}$$

Hence for $\sigma \geqslant 2$ we may write

$$\log\zeta(s) = \log|\zeta(s)|+i\arg\zeta(s),$$

where $\arg\zeta(s)$ is that value of $\arctan\{\mathbf{I}\zeta(s)/\mathbf{R}\zeta(s)\}$ which lies between $-\frac{1}{2}\pi$ and $\frac{1}{2}\pi$. It is clear that

$$|\log\zeta(s)| < A \quad (\sigma \geqslant 2). \tag{9.1.4}$$

For $\sigma < 2$, $t \neq 0$, we define $\log\zeta(s)$ as the analytic continuation of the above function along the straight line $(\sigma+it, 2+it)$, provided that $\zeta(s) \neq 0$ on this segment of line.

Now consider a system of four concentric circles C_1, C_2, C_3, C_4, with centre $3+iT$ and radii 1, 4, 5, and 6 respectively. Suppose that $\zeta(s) \neq 0$ in or on C_4. Then $\log\zeta(s)$, defined as above, is regular in C_4. Let M_1, M_2, M_3 be its maximum modulus on C_1, C_2, and C_3 respectively.

Since $\zeta(s) = O(t^A)$, $\mathbf{R}\{\log\zeta(s)\} < A\log T$ in C_4, and the Borel–Carathéodory theorem gives

$$M_3 \leqslant \frac{2.5}{6-5}A\log T+\frac{6+5}{6-5}\log|\zeta(3+iT)| < A\log T.$$

Also $M_1 < A$, by (9.1.4). Hence Hadamard's three-circles theorem, applied to the circles C_1, C_2, C_3, gives

$$M_2 \leqslant M_1^\alpha M_3^\beta < A\log^\beta T,$$

where

$$1-\alpha = \beta = \log 4/\log 5 < 1.$$

Hence $$\zeta(-1+iT) = O\{\exp(\log^{\beta} T)\} = O(T^{\epsilon}).$$

But by (9.1.2), and the functional equation (2.1.1) with $\sigma = 2$,

$$|\zeta(-1+iT)| > AT^{\frac{3}{2}}.$$

We have thus obtained a contradiction. Hence every such circle C_4 contains at least one zero of $\zeta(s)$, and so there are an infinity of zeros. The argument also shows that the gaps between the ordinates of successive zeros are bounded.

9.2. The function $N(T)$. Let $T > 0$, and let $N(T)$ denote the number of zeros of the function $\zeta(s)$ in the region $0 \leqslant \sigma \leqslant 1, 0 < t \leqslant T$. The distribution of the ordinates of the zeros can then be studied by means of formulae involving $N(T)$.

The most easily proved result is

THEOREM 9.2. *As $T \to \infty$*

$$N(T+1)-N(T) = O(\log T). \tag{9.2.1}$$

For it is easily seen that

$$N(T+1)-N(T) \leqslant n(\surd 5),$$

where $n(r)$ is the number of zeros of $\zeta(s)$ in the circle with centre $2+iT$ and radius r. Now, by Jensen's theorem,

$$\int\limits_0^3 \frac{n(r)}{r}\,dr = \frac{1}{2\pi}\int\limits_0^{2\pi} \log|\zeta(2+iT+3e^{i\theta})|\,d\theta - \log|\zeta(2+iT)|.$$

Since $|\zeta(s)| < t^A$ for $-1 \leqslant \sigma \leqslant 5$, we have

$$\log|\zeta(2+iT+3e^{i\theta})| < A\log T.$$

Hence $$\int\limits_0^3 \frac{n(r)}{r}\,dr < A\log T + A < A\log T.$$

Since $$\int\limits_0^3 \frac{n(r)}{r}\,dr \geqslant \int\limits_{\surd 5}^3 \frac{n(r)}{r}\,dr \geqslant n(\surd 5)\int\limits_{\surd 5}^3 \frac{dr}{r} = An(\surd 5),$$

the result (9.2.1) follows.

Naturally it also follows that

$$N(T+h)-N(T) = O(\log T)$$

for any fixed value of h. In particular, the multiplicity of a multiple zero of $\zeta(s)$ in the region considered is at most $O(\log T)$.

9.3. The closer study of $N(T)$ depends on the following theorem.† If T is not the ordinate of a zero, let $S(T)$ denote the value of

$$\pi^{-1}\arg\zeta(\tfrac{1}{2}+iT)$$

obtained by continuous variation along the straight lines joining 2, $2+iT$, $\frac{1}{2}+iT$, starting with the value 0. If T is the ordinate of a zero, let $S(T) = S(T+0)$. Let

$$L(T) = \frac{1}{2\pi}T\log T - \frac{1+\log 2\pi}{2\pi}T + \frac{7}{8}. \tag{9.3.1}$$

THEOREM 9.3. *As* $T \to \infty$

$$N(T) = L(T)+S(T)+O(1/T). \tag{9.3.2}$$

The number of zeros of the function $\Xi(z)$ (see § 2.1) in the rectangle with vertices at $z = \pm T \pm \frac{3}{2}i$ is $2N(T)$, so that

$$2N(T) = \frac{1}{2\pi i}\int \frac{\Xi'(z)}{\Xi(z)}\,dz$$

taken round the rectangle. Since $\Xi(z)$ is even and real for real z, this is equal to

$$\frac{2}{\pi i}\left(\int\limits_{T}^{T+\frac{3}{2}i} + \int\limits_{T+\frac{3}{2}i}^{\frac{3}{2}i}\right)\frac{\Xi'(z)}{\Xi(z)}\,dz = \frac{2}{\pi i}\left(\int\limits_{2}^{2+iT} + \int\limits_{2+iT}^{\frac{1}{2}+iT}\right)\frac{\xi'(s)}{\xi(s)}\,ds$$

$$= \frac{2}{\pi}\Delta\arg\xi(s),$$

where Δ denotes the variation from 2 to $2+iT$, and thence to $\frac{1}{2}+iT$, along straight lines. Recalling that

$$\xi(s) = \tfrac{1}{2}s(s-1)\pi^{-\frac{1}{2}s}\Gamma(\tfrac{1}{2}s)\zeta(s),$$

we obtain

$$\pi N(T) = \Delta\arg s(s-1)+\Delta\arg\pi^{-\frac{1}{2}s}+\Delta\arg\Gamma(\tfrac{1}{2}s)+\Delta\arg\zeta(s).$$

Now

$$\Delta\arg s(s-1) = \arg(-\tfrac{1}{4}-T^2) = \pi,$$

$$\Delta\arg\pi^{-\frac{1}{2}s} = \Delta\arg e^{-\frac{1}{2}s\log\pi} = -\tfrac{1}{2}T\log\pi,$$

and by (4.12.1)

$$\begin{aligned}\Delta\arg\Gamma(\tfrac{1}{2}s) &= \mathbf{I}\log\Gamma(\tfrac{1}{4}+\tfrac{1}{2}iT)\\ &= \mathbf{I}\{(-\tfrac{1}{4}+\tfrac{1}{2}iT)\log(\tfrac{1}{2}iT)-\tfrac{1}{2}iT+O(1/T)\}\\ &= \tfrac{1}{2}T\log\tfrac{1}{2}T-\tfrac{1}{8}\pi-\tfrac{1}{2}T+O(1/T).\end{aligned}$$

Adding these results, we obtain the theorem, provided that T is not the ordinate of a zero. If T is the ordinate of a zero, the result follows from

† Backlund (2), (3).

the definitions and what has already been proved, the term $O(1/T)$ being continuous.

The problem of the behaviour of $N(T)$ is thus reduced to that of $S(T)$.

9.4. We shall now prove the following lemma.

LEMMA. *Let $0 \leqslant \alpha < \beta < 2$. Let $f(s)$ be an analytic function, real for real s, regular for $\sigma \geqslant \alpha$ except at $s = 1$; let*

$$|\mathbf{R}f(2+it)| \geqslant m > 0$$

and
$$|f(\sigma'+it')| \leqslant M_{\sigma,t} \quad (\sigma' \geqslant \sigma,\ 1 \leqslant t' \leqslant t).$$

Then if T is not the ordinate of a zero of $f(s)$

$$|\arg f(\sigma+iT)| \leqslant \frac{\pi}{\log\{(2-\alpha)/(2-\beta)\}}\left(\log M_{\alpha,T+2}+\log\frac{1}{m}\right)+\frac{3\pi}{2} \tag{9.4.1}$$

for $\sigma \geqslant \beta$.

Since $\arg f(2) = 0$, and

$$\arg f(s) = \arctan\left\{\frac{\mathbf{I}f(s)}{\mathbf{R}f(s)}\right\},$$

where $\mathbf{R}f(s)$ does not vanish on $\sigma = 2$, we have

$$|\arg f(2+iT)| < \tfrac{1}{2}\pi.$$

Now if $\mathbf{R}f(s)$ vanishes q times between $2+iT$ and $\beta+iT$, this interval is divided into $q+1$ parts, throughout each of which $\mathbf{R}\{f(s)\} \geqslant 0$ or $\mathbf{R}\{f(s)\} \leqslant 0$. Hence in each part the variation of $\arg f(s)$ does not exceed π. Hence

$$|\arg f(s)| \leqslant (q+\tfrac{3}{2})\pi \quad (\sigma \geqslant \beta).$$

Now q is the number of zeros of the function

$$g(z) = \tfrac{1}{2}\{f(z+iT)+f(z-iT)\}$$

for $\mathbf{I}(z) = 0$, $\beta \leqslant \mathbf{R}(z) \leqslant 2$; hence $q \leqslant n(2-\beta)$, where $n(r)$ denotes the number of zeros of $g(z)$ for $|z-2| \leqslant r$. Also

$$\int_0^{2-\alpha} \frac{n(r)}{r}\,dr \geqslant \int_{2-\beta}^{2-\alpha} \frac{n(r)}{r}\,dr \geqslant n(2-\beta)\log\frac{2-\alpha}{2-\beta},$$

and by Jensen's theorem

$$\int_0^{2-\alpha} \frac{n(r)}{r}\,dr = \frac{1}{2\pi}\int_0^{2\pi} \log|g\{2+(2-\alpha)e^{i\theta}\}|\,d\theta-\log|g(2)|$$

$$\leqslant \log M_{\alpha,T+2}+\log 1/m.$$

This proves the lemma.

We deduce

THEOREM 9.4. *As* $T \to \infty$

$$S(T) = O(\log T), \tag{9.4.2}$$

i.e.

$$N(T) = \frac{1}{2\pi} T \log T - \frac{1+\log 2\pi}{2\pi} T + O(\log T). \tag{9.4.3}$$

We apply the lemma with $f(s) = \zeta(s)$, $\alpha = 0$, $\beta = \frac{1}{2}$, and (9.4.2) follows, since $\zeta(s) = O(t^A)$. Then (9.4.3) follows from (9.3.2).

Theorem 9.4 has a number of interesting consequences. It gives another proof of Theorem 9.2, since $(0 < \theta < 1)$

$$L(T+1) - L(T) = L'(T+\theta) = O(\log T).$$

We can also prove the following result.

If the zeros $\beta+i\gamma$ *of* $\zeta(s)$ *with* $\gamma > 0$ *are arranged in a sequence* $\rho_n = \beta_n + i\gamma_n$ *so that* $\gamma_{n+1} \geqslant \gamma_n$, *then as* $n \to \infty$

$$|\rho_n| \sim \gamma_n \sim \frac{2\pi n}{\log n}. \tag{9.4.4}$$

We have

$$N(T) \sim \frac{1}{2\pi} T \log T.$$

Hence

$$2\pi N(\gamma_n \pm 1) \sim (\gamma_n \pm 1)\log(\gamma_n \pm 1) \sim \gamma_n \log \gamma_n.$$

Also

$$N(\gamma_n - 1) \leqslant n \leqslant N(\gamma_n + 1).$$

Hence

$$2\pi n \sim \gamma_n \log \gamma_n.$$

Hence

$$\log n \sim \log \gamma_n,$$

and so

$$\gamma_n \sim \frac{2\pi n}{\log n}.$$

Also $|\rho_n| \sim \gamma_n$, since $\beta_n = O(1)$.

We can also deduce the result of § 9.1, that the gaps between the ordinates of successive zeros are bounded. For if $|S(t)| \leqslant C \log t$ $(t \geqslant 2)$,

$$N(T+H) - N(T) = \frac{1}{2\pi} \int_T^{T+H} \log \frac{t}{2\pi}\, dt + S(T+H) - S(T) + O\left(\frac{1}{T}\right)$$

$$\geqslant \frac{H}{2\pi} \log \frac{T}{2\pi} - C\{\log(T+H) + \log T\} + O\left(\frac{1}{T}\right),$$

which is ultimately positive if H is a constant greater than $4\pi C$.

The behaviour of the function $S(T)$ appears to be very complicated. It must have a discontinuity k where T passes through the ordinate of a zero of $\zeta(s)$ of order k (since the term $O(1/T)$ in the above theorem is in fact continuous). Between the zeros, $N(T)$ is constant, so that the variation of $S(T)$ must just neutralize that of the other terms. In the formula (9.3.1), the term $\frac{7}{8}$ is presumably overwhelmed by the variations of $S(T)$. On the other hand, in the integrated formula

$$\int_0^T N(t)\,dt = \int_0^T L(t)\,dt + \int_0^T S(t)\,dt + O(\log T)$$

the term in $S(T)$ certainly plays a much smaller part, since, as we shall presently prove, the integral of $S(t)$ over $(0, T)$ is still only $O(\log T)$. Presumably this is due to frequent variations in the sign of $S(t)$. Actually we shall show that $S(t)$ changes sign an infinity of times.

9.5. A problem of analytic continuation. The above theorems on the zeros of $\zeta(s)$ lead to the solution of a curious subsidiary problem of analytic continuation.† Consider the function

$$P(s) = \sum_p \frac{1}{p^s}. \tag{9.5.1}$$

This is an analytic function of s, regular for $\sigma > 1$. Now by (1.6.1)

$$P(s) = \sum_{n=1}^{\infty} \frac{\mu(n)}{n} \log \zeta(ns). \tag{9.5.2}$$

As $n \to \infty$, $\log \zeta(ns) \sim 2^{-ns}$. Hence the right-hand side represents an analytic function of s, regular for $\sigma > 0$, except at the singularities of individual terms. These are branch-points arising from the poles and zeros of the functions $\zeta(ns)$; there are an infinity of such points, but they have no limit-point in the region $\sigma > 0$. Hence $P(s)$ is regular for $\sigma > 0$, except at certain branch-points.

Similarly, the function

$$Q(s) = -P'(s) = -\sum_{n=1}^{\infty} \mu(n) \frac{\zeta'(ns)}{\zeta(ns)} \tag{9.5.3}$$

is regular for $\sigma > 0$, except at certain simple poles.

We shall now prove that *the line $\sigma = 0$ is a natural boundary of the functions $P(s)$ and $Q(s)$.*

We shall in fact prove that every point of $\sigma = 0$ is a limit-point of poles of $Q(s)$. By symmetry, it is sufficient to consider the upper half-line. Thus it is sufficient to prove that for every $u > 0$, $\delta > 0$, the square

$$0 < \sigma < \delta, \qquad u < t \leqslant u + \delta \tag{9.5.4}$$

contains at least one pole of $Q(s)$.

† Landau and Walfisz (1).

As $p \to \infty$ through primes,

$$N\{p(u+\delta)\} \sim \frac{1}{2\pi}(u+\delta)p\log p, \qquad N(pu) \sim \frac{1}{2\pi}up\log p,$$

by Theorem 9.4. Hence for all $p \geqslant p_0(\delta, u)$

$$N\{p(u+\delta)\}-N(pu) > 0. \tag{9.5.5}$$

Also, by Theorem 9.2, the multiplicity $v(\rho)$ of each zero $\rho = \beta+i\gamma$ with ordinate $\gamma \geqslant 2$ is less than $A\log\gamma$, where A is an absolute constant.

Now choose $p = p(\delta, u)$ satisfying the conditions

$$p > \frac{1}{\delta}, \qquad p \geqslant \frac{2}{u}, \qquad p \geqslant p_0(\delta, u), \qquad p > A\log\{p(u+\delta)\}.$$

There is then, by (9.5.5), a zero ρ of $\zeta(s)$ in the rectangle

$$\tfrac{1}{2} \leqslant \sigma < 1, \qquad pu < t \leqslant p(u+\delta). \tag{9.5.6}$$

Since $\gamma > pu \geqslant 2$, its multiplicity $v(\rho)$ satisfies

$$v(\rho) < A\log\gamma \leqslant A\log\{p(u+\delta)\} < p,$$

and so is not divisible by p.

The point ρ/p belongs to the square (9.5.4). We shall show that this point is a pole of $Q(s)$. Let m run through the positive integers (finite in number) for which $\zeta(m\rho/p) = 0$. Then we have to prove that

$$\sum \frac{\mu(m)}{m} v\left(\frac{m\rho}{p}\right) \neq 0. \tag{9.5.7}$$

The term of this sum corresponding to $m = p$ is $-v(\rho)/p$. No other m occurring in the sum is divisible by p, since for $m \geqslant 2p$

$$\mathbf{R}\left(\frac{m\rho}{p}\right) = \frac{m\beta}{p} \geqslant \frac{2p \cdot \frac{1}{2}}{p} = 1.$$

Hence

$$\sum \frac{\mu(m)}{m} v\left(\frac{m\rho}{p}\right) = \frac{a}{b} - \frac{v(\rho)}{p},$$

where a and b are integers, and p is not a factor of b. Since p is also not a factor of $v(\rho)$, $ap \neq bv(\rho)$, and (9.5.7) follows.

There are various other functions with similar properties. For example,† let

$$f_{l,k}(s) = \sum_{n=1}^{\infty} \frac{\{d_k(n)\}^l}{n^s},$$

where k and l are positive integers, $k \geqslant 2$. By (1.2.2) and (1.2.10), $f_{l,k}(s)$ is a meromorphic function of s if $l = 1$, or if $l = 2$ and $k = 2$. *For all other values of l and k, $f_{l,k}(s)$ has $\sigma = 0$ as a natural boundary, and it has no singularities other than poles in the half-plane $\sigma > 0$.*

† Estermann (1).

9.6. An approximate formula for $\zeta'(s)/\zeta(s)$. The following approximate formula for $\zeta'(s)/\zeta(s)$ in terms of the zeros near to s is often useful.

THEOREM 9.6 (A). *If $\rho = \beta+i\gamma$ runs through zeros of $\zeta(s)$,*

$$\frac{\zeta'(s)}{\zeta(s)} = \sum_{|t-\gamma|\leqslant 1} \frac{1}{s-\rho}+O(\log t), \tag{9.6.1}$$

uniformly for $-1 \leqslant \sigma \leqslant 2$.

Take $f(s) = \zeta(s)$, $s_0 = 2+iT$, $r = 12$ in Lemma α of § 3.9. Then $M = A\log T$, and we obtain

$$\frac{\zeta'(s)}{\zeta(s)} = \sum_{|\rho-s_0|\leqslant 6} \frac{1}{s-\rho}+O(\log T) \tag{9.6.2}$$

for $|s-s_0| \leqslant 3$, and so in particular for $-1 \leqslant \sigma \leqslant 2, t = T$. Replacing T by t in the particular case, we obtain (9.6.2) with error $O(\log t)$, and $-1 \leqslant \sigma \leqslant 2$. Finally any term occurring in (9.6.2) but not in (9.6.1) is bounded, and the number of such terms does not exceed

$$N(t+6)-N(t-6) = O(\log t)$$

by Theorem 9.2. This proves (9.6.1).

Another proof depends on (2.12.7), which, by a known property of the Γ-function, gives

$$\frac{\zeta'(s)}{\zeta(s)} = \sum_{\rho}\left(\frac{1}{s-\rho}+\frac{1}{\rho}\right)+O(\log t).$$

Replacing s by $2+it$ and subtracting,

$$\frac{\zeta'(s)}{\zeta(s)} = \sum_{\rho}\left(\frac{1}{s-\rho}-\frac{1}{2+it-\rho}\right)+O(\log t),$$

since $\zeta'(2+it)/\zeta(2+it) = O(1)$.

Now

$$\sum_{|t-\gamma|\leqslant 1}\frac{1}{2+it-\rho} = \sum_{|t-\gamma|\leqslant 1} O(1) = O(\log t)$$

by Theorem 9.2. Also

$$\sum_{t+n<\gamma\leqslant t+n+1}\left(\frac{1}{s-\rho}-\frac{1}{2+it-\rho}\right) = \sum_{t+n<\gamma\leqslant t+n+1}\frac{2-\sigma}{(s-\rho)(2+it-\rho)}$$

$$= \sum_{t+n<\gamma\leqslant t+n+1} O\left\{\frac{1}{(\gamma-t)^2}\right\} = \sum_{t+n<\gamma\leqslant t+n+1} O\left(\frac{1}{n^2}\right) = O\left\{\frac{\log(t+n)}{n^2}\right\},$$

again by Theorem 9.2. Since

$$\sum_{n=1}^{\infty}\frac{\log(t+n)}{n^2} < \sum_{n\leqslant t}\frac{\log 2t}{n^2}+\sum_{n>t}\frac{\log 2n}{n^2} = O(\log t),$$

it follows that $$\sum_{\gamma>t+1}\left(\frac{1}{s-\rho}-\frac{1}{2+it-\rho}\right)=O(\log t).$$

Similarly $$\sum_{\gamma<t-1}\left(\frac{1}{s-\rho}-\frac{1}{2+it-\rho}\right)=O(\log t)$$

and the result follows again.

The corresponding formula for $\log\zeta(s)$ is given by

THEOREM 9.6 (B). *We have*

$$\log\zeta(s)=\sum_{|t-\gamma|\leqslant 1}\log(s-\rho)+O(\log t) \tag{9.6.3}$$

uniformly for $-1\leqslant\sigma\leqslant 2$, *where* $\log\zeta(s)$ *has its usual meaning, and* $-\pi<\mathbf{I}\log(s-\rho)\leqslant\pi$.

Integrating (9.6.1) from s to $2+it$, and supposing that t is not equal to the ordinate of any zero, we obtain

$$\log\zeta(s)-\log\zeta(2+it)=\sum_{|t-\gamma|\leqslant 1}\{\log(s-\rho)-\log(2+it-\rho)\}+O(\log t).$$

Now $\log\zeta(2+it)$ is bounded; also $\log(2+it-\rho)$ is bounded, and there are $O(\log t)$ such terms. Their sum is therefore $O(\log t)$. The result therefore follows for such values of t, and then by continuity for all values of s in the strip other than the zeros.

9.7. As an application of Theorem 9.6 (B) we shall prove the following theorem on the minimum value of $\zeta(s)$ in certain parts of the critical strip. We know from Theorem 8.12 that $|\zeta(s)|$ is sometimes large in the critical strip, but we can prove little about the distribution of the values of t for which it is large. The following result† states a much weaker inequality, but states it for many more values of t.

THEOREM 9.7. *There is a constant A such that each interval* $(T, T+1)$ *contains a value of t for which*

$$|\zeta(s)|>t^{-A}\quad(-1\leqslant\sigma\leqslant 2). \tag{9.7.1}$$

Further, if H is any number greater than unity, then

$$|\zeta(s)|>T^{-AH} \tag{9.7.2}$$

for $-1\leqslant\sigma\leqslant 2$, $T\leqslant t\leqslant T+1$, *except possibly for a set of values of t of measure* $1/H$.

Taking real parts in (9.6.3),

$$\log|\zeta(s)|=\sum_{|t-\gamma|\leqslant 1}\log|s-\rho|+O(\log t)$$
$$\geqslant\sum_{|t-\gamma|\leqslant 1}\log|t-\gamma|+O(\log t). \tag{9.7.3}$$

† Valiron (1), Landau (8), (18), Hoheisel (3).

Now

$$\int_{T}^{T+1}\sum_{|t-\gamma|\leqslant 1}\log|t-\gamma|\,dt = \sum_{T-1\leqslant\gamma\leqslant T+2}\int_{\max(\gamma-1,T)}^{\min(\gamma+1,T+1)}\log|t-\gamma|\,dt$$

$$\geqslant \sum_{T-1\leqslant\gamma\leqslant T+2}\int_{\gamma-1}^{\gamma+1}\log|t-\gamma|\,dt$$

$$= \sum_{T-1\leqslant\gamma\leqslant T+2}(-2) > -A\log T.$$

Hence
$$\sum_{|t-\gamma|\leqslant 1}\log|t-\gamma| > -A\log T$$
for some t in $(T, T+1)$.

Hence $\log|\zeta(s)| > -A\log T$ for some t in $(T, T+1)$ and all σ in $-1\leqslant\sigma\leqslant 2$; and

$$\log|\zeta(s)| > -AH\log T$$

except in a set of measure $1/H$. This proves the theorem.

The exceptional values of t are, of course, those in the neighbourhood of ordinates of zeros of $\zeta(s)$.

9.8. Application to a formula of Ramanujan.† Let a and b be positive numbers such that $ab = \pi$, and consider the integral

$$\frac{1}{2\pi i}\int a^{-2s}\frac{\Gamma(s)}{\zeta(1-2s)}\,ds = \frac{1}{2\pi i}\int\frac{b^{2s}}{\sqrt{\pi}}\frac{\Gamma(\frac{1}{2}-s)}{\zeta(2s)}\,ds$$

taken round the rectangle $(1\pm iT, -\frac{1}{2}\pm iT)$. The two forms are equivalent on account of the functional equation.

Let $T\to\infty$ through values such that $|T-\gamma| > \exp(-A_1\gamma/\log\gamma)$ for every ordinate γ of a zero of $\zeta(s)$. Then by (9.7.3)

$$\log|\zeta(\sigma+iT)| \geqslant -\sum_{|T-\gamma|\leqslant 1}A_1\gamma/\log\gamma+O(\log T) > -A_2 T$$

where $A_2 < \frac{1}{4}\pi$ if A_1 is small enough, and $T > T_0$. It now follows from the asymptotic formula for the Γ-function that the integrals along the horizontal sides of the contour tend to zero as $T\to\infty$ through the above values. Hence by the theorem of residues‡

$$\frac{1}{2\pi i}\int_{-\frac{1}{2}-i\infty}^{-\frac{1}{2}+i\infty}a^{-2s}\frac{\Gamma(s)}{\zeta(1-2s)}\,ds-\frac{1}{2\pi i}\int_{1-i\infty}^{1+i\infty}\frac{b^{2s}}{\sqrt{\pi}}\frac{\Gamma(\frac{1}{2}-s)}{\zeta(2s)}\,ds$$

$$= -\frac{1}{2\sqrt{\pi}}\sum_{\rho}b^{\rho}\frac{\Gamma(\frac{1}{2}-\frac{1}{2}\rho)}{\zeta'(\rho)}.$$

† Hardy and Littlewood (2), 156–9.

‡ In forming the series of residues we have supposed for simplicity that the zeros of $\zeta(s)$ are all simple.

The first term on the left is equal to

$$\sum_{n=1}^{\infty} \frac{\mu(n)}{n} \frac{1}{2\pi i} \int_{-\frac{1}{2}-i\infty}^{-\frac{1}{2}+i\infty} \left(\frac{n}{a}\right)^{2s} \Gamma(s)\,ds = -\sum_{n=1}^{\infty} \frac{\mu(n)}{n} \{1-e^{-(a/n)^2}\}$$

$$= \sum_{n=1}^{\infty} \frac{\mu(n)}{n} e^{-(a/n)^2}.$$

Evaluating the other integral in the same way, and multiplying through by $\surd a$, we obtain Ramanujan's result

$$\surd a \sum_{n=1}^{\infty} \frac{\mu(n)}{n} e^{-(a/n)^2} - \surd b \sum_{n=1}^{\infty} \frac{\mu(n)}{n} e^{-(b/n)^2} = -\frac{1}{2\surd b} \sum b^{\rho} \frac{\Gamma(\frac{1}{2}-\frac{1}{2}\rho)}{\zeta'(\rho)}. \tag{9.8.1}$$

We have, of course, not proved that the series on the right is convergent in the ordinary sense. We have merely proved that it is convergent if the terms are bracketed in such a way that two terms for which

$$|\gamma-\gamma'| < \exp(-A_1\gamma/\log\gamma)+\exp(-A_1\gamma'/\log\gamma')$$

are included in the same bracket. Of course the zeros are, on the average, much farther apart than this, and it is quite possible that the series may converge without any bracketing. But we are unable to prove this, even on the Riemann hypothesis.

9.9. We next prove a general formula concerning the zeros of an analytic function in a rectangle.† Suppose that $\phi(s)$ is meromorphic in and upon the boundary of a rectangle bounded by the lines $t = 0$, $t = T$, $\sigma = \alpha$, $\sigma = \beta$ $(\beta > \alpha)$, and regular and not zero on $\sigma = \beta$. The function $\log\phi(s)$ is regular in the neighbourhood of $\sigma = \beta$, and here, starting with any one value of the logarithm, we define $F(s) = \log\phi(s)$. For other points s of the rectangle, we define $F(s)$ to be the value obtained from $\log\phi(\beta+it)$ by continuous variation along $t =$ constant from $\beta+it$ to $\sigma+it$, provided that the path does not cross a zero or pole of $\phi(s)$; if it does, we put

$$F(s) = \lim_{\epsilon\to+0} F(\sigma+it+i\epsilon).$$

Let $\nu(\sigma', T)$ denote the excess of the number of zeros over the number of poles in the part of the rectangle for which $\sigma > \sigma'$, including zeros or poles on $t = T$, but not those on $t = 0$.

Then
$$\int F(s)\,ds = -2\pi i \int_{\alpha}^{\beta} \nu(\sigma, T)\,d\sigma, \tag{9.9.1}$$
the integral on the left being taken round the rectangle in the positive direction.

† Littlewood (4).

We may suppose $t = 0$ and $t = T$ to be free from zeros and poles of $\phi(s)$; it is easily verified that our conventions then ensure the truth of the theorem in the general case.

We have

$$\int F(s)\,ds = \int_{\alpha}^{\beta} F(\sigma)\,d\sigma - \int_{\alpha}^{\beta} F(\sigma+iT)\,d\sigma + \int_{0}^{T} \{F(\beta+it)-F(\alpha+it)\}\,i\,dt. \tag{9.9.2}$$

The last term is equal to

$$\int_{0}^{T} i\,dt \int_{\alpha}^{\beta} \frac{\phi'(\sigma+it)}{\phi(\sigma+it)}\,d\sigma = \int_{\alpha}^{\beta} d\sigma \int_{\sigma}^{\sigma+iT} \frac{\phi'(s)}{\phi(s)}\,ds,$$

and by the theorem of residues

$$\int_{\sigma}^{\sigma+iT} \frac{\phi'(s)}{\phi(s)}\,ds = \left(\int_{\sigma}^{\beta} + \int_{\beta}^{\beta+iT} - \int_{\sigma+iT}^{\beta+iT}\right) \frac{\phi'(s)}{\phi(s)}\,ds - 2\pi i\nu(\sigma, T)$$

$$= F(\sigma+iT)-F(\sigma)-2\pi i\nu(\sigma, T).$$

Substituting this in (9.9.2), we obtain (9.9.1).

We deduce

THEOREM 9.9. *If* $\quad S_1(T) = \int_{0}^{T} S(t)\,dt,$

then
$$S_1(T) = \frac{1}{\pi} \int_{\frac{1}{2}}^{2} \log|\zeta(\sigma+iT)|\,d\sigma + O(1). \tag{9.9.3}$$

Take $\phi(s) = \zeta(s)$, $\alpha = \frac{1}{2}$, in the above formula, and take the real part. We obtain

$$\int_{\frac{1}{2}}^{\beta} \log|\zeta(\sigma)|\,d\sigma - \int_{0}^{T} \arg\zeta(\beta+it)\,dt - \int_{\frac{1}{2}}^{\beta} \log|\zeta(\sigma+iT)|\,d\sigma + {}$$
$$+ \int_{0}^{T} \arg\zeta(\tfrac{1}{2}+it)\,dt = 0, \tag{9.9.4}$$

the term in $\nu(\sigma, T)$, being purely imaginary, disappearing. Now make $\beta \to \infty$. We have

$$\log\zeta(s) = \log\left(1+\frac{1}{2^s}+\ldots\right) = O(2^{-\sigma})$$

as $\sigma \to \infty$, uniformly with respect to t. Hence $\arg\zeta(s) = O(2^{-\sigma})$, so that the second integral tends to 0 as $\beta \to \infty$. Also the first integral is a constant, and

$$\int_{2}^{\beta} \log|\zeta(\sigma+iT)|\,d\sigma = \int_{2}^{\beta} O(2^{-\sigma})\,d\sigma = O(1).$$

Hence the result.

THEOREM 9.9 (A). $S_1(T) = O(\log T)$.

By Theorem 9.6 (B)

$$\int_{\frac{1}{2}}^{2} \log|\zeta(s)|\, d\sigma = \sum_{|t-\gamma|\leqslant 1} \int_{\frac{1}{2}}^{2} \log|s-\rho|\, d\sigma + O(\log t).$$

The terms of the last sum are bounded, since

$$\tfrac{3}{2}\log(\tfrac{9}{4}+1) \geqslant \int_{\frac{1}{2}}^{2} \log\{(\sigma-\beta)^2+(\gamma-t)^2\}\, d\sigma \geqslant 2\int_{\frac{1}{2}}^{2} \log|\sigma-\beta|\, d\sigma > -A.$$

Hence
$$\int_{\frac{1}{2}}^{2} \log|\zeta(s)|\, d\sigma = O(\log t), \tag{9.9.5}$$

and the result follows from the previous theorem.

It was proved by F. and R. Nevanlinna (1), (2) that

$$\int_{0}^{T} \frac{S(t)}{t}\, dt = A + O\left(\frac{\log T}{T}\right). \tag{9.9.6}$$

This follows from the previous result by integration by parts; for

$$\int_{1}^{T} \frac{S(t)}{t}\, dt = \left[\frac{S_1(t)}{t}\right]_1^T + \int_{1}^{T} \frac{S_1(t)}{t^2}\, dt = A + \frac{S_1(T)}{T} - \int_{T}^{\infty} \frac{S_1(t)}{t^2}\, dt.$$

Since $S_1(T) = O(\log T)$, the middle term is $O(T^{-1}\log T)$, and the last term is

$$O\left(\int_{T}^{\infty} \frac{\log t}{t^2}\, dt\right) = O\left(-\left[\frac{\log t}{t}\right]_T^\infty + \int_{T}^{\infty} \frac{dt}{t^2}\right) = O\left(\frac{\log T}{T}\right).$$

Hence the result follows. A similar result clearly holds for

$$\int_{1}^{T} \frac{S(t)}{t^\alpha}\, dt \quad (0 < \alpha < 1).$$

It has recently been proved by A. Selberg (5) that

$$S(t) = \Omega_{\pm}\{(\log t)^{\frac{1}{3}}(\log\log t)^{-\frac{7}{3}}\} \tag{9.9.7}$$

with a similar result for $S_1(t)$; and that

$$S_1(t) = \Omega_{+}\{(\log t)^{\frac{1}{2}}(\log\log t)^{-4}\}. \tag{9.9.8}$$

9.10. THEOREM 9.10.† *$S(t)$ has an infinity of changes of sign.*

Consider the interval (γ_n, γ_{n+1}) in which $N(t) = n$. Let $l(t)$ be the

† Titchmarsh (17).

linear function of t such that $l(\gamma_n) = S(\gamma_n)$, $l(\gamma_{n+1}) = S(\gamma_{n+1}-0)$. Then for $\gamma_n < t < \gamma_{n+1}$

$$l(t)-S(t) = \{S(\gamma_{n+1}-0)-S(\gamma_n)\}\frac{t-\gamma_n}{\gamma_{n+1}-\gamma_n}-\{S(t)-S(\gamma_n)\}$$

$$= -\{L(\gamma_{n+1})-L(\gamma_n)\}\frac{t-\gamma_n}{\gamma_{n+1}-\gamma_n}+\{L(t)-L(\gamma_n)\}+O\left(\frac{1}{\gamma_n}\right),$$

using (9.3.2) and the fact that $N(t)$ is constant in the interval. The first two terms on the right give

$$\begin{aligned} -L'(\xi)(t-\gamma_n)+L'(\eta)(t-\gamma_n) & \quad (\gamma_n < \eta < t,\ \gamma_n < \xi < \gamma_{n+1}) \\ = L''(\xi_1)(\eta-\xi)(t-\gamma_n) & \quad (\xi_1 \text{ between } \xi \text{ and } \eta) \\ = O(1/\gamma_n) & \end{aligned}$$

since $\gamma_{n+1}-\gamma_n = O(1)$. Hence

$$\int_{\gamma_n}^{\gamma_{n+1}} S(t)\,dt = \int_{\gamma_n}^{\gamma_{n+1}} l(t)\,dt+O\left(\frac{\gamma_{n+1}-\gamma_n}{\gamma_n}\right)$$

$$= \tfrac{1}{2}(\gamma_{n+1}-\gamma_n)\{S(\gamma_n)+S(\gamma_{n+1}-0)\}+O\left(\frac{\gamma_{n+1}-\gamma_n}{\gamma_n}\right).$$

Suppose that $S(t) \geqslant 0$ for $t > t_0$. Then

$$N(\gamma_n) \geqslant N(\gamma_n-0)+1$$

gives

$$S(\gamma_n) \geqslant S(\gamma_n-0)+1 \geqslant 1.$$

Hence

$$\int_{\gamma_n}^{\gamma_{n+1}} S(t)\,dt \geqslant \tfrac{1}{2}(\gamma_{n+1}-\gamma_n)+O\left(\frac{\gamma_{n+1}-\gamma_n}{\gamma_n}\right)$$

$$\geqslant \tfrac{1}{4}(\gamma_{n+1}-\gamma_n) \quad (n \geqslant n_0).$$

Hence

$$\int_{\gamma_{n_0}}^{\gamma_N} S(t)\,dt \geqslant \tfrac{1}{4}(\gamma_N-\gamma_{n_0}),$$

contrary to Theorem 9.9 (A). Similarly the hypothesis $S(t) \leqslant 0$ for $t > t_0$ can be shown to lead to a contradiction.

It has been proved by A. Selberg (5) that $S(t)$ changes sign at least

$$T(\log T)^{\frac{1}{3}}e^{-A\sqrt{\log\log T}}$$

times in the interval $(0, T)$.

9.11. At the present time no improvement on the result

$$S(T) = O(\log T)$$

is known. But it is possible to prove directly some of the results which would follow from such an improvement. We shall first prove†

THEOREM 9.11. *The gaps between the ordinates of successive zeros of $\zeta(s)$ tend to* 0.

† Littlewood (3).

This would follow at once from (9.3.2) if it were possible to prove that $S(t) = o(\log t)$.

The argument given in § 9.1 shows that the gaps are bounded. Here we have to apply a similar argument to the strip $T-\delta \leqslant t \leqslant T+\delta$, where δ is arbitrarily small, and it is clear that we cannot use four concentric circles. But the ideas of the theorems of Borel–Carathéodory and Hadamard are in no way essentially bound up with sets of concentric circles, and the difficulty can be surmounted by using suitable elongated curves instead.

Let D_4 be the rectangle with centre $3+iT$ and a corner at $-3+i(T+\delta)$, the sides being parallel to the axes. We represent D_4 conformally on the unit circle D_4' in the z-plane, so that its centre $3+iT$ corresponds to $z = 0$. By this representation a set of concentric circles $|z| = r$ inside D_4' will correspond to a set of convex curves inside D_4, such that as $r \to 0$ the curve shrinks upon the point $3+iT$, while as $r \to 1$ it tends to coincidence with D_4. Let D_1', D_2', D_3' be circles (independent, of course, of T) for which the corresponding curves D_1, D_2, D_3 in the s-plane pass through the points $2+iT$, $-1+iT$, $-2+iT$ respectively.

The proof now proceeds as before. We consider the function

$$f(z) = \log \zeta\{s(z)\},$$

where $s = s(z)$ is the analytic function corresponding to the conformal representation; and we apply the theorems of Borel–Carathéodory and Hadamard in the same way as before.

9.12. We shall now obtain a more precise result of the same kind.†

THEOREM 9.12. *For every large positive T, $\zeta(s)$ has a zero $\beta+i\gamma$ satisfying*

$$|\gamma-T| < \frac{A}{\log\log\log T}.$$

This was first proved by Littlewood by a detailed study of the conformal representation used in the previous proof. This involves rather complicated calculations with elliptic functions. We shall give here two proofs which avoid these calculations.

In the first, we replace the rectangles by a succession of circles. Let T be a large positive number, and suppose that $\zeta(s)$ has no zero $\beta+i\gamma$ such that $T-\delta \leqslant \gamma \leqslant T+\delta$, where $\delta < \frac{1}{2}$. Then the function

$$f(s) = \log \zeta(s),$$

where the logarithm has its principal value for $\sigma > 2$, is regular in the rectangle

$$-2 \leqslant \sigma \leqslant 3, \qquad T-\delta \leqslant t \leqslant T+\delta.$$

† Littlewood (3); proofs given here by Titchmarsh (13), Kramaschke (1).

Let c_ν, C_ν, $\mathbf{C}_\nu$, Γ_ν be four concentric circles, with centre $2-\frac{1}{4}\nu\delta+iT$, and radii $\frac{1}{4}\delta$, $\frac{1}{2}\delta$, $\frac{3}{4}\delta$, and δ respectively. Consider these sets of circles for $\nu = 0, 1, \ldots, n$, where $n = [12/\delta]+1$, so that $2-\frac{1}{4}n\delta \leqslant -1$, i.e. the centre of the last circle lies on, or to the left of, $\sigma = -1$. Let m_ν, M_ν, and $\mathbf{M}_\nu$ denote the maxima of $|f(s)|$ on c_ν, C_ν, and $\mathbf{C}_\nu$ respectively.

Let $A_1, A_2, \ldots$ denote absolute constants (it is convenient to preserve their identity throughout the proof). We have $\mathbf{R}\{f(s)\} < A_1 \log T$ on all the circles, and $|f(2+iT)| < A_2$. Hence the Borel–Carathéodory theorem for the circles $\mathbf{C}_0$ and Γ_0 gives

$$\mathbf{M}_0 < \frac{\delta+\frac{3}{4}\delta}{\delta-\frac{3}{4}\delta}(A_1 \log T+A_2) = 7(A_1 \log T+A_2),$$

and in particular

$$|f(2-\tfrac{1}{4}\delta+iT)| < 7(A_1 \log T+A_2).$$

Hence, applying the Borel–Carathéodory theorem to $\mathbf{C}_1$ and Γ_1,

$$\mathbf{M}_1 < 7\{A_1 \log T+|f(2-\tfrac{1}{4}\delta+iT)|\} < (7+7^2)A_1 \log T+7^2 A_2.$$

So generally $\qquad \mathbf{M}_\nu < (7+\ldots+7^{\nu+1})A_1 \log T+7^{\nu+1}A_2,$

or, say,

$$\mathbf{M}_\nu < 7^\nu A_3 \log T. \tag{9.12.1}$$

Now by Hadamard's three-circles theorem

$$M_\nu \leqslant m_\nu^a \mathbf{M}_\nu^b,$$

where a and b are positive constants such that $a+b = 1$; in fact $a = \log\frac{3}{2}/\log 3$, $b = \log 2/\log 3$. Also, since the circle $C_{\nu-1}$ includes the circle c_ν, $m_\nu \leqslant M_{\nu-1}$. Hence

$$M_\nu \leqslant M_{\nu-1}^a \mathbf{M}_\nu^b \quad (\nu = 1, 2, \ldots, n).$$

Thus $\qquad M_1 \leqslant M_0^a \mathbf{M}_1^b, \qquad M_2 \leqslant M_1^a \mathbf{M}_2^b \leqslant M_0^{a^2}\mathbf{M}_1^{ab}\mathbf{M}_2^b,$

and so on, giving finally

$$M_n \leqslant M_0^{a^n}\mathbf{M}_1^{a^{n-1}b}\mathbf{M}_2^{a^{n-2}b}\ldots\mathbf{M}_n^b.$$

Hence, by (9.12.1),

$$M_n \leqslant M_0^{a^n} 7^{a^{n-1}b+2a^{n-2}b+\ldots+nb}(A_3 \log T)^{a^{n-1}b+a^{n-2}b+\ldots+b}.$$

Now

$$a^{n-1}b+2a^{n-2}b+\ldots+nb < n^2,$$

$$a^{n-1}b+a^{n-2}b+\ldots+b = b(1-a^n)/(1-a) = 1-a^n.$$

Hence $\qquad M_n \leqslant M_0^{a^n} 7^{n^2}(A_3 \log T)^{1-a^n} < A_4\, 7^{n^2}(\log T)^{1-a^n},$

since M_0 is bounded as $T \to \infty$.

But $|\zeta(s)| > t^{A_5}$ for $\sigma \leqslant -1$, $t > t_0$, so that $M_n > A_5 \log T$. Hence

$$A_5 < A_4\, 7^{n^2}(\log T)^{-a^n},$$

$$\log\log T < \left(\frac{1}{a}\right)^n\left(n^2 \log 7-\log\frac{A_5}{A_4}\right),$$

$$\log\log\log T < n \log\frac{1}{a}+A_6 \log n,$$

so that
$$\delta < \frac{12}{n-1} < \frac{A}{\log\log\log T},$$
and the result follows.

9.13. Second Proof. Consider the angular region in the s-plane with vertex at $s = -3+iT$, bounded by straight lines making angles $\pm\frac{1}{2}\alpha (0 < \alpha < \pi)$ with the real axis.

Let
$$w = (s+3-iT)^{\pi/\alpha}.$$
Then the angular region is mapped on the half-plane $\mathbf{R}(w) \geqslant 0$. The point $s = 2+iT$ corresponds to
$$w = 5^{\pi/\alpha}.$$
Let
$$z = \frac{w-5^{\pi/\alpha}}{w+5^{\pi/\alpha}}.$$
Then the angular region corresponds to the unit circle in the z-plane, and $s = 2+iT$ corresponds to its centre $z = 0$. If $s = \sigma+iT$ corresponds to $z = -r$, then
$$(\sigma+3)^{\pi/\alpha} = w = 5^{\pi/\alpha}\frac{1-r}{1+r},$$
i.e.
$$r = \left\{1-\left(\frac{\sigma+3}{5}\right)^{\pi/\alpha}\right\}\Big/\left\{1+\left(\frac{\sigma+3}{5}\right)^{\pi/\alpha}\right\}.$$

Suppose that $\zeta(s)$ has no zeros in the angular region, so that $\log\zeta(s)$ is regular in it.

Let $s = \frac{3}{2}+iT$, $-1+iT$, $-2+iT$ correspond to $z = -r_1, -r_2, -r_3$ respectively. Let M_1, M_2, M_3 be the maxima of $|\log\zeta(s)|$ on the s-curves corresponding to $|z| = r_1, r_2, r_3$. Then Hadamard's three-circles theorem gives
$$\log M_2 \leqslant \frac{\log r_3/r_2}{\log r_3/r_1}\log M_1 + \frac{\log r_2/r_1}{\log r_3/r_1}\log M_3.$$

It is easily verified that, on the curve corresponding to $|z| = r_1$, $\sigma \geqslant \frac{3}{2}$. For if $w = \xi+i\eta$, then
$$\sigma = -3+(\xi^2+\eta^2)^{\alpha/2\pi}\cos\left(\frac{\alpha}{\pi}\arctan\frac{\eta}{\xi}\right),$$
which is a minimum at $\eta = 0$, for given ξ, if $0 < \alpha < \frac{1}{2}\pi$; and the minimum is $-3+\xi^{\alpha/\pi}$, which, as a function of ξ, is a minimum when ξ is a minimum, i.e. when $z = -r_1$. It therefore follows that $\log M_1 < A$.

Since $\mathbf{R}\{\log\zeta(s)\} < A\log T$ in the angle, it follows from the Borel–Carathéodory theorem that
$$M_3 < \frac{2}{1-r_3}(A\log T+A) < \frac{A\log T}{1-r_3}.$$

Hence $$\log M_2 \leqslant A+\frac{\log r_2/r_1}{\log r_3/r_1}\log\left(\frac{A\log T}{1-r_3}\right).$$

Now if r_1, r_2, and r_3 are sufficiently near to 1, i.e. if α is sufficiently small,

$$\frac{\log r_2/r_1}{\log r_3/r_1} = \frac{\log\left(1+\dfrac{r_2-r_1}{r_1}\right)}{\log\left(1+\dfrac{r_3-r_1}{r_1}\right)} \leqslant \left(\frac{r_2-r_1}{r_3-r_1}\right)^{\frac{1}{2}},$$

and $$\frac{r_2-r_1}{r_3-r_1} = \frac{\dfrac{1-r_1}{1+r_1}-\dfrac{1-r_2}{1+r_2}}{\dfrac{1-r_1}{1+r_1}-\dfrac{1-r_3}{1+r_3}}\frac{1+r_2}{1+r_3} < \frac{(\frac{9}{10})^{\pi/\alpha}-(\frac{2}{5})^{\pi/\alpha}}{(\frac{9}{10})^{\pi/\alpha}-(\frac{1}{5})^{\pi/\alpha}}$$
$$< 1-A(\tfrac{4}{9})^{\pi/\alpha}.$$

Hence $$\frac{\log r_2/r_1}{\log r_3/r_1} < 1-A(\tfrac{4}{9})^{\pi/\alpha}.$$

Also $$1/(1-r_3) < A5^{\pi/\alpha}.$$

Hence $$\log M_2 < A+\{1-A(\tfrac{4}{9})^{\pi/\alpha}\}\left\{\log\log T+\frac{\pi}{\alpha}\log 5+A\right\}.$$

Let $\alpha = \pi/(c\log\log\log T)$. Then

$$\log M_2 < A+\{1-A(\log\log T)^{-c\log\frac{9}{4}}\}\{\log\log T+c\log 5\log\log\log T+A\}$$
$$< \log\log T-(\log\log T)^{\frac{1}{2}}$$

if $c\log\frac{9}{4} < \frac{1}{2}$ and T is large enough. Hence

$$M_2 < \log T\, e^{-(\log\log T)^{\frac{1}{2}}} < \epsilon\log T \quad (T > T_0(\epsilon)).$$

In particular
$$\log|\zeta(-1+iT)| < \epsilon\log T,$$
$$|\zeta(-1+iT)| < T^{\epsilon}.$$

But $$|\zeta(-1+iT)| = |\chi(-1+iT)\zeta(2-iT)| > KT^{\frac{3}{2}}.$$

We thus obtain a contradiction, and the result follows.

9.14. Another result† in the same order of ideas is

THEOREM 9.14. *For any fixed h, however small,*

$$N(T+h)-N(T) > K\log T$$

for $K = K(h)$, $T > T_0$.

This result is not a consequence of Theorem 9.4 if h is less than a certain value.

Consider the same angular region as before, with a new α such that

† Not previously published.

$\tan\alpha \leqslant \frac{1}{4}$, and suppose now that $\zeta(s)$ has zeros $\rho_1, \rho_2, \ldots, \rho_n$ in the angular region. Let

$$F(s) = \frac{\zeta(s)}{(s-\rho_1)\ldots(s-\rho_n)}.$$

Let C be the circle with centre $\frac{1}{2}+iT$ and radius 3. Then $|s-\rho_\nu| \geqslant 1$ on C. Hence

$$|F(s)| \leqslant |\zeta(s)| < T^A$$

on C, and so also inside C.

Let $f(s) = \log F(s)$. Then $f(s)$ is regular in the angle, and

$$\mathbf{R}f(s) < A\log T.$$

Also

$$f(2+iT) = \log\zeta(2+iT) - \sum_{\nu=1}^{n}\log(2+iT-\rho_\nu)$$
$$= O(1)+\sum_{\nu=1}^{n} O(1) = O(n).$$

Let M_1, M_2, and M_3 now denote the maxima of $|f(s)|$ on the three s-curves. Then

$$M_3 < \frac{A}{1-r_3}(\log T+n).$$

Also $M_1 < An$, as for $f(2+iT)$. Hence

$$\log|f(-1+iT)| \leqslant \log M_2$$
$$< \frac{\log r_3/r_2}{\log r_3/r_1}(A+\log n)+\frac{\log r_2/r_1}{\log r_3/r_1}\log\left\{\frac{A(n+\log T)}{1-r_3}\right\}$$
$$< A+\log n+\frac{\log r_2/r_1}{\log r_3/r_1}\left\{\log\frac{1}{1-r_3}+\log\left(\frac{\log T}{n}\right)\right\}$$
$$< A+\log n+\{1-A(\tfrac{4}{9})^{\pi/\alpha}\}\left\{\frac{\pi}{\alpha}\log 5+\log\left(\frac{\log T}{n}\right)\right\}$$

as before. But

$$|f(-1+iT)| = \left|\log\zeta(-1+iT)-\sum_{\nu=1}^{n}\log(-1+iT-\rho_\nu)\right|$$
$$\geqslant \log|\zeta(-1+iT)|-\sum_{\nu=1}^{n} O(1)$$
$$> A_1\log T-A_2 n,$$

say. If $n > \frac{1}{2}(A_1/A_2)\log T$ the theorem follows at once. Otherwise

$$|f(-1+iT)| > \tfrac{1}{2}A_1\log T,$$

and we obtain

$$\log\left(\frac{\log T}{n}\right) < A+\{1-A(\tfrac{4}{9})^{\pi/\alpha}\}\left\{\frac{\pi}{\alpha}\log 5+\log\left(\frac{\log T}{n}\right)\right\},$$

$$A(\tfrac{4}{9})^{\pi/\alpha}\log\left(\frac{\log T}{n}\right) < A+\{1-A(\tfrac{4}{9})^{\pi/\alpha}\}\frac{\pi}{\alpha}\log 5,$$

and hence $$\log\log\left(\frac{\log T}{n}\right) < \frac{\pi}{\alpha}\log\frac{9}{4}+\log\frac{1}{\alpha}+A < \frac{A}{\alpha},$$
$$n > e^{-e^{A/\alpha}}\log T.$$
This proves the theorem.

9.15. The function $N(\sigma, T)$. We define $N(\sigma, T)$ to be the number of zeros $\beta+i\gamma$ of the zeta-function such that $\beta > \sigma$, $0 < t \leqslant T$. For each T, $N(\sigma, T)$ is a non-increasing function of σ, and is 0 for $\sigma \geqslant 1$. On the Riemann hypothesis, $N(\sigma, T) = 0$ for $\sigma > \frac{1}{2}$. Without any hypothesis, all that we can say so far is that
$$N(\sigma, T) \leqslant N(T) < AT\log T$$
for $\frac{1}{2} < \sigma < 1$.

The object of the next few sections is to improve upon this inequality for values of σ between $\frac{1}{2}$ and 1.

We return to the formula (9.9.1). Let $\phi(s) = \zeta(s)$, $\alpha = \sigma_0$, $\beta = 2$, and this time take the imaginary part. We have
$$\nu(\sigma, T) = N(\sigma, T) \quad (\sigma < 1), \qquad \nu(\sigma, T) = 0 \quad (\sigma \geqslant 1).$$
We obtain, if T is not the ordinate of a zero,
$$2\pi\int\limits_{\sigma_0}^{1} N(\sigma, T)\,d\sigma = \int\limits_0^T \log|\zeta(\sigma_0+it)|\,dt - \int\limits_0^T \log|\zeta(2+it)|\,dt +$$
$$+\int\limits_{\sigma_0}^{2} \arg\zeta(\sigma+iT)\,d\sigma + K(\sigma_0),$$
where $K(\sigma_0)$ is independent of T. We deduce†

THEOREM 9.15. *If* $\frac{1}{2} \leqslant \sigma_0 \leqslant 1$, *and* $T \to \infty$,
$$2\pi\int\limits_{\sigma_0}^{1} N(\sigma, T)\,d\sigma = \int\limits_0^T \log|\zeta(\sigma_0+it)|\,dt + O(\log T).$$

We have
$$\int\limits_0^T \log|\zeta(2+it)|\,dt = \mathbf{R}\sum_{n=2}^{\infty}\frac{\Lambda_1(n)}{n^2}\frac{n^{-iT}-1}{-i\log n} = O(1).$$

Also, by § 9.4, $\arg\zeta(\sigma+iT) = O(\log T)$ uniformly for $\sigma \geqslant \frac{1}{2}$, if T is not the ordinate of a zero. Hence the integral involving $\arg\zeta(\sigma+iT)$ is $O(\log T)$. The result follows if T is not the ordinate of a zero, and this restriction can then be removed from considerations of continuity.

† Littlewood (4).

THEOREM 9.15 (A).† *For any fixed σ greater than $\frac{1}{2}$,*

$$N(\sigma, T) = O(T).$$

For any non-negative continuous $f(t)$

$$\frac{1}{b-a}\int_a^b \log f(t)\,dt \leqslant \log\left\{\frac{1}{b-a}\int_a^b f(t)\,dt\right\}.$$

Thus, for $\frac{1}{2} < \sigma < 1$,

$$\int_0^T \log|\zeta(\sigma+it)|\,dt = \tfrac{1}{2}\int_0^T \log|\zeta(\sigma+it)|^2\,dt$$

$$\leqslant \tfrac{1}{2}T\log\left\{\frac{1}{T}\int_0^T |\zeta(\sigma+it)|^2\,dt\right\} = O(T)$$

by Theorem 7.2. Hence, by Theorem 9.15,

$$\int_{\sigma_0}^1 N(\sigma, T)\,d\sigma = O(T)$$

for $\sigma_0 > \frac{1}{2}$. Hence, if $\sigma_1 = \frac{1}{2}+\frac{1}{2}(\sigma_0-\frac{1}{2})$,

$$N(\sigma_0, T) \leqslant \frac{1}{\sigma_0-\sigma_1}\int_{\sigma_1}^{\sigma_0} N(\sigma, T)\,d\sigma \leqslant \frac{2}{\sigma_0-\frac{1}{2}}\int_{\sigma_1}^{1} N(\sigma, T)\,d\sigma = O(T),$$

the required result.

From this theorem, and the fact that $N(T) \sim AT\log T$, it follows that *all but an infinitesimal proportion of the zeros of $\zeta(s)$ lie in the strip $\frac{1}{2}-\delta < \sigma < \frac{1}{2}+\delta$, however small δ may be.*

9.16. We shall next prove a number of theorems in which the $O(T)$ of Theorem 9.15 (A) is replaced by $O(T^\theta)$, where $\theta < 1$.‡ We do this by applying the above methods, not to $\zeta(s)$ itself, but to the function

$$\zeta(s)M_X(s) = \zeta(s)\sum_{n<X}\frac{\mu(n)}{n^s}.$$

The zeros of $\zeta(s)$ are zeros of $\zeta(s)M_X(s)$. If $\sigma > 1$, $M_X(s) \to 1/\zeta(s)$ as $X \to \infty$, so that $\zeta(s)M_X(s) \to 1$. On the Riemann hypothesis this is also true for $\frac{1}{2} < \sigma \leqslant 1$. Of course we cannot prove this without any hypothesis; but we can choose X so that the additional factor neutralizes to a certain extent the peculiarities of $\zeta(s)$, even for values of σ less than 1.

Let $$f_X(s) = \zeta(s)M_X(s)-1.$$

† Bohr and Landau (4), Littlewood (4).

‡ Bohr and Landau (5), Carlson (1), Landau (12), Titchmarsh (5), Ingham (5).

We shall first prove

THEOREM 9.16. *If for some* $X = X(\sigma, T)$, $T^{1-l(\sigma)} \leqslant X < T^A$,

$$\int\limits_{\frac{1}{2}T}^{T} |f_X(s)|^2\, dt = O(T^{l(\sigma)} \log^m T)$$

as $T \to \infty$, *uniformly for* $\sigma \geqslant \alpha$, *where* $l(\sigma)$ *is a positive non-increasing function with a bounded derivative, and* m *is a constant* $\geqslant 0$, *then*

$$N(\sigma, T) = O(T^{l(\sigma)} \log^{m+1} T)$$

uniformly for $\sigma \geqslant \alpha + 1/\log T$.

We have
$$f_X(s) = \zeta(s) \sum_{n<X} \frac{\mu(n)}{n^s} - 1 = \sum \frac{a_n(X)}{n^s},$$
where $a_1(X) = 0$,
$$a_n(X) = \sum_{d|n} \mu(d) = 0 \quad (n < X),$$
and
$$|a_n(X)| = \left| \sum_{\substack{d|n \\ d<X}} \mu(d) \right| \leqslant d(n)$$
for all n and X.

Let
$$1 - f_X^2 = \zeta M_X(2 - \zeta M_X) = \zeta(s) g(s) = h(s)$$
say, where $g(s) = g_X(s)$ and $h(s) = h_X(s)$ are regular except at $s = 1$. Now for $\sigma \geqslant 2$, $X > X_0$,
$$|f_X(s)|^2 \leqslant \left(\sum_{n \geqslant X} \frac{d(n)}{n^2} \right)^2 = O(X^{2\epsilon - 2}) < \frac{1}{2X} < \frac{1}{2},$$
so that $h(s) \neq 0$. Applying (9.9.1) to $h(s)$, and writing
$$\nu(\sigma, T_1, T_2) = \nu(\sigma, T_2) - \nu(\sigma, T_1),$$
we obtain
$$2\pi \int\limits_{\sigma_0}^{2} \nu(\sigma, \tfrac{1}{2}T, T)\, d\sigma = \int\limits_{\frac{1}{2}T}^{T} \{\log|h(\sigma_0 + it)| - \log|h(2+it)|\}\, dt +$$
$$+ \int\limits_{\sigma_0}^{2} \{\arg h(\sigma + iT) - \arg h(\sigma + \tfrac{1}{2}iT)\}\, d\sigma.$$

Now
$$\log|h(s)| \leqslant \log\{1 + |f_X(s)|^2\} \leqslant |f_X(s)|^2,$$
so that, if $\sigma_0 \geqslant \alpha$,
$$\int\limits_{\frac{1}{2}T}^{T} \log|h(\sigma_0 + it)|\, dt \leqslant \int\limits_{\frac{1}{2}T}^{T} |f_X(\sigma_0 + it)|^2\, dt = O(T^{l(\sigma_0)} \log^m T).$$

Next
$$-\log|h(2+it)| \leqslant -\log\{1 - |f_X(2+it)|^2\} \leqslant 2|f_X(2+it)|^2 < X^{-1}$$

so that $$-\int_{\frac{1}{2}T}^{T} \log|h(2+it)|\,dt < \frac{T}{2X} = O(T^{l(\sigma_0)}).$$

Also we can apply the lemma of § 9.4 to $h(s)$, with $\alpha = 0$, $\beta \geqslant \frac{1}{2}$, $m \geqslant \frac{1}{2}$, and $M_{\sigma,t} = O(X^A T^A)$. We obtain
$$\arg h(s) = O(\log X + \log t)$$
for $\sigma \geqslant \frac{1}{2}$. Hence
$$\int_{\sigma_0}^{\frac{1}{2}} \{\arg h(\sigma+iT) - \arg h(\sigma+\tfrac{1}{2}iT)\}\,d\sigma = O(\log X+\log T) = O(\log T).$$

Hence $$\int_{\sigma_0}^{2} \nu(\sigma, \tfrac{1}{2}T, T)\,d\sigma = O(T^{l(\sigma_0)}\log^m T).$$

Also
$$\int_{\sigma_0}^{2} \nu(\sigma, \tfrac{1}{2}T, T)\,d\sigma \geqslant \int_{\sigma_0}^{2} N(\sigma, \tfrac{1}{2}T, T)\,d\sigma \geqslant (\sigma_1-\sigma_0)N(\sigma_1, \tfrac{1}{2}T, T)$$
if $\sigma_0 < \sigma_1 \leqslant 2$. Taking $\sigma_1 = \sigma_0 + 1/\log T$, we have
$$T^{l(\sigma_0)} = T^{l(\sigma_1)+O(\sigma_1-\sigma_0)} = O(T^{l(\sigma_1)}).$$

Hence $$N(\sigma_1, \tfrac{1}{2}T, T) = O(T^{l(\sigma_1)}\log^{m+1}T).$$

Replacing T by $\frac{1}{2}T$, $\frac{1}{4}T$,... and adding, the result follows.

9.17. The simplest application is

THEOREM 9.17. *For any fixed σ in $\frac{1}{2} < \sigma < 1$,*
$$N(\sigma, T) = O(T^{4\sigma(1-\sigma)+\epsilon}).$$

We use Theorem 4.11 with $x = T$, and obtain
$$\begin{aligned} f_X(s) &= \sum_{m<T} \frac{1}{m^s} \sum_{n<X} \frac{\mu(n)}{n^s} - 1 + O(T^{-\sigma}|M_X(s)|) \\ &= \sum \frac{b_n(X)}{n^s} + O(T^{-\sigma}X^{1-\sigma}), \end{aligned} \tag{9.17.1}$$

where, if $X < T$, $b_n(X) = 0$ for $n < X$ and for $n > XT$; and, as for a_n, $|b_n(X)| \leqslant d(n) = O(n^\epsilon)$. Hence
$$\begin{aligned} \int_{\frac{1}{2}T}^{T} \left|\sum \frac{b_n(X)}{n^s}\right|^2 dt &= \tfrac{1}{2}T \sum \frac{|b_n(X)|^2}{n^{2\sigma}} + \sum\sum \frac{b_m b_n}{(mn)^\sigma} \int_{\frac{1}{2}T}^{T} \left(\frac{n}{m}\right)^{it} dt \\ &= O\Big(T \sum_{n\geqslant X} \frac{1}{n^{2\sigma-\epsilon}}\Big) + O\Big(\sum_{n<m<XT}\sum \frac{1}{(mn)^{\sigma-\epsilon}\log m/n}\Big) \\ &= O(TX^{1-2\sigma+\epsilon}) + O\{(XT)^{2-2\sigma+\epsilon}\} \end{aligned}$$

by (7.2.1). These terms are of the same order (apart from ϵ's) if $X = T^{2\sigma-1}$, and then

$$\int_{\frac{1}{2}T}^{T} \left|\sum \frac{b_n(X)}{n^s}\right|^2 dt = O(T^{4\sigma(1-\sigma)+\epsilon}).$$

The O-term in (9.17.1) gives

$$O(T^{1-2\sigma}X^{2-2\sigma}) = O(T^{1-2\sigma}X) = O(1).$$

The result therefore follows from Theorem 9.16.

9.18. The main instrument used in obtaining still better results for $N(\sigma, T)$ is the convexity theorem for mean values of analytic functions proved in § 7.8. We require, however, some slight extensions of the theorem. If the right-hand sides of (7.8.1) and (7.8.2) are replaced by finite sums

$$\sum C(T^a+1), \qquad \sum C'(T^b+1),$$

then the right-hand side of (7.8.3) is clearly to be replaced by

$$K\sum\sum (CT^a)^{(\beta-\sigma)/(\beta-\alpha)}(C'T^b)^{(\sigma-\alpha)/(\beta-\alpha)}.$$

In one of the applications a term $T^a\log^4 T$ occurs in the data instead of the above T^a. This produces the same change in the result. The only change in the proof is that, instead of the term

$$\int_0^\infty \left(\frac{u}{\delta}\right)^{a+2\alpha-1} e^{-2u}\,du = \frac{K}{\delta^{a+2\alpha-1}},$$

we obtain a term

$$\int_0^\infty \left(\frac{u}{\delta}\right)^{a+2\alpha-1} \log^4\frac{u}{\delta}\, e^{-2u}\,du$$

$$= \int_0^\infty \left(\frac{u}{\delta}\right)^{a+2\alpha-1} \left\{\log^4\frac{1}{\delta}+4\log^3\frac{1}{\delta}\log u+\ldots\right\} e^{-2u}\,du < \frac{K}{\delta^{a+2\alpha-1}}\log^4\frac{1}{\delta}.$$

THEOREM 9.18. *If* $\zeta(\frac{1}{2}+it) = O(t^c\log^{c'}t)$, *where* $c' \leqslant \frac{3}{2}$, *then*

$$N(\sigma, T) = O(T^{2(1+2c)(1-\sigma)}\log^5 T)$$

uniformly for $\frac{1}{2} \leqslant \sigma \leqslant 1$.

If $0 < \delta < 1$,

$$\int_0^T |f_X(1+\delta+it)|^2\,dt = \sum_{m\geqslant X}\sum_{n\geqslant X} \frac{a_X(m)a_X(n)}{m^{1+\delta}n^{1+\delta}} \int_0^T \left(\frac{m}{n}\right)^{it} dt$$

$$= T\sum_{n\geqslant X}\frac{a_X^2(n)}{n^{2+2\delta}} + 2\sum_{X\leqslant m<n}\sum \frac{a_X(m)a_X(n)}{m^{1+\delta}n^{1+\delta}}\frac{\sin(T\log m/n)}{\log m/n}$$

$$\leqslant T\sum_{n\geqslant X}\frac{d^2(n)}{n^{2+2\delta}} + 2\sum_{X\leqslant m<n}\sum \frac{d(m)d(n)}{m^{1+\delta}n^{1+\delta}}.$$

Now† $\sum_{n\leqslant x} d^2(n) < Ax\log^3 x$, $\quad \sum\sum_{m<n\leqslant x} \frac{d(m)d(n)}{(mn)^{\frac{1}{2}}\log n/m} < Ax\log^3 x$.

Hence

$$\sum_{n\geqslant X} \frac{d^2(n)}{n^{1+\xi}} = \sum_{n\geqslant X} d^2(n) \int_n^\infty \frac{1+\xi}{x^{2+\xi}}\,dx = \int_X^\infty \frac{1+\xi}{x^{2+\xi}} \sum_{X\leqslant n\leqslant x} d^2(n)\,dx$$

$$< \int_X^\infty \frac{(1+\xi)A\log^3 x}{x^{1+\xi}}\,dx = \frac{A(1+1/\xi)}{X^\xi}\int_1^\infty \frac{\log^3(Xy^{1/\xi})}{y^2}\,dy$$

(putting $x = Xy^{1/\xi}$) $\quad < \frac{A}{\xi X^\xi}\left(\log X+\frac{1}{\xi}\right)^3.$

Hence $\quad \sum_{n\geqslant X} \frac{d^2(n)}{n^{2+2\delta}} < \frac{A\log^3 X}{X^{1+2\delta}} < \frac{A}{X\delta^3}$

since $X^{2\delta} = e^{2\delta\log X} > \frac{1}{6}(2\delta\log X)^3$.

Also, since $\quad 1 < \log\lambda+\lambda^{-1} < \log\lambda+\lambda^{-\frac{1}{2}}$

for $\lambda > 1$,

$$\sum\sum_{X\leqslant m<n} \frac{d(m)d(n)}{(mn)^{1+\xi}\log n/m} < \sum\sum_{X\leqslant m<n} \frac{d(m)d(n)}{(mn)^{1+\xi}} + \sum\sum_{X\leqslant m<n} \frac{d(m)d(n)}{m^\xi n^{1+\xi}(mn)^{\frac{1}{2}}\log n/m}$$

$$< \left(\sum_{n=1}^\infty \frac{d(n)}{n^{1+\xi}}\right)^2 + \sum\sum_{1\leqslant m<n} \frac{d(m)d(n)}{(mn)^{\frac{1}{2}}\log n/m} \int_n^\infty \frac{1+\xi}{x^{2+\xi}}\,dx$$

$$< \zeta^4(1+\xi) + \int_1^\infty \frac{1+\xi}{x^{2+\xi}} \sum\sum_{m<n\leqslant x} \frac{d(m)d(n)}{(mn)^{\frac{1}{2}}\log n/m}\,dx$$

$$< \zeta^4(1+\xi) + \int_1^\infty \frac{(1+\xi)A\log^3 x}{x^{1+\xi}}\,dx < \frac{A}{\xi^4}.$$

Hence $$\int_0^T |f_X(1+\delta+it)|^2\,dt < A\left(\frac{T}{X}+1\right)\delta^{-4}. \tag{9.18.1}$$

For $\sigma = \frac{1}{2}$ we use the inequalities

$$|f_X|^2 \leqslant 2(|\zeta|^2|M_X|^2+1),$$

$$\int_0^T |M_X(\tfrac{1}{2}+it)|^2\,dt \leqslant T\sum_{n<X} \frac{\mu^2(n)}{n} + 2\sum\sum_{m<n<X} \frac{|\mu(m)\mu(n)|}{(mn)^{\frac{1}{2}}\log n/m}$$

$$\leqslant T\sum_{n<X}\frac{1}{n} + 2\sum\sum_{m<n<X} \frac{1}{(mn)^{\frac{1}{2}}\log n/m}$$

$$< A(T+X)\log X,$$

by (7.2.1).

† The first result follows easily from (7.16.3); for the second, see Ingham (1); the argument of § 7.21, and the first result, give an extra $\log x$.

Hence $$\int_0^T |f_X(\tfrac{1}{2}+it)|^2\,dt < AT^{2c}(T+X)\log^{2c'}(T+2)\log X. \tag{9.18.2}$$

The convexity theorem therefore gives

$$\int_{\frac{1}{2}T}^{T} |f_X(\sigma+it)|^2\,dt$$
$$= O\left\{\left(\frac{T}{X}+1\right)\delta^{-4}\right\}^{(\sigma-\frac{1}{2})/(\frac{1}{2}+\delta)}\{T^{2c}(T+X)\log^{2c'}(T+2)\log X\}^{(1+\delta-\sigma)/(\frac{1}{2}+\delta)}$$
$$= O\left\{\frac{T+X}{\delta^4}\,\frac{T^{4c(1-\sigma)}}{X^{2\sigma-1}}\,(XT^{2c})^{\{(2\sigma-1)\delta\}/(\frac{1}{2}+\delta)}(\delta^4\log^3(T+2)\log X)^{(1+\delta-\sigma)/(\frac{1}{2}+\delta)}\right\}.$$

Taking $\delta = 1/\log(T+X)$, we obtain
$$O\{(T+X)T^{4c(1-\sigma)}X^{1-2\sigma}\log^4(T+X)\}.$$
If $X = T$, the result follows from Theorem 9.16.

For example, by Theorem 5.5 we may take $c = \frac{1}{6}$, $c' = \frac{3}{2}$. Hence
$$N(\sigma, T) = O(T^{\frac{8}{3}(1-\sigma)}\log^5 T). \tag{9.18.3}$$
This is an improvement on Theorem 9.17 if $\sigma > \frac{2}{3}$.

On the unproved Lindelöf hypothesis that $\zeta(\frac{1}{2}+it) = O(t^\epsilon)$, Theorem 9.18 gives
$$N(\sigma, T) = O(T^{2(1-\sigma)+\epsilon}).$$

9.19. An improvement on Theorem 9.17 for all values of σ in $\frac{1}{2} < \sigma < 1$ is effected by combining (9.18.3) with

THEOREM 9.19 (A). *$N(\sigma, T) = O(T^{\frac{3}{2}-\sigma}\log^5 T)$.*

We have
$$\int_0^T |f_X(\tfrac{1}{2}+it)|^2\,dt < A\int_0^T |\zeta(\tfrac{1}{2}+it)|^2|M_X(\tfrac{1}{2}+it)|^2\,dt+AT$$
$$< A\left\{\int_0^T |\zeta(\tfrac{1}{2}+it)|^4\,dt\int_0^T |M_X(\tfrac{1}{2}+it)|^4\,dt\right\}^{\frac{1}{2}}+AT.$$

Now $$M_X^2(s) = \sum_{n<X^2}\frac{c_n}{n^s}, \qquad |c_n| \leqslant d(n).$$

Hence
$$\int_0^T |M_X(\tfrac{1}{2}+it)|^4\,dt \leqslant T\sum_{n<X^2}\frac{d^2(n)}{n}+2\sum\sum_{m<n<X^2}\frac{d(m)d(n)}{(mn)^{\frac{1}{2}}\log n/m}$$
$$< AT\log^4 X+AX^2\log^3 X.$$

Hence $$\int_0^T |f_X(\tfrac{1}{2}+it)|^2\,dt < AT^{\frac{1}{2}}(T+X^2)^{\frac{1}{2}}\log^2(T+2)\log^2 X. \tag{9.19.1}$$

From (9.18.1), (9.19.1), and the convexity theorem, we obtain

$$\int_{\frac{1}{2}T}^{T} |f_X(\sigma+it)|^2\,dt$$
$$= O\left\{\left(\left(\frac{T}{X}+1\right)\delta^{-4}\right)^{(\sigma-\frac{1}{2})/(\frac{1}{2}+\delta)}\{T^{\frac{1}{2}}(T+X^2)^{\frac{1}{2}}\log^2(T+2)\log^2 X\}^{(1+\delta-\sigma)/(\frac{1}{2}+\delta)}\right\}.$$

If $X = T^{\frac{1}{2}}$, $\delta = 1/\log(T+2)$, the result follows as before.

This is an improvement on Theorem 9.17 if $\frac{1}{2} < \sigma < \frac{3}{4}$.

Various results of this type have been obtained,† the most successful‡ being

THEOREM 9.19 (B). $N(\sigma, T) = O(T^{3(1-\sigma)/(2-\sigma)}\log^5 T)$.

This depends on a two-variable convexity theorem;§ if

$$J(\sigma,\lambda) = \left\{\int_0^T |f(\sigma+it)|^{1/\lambda}\,dt\right\}^{\lambda},$$

then $$J(\sigma, p\lambda+q\mu) = O\{J^p(\alpha,\lambda)J^q(\beta,\mu)\} \quad (\alpha<\sigma<\beta),$$

where $$p = \frac{\beta-\sigma}{\beta-\alpha}, \qquad q = \frac{\sigma-\alpha}{\beta-\alpha}.$$

We have

$$\int_0^T |f_X(\tfrac{1}{2}+it)|^{\frac{4}{3}}\,dt < A\int_0^T |\zeta(\tfrac{1}{2}+it)|^{\frac{4}{3}}|M_X(\tfrac{1}{2}+it)|^{\frac{4}{3}}\,dt + AT$$
$$< A\left\{\int_0^T |\zeta(\tfrac{1}{2}+it)|^4\,dt\right\}^{\frac{1}{3}}\left\{\int_0^T |M_X(\tfrac{1}{2}+it)|^2\,dt\right\}^{\frac{2}{3}} + AT$$
$$< A\{T\log^4(T+2)\}^{\frac{1}{3}}\{(T+X)\log X\}^{\frac{2}{3}} + AT$$
$$< A(T+X)\log^2(T+X). \tag{9.19.2}$$

In the two-variable convexity theorem, take $\alpha = \frac{1}{2}$, $\beta = 1+\delta$, $\lambda = \frac{3}{4}$, $\mu = \frac{1}{2}$, and use (9.18.1) and (9.19.2). We obtain

$$\int_0^T |f_X(\sigma+it)|^{1/K}\,dt$$
$$< A\{(T+X)\log^2(T+X)\}^{\frac{3}{2}(1-\sigma+\delta)/(1-\frac{1}{2}\sigma+\frac{3}{2}\delta)}\left\{\left(\frac{T}{X}+1\right)\delta^{-4}\right\}^{(\sigma-\frac{1}{2})/(1-\frac{1}{2}\sigma+\frac{3}{2}\delta)},$$

where $K = p\lambda+q\mu$ lies between $\frac{1}{2}$ and $\frac{3}{4}$. Taking $X = T$, $\delta = 1/\log T$, we obtain

$$\int_0^T |f_X(\sigma+it)|^{1/K}\,dt < AT^{3(1-\sigma)/(2-\sigma)}\log^4 T.$$

† Titchmarsh (5), Ingham (5), (6). ‡ Ingham (6). § Gabriel (1).

The result now follows from a modified form of Theorem 9.16, since

$$\log|1-f_X^2| \leqslant \log(1+|f_X|^2) < A|f_X|^{1/K}.$$

A. Selberg† has recently proved

THEOREM 9.19 (C). $N(\sigma, T) = O(T^{1-\frac{1}{4}(\sigma-\frac{1}{2})}\log T)$

uniformly for $\frac{1}{2} \leqslant \sigma \leqslant 1.$

This is an improvement on the previous theorem if σ is a function of T such that $\sigma-\frac{1}{2}$ is sufficiently small.

9.20. The corresponding problems with σ equal or nearly equal to $\frac{1}{2}$ are naturally more difficult. Here the most interesting question is that of the behaviour of

$$\int_{\frac{1}{2}}^{1} N(\sigma, T)\, d\sigma \tag{9.20.1}$$

as $T \to \infty$. If the zeros of $\zeta(s)$ are $\beta+i\gamma$, this is equal to

$$\int_{\frac{1}{2}}^{1} \Big(\sum_{\beta>\sigma, 0<\gamma\leqslant T} 1\Big)\, d\sigma = \sum_{\beta>\frac{1}{2}, 0<\gamma\leqslant T} \int_{\frac{1}{2}}^{\beta} d\sigma = \sum_{\beta>\frac{1}{2}, 0<\gamma\leqslant T} (\beta-\tfrac{1}{2}).$$

Hence an equivalent problem is that of the sum

$$\sum_{0<\gamma\leqslant T} |\beta-\tfrac{1}{2}|. \tag{9.20.2}$$

There are some immediate results.‡ If we apply the above argument, but use Theorem 7.2 (A) instead of Theorem 7.2, we obtain at once

$$\int_{\sigma_0}^{1} N(\sigma, T)\, d\sigma < AT\log\left\{\min\left(\log T, \log\frac{1}{\sigma_0-\frac{1}{2}}\right)\right\} \tag{9.20.3}$$

for $\frac{1}{2} \leqslant \sigma_0 \leqslant 1$; and in particular

$$\int_{\frac{1}{2}}^{1} N(\sigma, T)\, d\sigma = O(T\log\log T). \tag{9.20.4}$$

These, however, are superseded by the following analysis, due to A. Selberg (2), the principal result of which is that

$$\int_{\frac{1}{2}}^{1} N(\sigma, T)\, d\sigma = O(T). \tag{9.20.5}$$

We consider the integral

$$\int_{T}^{T+U} |\zeta(\tfrac{1}{2}+it)\psi(\tfrac{1}{2}+it)|^2\, dt,$$

† Selberg (5). ‡ Littlewood (4).

where $0 < U \leqslant T$ and ψ is a function to be specified later. We use the formulae of §4.17. Since

$$e^{i\vartheta} = \{\chi(\tfrac{1}{2}+it)\}^{-\frac{1}{2}} = \left(\frac{t}{2\pi e}\right)^{\frac{1}{2}it} e^{-\frac{1}{8}\pi i}\left\{1+O\left(\frac{1}{t}\right)\right\},$$

we have

$$Z(t) = z(t)+\bar{z}(t)+O(t^{-\frac{1}{4}}), \tag{9.20.6}$$

where

$$z(t) = \left(\frac{t}{2\pi e}\right)^{\frac{1}{2}it} e^{-\frac{1}{8}\pi i} \sum_{n\leqslant x} n^{-\frac{1}{2}-it}$$

and $x = (t/2\pi)^{\frac{1}{2}}$. Let $T \leqslant t \leqslant T+U$, $\tau = (T/2\pi)^{\frac{1}{2}}$, $\tau' = \{(T+U)/2\pi\}^{\frac{1}{2}}$. Let

$$z_1(t) = \left(\frac{t}{2\pi e}\right)^{\frac{1}{2}it} e^{-\frac{1}{8}\pi i} \sum_{n\leqslant \tau} n^{-\frac{1}{2}-it}.$$

Proceeding as in § 7.3, we have

$$\begin{aligned}\int\limits_T^{T+U} |z(t)-z_1(t)|^2\,dt &= O\left(U \sum_{\tau<n\leqslant\tau'} \frac{1}{n}\right)+O(T^{\frac{1}{2}}\log T)\\ &= O\left(U\frac{\tau'-\tau}{\tau}\right)+O(T^{\frac{1}{2}}\log T)\\ &= O(U^2/T)+O(T^{\frac{1}{2}}\log T).\end{aligned} \tag{9.20.7}$$

9.21. LEMMA 9.21. *Let m and n be positive integers, $(m,n)=1$, $M=\max(m,n)$. Then*

$$\int\limits_T^{T+U} z_1(t)\bar{z}_1(t)\left(\frac{n}{m}\right)^{it} dt = \frac{U}{(mn)^{\frac{1}{2}}} \sum_{r<\tau/M} \frac{1}{r}+O\{T^{\frac{1}{2}}M^2\log(MT)\}.$$

The integral is

$$\sum_{\mu\leqslant\tau}\sum_{\nu\leqslant\tau} \frac{1}{(\mu\nu)^{\frac{1}{2}}} \int\limits_T^{T+U} \left(\frac{n\nu}{m\mu}\right)^{it} dt.$$

The terms with $m\mu = n\nu$ contribute

$$U\sum_{m\mu=n\nu}\sum \frac{1}{(\mu\nu)^{\frac{1}{2}}} = U\sum_{rn\leqslant\tau,\, rm\leqslant\tau} \frac{1}{(rn\,.\,rm)^{\frac{1}{2}}} = \frac{U}{(mn)^{\frac{1}{2}}}\sum_{r\leqslant\tau/M}\frac{1}{r}.$$

The remaining terms are

$$\begin{aligned}O\left\{\sum_{m\mu\neq n\nu}\sum \frac{1}{(\mu\nu)^{\frac{1}{2}}|\log(n\nu/m\mu)|}\right\} &= O\left\{\sum_{m\mu\neq n\nu}\sum \frac{M}{(m\mu n\nu)^{\frac{1}{2}}|\log(n\nu/m\mu)|}\right\}\\ &= O\left\{M \sum_{\kappa\leqslant M\tau,\,\lambda\leqslant M\tau}\sum \frac{1}{(\kappa\lambda)^{\frac{1}{2}}|\log\lambda/\kappa|}\right\} = O\{M^2\tau\log(M\tau)\},\end{aligned}$$

and the result follows.

9.22. LEMMA 9.22. *Defining m, n, M as before, and supposing*

$$T^{\frac{4}{5}} < U \leqslant T,$$

$$\int_T^{T+U} z_1^2(t)\left(\frac{n}{m}\right)^{it} dt = \frac{U}{(mn)^{\frac{1}{2}}} \sum_{\tau/m \leqslant r \leqslant \tau/n} \frac{1}{r} + O(MT^{\frac{1}{2}}) + O(U^2/T) + O(T^{\frac{9}{10}}) \tag{9.22.1}$$

if $n \leqslant m$. If $m < n$, the first term on the right-hand side is to be omitted.

The left-hand side is

$$e^{-\frac{1}{4}\pi i} \sum_{\mu \leqslant \tau} \sum_{\nu \leqslant \tau} \frac{1}{(\mu\nu)^{\frac{1}{2}}} \int_T^{T+U} \left(\frac{t}{2\pi e} \frac{n}{\mu\nu m}\right)^{it} dt.$$

The integral is of the form considered in § 4.6, with

$$F(t) = t \log \frac{t}{ec}, \qquad c = \frac{2\pi\mu\nu m}{n}.$$

Hence by (4.6.5), with $\lambda_2 = (T+U)^{-1}$, $\lambda_3 = (T+U)^{-2}$, it is equal to

$$(2\pi c)^{\frac{1}{2}} e^{\frac{1}{4}\pi i - ic} + O(T^{\frac{2}{5}}) + O\left\{\min\left(\frac{1}{|\log c/T|}, T^{\frac{1}{2}}\right)\right\}$$

$$+ O\left\{\min\left(\frac{1}{\log|(T+U)/c|}, T^{\frac{1}{2}}\right)\right\}, \tag{9.22.2}$$

with the leading term present only when $T \leqslant c \leqslant T+U$. We therefore obtain a main term

$$2\pi \left(\frac{m}{n}\right)^{\frac{1}{2}} \sum_{\mu \leqslant \tau} \sum_{\nu \leqslant \tau} e^{-2\pi i \mu\nu m/n} \tag{9.22.3}$$

where μ and ν also satisfy

$$\tau^2 n/m \leqslant \mu\nu \leqslant \tau'^2 n/m.$$

The double sum is clearly zero unless $n \leqslant m$, as we now suppose. The ν-summation runs over the range $\nu_1 \leqslant \nu \leqslant \nu_2$, where $\nu_1 = \tau^2 n/m\mu$ and $\nu_2 = \min(\tau'^2 n/m\mu, \tau)$, and μ runs over $\tau n/m \leqslant \mu \leqslant \tau$. The inner sum is therefore $\nu_2 - \nu_1 + O(n)$ if $n|\mu$, and $O(n)$ otherwise. The error term $O(n)$ contributes $O\{(mn)^{\frac{1}{2}}\tau\} = O(MT^{\frac{1}{2}})$ in (9.22.1). On writing $\mu = nr$ we are left with

$$2\pi \left(\frac{m}{n}\right)^{\frac{1}{2}} \sum_{\tau/m \leqslant r \leqslant \tau/n} (\nu_2 - \nu_1).$$

Let $\nu_3 = \tau'^2/mr$. Then $\nu_2 = \nu_3$ unless $r < \tau'^2/m\tau$. Hence the error on replacing ν_2 by ν_3 is

$$O\left\{\left(\frac{m}{n}\right)^{\frac{1}{2}}\sum_{\tau/m\leqslant r<\tau'^2/m\tau}\left(\frac{\tau'^2}{mr}-\tau\right)\right\}=O\left\{\left(\frac{m}{n}\right)^{\frac{1}{2}}\left(\frac{\tau'^2}{m\tau}-\frac{\tau}{m}+1\right)\left(\frac{\tau'^2}{\tau}-\tau\right)\right\}$$

$$=O\left\{(mn)^{-\frac{1}{2}}\left(\frac{\tau'^2-\tau^2}{\tau}\right)^2\right\}+O\left\{\left(\frac{m}{n}\right)^{\frac{1}{2}}\left(\frac{\tau'^2-\tau^2}{\tau}\right)\right\}$$

$$=O(U^2T^{-1})+O(M^{\frac{1}{2}}UT^{-\frac{1}{2}}).$$

Finally there remains

$$2\pi\left(\frac{m}{n}\right)^{\frac{1}{2}}\sum_{\tau/m\leqslant r\leqslant\tau/n}(\nu_3-\nu_1)=2\pi\left(\frac{m}{n}\right)^{\frac{1}{2}}\sum_{\tau/m\leqslant r\leqslant\tau/n}\left(\frac{\tau'^2}{mr}-\frac{\tau^2}{mr}\right)$$

$$=\frac{U}{(mn)^{\frac{1}{2}}}\sum_{\tau/m\leqslant r\leqslant\tau/n}\frac{1}{r}.$$

Now consider the O-terms arising from (9.22.2). The term $O(T^{\frac{2}{5}})$ gives

$$O\left\{T^{\frac{2}{5}}\sum_{\mu\leqslant\tau}\sum_{\nu\leqslant\tau}\frac{1}{(\mu\nu)^{\frac{1}{2}}}\right\}=O(T^{\frac{2}{5}}\tau)=O(T^{\frac{9}{10}}).$$

Next

$$\sum_{\mu\leqslant\tau}\sum_{\nu\leqslant\tau}\frac{1}{(\mu\nu)^{\frac{1}{2}}}\min\left(\frac{1}{|\log(2\pi\mu\nu m/nT)|},T^{\frac{1}{2}}\right)$$

$$=O\left\{T^{\epsilon}\sum_{r\leqslant\tau^2}\frac{1}{r^{\frac{1}{2}}}\min\left(\frac{1}{|\log(rm/n\tau^2)|},T^{\frac{1}{2}}\right)\right\}.$$

Suppose, for example, that $n<m$. Then the terms with $r<\frac{1}{2}n\tau^2/m$ or $r>2n\tau^2/m$ are

$$O\left(T^{\epsilon}\sum_{r\leqslant\tau^2}\frac{1}{r^{\frac{1}{2}}}\right)=O(T^{\epsilon}\tau)=O(T^{\frac{1}{2}+\epsilon}).$$

In the other terms, let $r=[n\tau^2/m]-r'$. We obtain

$$O\left\{T^{\epsilon}\sum_{r'}\frac{1}{(n\tau^2/m)^{\frac{1}{2}}}\frac{1}{|r'-\theta|/(n\tau^2/m)}\right\}\quad(|\theta|<1)$$

$$=O\left\{T^{\epsilon}\left(\frac{n\tau^2}{m}\right)^{\frac{1}{2}}\log T\right\}=O(T^{\frac{1}{2}+\epsilon}),$$

omitting the terms $r'=-1,0,1$; and these are $O(T^{\frac{1}{2}+\epsilon})$.

A similar argument applies in the other cases.

9.23. LEMMA 9.23. *Let* $(m,n)=1$ *with* $m,n\leqslant X\leqslant T^{\frac{1}{5}}$. *If* $T^{\frac{14}{15}}\leqslant U\leqslant T$, *then*

$$\int_T^{T+U} Z^2(t)\left(\frac{n}{m}\right)^{it} dt = \frac{U}{(mn)^{\frac{1}{2}}}\left\{\log\frac{T}{2\pi mn}+2\gamma\right\}+O(U^{\frac{3}{2}}T^{-\frac{1}{2}}\log T).$$

Let $Z(t) = z_1(t)+\overline{z_1(t)}+e(t)$. Then

$$\int_T^{T+U} \{z_1(t)+\overline{z_1(t)}\}^2\left(\frac{n}{m}\right)^{it} dt$$

$$= \int_T^{T+U} Z(t)^2\left(\frac{n}{m}\right)^{it} dt + O\left(\int_T^{T+U} |Z(t)e(t)|\,dt\right)+O\left(\int_T^{T+U} |e(t)|^2 dt\right).$$

We have

$$\int_T^{T+U} |e(t)|^2\,dt = O(U^2/T)+O(T^{\frac{1}{2}}\log T) = O(U^2/T)$$

by (9.20.7), and

$$\int_T^{T+U} |Z(t)|^2\,dt = O(U\log T)+O(T^{\frac{1}{2}+\varepsilon}) = O(U\log T),$$

by Theorem 7.4. Hence

$$\int_T^{T+U} |Z(t)e(t)|\,dt = O\{(U^2/T)^{\frac{1}{2}}(U\log T)^{\frac{1}{2}}\}$$

by Cauchy's inequality. It follows that

$$\int_T^{T+U} Z(t)^2\left(\frac{n}{m}\right)^{it} dt$$

$$= \int_T^{T+U} \{z_1(t)^2+\overline{z_1(t)^2}+2z_1(t)\overline{z_1(t)}\}\left(\frac{n}{m}\right)^{it} dt + O(U^{\frac{3}{2}}T^{-\frac{1}{2}}\log^{\frac{1}{2}}T).$$

By Lemmas 9.21 and 9.22 the main integral on the right is

$$\frac{U}{(mn)^{\frac{1}{2}}}\left(\sum_{r\leqslant\tau/n}\frac{1}{r}+\sum_{r\leqslant\tau/m}\frac{1}{r}\right)+O\{T^{\frac{1}{2}}X^2\log(XT)\}+O(XT^{\frac{1}{2}})+$$
$$+O(U^2/T)+O(T^{\frac{9}{10}})$$

whether $n \leqslant m$ or not. The result then follows, since

$$\sum_{r \leqslant \tau/n} \frac{1}{r} + \sum_{r \leqslant \tau/m} \frac{1}{r} = \log\frac{\tau^2}{mn} + 2\gamma + O\left(\frac{X}{\tau}\right),$$

and since the error terms $O\{T^{\frac{1}{2}}X^2\log(XT)\}$, $O(XT^{\frac{1}{2}})$, $O(U^2/T)$, $O(T^{\frac{9}{10}})$ and $O(UXT^{-\frac{1}{2}})$ are all $O(U^{\frac{3}{2}}T^{-\frac{1}{2}}\log T)$.

9.24. Theorem 9.24.

$$\int\limits_{\frac{1}{2}}^{1} N(\sigma, T)\, d\sigma = O(T). \tag{9.24.1}$$

Consider the integral

$$I = \int\limits_{T}^{T+U} |\zeta(\tfrac{1}{2}+it)\psi(\tfrac{1}{2}+it)|^2\, dt = \int\limits_{T}^{T+U} Z^2(t)|\psi(\tfrac{1}{2}+it)|^2\, dt,$$

where

$$\psi(s) = \sum_{r<X} \delta_r r^{1-s}$$

and

$$\delta_r = \frac{\sum\limits_{\rho r<X} \mu(\rho r)\mu(\rho)/\phi(\rho r)}{\sum\limits_{\rho<X} \mu^2(\rho)/\phi(\rho)} = \frac{\mu(r)}{\phi(r)} \frac{\sum\limits_{\rho r<X, (\rho,r)=1} \mu^2(\rho)/\phi(\rho)}{\sum\limits_{\rho<X} \mu^2(\rho)/\phi(\rho)}.$$

Clearly

$$|\delta_r| \leqslant \frac{1}{\phi(r)}$$

for all values of r. Now

$$I = \sum_{q<X} \sum_{r<X} \delta_q \delta_r q^{\frac{1}{2}} r^{\frac{1}{2}} \int\limits_{T}^{T+U} Z^2(t)\left(\frac{n}{m}\right)^{it} dt,$$

where $m = q/(q,r)$, $n = r/(q,r)$. Using Lemma 9.23, the main term contributes to this

$$\sum_{q<X} \sum_{r<X} \delta_q \delta_r q^{\frac{1}{2}} r^{\frac{1}{2}} \frac{U}{(mn)^{\frac{1}{2}}} \log\frac{Te^{2\gamma}}{2\pi mn} = U \sum_{q<X} \sum_{r<X} \delta_q \delta_r (q,r) \log\frac{Te^{2\gamma}(q,r)^2}{2\pi qr}$$

$$= U \log\frac{Te^{2\gamma}}{2\pi} \sum_{q<X} \sum_{r<X} \delta_q \delta_r (q,r) - 2U \sum_{q<X} \sum_{r<X} \delta_q \delta_r (q,r) \log q +$$

$$+ 2U \sum_{q<X} \sum_{r<X} \delta_q \delta_r (q,r) \log (q,r).$$

For a fixed $q < X$,

$$\sum_{r<X} (q,r)\, \delta_r = \left\{\sum_{\rho<X} \frac{\mu^2(\rho)}{\phi(\rho)}\right\}^{-1} \sum_{r<X, \rho r<X} \frac{(q,r)\mu(\rho r)\mu(\rho)}{\phi(\rho r)}.$$

Now

$$(q,r) = \sum_{\nu|(q,r)} \phi(\nu) = \sum_{\nu|q, \nu|r} \phi(\nu).$$

Hence the second factor on the right is

$$\sum_{r<X,\rho r<X}\sum \frac{\mu(\rho r)\mu(\rho)}{\phi(\rho r)} \sum_{\nu|q,\nu|r} \phi(\nu) = \sum_{\nu|q} \phi(\nu) \sum_{\substack{r<X,\rho r<X \\ \nu|r}}\sum \frac{\mu(\rho r)\mu(\rho)}{\phi(\rho r)}.$$

Put $\rho r = l$. Then $\rho\nu|\rho r$, $\rho\nu|l$, i.e. $\rho|(l/\nu)$. Hence we get

$$\sum_{\nu|q} \phi(\nu) \sum_{\substack{l<X \\ \nu|l}} \frac{\mu(l)}{\phi(l)} \sum_{\rho|(l/\nu)} \mu(\rho).$$

The ρ-sum is 0 unless $l = \nu$, when it is 1. Hence we get

$$\sum_{\nu|q} \phi(\nu)\frac{\mu(\nu)}{\phi(\nu)} = \sum_{\nu|q} \mu(\nu) = \begin{cases} 1 & (q=1), \\ 0 & (q>1). \end{cases}$$

Hence

$$\sum_{q<X} \sum_{r<X} \delta_q \delta_r(q,r) = \left\{\sum_{\rho<X} \frac{\mu^2(\rho)}{\phi(\rho)}\right\}^{-1} \delta_1 = \left\{\sum_{\rho<X} \frac{\mu^2(\rho)}{\phi(\rho)}\right\}^{-1}$$

and $$\sum_{q<X} \sum_{r<X} \delta_q \delta_r(q,r)\log q = \left\{\sum_{\rho<X} \frac{\mu^2(\rho)}{\phi(\rho)}\right\}^{-1} \delta_1 \log 1 = 0.$$

Let $\phi_a(n)$ be defined by

$$\sum_{n=1}^{\infty} \frac{\phi_a(n)}{n^s} = \frac{\zeta(s-a-1)}{\zeta(s)},$$

so that $$\phi_a(n) = n^{1+a} \sum_{m|n} \frac{\mu(m)}{m^{1+a}} = n^{1+a} \prod_{p|n} \left(1-\frac{1}{p^{1+a}}\right).$$

Let $\psi(n)$ be defined by

$$\sum_{n=1}^{\infty} \frac{\psi(n)}{n^s} = -\frac{\zeta'(s-1)}{\zeta(s)}.$$

Then $$-\zeta'(s-1) = \zeta(s) \sum_{n=1}^{\infty} \frac{\psi(n)}{n^s},$$

and hence $$n \log n = \sum_{d|n} \psi(d).$$

Hence $$(q,r)\log(q,r) = \sum_{d|q,d|r} \psi(d)$$

and $$\sum_{q<X} \sum_{r<X} \delta_q \delta_r(q,r)\log(q,r) = \sum_{d<X} \psi(d) \sum_{\substack{d|q,d|r \\ q<X,r<X}} \delta_q \delta_r$$

$$= \sum_{d<X} \psi(d) \Big(\sum_{d|q,q<X} \delta_q\Big)^2.$$

Now $$\psi(n) = \left[\frac{\partial}{\partial a}\phi_a(n)\right]_{a=0} = \phi(n)\left(\log n + \sum_{p|n} \frac{\log p}{p-1}\right),$$

$$\psi(n) \leqslant \phi(n)\left(\log n + \sum_{p|n} \log p\right) \leqslant 2\phi(n)\log n.$$

Also

$$\sum_{\substack{d|q \\ q<X}} \delta_q = \left\{\sum_{\rho<X} \frac{\mu^2(\rho)}{\phi(\rho)}\right\}^{-1} \sum\sum_{\substack{\rho q<X \\ d|q}} \frac{\mu(\rho q)\mu(\rho)}{\phi(\rho q)}$$

$$= \left\{\sum_{\rho<X} \frac{\mu^2(\rho)}{\phi(\rho)}\right\}^{-1} \sum_{\substack{n<X \\ d|n}} \frac{\mu(n)}{\phi(n)} \sum_{\rho|n/d} \mu(\rho) = \left\{\sum_{\rho<X} \frac{\mu^2(\rho)}{\phi(\rho)}\right\}^{-1} \frac{\mu(d)}{\phi(d)}.$$

Hence $$\sum_{q<X} \sum_{r<X} \delta_q \delta_r (q,r) \log(q,r) \leqslant 2\log X \left\{\sum_{\rho<X} \frac{\mu^2(\rho)}{\phi(\rho)}\right\}^{-1}.$$

Since

$$\sum_{n=1}^{\infty} \frac{\mu^2(n)}{\phi(n)n^s} = \prod_p \left(1 + \frac{\mu^2(p)}{\phi(p)p^s}\right) = \prod_p \left(1 + \frac{1}{(p-1)p^s}\right)$$

$$= \zeta(s+1) \prod_p \left(1 - \frac{1}{p^{s+1}}\right)\left(1 + \frac{1}{(p-1)p^s}\right),$$

we have $$\sum_{\rho<X} \frac{\mu^2(\rho)}{\phi(\rho)} \sim A \log X.$$

The contribution of all the above terms to I is therefore

$$O\left(U \frac{\log T}{\log X}\right) + O(U) = O(U)$$

on taking, say, $X = T^{\frac{1}{100}}$.

The O-term in Lemma 9.23 gives

$$O(U^{\frac{3}{2}} T^{-\frac{1}{2}} \log T) \sum_{q<X} \sum_{r<X} \frac{q^{\frac{1}{2}} r^{\frac{1}{2}}}{\phi(q)\phi(r)}$$

$$= O(U^{\frac{3}{2}} T^{-\frac{1}{2}} \log T) O(X)$$

$$= O(U^{\frac{3}{2}} T^{-\frac{49}{100}} \log T).$$

Taking say $U = T^{\frac{14}{15}}$, this is $O(U)$. Hence $I = O(U)$.

By an argument similar to that of § 9.16, it follows that

$$\int_{\frac{1}{2}}^{1} \{N(\sigma, T+U) - N(\sigma, T)\}\, d\sigma = O(U).$$

Replacing T by $T+U$ $T+2U$,... and adding, $O(T/U)$ terms, we obtain

$$\int_{\frac{1}{2}}^{1} \{N(\sigma, 2T)-N(\sigma, T)\}\, d\sigma = O(T).$$

Replacing T by $\frac{1}{2}T$, $\frac{1}{4}T$,... and adding, the theorem follows.

It also follows that, if $\frac{1}{2} < \sigma \leqslant 1$,

$$N(\sigma, T) = \frac{2}{\sigma-\frac{1}{2}} \int_{\frac{1}{2}\sigma+\frac{1}{4}}^{\sigma} N(\sigma', T)\, d\sigma'$$

$$\leqslant \frac{2}{\sigma-\frac{1}{2}} \int_{\frac{1}{2}}^{1} N(\sigma, T)\, d\sigma = O\left(\frac{T}{\sigma-\frac{1}{2}}\right). \tag{9.24.2}$$

Lastly, *if $\phi(t)$ is positive and increases to infinity with t, all but an infinitesimal proportion of the zeros of $\zeta(s)$ in the upper half-plane lie in the region*

$$|\sigma-\tfrac{1}{2}| < \frac{\phi(t)}{\log t}.$$

The curved boundary of the region

$$\sigma = \tfrac{1}{2}+\frac{\phi(t)}{\log t}, \qquad T^{\frac{1}{2}} < t < T$$

lies to the right of
$$\sigma = \sigma_1 = \tfrac{1}{2}+\frac{\phi(T^{\frac{1}{2}})}{\log T},$$

and
$$N(\sigma_1, T) = O\left(\frac{T}{\sigma_1-\frac{1}{2}}\right) = O\left(\frac{T\log T}{\phi(T^{\frac{1}{2}})}\right) = o\,(T\log T).$$

Hence the number of zeros outside the region specified is $o\,(T\log T)$, and the result follows.

NOTES FOR CHAPTER 9

9.25. The mean value of $S(t)$ has been investigated by Selberg (5). One has

$$\int_{0}^{T} |S(t)|^{2k}\, dt \sim \frac{(2k)!}{k!(2\pi)^{2k}}\, T(\log\log T)^{k} \tag{9.25.1}$$

for every positive integer k. Selberg's earlier conditional treatment (4) is discussed in §§ 14.20–24, the key feature used in (5) to deal with zeros off the critical line being the estimate given in Theorem 9.19(C). Selberg (5) also gave an unconditional proof of Theorem 14.19, which had previously been established on the Riemann hypothesis by Littlewood.

These results have been investigated further by Fujii [1], [2] and Ghosh [1], [2], who give results which are uniform in k.

It follows in particular from Fujii [1] that

$$\int_0^T |S(t+h)-S(t)|^2\,dt = \pi^{-2}T\log(3+h\log T)+O[T\{\log(3+h\log T)\}^{\frac{1}{2}}] \tag{9.25.2}$$

and

$$\int_0^T |S(t+h)-S(t)|^{2k}\,dt \ll T\{Ak^4\log(3+h\log T)\}^k \tag{9.25.3}$$

uniformly for $0 \leqslant h \leqslant \frac{1}{2}T$. One may readily deduce that

$$N_j(T) \ll N(T)\,\mathrm{e}^{-A\sqrt{j}},$$

where $N_j(T)$ denotes the number of zeros $\beta+i\gamma$ of multiplicity exactly j, in the range $0<\gamma\leqslant T$. Moreover one finds that

$$\#\{n: 0<\gamma_n\leqslant T, \gamma_{n+1}-\gamma_n \geqslant \lambda/\log T\} \ll N(T)\exp\{-A\lambda^{\frac{1}{2}}(\log\lambda)^{-\frac{1}{4}}\},$$

uniformly for $\lambda \geqslant 2$, whence, in particular,

$$\sum_{0<\gamma_n\leqslant T} (\gamma_{n+1}-\gamma_n)^k \ll \frac{N(T)}{(\log T)^k}, \tag{9.25.4}$$

for any fixed $k \geqslant 0$. Fujii [2] also states that there exist constants $\lambda > 1$ and $\mu < 1$ such that

$$\frac{\gamma_{n+1}-\gamma_n}{2\pi/\log\gamma_n} \geqslant \lambda \tag{9.25.5}$$

and

$$\frac{\gamma_{n+1}-\gamma_n}{2\pi/\log\gamma_n} \leqslant \mu \tag{9.25.6}$$

each hold for a positive proportion of n (i.e. the number of n for which $0<\gamma_n\leqslant T$ is at least $AN(T)$ if $T\geqslant T_0$). Note that $2\pi/\log\gamma_n$ is the average spacing between zeros. The possibility of results such as (9.25.5) and (9.25.6) was first observed by Selberg [6].

9.26. Since the deduction of the results (9.25.5) and (9.25.6) is not obvious, we give a sketch. If M is a sufficiently large integer constant, then (9.25.2) and (9.25.3) yield

$$\int_T^{2T} |S(t+h)-S(t)|^2 dt \gg T$$

and

$$\int_T^{2T} |S(t+h)-S(t)|^4 dt \ll T$$

uniformly for

$$\frac{2\pi M}{\log T} \leqslant h \leqslant \frac{4\pi M}{\log T}.$$

By Hölder's inequality we have

$$\int_T^{2T} |S(t+h)-S(t)|^2 dt \leqslant \left(\int_T^{2T} |S(t+h)-S(t)| dt\right)^{\frac{2}{3}} \times \left(\int_T^{2T} |S(t+h)-S(t)|^4 dt\right)^{\frac{1}{3}},$$

so that

$$\int_T^{2T} |S(t+h)-S(t)| dt \gg T.$$

We now observe that

$$S(t+h)-S(t) = N(t+h)-N(t)-\frac{h\log T}{2\pi}+O\left(\frac{1}{\log T}\right),$$

for $T \leqslant t \leqslant 2T$, whence

$$\int_T^{2T} \left|N(t+h)-N(t)-\frac{h\log T}{2\pi}\right| dt \gg T.$$

We proceed to write $h = 2\pi M\lambda/\log T$ and

$$\delta(t,\lambda) = N\left(t+\frac{2\pi\lambda}{\log T}\right)-N(t)-\lambda,$$

so that

$$N(t+h)-N(t)-\frac{h\log T}{2\pi}=\sum_{m=0}^{M-1}\delta\left(t+\frac{2\pi m\lambda}{\log T},\lambda\right).$$

Thus

$$T \ll \sum_{m=0}^{M-1}\int_{T+2\pi m\lambda/\log T}^{2T+2\pi m\lambda/\log T}|\delta(t,\lambda)|dt$$

$$= M\int_{T}^{2T}|\delta(t,\lambda)|dt+O(1),$$

and hence

$$\int_{T}^{2T}|\delta(t,\lambda)|dt \gg T \tag{9.26.1}$$

uniformly for $1 \leqslant \lambda \leqslant 2$, since M is constant.

Now, if I is the subset of $[T,2T]$ on which $N\left(t+\dfrac{2\pi\lambda}{\log T}\right)=N(t)$, then

$$|\delta(t,\lambda)| \leqslant \begin{cases}\delta(t,\lambda)+2\lambda & (t\in I),\\ \delta(t,\lambda)+2\lambda-2 & (t\in[T,2T]-I),\end{cases}$$

so that (9.26.1) yields

$$T \ll \int_{T}^{2T}\delta(t,\lambda)dt+(2\lambda-2)T+2m(I),$$

where $m(I)$ is the measure of I. However

$$\int_{T}^{2T}\delta(t,\lambda)\,dt=O\left(\frac{T}{\log T}\right),$$

whence $m(I) \gg T$, if $\lambda>1$ is chosen sufficiently close to 1. Thus, if

$$S=\left\{n\colon T\leqslant\gamma_n\leqslant 2T,\,\gamma_{n+1}-\gamma_n\geqslant\frac{2\pi\lambda}{\log T}\right\},$$

then

$$T \ll m(I) \ll \sum_{n\in S}(\gamma_{n+1}-\gamma_n)+O(1),$$

so that

$$T^2 \ll \left\{ \sum_{n \in S} (\gamma_{n+1} - \gamma_n) \right\}^2 \leqslant (\# S)\left(\sum_{n \in S} (\gamma_{n+1} - \gamma_n)^2 \right)$$

$$\ll \# S \frac{T}{\log T},$$

by (9.25.4) with $k = 2$. It follows that

$$\# S \gg N(T), \tag{9.26.2}$$

proving that (9.25.5) holds for a positive proportion of n.

Now suppose that μ is a constant in the range $0 < \mu < 1$, and put

$$U = \{n \colon T \leqslant \gamma_n \leqslant 2T\},$$

and

$$V = \left\{ n \in U \colon \gamma_{n+1} - \gamma_n \leqslant \frac{2\pi\mu}{\log T} \right\},$$

whence $\# U = \dfrac{T}{2\pi} \log T + O(T)$. Then

$$T = \sum_{n \in U} (\gamma_{n+1} - \gamma_n) + O(1)$$

$$\geqslant \sum_{n \in U - V} (\gamma_{n+1} - \gamma_n) + O(1)$$

$$\geqslant \frac{2\pi\mu}{\log T} (\# U - \# V - \# S) + \frac{2\pi\lambda}{\log T} S + O(1)$$

$$= \frac{2\pi\mu}{\log T} \left(\frac{T}{2\pi} \log T - \# V \right) + \frac{2\pi(\lambda - \mu)}{\log T} \# S + O\left(\frac{T}{\log T} \right).$$

If the implied constant in (9.26.2) is η, it follows that $\# V \gg N(T)$, on taking $\mu = 1 - \nu$, with $0 < \nu < \eta(\lambda - 1)/(1 - \eta)$. Thus (9.25.6) also holds for a positive proportion of n.

9.27. Ghosh [**1**] was able to sharpen the result of Selberg mentioned at the end of §9.10, to show that $S(t)$ has at least

$$T(\log T) \exp\left(- \frac{A \operatorname{loglog} T}{(\operatorname{logloglog} T)^{\frac{1}{2} - \delta}} \right)$$

sign changes in the range $0 \leqslant t \leqslant T$, for any positive δ, and $A = A(\delta)$, $T \geqslant T(\delta)$. He also proved (Ghosh [**2**]) that the asymptotic formula (9.25.1) holds for any positive real k, with the constant on the right hand side replaced by $\Gamma(2k+1)/\Gamma(k+1)(2\pi)^{2k}$. Moreover he showed (Ghosh [**2**]) that

$$\frac{|S(t)|}{\sqrt{(\log\log t)}} = f(t),$$

say, has a limiting distribution

$$P(\sigma) = 2\pi^{\frac{1}{2}} \int_0^{\sigma} e^{-\pi^2 z^2}\, dz,$$

in the sense that, for any $\sigma > 0$, the measure of the set of $t \in [0, T]$ for which $f(t) \leqslant \sigma$, is asymptotically $TP(\sigma)$. (A minor error in Ghosh's statement of the result has been corrected here.)

9.28. A great deal of work has been done on the 'zero-density estimates' of §§9.15–19, using an idea which originates with Halász [**1**]. However it is not possible to combine this with the method of §9.16, based on Littlewood's formula (9.9.1). Instead one argues as follows (Montgomery [**1**; Chapter 12]). Let

$$M_X(s)\zeta(s) = \sum_1^{\infty} a_n n^{-s},$$

so that $a_n = 0$ for $2 \leqslant n \leqslant X$. If $\zeta(\rho) = 0$, where $\rho = \beta + i\gamma$ and $\beta > \frac{1}{2}$, then we have

$$e^{-1/Y} + \sum_{n>X} a_n n^{-\rho} e^{-n/Y} = \sum_{n=1}^{\infty} a_n n^{-\rho} e^{-n/Y}$$

$$= \frac{1}{2\pi i} \int_{2-i\infty}^{2+i\infty} M_X(s+\rho)\zeta(s+\rho)\Gamma(s)\, Y^s\, ds,$$

by the lemma of §7.9. On moving the line of integration to $\mathbf{R}(s) = \frac{1}{2} - \beta$ this yields

$$M_X(1)\Gamma(1-\rho)\, Y^{1-\rho} +$$

$$+ \frac{1}{2\pi i} \int_{-\infty}^{\infty} M_X(\tfrac{1}{2}+it)\,\zeta(\tfrac{1}{2}+it)\Gamma(\tfrac{1}{2}-\beta+i(t-\gamma))\, Y^{\frac{1}{2}-\beta+i(t-\gamma)}\, dt,$$

since the pole of $\Gamma(s)$ at $s = 0$ is cancelled by the zero of $\zeta(s+\rho)$. If we now assume that $\log^2 T \leqslant \gamma \leqslant T$, and that $\log T \ll \log X$, $\log Y \ll \log T$, then $e^{-1/Y} \gg 1$ and

$$M_X(1)\Gamma(1-\rho)Y^{1-\rho} = o(1),$$

whence either

$$\left|\sum_{n>X} a_n n^{-\rho} e^{-n/Y}\right| \gg 1$$

or

$$\int_{-\infty}^{\infty} |M_X(\tfrac{1}{2}+it)\zeta(\tfrac{1}{2}+it)\Gamma(\tfrac{1}{2}-\beta+i(t-\gamma))|\,dt \gg Y^{\beta-\frac{1}{2}}.$$

In the latter case one has

$$|M_X(\tfrac{1}{2}+it_\rho)\zeta(\tfrac{1}{2}+it_\rho)| \gg (\beta-\tfrac{1}{2})\,Y^{\beta-\frac{1}{2}}$$

for some t_ρ in the range $|t_\rho-\gamma| \leqslant \log^2 T$. The problem therefore reduces to that of counting discrete points at which one of the Dirichlet series $\Sigma a_n n^{-s}e^{-n/Y}$, $M_X(s)$, and $\zeta(s)$ is large. In practice it is more convenient to take finite Dirichlet polynomials approximating to these.

The methods given in §§9.17–19 correspond to the use of a mean-value bound. Thus Montgomery [**1**; Chapter 7] showed that

$$\sum_{r=1}^{R}\left|\sum_{n=1}^{N} a_n n^{-s_r}\right|^2 \ll (T+N)(\log N)^2 \sum_{n=1}^{N} |a_n|^2 n^{-2\sigma} \qquad (9.28.1)$$

for any points s_r satisfying

$$\mathbf{R}(s_r) \geqslant \sigma, \qquad |\mathbf{I}(s_r)| \leqslant T, \qquad \mathbf{I}(s_{r+1}-s_r) \geqslant 1, \qquad (9.28.2)$$

and any complex a_n. Theorems 9.17, 9.18, 9.19(A), and 9.19(B) may all be recovered from this (except possibly for worse powers of $\log T$). However one may also use Halász's lemma. One simple form of this (Montgomery [**1**; Theorem 8.2]) gives

$$\sum_{r=1}^{R}\left|\sum_{n=1}^{N} a_n n^{-s_r}\right|^2 \ll (N+RT^{\frac{1}{2}})(\log T) \sum_{n=1}^{N} |a_n|^2 n^{-2\sigma} \quad (9.28.3)$$

for any points s_r satisfying (9.28.2). Under suitable circumstances this implies a sharper bound for R than does (9.28.1). Under the Lindelöf hypothesis one may replace the term $RT^{\frac{1}{2}}$ in (9.28.3) by $RT^{\varepsilon}N^{\frac{1}{2}}$, which is superior, since one invariably takes $N \leqslant T$ in applying the Halász lemma. (If $N \geqslant T$ it would be better to use (9.28.1).) Moreover Montgomery [**1**; Chapter 9] makes the conjecture (the Large Values

Conjecture):

$$\sum_{r=1}^{R}\left|\sum_{n=1}^{N} a_n n^{-s_r}\right|^2 \ll (N+RT^{\varepsilon})\sum_{n=1}^{N}|a_n|^2 n^{-2\sigma}$$

for points s_r satisfying (9.28.2). Using the Halász lemma with the Lindelöf hypothesis one obtains

$$N(\sigma, T) \ll T^{\varepsilon}, \qquad \tfrac{3}{4}+\varepsilon \leqslant \sigma \leqslant 1, \tag{9.28.4}$$

(Halász and Turán [**1**], Montgomery [**1**; Theorem 12.3]). If the Large Values Conjecture is true then the Lindelöf hypothesis gives the wider range $\frac{1}{2}+\varepsilon \leqslant \sigma \leqslant 1$ for (9.28.4)

9.29. The picture for unconditional estimates is more complex. At present it seems that the Halász method is only useful for $\sigma \geqslant \frac{3}{4}$. Thus Ingham's result, Theorem 9.19(B), is still the best known for $\frac{1}{2} < \sigma \leqslant \frac{3}{4}$. Using (9.28.3), Montgomery [**1**; Theorem 12.1] showed that

$$N(\sigma, T) \ll T^{2(1-\sigma)/\sigma}(\log T)^{14} \qquad (\tfrac{4}{5} \leqslant \sigma \leqslant 1),$$

which is superior to Theorem 9.19(B). This was improved by Huxley [**1**] to give

$$N(\sigma, T) \ll T^{3(1-\sigma)/(3\sigma-1)}(\log T)^{44} \quad (\tfrac{3}{4} \leqslant \sigma \leqslant 1). \tag{9.29.1}$$

Huxley used the Halász lemma in the form

$$R \ll \left\{NV^{-2}\sum_{n=1}^{N}|a_n|^2 n^{-2\sigma} + TNV^{-6}\left(\sum_{n=1}^{N}|a_n|^2 n^{-2\sigma}\right)^3\right\}(\log T)^2,$$

for points s_r satisfying (9.28.2) and the condition

$$\left|\sum_{n=1}^{N} a_n n^{-s_r}\right| \geqslant V.$$

In conjunction with Theorem 9.19(B), Huxley's result yields

$$N(\sigma, T) \ll T^{12/5(1-\sigma)}(\log T)^{44} \quad (\tfrac{1}{2} \leqslant \sigma \leqslant 1),$$

(c.f. (9.18.3)). A considerable number of other estimates have been given, for which the interested reader is referred to Ivić [**3**; Chapter 11]. We mention only a few of the most significant. Ivić [**2**] showed that

$$N(\sigma, T) \ll \begin{cases} T^{(3-3\sigma)/(7\sigma-4)+\varepsilon} & (\frac{3}{4} \leqslant \sigma \leqslant \frac{10}{13}) \\ T^{(9-9\sigma)/(8\sigma-2)+\varepsilon} & (\frac{10}{13} \leqslant \sigma \leqslant 1), \end{cases}$$

which supersede Huxley's result (9.29.1) throughout the range $\frac{3}{4} < \sigma < 1$. Jutila [**1**] gave a more powerful, but more complicated, result, which has a similar effect. His bounds also imply the 'Density hypothesis' $N(\sigma, T) \ll T^{2-2\sigma+\varepsilon}$, for $\frac{11}{14} \leqslant \sigma \leqslant 1$. Heath-Brown [**6**] improved this by giving

$$N(\sigma, T) \ll T^{(9-9\sigma)/(7\sigma-1)+\varepsilon} \quad (\tfrac{11}{14} \leqslant \sigma \leqslant 1).$$

When σ is very close to 1 one can use the Vinogradov–Korobov exponential sum estimates, as described in Chapter 6. These lead to

$$N(\sigma, T) \ll T^{A(1-\sigma)^{\frac{3}{2}}}(\log T)^{A'},$$

for suitable numerical constants A and A', (see Montgomery [**1**; Corollary 12.5], who gives $A = 1334$, after correction of a numerical error).

Selberg's estimate given in Theorem 9.19(C) has been improved by Jutila [**2**] to give

$$N(\sigma, T) \ll T^{1-(1-\delta)(\sigma-\frac{1}{2})} \log T$$

uniformly for $\frac{1}{2} \leqslant \sigma \leqslant 1$, for any fixed $\delta > 0$.

9.30. Of course Theorem 19.24 is an immediate consequence of Theorem 19.9(C), but the proof is a little easier. The coefficients δ_r used in § 9.24 are essentially

$$\mu(r) r^{-1} \frac{\log X/r}{\log X},$$

and indeed a more careful analysis yields

$$\int_0^T |\zeta(\tfrac{1}{2}+it)|^2 \left| \sum_{r \leqslant X} \mu(r) \frac{\log X/r}{\log X} r^{-\frac{1}{2}-it} \right|^2 dt \sim T\left(1 + \frac{\log T}{\log X}\right).$$

Here one can take $X \leqslant T^{\frac{1}{2}-\varepsilon}$ using fairly standard techniques, or $X \leqslant T^{\frac{9}{17}-\varepsilon}$ by employing estimates for Kloosterman sums (see Balasubramanian, Conrey and Heath-Brown [**1**]). The latter result yields (9.24.1) with the implied constant 0·0845.

X

THE ZEROS ON THE CRITICAL LINE

10.1. General discussion. The memoir in which Riemann first considered the zeta-function has become famous for the number of ideas it contains which have since proved fruitful, and it is by no means certain that these are even now exhausted. The analysis which precedes his observations on the zeros is particularly interesting. He obtains, as in § 2.6, the formula

$$\Gamma(\tfrac{1}{2}s)\pi^{-\frac{1}{2}s}\zeta(s) = \frac{1}{s(s-1)} + \int_1^\infty \psi(x)(x^{\frac{1}{2}s-1}+x^{-\frac{1}{2}-\frac{1}{2}s})\,dx,$$

where

$$\psi(x) = \sum_{n=1}^\infty e^{-n^2\pi x}.$$

Multiplying by $\frac{1}{2}s(s-1)$, and putting $s = \frac{1}{2}+it$, we obtain

$$\Xi(t) = \tfrac{1}{2}-(t^2+\tfrac{1}{4})\int_1^\infty \psi(x)x^{-\frac{3}{4}}\cos(\tfrac{1}{2}t\log x)\,dx. \tag{10.1.1}$$

Integrating by parts, and using the relation

$$4\psi'(1)+\psi(1) = -\tfrac{1}{2},$$

which follows at once from (2.6.3), we obtain

$$\Xi(t) = 4\int_1^\infty \frac{d}{dx}\{x^{\frac{3}{2}}\psi'(x)\}x^{-\frac{1}{4}}\cos(\tfrac{1}{2}t\log x)\,dx. \tag{10.1.2}$$

Riemann then observes:

'Diese Function ist für alle endlichen Werthe von t endlich, und lässt sich nach Potenzen von tt in eine sehr schnell convergirende Reihe entwickeln. Da für einen Werth von s, dessen reeller Bestandtheil grösser als 1 ist, $\log\zeta(s) = -\sum\log(1-p^{-s})$ endlich bleibt, und von den Logarithmen der übrigen Factoren von $\Xi(t)$ dasselbe gilt, so kann die Function $\Xi(t)$ nur verschwinden, wenn der imaginäre Theil von t zwischen $\frac{1}{2}i$ und $-\frac{1}{2}i$ liegt. Die Anzahl der Wurzeln von $\Xi(t) = 0$, deren reeller Theil zwischen 0 und T liegt, ist etwa

$$= \frac{T}{2\pi}\log\frac{T}{2\pi}-\frac{T}{2\pi};$$

denn dass Integral $\int d\log\Xi(t)$ positive um den Inbegriff der Werthe von t erstreckt, deren imaginäre Theil zwischen $\frac{1}{2}i$ und $-\frac{1}{2}i$, und deren reeller Theil zwischen 0 und T liegt, ist (bis auf einen Bruchtheil von der Ordnung der Grösse $1/T$) gleich $\{T\log(T/2\pi)-T\}i$; dieses Integral aber ist gleich der Anzahl der in diesem Gebiet liegenden Wurzeln von $\Xi(t) = 0$, multiplicirt mit $2\pi i$. Man findet nun in der That etwa so viel reelle Wurzeln innerhalb dieser Grenzen, und es ist sehr wahrscheinlich, dass alle Wurzeln reelle sind.'

This statement, that all the zeros of $\Xi(t)$ are real, is the famous 'Riemann hypothesis', which remains unproved to this day. The memoir goes on:

'Hiervon wäre allerdings ein strenger Beweis zu wünschen; ich habe indess die Aufsuchung desselben nach einigen flüchtigen vergeblichen Versuchen vorläufig bei Seite gelassen, da er für den nächsten Zweck meiner Untersuchung [i.e. the explicit formula for $\pi(x)$] entbehrlich schien.'

In the approximate formula for $N(T)$, Riemann's $1/T$ may be a mistake for $\log T$; for, since $N(T)$ has an infinity of discontinuities at least equal to 1, the remainder cannot tend to zero. With this correction, Riemann's first statement is Theorem 9.4, which was proved by von Mangoldt many years later.

Riemann's second statement, on the real zeros of $\Xi(t)$, is more obscure, and his exact meaning cannot now be known. It is, however, possible that anyone encountering the subject for the first time might argue as follows. We can write (10.1.2) in the form

$$\Xi(t) = 2\int_0^\infty \Phi(u)\cos ut\,du, \tag{10.1.3}$$

where

$$\Phi(u) = 2\sum_{n=1}^\infty (2n^4\pi^2 e^{\frac{9}{2}u} - 3n^2\pi e^{\frac{5}{2}u})e^{-n^2\pi e^{2u}}. \tag{10.1.4}$$

This series converges very rapidly, and one might suppose that an approximation to the truth could be obtained by replacing it by its first term; or perhaps better by

$$\Phi^*(u) = 2\pi^2\cosh\tfrac{9}{2}u\,e^{-2\pi\cosh 2u},$$

since this, like $\Phi(u)$, is an even function of u, which is asymptotically equivalent to $\Phi(u)$. We should thus replace $\Xi(t)$ by

$$\Xi^*(t) = 4\pi^2\int_0^\infty \cosh\tfrac{9}{2}u\,e^{-2\pi\cosh 2u}\cos ut\,du.$$

The asymptotic behaviour of $\Xi^*(t)$ can be found by the method of steepest descents. To avoid the calculation we shall quote known Bessel-function formulae. We have†

$$K_z(a) = \int_0^\infty e^{-a\cosh u}\cosh zu\,du,$$

and hence

$$\Xi^*(t) = \pi^2\{K_{\frac{9}{4}+\frac{1}{2}it}(2\pi)+K_{\frac{9}{4}-\frac{1}{2}it}(2\pi)\}.$$

For fixed z, as $\nu\to\infty$

$$I_\nu(z) \sim (\tfrac{1}{2}z)^\nu/\Gamma(\nu+1).$$

† Watson, *Theory of Bessel Functions*, 6.22 (5).

Hence

$$I_{-\frac{9}{4}-\frac{1}{2}it}(2\pi) \sim \frac{\pi^{-\frac{9}{4}-\frac{1}{2}it}}{\Gamma(-\frac{5}{4}-\frac{1}{2}it)} \sim \frac{1}{\pi\sqrt{2}} e^{\frac{1}{4}\pi t}\left(\frac{t}{2\pi}\right)^{\frac{7}{4}}\left(\frac{t}{2\pi e}\right)^{\frac{1}{2}it} e^{-\frac{7}{8}i\pi},$$

$$I_{\frac{9}{4}+\frac{1}{2}it}(2\pi) \sim \frac{\pi^{\frac{9}{4}+\frac{1}{2}it}}{\Gamma(\frac{13}{4}+\frac{1}{2}it)} = O(e^{\frac{1}{4}\pi t}t^{-\frac{11}{4}}),$$

$$K_{\frac{9}{4}+\frac{1}{2}it}(2\pi) = \tfrac{1}{2}\pi \operatorname{cosec} \pi(\tfrac{9}{4}+\tfrac{1}{2}it)\{I_{-\frac{9}{4}-\frac{1}{2}it}(2\pi) - I_{\frac{9}{4}+\frac{1}{2}it}(2\pi)\}$$

$$\sim \frac{1}{\sqrt{2}} e^{-\frac{1}{4}\pi t}\left(\frac{t}{2\pi}\right)^{\frac{7}{4}}\left(\frac{t}{2\pi e}\right)^{\frac{1}{2}it} e^{\frac{7}{8}i\pi}.$$

Hence $$\Xi^*(t) \sim \pi^{\frac{1}{4}}2^{-\frac{5}{4}}t^{\frac{7}{4}}e^{-\frac{1}{4}\pi t}\cos\left(\tfrac{1}{2}t\log\frac{t}{2\pi e}+\tfrac{7}{8}\pi\right).$$

The right-hand side has zeros at

$$\tfrac{1}{2}t\log\frac{t}{2\pi e}+\tfrac{7}{8}\pi = (n+\tfrac{1}{2})\pi,$$

and the number of these in the interval $(0, T)$ is

$$\frac{T}{2\pi}\log\frac{T}{2\pi}-\frac{T}{2\pi}+O(1).$$

The similarity to the formula for $N(T)$ is indeed striking.

However, if we try to work on this suggestion, difficulties at once appear. We can write

$$\Xi(t)-\Xi^*(t) = \int_{-\infty}^{\infty} \{\Phi(u)-\Phi^*(u)\}e^{iut}\,du.$$

To show that this is small compared with $\Xi(t)$ we should want to move the line of integration into the upper half-plane, at least as far as $\mathbf{I}(u) = \frac{1}{4}\pi$; and this is just where the series for $\Phi(u)$ ceases to converge. Actually

$$|\Xi(t)| > At^{\frac{7}{4}}e^{-\frac{1}{4}\pi t}|\zeta(\tfrac{1}{2}+it)|,$$

and $|\zeta(\frac{1}{2}+it)|$ is unbounded, so that the suggestion that $\Xi^*(t)$ is an approximation to $\Xi(t)$ is false, at any rate if it is taken in the most obvious sense.

10.2. Although every attempt to prove the Riemann hypothesis, that all the complex zeros of $\zeta(s)$ lie on $\sigma = \frac{1}{2}$, has failed, it is known *that* $\zeta(s)$ *has an infinity of zeros on* $\sigma = \frac{1}{2}$. This was first proved by Hardy in 1914. We shall give here a number of different proofs of this theorem.

First method.† We have

$$\Xi(t) = -\tfrac{1}{2}(t^2+\tfrac{1}{4})\pi^{-\frac{1}{4}-\frac{1}{2}it}\Gamma(\tfrac{1}{4}+\tfrac{1}{2}it)\zeta(\tfrac{1}{2}+it),$$

where $\Xi(t)$ is an even integral function of t, and is real for real t. A zero

† Hardy (1).

of $\zeta(s)$ on $\sigma = \frac{1}{2}$ therefore corresponds to a real zero of $\Xi(t)$, and it is a question of proving that $\Xi(t)$ has an infinity of real zeros.

Putting $x = -i\alpha$ in (2.16.2), we have

$$\begin{aligned}\frac{2}{\pi}\int_0^\infty \frac{\Xi(t)}{t^2+\frac{1}{4}}\cosh\alpha t\,dt &= e^{-\frac{1}{2}i\alpha}-2e^{\frac{1}{2}i\alpha}\psi(e^{2i\alpha}) \\ &= 2\cos\tfrac{1}{2}\alpha-2e^{\frac{1}{2}i\alpha}\{\tfrac{1}{2}+\psi(e^{2i\alpha})\}.\end{aligned} \tag{10.2.1}$$

Since $\zeta(\frac{1}{2}+it) = O(t^A)$, $\Xi(t) = O(t^A e^{-\frac{1}{4}\pi t})$, and the above integral may be differentiated with respect to α any number of times provided that $\alpha < \frac{1}{4}\pi$. Thus

$$\frac{2}{\pi}\int_0^\infty \frac{\Xi(t)}{t^2+\frac{1}{4}}t^{2n}\cosh\alpha t\,dt = \frac{(-1)^n\cos\frac{1}{2}\alpha}{2^{2n-1}}-2\left(\frac{d}{d\alpha}\right)^{2n}e^{\frac{1}{2}i\alpha}\{\tfrac{1}{2}+\psi(e^{2i\alpha})\}.$$

We next prove that the last term tends to 0 as $\alpha \to \frac{1}{4}\pi$, for every fixed n. The equation (2.6.3) gives at once the functional equation

$$x^{-\frac{1}{4}}-2x^{\frac{1}{4}}\psi(x) = x^{\frac{1}{4}}-2x^{-\frac{1}{4}}\psi\left(\frac{1}{x}\right),$$

or

$$\psi(x) = x^{-\frac{1}{2}}\psi\left(\frac{1}{x}\right)+\tfrac{1}{2}x^{-\frac{1}{2}}-\tfrac{1}{2}.$$

Hence

$$\begin{aligned}\psi(i+\delta) &= \sum_1^\infty e^{-n^2\pi(i+\delta)} = \sum_1^\infty (-1)^n e^{-n^2\pi\delta} \\ &= 2\psi(4\delta)-\psi(\delta) \\ &= \frac{1}{\sqrt{\delta}}\psi\left(\frac{1}{4\delta}\right)-\frac{1}{\sqrt{\delta}}\psi\left(\frac{1}{\delta}\right)-\frac{1}{2}.\end{aligned}$$

It is easily seen from this that $\frac{1}{2}+\psi(x)$ and all its derivatives tend to zero as $x \to i$ along any route in an angle $|\arg(x-i)| < \frac{1}{2}\pi$.

We have thus proved that

$$\lim_{\alpha\to\frac{1}{4}\pi}\int_0^\infty \frac{\Xi(t)}{t^2+\frac{1}{4}}t^{2n}\cosh\alpha t\,dt = \frac{(-1)^n\pi\cos\frac{1}{8}\pi}{2^{2n}}. \tag{10.2.2}$$

Suppose now that $\Xi(t)$ were ultimately of one sign, say, for example, positive for $t \geqslant T$. Then

$$\lim_{\alpha\to\frac{1}{4}\pi}\int_T^\infty \frac{\Xi(t)}{t^2+\frac{1}{4}}t^{2n}\cosh\alpha t\,dt = L,$$

say. Hence

$$\int_T^{T'} \frac{\Xi(t)}{t^2+\frac{1}{4}}t^{2n}\cosh\alpha t\,dt \leqslant L$$

for all $\alpha < \frac{1}{4}\pi$ and $T' > T$. Hence, making $\alpha \to \frac{1}{4}\pi$,

$$\int_T^{T'} \frac{\Xi(t)}{t^2+\frac{1}{4}} t^{2n} \cosh \tfrac{1}{4}\pi t \, dt \leqslant L.$$

Hence the integral
$$\int_0^{\infty} \frac{\Xi(t)}{t^2+\frac{1}{4}} t^{2n} \cosh \tfrac{1}{4}\pi t \, dt$$

is convergent. The integral on the left of (10.2.2) is therefore uniformly convergent with respect to α for $0 \leqslant \alpha \leqslant \frac{1}{4}\pi$, and it follows that

$$\int_0^{\infty} \frac{\Xi(t)}{t^2+\frac{1}{4}} t^{2n} \cosh \tfrac{1}{4}\pi t \, dt = \frac{(-1)^n \pi \cos \frac{1}{8}\pi}{2^{2n}}$$

for every n.

This, however, is impossible; for, taking n odd, the right-hand side is negative, and hence

$$\int_T^{\infty} \frac{\Xi(t)}{t^2+\frac{1}{4}} t^{2n} \cosh \tfrac{1}{4}\pi t \, dt < -\int_0^{T} \frac{\Xi(t)}{t^2+\frac{1}{4}} t^{2n} \cosh \tfrac{1}{4}\pi t \, dt$$
$$< KT^{2n},$$

where K is independent of n. But by hypothesis there is a positive $m = m(T)$ such that $\Xi(t)/(t^2+\frac{1}{4}) \geqslant m$ for $2T \leqslant t \leqslant 2T+1$. Hence

$$\int_T^{\infty} \frac{\Xi(t)}{t^2+\frac{1}{4}} t^{2n} \cosh \tfrac{1}{4}\pi t \, dt \geqslant \int_{2T}^{2T+1} mt^{2n} \, dt \geqslant m(2T)^{2n}.$$

Hence
$$m2^{2n} < K,$$

which is false for sufficiently large n. This proves the theorem.

10.3. A variant of the above proof depends on the following theorem of Fejér:†

Let n be any positive integer. Then the number of changes in sign in the interval $(0, a)$ of a continuous function $f(x)$ is not less than the number of changes in sign of the sequence

$$f(0), \quad \int_0^a f(t) \, dt, \quad ..., \quad \int_0^a f(t) t^n \, dt. \tag{10.3.1}$$

We deduce this from the following theorem of Fekete:‡

† Fejér (1). ‡ Fekete (1).

The number of changes in sign in the interval $(0, a)$ *of a continuous function* $f(x)$ *is not less than the number of changes in sign of the sequence*

$$f(a),\quad f_1(a),\quad \ldots,\quad f_n(a), \tag{10.3.2}$$

where

$$f_\nu(x) = \int_0^x f_{\nu-1}(t)\,dt \qquad (\nu = 1, 2,\ldots, n), \qquad f_0(x) = f(x).$$

To prove Fekete's theorem, suppose first that $n = 1$. Consider the curve $y = f_1(x)$. Now $f_1(0) = 0$, and, if $f(a)$ and $f_1(a)$ have opposite signs, y is positive decreasing or negative increasing at $x = a$. Hence $f(x)$ has at least one zero.

Now assume the theorem for $n-1$. Suppose that there are k changes of sign in the sequence $f_1(x),\ldots, f_n(x)$. Then $f_1(x)$ has at least k changes of sign. We have then to prove that

(i) if $f(a)$ and $f_1(a)$ have the same sign, $f(x)$ has at least k changes of sign,

(ii) if $f(a)$ and $f_1(a)$ have opposite signs, $f(x)$ has at least $k+1$ changes of sign.

Each of these cases is easily verified by considering the curve $y = f_1(x)$. This proves Fekete's theorem.

To deduce Fejér's theorem, we have

$$f_\nu(x) = \frac{1}{(\nu-1)!}\int_0^x (x-t)^{\nu-1} f(t)\,dt,$$

and hence

$$f_\nu(a) = \frac{1}{(\nu-1)!}\int_0^a (a-t)^{\nu-1} f(t)\,dt = \frac{1}{(\nu-1)!}\int_0^a f(a-t)t^{\nu-1}\,dt.$$

We may therefore replace the sequence (10.3.2) by the sequence

$$f(a),\quad \int_0^a f(a-t)\,dt,\quad \ldots,\quad \int_0^a f(a-t)t^{n-1}\,dt. \tag{10.3.3}$$

Since the number of changes of sign of $f(t)$ is the same as the number of changes of sign of $f(a-t)$, we can replace $f(t)$ by $f(a-t)$. This proves Fejér's theorem.

To prove that there are an infinity of zeros of $\zeta(s)$ on the critical line, we prove as before that

$$\lim_{\alpha\to\frac{1}{4}\pi}\int_0^\infty \frac{\Xi(t)}{t^2+\frac{1}{4}} t^{2n}\cosh\alpha t\,dt = \frac{(-1)^n\pi\cos\frac{1}{8}\pi}{2^{2n}}.$$

Hence

$$\int_0^a \frac{\Xi(t)}{t^2+\frac{1}{4}} t^{2n}\cosh\alpha t\,dt$$

has the same sign as $(-1)^n$ for $n = 0, 1,..., N$, if $a = a(N)$ is large enough and $\alpha = \alpha(N)$ is near enough to $\frac{1}{4}\pi$. Hence $\Xi(t)$ has at least N changes of sign in $(0, a)$, and the result follows.†

10.4. Another method‡ is based on Riemann's formula (10.1.2).

Putting $x = e^{2u}$ in (10.1.2), we have

$$\Xi(t) = 4\int_0^\infty \frac{d}{du}\{e^{3u}\psi'(e^{2u})\}e^{-\frac{1}{2}u}\cos ut\, du$$

$$= 2\int_0^\infty \Phi(u)\cos ut\, du,$$

say. Then, by Fourier's integral theorem,

$$\Phi(u) = \frac{1}{\pi}\int_0^\infty \Xi(t)\cos ut\, dt,$$

and hence also

$$\Phi^{(2n)}(u) = \frac{(-1)^n}{\pi}\int_0^\infty \Xi(t)t^{2n}\cos ut\, dt.$$

Since $\psi(x)$ is regular for $\mathbf{R}(x) > 0$, $\Phi(u)$ is regular for $-\frac{1}{4}\pi < \mathbf{I}(u) < \frac{1}{4}\pi$. Let

$$\Phi(iu) = c_0+c_1 u^2+c_2 u^4+... \quad (|u| < \tfrac{1}{4}\pi).$$

Then

$$(2n)!\, c_n = (-1)^n\Phi^{(2n)}(0) = \frac{1}{\pi}\int_0^\infty \Xi(t)t^{2n}\, dt.$$

Suppose now that $\Xi(t)$ is of one sign, say $\Xi(t) > 0$, for $t > T$. Then $c_n > 0$ for $n > n_0$, since

$$\int_0^\infty \Xi(t)t^{2n}\, dt > \int_{T+1}^{T+2} \Xi(t)t^{2n}\, dt - \int_0^T |\Xi(t)|t^{2n}\, dt$$

$$> (T+1)^{2n}\int_{T+1}^{T+2} \Xi(t)\, dt - T^{2n}\int_0^T |\Xi(t)|\, dt.$$

It follows that $\Phi^{(n)}(iu)$ increases steadily with u if $n > 2n_0$. But in fact $\Phi(u)$ and all its derivatives tend to 0 as $u \to \frac{1}{4}i\pi$ along the imaginary axis, by the properties of $\psi(x)$ obtained in § 10.2. The theorem therefore follows again.

10.5. The above proofs of Hardy's theorem are all similar in that they depend on the consideration of 'moments' $\int f(t)t^n\, dt$. The following

† Fekete (2). ‡ Pólya (3).

method† depends on a contrast between the asymptotic behaviour of the integrals

$$\int_T^{2T} Z(t)\,dt, \qquad \int_T^{2T} |Z(t)|\,dt,$$

where $Z(t)$ is the function defined in § 4.17. If $Z(t)$ were ultimately of one sign, these integrals would be ultimately equal, apart possibly from sign. But we shall see that in fact they behave quite differently.

Consider the integral

$$\int \{\chi(s)\}^{-\frac{1}{2}}\zeta(s)\,ds,$$

where the integrand is the function which reduces to $Z(t)$ on $\sigma = \frac{1}{2}$, taken round the rectangle with sides $\sigma = \frac{1}{2}$, $\sigma = \frac{5}{4}$, $t = T$, $t = 2T$. This integral is zero, by Cauchy's theorem. Now

$$\int_{\frac{1}{2}+iT}^{\frac{1}{2}+2iT} \{\chi(s)\}^{-\frac{1}{2}}\zeta(s)\,ds = i\int_T^{2T} Z(t)\,dt.$$

Also by (4.12.3)

$$\{\chi(s)\}^{-\frac{1}{2}} = \left(\frac{t}{2\pi}\right)^{\frac{1}{2}\sigma-\frac{1}{4}+\frac{1}{2}it} e^{-\frac{1}{2}it-\frac{1}{8}i\pi}\left\{1+O\left(\frac{1}{t}\right)\right\}.$$

Hence, by (5.1.2) and (5.1.4),

$$\begin{aligned}\{\chi(s)\}^{-\frac{1}{2}}\zeta(s) &= O(t^{\frac{1}{2}\sigma-\frac{1}{4}}.t^{\frac{1}{2}-\frac{1}{2}\sigma+\epsilon}) = O(t^{\frac{1}{4}+\epsilon}) \quad (\tfrac{1}{2} \leqslant \sigma \leqslant 1),\\ &= O(t^{\frac{1}{2}\sigma-\frac{1}{4}+\epsilon}) = O(t^{\frac{3}{8}+\epsilon}) \quad (1 < \sigma \leqslant \tfrac{5}{4}).\end{aligned}$$

The integrals along the sides $t = T$, $t = 2T$ are therefore $O(T^{\frac{3}{8}+\epsilon})$.

The integral along the right-hand side is

$$\int_T^{2T} \left(\frac{t}{2\pi}\right)^{\frac{3}{8}+\frac{1}{2}it} e^{-\frac{1}{2}it-\frac{1}{8}i\pi}\left\{1+O\left(\frac{1}{t}\right)\right\}\zeta(\tfrac{5}{4}+it)i\,dt.$$

The contribution of the O-term is

$$\int_T^{2T} O(t^{-\frac{5}{8}})\,dt = O(T^{\frac{3}{8}}).$$

The other term is a constant multiple of

$$\sum_{n=1}^{\infty} n^{-\frac{5}{4}} \int_T^{2T} \left(\frac{t}{2\pi}\right)^{\frac{3}{8}+\frac{1}{2}it} e^{-\frac{1}{2}it-it\log n}\,dt.$$

Now
$$\frac{d^2}{dt^2}\left(\tfrac{1}{2}t\log\frac{t}{2\pi}-\tfrac{1}{2}t-t\log n\right) = \frac{1}{2t}.$$

Hence, by Lemma 4.5, the integral in the above sum is $O(T^{\frac{7}{8}})$, uniformly with respect to n, so that the whole sum is also $O(T^{\frac{7}{8}})$.

† See Landau, *Vorlesungen*, ii. 78–85.

Combining all these results, we obtain

$$\int_T^{2T} Z(t)\,dt = O(T^{\frac{7}{8}}). \tag{10.5.1}$$

On the other hand,

$$\int_T^{2T} |Z(t)|\,dt = \int_T^{2T} |\zeta(\tfrac{1}{2}+it)|\,dt \geqslant \left|\int_T^{2T} \zeta(\tfrac{1}{2}+it)\,dt\right|.$$

But

$$i\int_T^{2T} \zeta(\tfrac{1}{2}+it)\,dt = \int_{\frac{1}{2}+iT}^{\frac{1}{2}+2iT} \zeta(s)\,ds = \int_{\frac{1}{2}+iT}^{2+iT} + \int_{2+iT}^{2+2iT} + \int_{2+2iT}^{\frac{1}{2}+2iT}$$

$$= \left[s - \sum_{n=2}^{\infty} \frac{1}{n^s \log n}\right]_{2+iT}^{2+2iT} + \int_{\frac{1}{2}}^{2} O(T^{\frac{1}{2}})\,d\sigma = iT + O(T^{\frac{1}{2}}).$$

Hence

$$\int_T^{2T} |Z(t)|\,dt > AT. \tag{10.5.2}$$

Hardy's theorem now follows from (10.5.1) and (10.5.2).

Another variant of this method is obtained by starting again from (10.2.1). Putting $\alpha = \frac{1}{4}\pi - \delta$, we obtain

$$\int_0^{\infty} \frac{\Xi(t)}{t^2+\frac{1}{4}} \cosh\{(\tfrac{1}{4}\pi-\delta)t\}\,dt = O(1) + O\left\{\sum_{n=1}^{\infty} \exp(-n^2\pi i e^{-2i\delta})\right\}$$

$$= O(1) + O\left(\sum_{n=1}^{\infty} e^{-n^2\pi \sin 2\delta}\right) = O(1) + O\left(\int_0^{\infty} e^{-x^2\pi \sin 2\delta}\,dx\right) = O(\delta^{-\frac{1}{2}})$$

as $\delta \to 0$. If, for example, $\Xi(t) > 0$ for $t > t_0$, it follows that for $T > t_0$

$$\int_T^{2T} |Z(t)|\,dt = \left|\int_T^{2T} Z(t)\,dt\right| < A \int_T^{2T} \frac{\Xi(t)}{t^2+\frac{1}{4}} t^{\frac{1}{4}} e^{\frac{1}{4}\pi t}\,dt$$

$$< AT^{\frac{1}{4}} \int_T^{2T} \frac{\Xi(t)}{t^2+\frac{1}{4}} e^{\frac{1}{4}\pi t - \frac{1}{2}t/T}\,dt < AT^{\frac{1}{4}} \int_{t_0}^{\infty} \frac{\Xi(t)}{t^2+\frac{1}{4}} \cosh\left\{\left(\tfrac{1}{4}\pi - \frac{1}{2T}\right)t\right\}\,dt$$

$$= O(T^{\frac{1}{4}}.T^{\frac{1}{2}}) = O(T^{\frac{3}{4}}).$$

This is inconsistent with (10.5.2), so that the theorem again follows.

10.6. Still another method† depends on the formula (4.17.4), viz.

$$Z(t) = 2\sum_{n \leqslant x} \frac{\cos(\vartheta - t\log n)}{\sqrt{n}} + O(t^{-\frac{1}{4}}),$$

† Titchmarsh (11).

where $x = \sqrt{(t/2\pi)}$. Here $\vartheta = \vartheta(t)$ is defined by

$$\chi(\tfrac{1}{2}+it) = e^{-2i\vartheta(t)},$$

so that

$$\vartheta'(t) = -\frac{1}{2}\frac{\chi'(\frac{1}{2}+it)}{\chi(\frac{1}{2}+it)} = -\frac{1}{2}\left\{\log\pi - \frac{1}{2}\frac{\Gamma'(\frac{1}{4}-\frac{1}{2}it)}{\Gamma(\frac{1}{4}-\frac{1}{2}it)} - \frac{1}{2}\frac{\Gamma'(\frac{1}{4}+\frac{1}{2}it)}{\Gamma(\frac{1}{4}+\frac{1}{2}it)}\right\}$$

$$= -\tfrac{1}{2}\log\pi + \tfrac{1}{4}\log(\tfrac{1}{16}+\tfrac{1}{4}t^2) - \frac{1}{1+4t^2} - \mathbf{R}\int_0^\infty \frac{u\,du}{\{u^2+(\frac{1}{4}+\frac{1}{2}it)^2\}(e^{2\pi u}-1)},$$

and we have

$$\vartheta'(t) = \tfrac{1}{2}\log t - \tfrac{1}{2}\log 2\pi + O(1/t),$$

$$\vartheta(t) \sim \tfrac{1}{2}t\log t, \qquad \vartheta''(t) \sim \frac{1}{2t}.$$

The function $\vartheta(t)$ is steadily increasing for $t \geqslant t_0$. If ν is any positive integer ($\geqslant \nu_0$), the equation $\vartheta(t) = \nu\pi$ therefore has just one solution, say t_ν, and $t_\nu \sim 2\pi\nu/\log\nu$. Now

$$Z(t_\nu) = 2(-1)^\nu \sum_{n\leqslant x} \frac{\cos(t_\nu \log n)}{\sqrt{n}} + O(t_\nu^{-\frac{1}{4}}).$$

The sum

$$g(t_\nu) = \sum_{n\leqslant x} \frac{\cos(t_\nu \log n)}{\sqrt{n}} = 1 + \frac{\cos(t_\nu \log 2)}{\sqrt{2}} + \dots$$

consists of the constant term unity and oscillatory terms; and the formula suggests that $g(t_\nu)$ will usually be positive, and hence that $Z(t)$ will usually change sign in the interval $(t_\nu, t_{\nu+1})$.

We shall prove

THEOREM 10.6. *As* $N \to \infty$

$$\sum_{\nu=\nu_0}^{N} Z(t_{2\nu}) \sim 2N, \qquad \sum_{\nu=\nu_0}^{N} Z(t_{2\nu+1}) \sim -2N.$$

It follows at once that $Z(t_{2\nu})$ is positive for an infinity of values of ν, and that $Z(t_{2\nu+1})$ is negative for an infinity of values of ν; and the existence of an infinity of real zeros of $Z(t)$, and so of $\Xi(t)$, again follows.

We have

$$\sum_{\nu=M+1}^{N} g(t_{2\nu}) = \sum_{\nu=M+1}^{N} \sum_{n\leqslant\sqrt{(t_{2\nu}/2\pi)}} \frac{\cos(t_{2\nu}\log n)}{\sqrt{n}}$$

$$= N - M + \sum_{2\leqslant n\leqslant\sqrt{(t_{2N}/2\pi)}} \frac{1}{\sqrt{n}} \sum_{\tau\leqslant t_{2\nu}\leqslant t_{2N}} \cos(t_{2\nu}\log n),$$

where $\tau = \max(t_{2M+2}, 2\pi n^2)$. The inner sum is of the form

$$\sum \cos\{2\pi\phi(\nu)\},$$

where
$$\phi(\nu) = \frac{t_{2\nu}\log n}{2\pi}.$$

We may define t_ν for all $\nu \geqslant \nu_0$ (not necessarily integral) by $\vartheta(t_\nu) = \nu\pi$. Then
$$\phi'(\nu) = \frac{\log n}{2\pi}\frac{dt_{2\nu}}{d\nu}, \qquad \vartheta'(t_{2\nu})\frac{dt_{2\nu}}{d\nu} = 2\pi,$$

so that
$$\phi'(\nu) = \frac{\log n}{\vartheta'(t_{2\nu})}.$$

Hence $\phi'(\nu)$ is positive and steadily decreasing, and, if ν is large enough,
$$\phi''(\nu) = -2\pi\log n\frac{\vartheta''(t_{2\nu})}{\{\vartheta'(t_\nu)\}^3} \sim -\frac{8\pi\log n}{t_{2\nu}\log^3 t_{2\nu}} < -A\frac{\log n}{t_{2N}\log^3 t_{2N}}.$$

Hence, by Theorem 5.9,
$$\sum_{\tau\leqslant t_{2\nu}\leqslant t_{2N}} \cos(t_{2\nu}\log n) = O\left(t_{2N}\frac{\log^{\frac{1}{2}}n}{t_{2N}^{\frac{1}{2}}\log^{\frac{3}{2}}t_{2N}}\right)+O\left(\frac{t_{2N}^{\frac{1}{2}}\log^{\frac{3}{2}}t_{2N}}{\log^{\frac{1}{2}}n}\right)$$
$$= O(t_{2N}^{\frac{1}{2}}\log^{\frac{3}{2}}t_{2N}).$$

Hence
$$\sum_{2\leqslant n\leqslant\sqrt{(t_{2N}/2\pi)}} \frac{1}{\sqrt{n}} \sum_{\tau\leqslant t_{2\nu}\leqslant t_{2N}} \cos(t_{2\nu}\log n) = O(t_{2N}^{\frac{3}{4}}\log^{\frac{3}{2}}t_{2N})$$
$$= O(N^{\frac{3}{4}}\log^{\frac{3}{4}}N).$$

Hence
$$\sum_{\nu=M+1}^{N} Z(t_{2\nu}) = 2N+O(N^{\frac{3}{4}}\log^{\frac{3}{4}}N),$$

and a similar argument applies to the other sum.

10.7. We denote by $N_0(T)$ the number of zeros of $\zeta(s)$ of the form $\frac{1}{2}+it$ $(0 < t \leqslant T)$. The theorem already proved shows that $N_0(T)$ tends to infinity with T. We can, however, prove much more than this.

THEOREM 10.7.† $N_0(T) > AT$.

Any of the above proofs can be put in a more precise form so as to give results in this direction. The most successful method is similar in principle to that of § 10.5, but is more elaborate. We contrast the behaviour of the integrals
$$I = \int_t^{t+H} \Xi(u)\frac{e^{\frac{1}{4}\pi u}}{u^2+\frac{1}{4}}e^{-u/T}\,du, \qquad J = \int_t^{t+H} |\Xi(u)|\frac{e^{\frac{1}{4}\pi u}}{u^2+\frac{1}{4}}e^{-u/T}\,du,$$

where $T \leqslant t \leqslant 2T$ and $T \to \infty$.

† Hardy and Littlewood (3).

We use the theory of Fourier transforms. Let $F(u), f(y)$ be functions related by the Fourier formulae

$$F(u) = \frac{1}{\sqrt{(2\pi)}} \int_{-\infty}^{\infty} f(y) e^{iyu}\, dy, \qquad f(y) = \frac{1}{\sqrt{(2\pi)}} \int_{-\infty}^{\infty} F(u) e^{-iyu}\, du.$$

Integrating over $(t, t+H)$, we obtain

$$\int_{t}^{t+H} F(u)\, du = \frac{1}{\sqrt{(2\pi)}} \int_{-\infty}^{\infty} f(y) \frac{e^{iyH}-1}{iy} e^{iyt}\, dy,$$

so that

$$\int_{t}^{t+H} F(u)\, du, \qquad f(y)\frac{e^{iyH}-1}{iy}$$

are Fourier transforms. Hence the Parseval formula gives

$$\int_{-\infty}^{\infty} \left| \int_{t}^{t+H} F(u)\, du \right|^2 dt = \int_{-\infty}^{\infty} |f(y)|^2 \frac{4\sin^2 \frac{1}{2}Hy}{y^2}\, dy.$$

If $F(u)$ is real, $|f(y)|$ is even, and we have

$$\begin{aligned} \int_{-\infty}^{\infty} \left| \int_{t}^{t+H} F(u)\, du \right|^2 dt &= 2\int_{0}^{\infty} |f(y)|^2 \frac{4\sin^2 \frac{1}{2}Hy}{y^2}\, dy \\ &\leqslant 2H^2 \int_{0}^{1/H} |f(y)|^2\, dy + 8\int_{1/H}^{\infty} \frac{|f(y)|^2}{y^2}\, dy. \qquad (10.7.1) \end{aligned}$$

Now (2.16.2) may be written

$$\frac{1}{2\pi} \int_{-\infty}^{\infty} \frac{\Xi(t)}{t^2+\frac{1}{4}} e^{i\xi t}\, dt = \tfrac{1}{2} e^{\frac{1}{2}\xi} - e^{-\frac{1}{2}\xi} \psi(e^{-2\xi}).$$

Putting $\xi = -i(\frac{1}{4}\pi - \frac{1}{2}\delta) - y$, it is seen that we may take

$$F(t) = \frac{1}{\sqrt{(2\pi)}} \frac{\Xi(t)}{t^2+\frac{1}{4}} e^{(\frac{1}{4}\pi-\frac{1}{2}\delta)t}, \qquad f(y) = \tfrac{1}{2} e^{-\frac{1}{2}i(\frac{1}{4}\pi-\frac{1}{2}\delta)-\frac{1}{2}y} - \\ - e^{\frac{1}{2}i(\frac{1}{4}\pi-\frac{1}{2}\delta)+\frac{1}{2}y} \psi(e^{i(\frac{1}{2}\pi-\delta)+2y}).$$

Let $H \geqslant 1$. The contribution of the first term in $f(y)$ to (10.7.1) is clearly $O(H)$. Putting $y = \log x$, $G = e^{1/H}$, we therefore obtain

$$\begin{aligned} \int_{-\infty}^{\infty} \left| \int_{t}^{t+H} F(u)\, du \right|^2 dt &= O\left\{ H^2 \int_{1}^{G} |\psi(e^{i(\frac{1}{2}\pi-\delta)}x^2)|^2\, dx \right\} + \\ &\quad + O\left\{ \int_{G}^{\infty} |\psi(e^{i(\frac{1}{2}\pi-\delta)}x^2)|^2 \frac{dx}{\log^2 x} \right\} + O(H). \qquad (10.7.2) \end{aligned}$$

Now

$$|\psi(e^{i(\frac{1}{2}\pi-\delta)}x^2)|^2 = \left|\sum_{n=1}^{\infty} e^{-n^2\pi x^2(\sin\delta+i\cos\delta)}\right|^2$$

$$= \sum_{n=1}^{\infty} e^{-2n^2\pi x^2\sin\delta} + \sum_{m\neq n}\sum e^{-(m^2+n^2)\pi x^2\sin\delta+i(m^2-n^2)\pi x^2\cos\delta}.$$

As in § 10.5, the first sum is $O(x^{-1}\delta^{-\frac{1}{2}})$, and its contribution to (10.7.2) is therefore

$$O\left(H^2\int_1^G x^{-1}\delta^{-\frac{1}{2}}\,dx\right)+O\left(\int_G^\infty \frac{\delta^{-\frac{1}{2}}\,dx}{x\log^2 x}\right)$$

$$= O\{H^2(G-1)\delta^{-\frac{1}{2}}\}+O(\delta^{-\frac{1}{2}}/\log G) = O(H\delta^{-\frac{1}{2}}).$$

The sum with $m \neq n$ contributes to the second term in (10.7.2) terms of the form

$$\int_G^\infty e^{-(m^2+n^2)\pi x^2\sin\delta+i(m^2-n^2)\pi x^2\cos\delta}\frac{dx}{\log^2 x} = O\left\{\frac{e^{-(m^2+n^2)\pi G^2\sin\delta}}{|m^2-n^2|\,G\log^2 G}\right\}$$

$$= O\left(\frac{H^2 e^{-(m^2+n^2)\pi\sin\delta}}{|m^2-n^2|}\right)$$

by Lemma 4.3. Hence the sum is

$$O\left(H^2\sum_{m=2}^{\infty}\sum_{n=1}^{m-1}\frac{e^{-(m^2+n^2)\pi\sin\delta}}{m^2-n^2}\right) = O\left(H^2\sum_{m=2}^{\infty}\frac{e^{-m^2\pi\sin\delta}}{m}\sum_{n=1}^{m-1}\frac{1}{m-n}\right)$$

$$= O\left(H^2\sum_{m=2}^{\infty}\frac{\log m}{m}e^{-m^2\pi\sin\delta}\right) = O\left\{H^2\left(\sum_{m\leqslant 1/\delta}\frac{\log m}{m}+\sum_{m>1/\delta}e^{-m^2\pi\sin\delta}\right)\right\}$$

$$= O\left(H^2\log^2\frac{1}{\delta}\right) = O(H\delta^{-\frac{1}{2}})$$

for $\delta < \delta_0(H)$. The first integral in (10.7.2) may be dealt with in the same way. Hence

$$\int_{-\infty}^{\infty}\left|\int_t^{t+H} F(u)\,du\right|^2 dt = O(H\delta^{-\frac{1}{2}}).$$

Taking $\delta = 1/T$ and $T > T_0(H)$, it follows that

$$\int_T^{2T} |I|^2\,dt = O(HT^{\frac{1}{2}}). \tag{10.7.3}$$

10.8. We next prove that

$$J > (AH+\Psi)T^{-\frac{1}{4}}, \tag{10.8.1}$$

where

$$\int_T^{2T} |\Psi|^2\,dt = O(T) \quad (0 < H < T). \tag{10.8.2}$$

We have, if $s = \frac{1}{2}+it$, $T \leqslant t \leqslant 2T$,

$$T^{\frac{1}{4}}\,|\Xi(t)|\frac{e^{\frac{1}{4}\pi t}}{t^2+\frac{1}{4}} > A\,|\zeta(\tfrac{1}{2}+it)|.$$

Hence

$$T^{\frac{1}{4}}J > A\int\limits_{t}^{t+H} |\zeta(\tfrac{1}{2}+iu)|\,du > A\left|\int\limits_{t}^{t+H} \zeta(\tfrac{1}{2}+iu)\,du\right|$$

$$= A\left|\int\limits_{t}^{t+H} \left\{\sum_{n<AT}\frac{1}{n^{\frac{1}{2}+iu}}+O(T^{-\frac{1}{2}})\right\}du\right|$$

$$= AH+O\left\{\left|\int\limits_{t}^{t+H}\sum_{2\leqslant n<AT}\frac{1}{n^{\frac{1}{2}+iu}}\,du\right|\right\}+O(HT^{-\frac{1}{2}})$$

$$= AH+O\left\{\left|\sum_{2\leqslant n<AT}\left(\frac{1}{n^{\frac{1}{2}+i(t+H)}\log n}-\frac{1}{n^{\frac{1}{2}+it}\log n}\right)\right|\right\}+O(HT^{-\frac{1}{2}}).$$

It is now sufficient to prove that

$$\int\limits_{T}^{2T}\left|\sum_{2\leqslant n<AT}\frac{1}{n^{\frac{1}{2}+it}\log n}\right|^2 dt = O(T),$$

and the calculations are similar to those of § 7.3, but with an extra factor $\log m\log n$ in the denominator.

To prove Theorem 10.7, let S be the sub-set of the interval $(T, 2T)$ where $I = J$. Then

$$\int\limits_{S} |I|\,dt = \int\limits_{S} J\,dt.$$

Now
$$\int\limits_{S} |I|\,dt \leqslant \int\limits_{T}^{2T} |I|\,dt \leqslant \left(T\int\limits_{T}^{2T} |I|^2\,dt\right)^{\frac{1}{2}} < AH^{\frac{1}{2}}T^{\frac{3}{4}}$$

by (10.7.3); and by (10.8.1) and (10.8.2)

$$\int\limits_{S} J\,dt > T^{-\frac{1}{4}}\int\limits_{S}(AH+\Psi)\,dt$$

$$> AT^{-\frac{1}{4}}Hm(S)-T^{-\frac{1}{4}}\int\limits_{T}^{2T}|\Psi|\,dt$$

$$> AT^{-\frac{1}{4}}Hm(S)-T^{-\frac{1}{4}}\left(T\int\limits_{T}^{2T}|\Psi|^2\,dt\right)^{\frac{1}{2}}$$

$$> AT^{-\frac{1}{4}}Hm(S)-AT^{\frac{3}{4}},$$

where $m(S)$ is the measure of S. Hence, for $H \geqslant 1$ and $T > T_0(H)$,

$$m(S) < ATH^{-\frac{1}{2}}.$$

Now divide the interval $(T, 2T)$ into $[T/2H]$ pairs of abutting intervals j_1, j_2, each, except the last j_2, of length H, and each j_2 lying to the right of the corresponding j_1. Then either j_1 or j_2 contains a zero of $\Xi(t)$ unless j_1 consists entirely of points of S. Suppose that the latter occurs for ν j_1's. Then

$$\nu H \leqslant m(S) < ATH^{-\frac{1}{2}}.$$

Hence there are, in $(T, 2T)$, at least

$$[T/2H] - \nu > \frac{T}{H}\left(\frac{1}{3} - \frac{A}{\sqrt{H}}\right) > \frac{T}{4H}$$

zeros if H is large enough. This proves the theorem.

10.9. For many years the above theorem of Hardy and Littlewood, that $N_0(T) > AT$, was the best that was known in this direction. Recently it has been proved by A. Selberg (2) that $N_0(T) > AT \log T$. This is a remarkable improvement, since it shows that a finite proportion of the zeros of $\zeta(s)$ lie on the critical line. On the Riemann hypothesis, of course,

$$N_0(T) = N(T) \sim \frac{1}{2\pi} T \log T.$$

The numerical value of the constant A in Selberg's theorem is very small.†

The essential idea of Selberg's proof is to modify the series for $\zeta(s)$ by multiplying it by the square of a partial sum of the series for $\{\zeta(s)\}^{-\frac{1}{2}}$. To this extent, it is similar to the proofs given in Chapter IX of theorems about the general distribution of the zeros.

We define α_ν by

$$\frac{1}{\sqrt{\zeta(s)}} = \sum_{\nu=1}^{\infty} \frac{\alpha_\nu}{\nu^s} \quad (\sigma > 1), \qquad \alpha_1 = 1.$$

It is seen from the Euler product that $\alpha_\mu \alpha_\nu = \alpha_{\mu\nu}$ if $(\mu, \nu) = 1$. Since the series for $(1-z)^{\frac{1}{2}}$ is majorized by that for $(1-z)^{-\frac{1}{2}}$, we see that, if

$$\sqrt{\zeta(s)} = \sum_{\nu=1}^{\infty} \frac{\alpha'_\nu}{\nu^s}, \qquad \alpha'_1 = 1,$$

then $|\alpha_\nu| \leqslant \alpha'_\nu \leqslant 1$.

Let
$$\beta_\nu = \alpha_\nu\left(1 - \frac{\log \nu}{\log X}\right) \quad (1 \leqslant \nu < X).$$

Then
$$|\beta_\nu| \leqslant 1.$$

† It was calculated in an Oxford dissertation by S. H. Min.

All sums involving β_ν run over $[1, X]$ (or we may suppose $\beta_\nu = 0$ for $\nu \geqslant X$). Let

$$\phi(s) = \sum \frac{\beta_\nu}{\nu^s}.$$

10.10. Let†

$$\Phi(z) = \frac{1}{4\pi i} \int_{c-i\infty}^{c+i\infty} \Gamma(\tfrac{1}{2}s)\pi^{-\frac{1}{2}s}\zeta(s)\phi(s)\phi(1-s)z^s\, ds$$

where $c > 1$. Moving the line of integration to $\sigma = \frac{1}{2}$, and evaluating the residue at $s = 1$, we obtain

$$\Phi(z) = \tfrac{1}{2}z\phi(1)\phi(0) + \frac{1}{4\pi i} \int_{\frac{1}{2}-i\infty}^{\frac{1}{2}+i\infty} \Gamma(\tfrac{1}{2}s)\pi^{-\frac{1}{2}s}\zeta(s)\phi(s)\phi(1-s)z^s\, ds$$

$$= \tfrac{1}{2}z\phi(1)\phi(0) - \frac{z^{\frac{1}{2}}}{2\pi} \int_{-\infty}^{\infty} \frac{\Xi(t)}{t^2+\frac{1}{4}} |\phi(\tfrac{1}{2}+it)|^2 z^{it}\, dt.$$

On the other hand,

$$\Phi(z) = \frac{1}{4\pi i} \sum_{n=1}^{\infty} \sum_{\mu} \sum_{\nu} \beta_\mu \beta_\nu \int_{c-i\infty}^{c+i\infty} \Gamma(\tfrac{1}{2}s)\pi^{-\frac{1}{2}s} \frac{z^s}{n^s\mu^s\nu^{1-s}}\, ds$$

$$= \sum_{n=1}^{\infty} \sum_{\mu} \sum_{\nu} \frac{\beta_\mu \beta_\nu}{\nu} \exp\left(-\frac{\pi n^2\mu^2}{z^2\nu^2}\right).$$

Putting $z = e^{-i(\frac{1}{4}\pi - \frac{1}{2}\delta) - y}$, it follows that the functions

$$F(t) = \frac{1}{\sqrt{(2\pi)}} \frac{\Xi(t)}{t^2+\frac{1}{4}} |\phi(\tfrac{1}{2}+it)|^2 e^{(\frac{1}{4}\pi - \frac{1}{2}\delta)t},$$

$$f(y) = \tfrac{1}{2}z^{\frac{1}{2}}\phi(1)\phi(0) - z^{-\frac{1}{2}} \sum_{n=1}^{\infty} \sum_{\mu} \sum_{\nu} \frac{\beta_\mu \beta_\nu}{\nu} \exp\left(-\frac{\pi n^2\mu^2}{z^2\nu^2}\right)$$

are Fourier transforms. Hence, as in § 10.7,

$$\int_{-\infty}^{\infty} \left\{\int_{t}^{t+h} F(u)\, du\right\}^2 dt \leqslant 2h^2 \int_{0}^{1/h} |f(y)|^2\, dy + 8 \int_{1/h}^{\infty} |f(y)|^2 y^{-2}\, dy \qquad (10.10.1)$$

where $h \leqslant 1$ is to be chosen later.

Putting $y = \log x$, $G = e^{1/h}$, the first integral on the right is equal to

$$\int_{1}^{G} \left| \frac{e^{-i(\frac{1}{4}\pi - \frac{1}{2}\delta)}}{2x} \phi(1)\phi(0) - \sum_{n=1}^{\infty} \sum_{\mu} \sum_{\nu} \frac{\beta_\mu \beta_\nu}{\nu} \exp\left(-\frac{\pi n^2\mu^2}{\nu^2} e^{i(\frac{1}{2}\pi - \delta)} x^2\right) \right|^2 dx.$$

† Titchmarsh (26).

Calling the triple sum $g(x)$, this is not greater than

$$2\int_1^G \frac{|\phi(1)\phi(0)|^2}{4x^2}\,dx+2\int_1^G |g(x)|^2\,dx < \tfrac{1}{2}|\phi(1)\phi(0)|^2+2\int_1^G |g(x)|^2\,dx.$$

Similarly the second integral in (10.10.1) does not exceed

$$\frac{|\phi(1)\phi(0)|^2}{2G\log^2 G}+2\int_G^\infty \frac{|g(x)|^2}{\log^2 x}\,dx.$$

10.11. We have to obtain upper bounds for these integrals as $\delta \to 0$, but it is more convenient to consider directly the integral

$$J(x,\theta) = \int_x^\infty |g(u)|^2 u^{-\theta}\,du \quad (0<\theta\leqslant \tfrac{1}{2},\ x\geqslant 1).$$

This is equal to

$$\sum_{m=1}^\infty \sum_{n=1}^\infty \sum_{\kappa\lambda\mu\nu} \frac{\beta_\kappa\beta_\lambda\beta_\mu\beta_\nu}{\lambda\nu} \int_x^\infty \exp\left\{-\pi\left(\frac{m^2\kappa^2}{\lambda^2}+\frac{n^2\mu^2}{\nu^2}\right)u^2\sin\delta+\right.$$

$$\left.+i\pi\left(\frac{m^2\kappa^2}{\lambda^2}-\frac{n^2\mu^2}{\nu^2}\right)u^2\cos\delta\right\}\frac{du}{u^\theta}.$$

Let Σ_1 denote the sum of those terms in which $m\kappa/\lambda = n\mu/\nu$, and Σ_2 the remainder. Let $(\kappa\nu, \lambda\mu) = q$, so that

$$\kappa\nu = aq, \qquad \lambda\mu = bq, \qquad (a,b)=1.$$

Then, in Σ_1, $ma = nb$, so that $n = ra$, $m = rb$ $(r = 1, 2,\ldots)$. Hence

$$\Sigma_1 = \sum_{\kappa\lambda\mu\nu} \frac{\beta_\kappa\beta_\lambda\beta_\mu\beta_\nu}{\lambda\nu} \sum_{r=1}^\infty \int_x^\infty \exp\left(-2\pi\frac{r^2\kappa^2\mu^2}{q^2}u^2\sin\delta\right)\frac{du}{u^\theta}.$$

Now

$$\sum_{r=1}^\infty \int_x^\infty e^{-r^2u^2\eta}\frac{du}{u^\theta} = \eta^{\frac{1}{2}\theta-\frac{1}{2}} \sum_{r=1}^\infty \frac{1}{r^{1-\theta}} \int_{xr\sqrt{\eta}}^\infty e^{-y^2}\frac{dy}{y^\theta}$$

$$= \eta^{\frac{1}{2}\theta-\frac{1}{2}} \int_{x\sqrt{\eta}}^\infty \frac{e^{-y^2}}{y^\theta}\left(\sum_{r\leqslant y/(x\sqrt{\eta})}\frac{1}{r^{1-\theta}}\right)dy.$$

The last r-sum is of the form

$$\frac{1}{\theta}\left(\frac{y}{x\sqrt{\eta}}\right)^\theta - \frac{1}{\theta} + K(\theta) + O\left\{\left(\frac{y}{x\sqrt{\eta}}\right)^{\theta-1}\right\},$$

where $K(\theta)$, and later $K_1(\theta)$, are bounded functions of θ. Hence we obtain

$$\frac{1}{\theta x^{\theta}\eta^{\frac{1}{2}}}\left\{\int_0^{\infty} e^{-y^2}\,dy+O(x\sqrt{\eta})\right\}-\frac{\eta^{\frac{1}{2}\theta-\frac{1}{2}}}{\theta}\left[\int_0^{\infty} e^{-y^2}y^{-\theta}\,dy+O\{(x\sqrt{\eta})^{1-\theta}\}\right]+$$

$$+\eta^{\frac{1}{2}\theta-\frac{1}{2}}K(\theta)\left[\int_0^{\infty} e^{-y^2}y^{-\theta}\,dy+O\{(x\sqrt{\eta})^{1-\theta}\}\right]+O\{x^{1-\theta}\log(2+\eta^{-1})\}$$

$$=\frac{\sqrt{\pi}}{2\theta x^{\theta}\eta^{\frac{1}{2}}}+\frac{K_1(\theta)\eta^{\frac{1}{2}\theta-\frac{1}{2}}}{\theta}+O\left\{\frac{x^{1-\theta}}{\theta}\log(2+\eta^{-1})\right\}.$$

Putting $\eta = 2\pi\kappa^2\mu^2q^{-2}\sin\delta$, it follows that

$$\Sigma_1=\frac{S(0)}{2(2\sin\delta)^{\frac{1}{2}}\theta x^{\theta}}+\frac{K_1(\theta)}{\theta}(2\pi\sin\delta)^{\frac{1}{2}\theta-\frac{1}{2}}S(\theta)+$$

$$+O\left\{\frac{x^{1-\theta}\log(2+\eta^{-1})}{\theta}\sum_{\kappa\lambda\mu\nu}\frac{|\beta_\kappa\beta_\lambda\beta_\mu\beta_\nu|}{\lambda\nu}\right\}, \tag{10.11.1}$$

where
$$S(\theta)=\sum_{\kappa\lambda\mu\nu}\left(\frac{q}{\kappa\mu}\right)^{1-\theta}\frac{\beta_\kappa\beta_\lambda\beta_\mu\beta_\nu}{\lambda\nu}.$$

Defining $\phi_a(n)$ as in § 9.24, we have

$$q^{1-\theta}=\sum_{\rho|q}\phi_{-\theta}(\rho)=\sum_{\rho|\kappa\nu,\rho|\lambda\mu}\phi_{-\theta}(\rho).$$

Hence
$$S(\theta)=\sum_{\rho<X^2}\phi_{-\theta}(\rho)\left(\sum_{\rho|\kappa\nu}\frac{\beta_\kappa\beta_\nu}{\kappa^{1-\theta}\nu}\right)^2.$$

Let d and d_1 denote positive integers whose prime factors divide ρ. Let $\kappa = d\kappa'$, $\nu = d_1\nu'$, where $(\kappa',\rho) = 1$, $(\nu',\rho) = 1$. Then

$$\sum_{\rho|\kappa\nu}\frac{\beta_\kappa\beta_\nu}{\kappa^{1-\theta}\nu}=\sum_{\rho|dd_1}\frac{1}{d^{1-\theta}d_1}\sum_{\kappa'}\frac{\beta_{d\kappa'}}{\kappa'^{1-\theta}}\sum_{\nu'}\frac{\beta_{d_1\nu'}}{\nu'}.$$

Now, for $(\kappa',\rho) = 1$, $\qquad \beta_{d\kappa'}=\dfrac{\alpha_d\,\alpha_{\kappa'}}{\log X}\log\dfrac{X}{d\kappa'}.$

Hence the above sum is equal to

$$\frac{1}{\log^2 X}\sum_{\rho|dd_1}\frac{\alpha_d\,\alpha_{d_1}}{d^{1-\theta}d_1}\sum_{\kappa'\leqslant X/d}\frac{\alpha_{\kappa'}}{\kappa'^{1-\theta}}\log\frac{X}{d\kappa'}\sum_{\nu'\leqslant X/d_1}\frac{\alpha_{\nu'}}{\nu'}\log\frac{X}{d_1\nu'}.$$

10.12. LEMMA 10.12. *We have*

$$\sum_{\kappa'\leqslant X/d}\frac{\alpha_{\kappa'}}{\kappa'^{1-\theta}}\log\frac{X}{d\kappa'}=O\left\{\left(\frac{X}{d}\right)^{\theta}\log^{\frac{1}{2}}\frac{X}{d}\prod_{p|\rho}\left(1+\frac{1}{p}\right)^{\frac{1}{2}}\right\} \tag{10.12.1}$$

uniformly with respect to θ.

We may suppose that $X \geqslant 2d$, since otherwise the lemma is trivial.

Now
$$\frac{1}{2\pi i}\int\limits_{1-i\infty}^{1+i\infty}\frac{x^s}{s^2}\,ds = 0 \quad (0 < x \leqslant 1), \quad \log x \quad (x > 1).$$

Also
$$\sum_{(\kappa',\rho)=1}\frac{\alpha_{\kappa'}}{\kappa'^{1-\theta+s}} = \prod_{(p,\rho)=1}\left(1-\frac{1}{p^{1-\theta+s}}\right)^{\frac{1}{2}} = \sum_{p|\rho}\left(1-\frac{1}{p^{1-\theta+s}}\right)^{-\frac{1}{2}}\frac{1}{\sqrt{\zeta(1-\theta+s)}}.$$

Hence the left-hand side of (10.12.1) is equal to
$$\frac{1}{2\pi i}\int\limits_{1-i\infty}^{1+i\infty}\frac{1}{s^2}\left(\frac{X}{d}\right)^s\prod_{p|\rho}\left(1-\frac{1}{p^{1-\theta+s}}\right)^{-\frac{1}{2}}\frac{ds}{\sqrt{\zeta(1-\theta+s)}}. \tag{10.12.2}$$

There are singularities at $s = 0$ and $s = \theta$. If $\theta \geqslant \{\log(X/d)\}^{-1}$, we can take the line of integration through $s = \theta$, the integral round a small indentation tending to zero. Now
$$\left|\frac{1}{\zeta(1+it)}\right| < A|t|$$
for all t (large or small). Also
$$\prod_{p|\rho}\left(1-\frac{1}{p^{1-\theta+s}}\right)^{-1} = O\left\{\left|\prod_{p|\rho}\left(1+\frac{1}{p^{1-\theta+s}}\right)\right|\right\} = O\left\{\prod_{p|\rho}\left(1+\frac{1}{p}\right)\right\}.$$

Hence (10.12.2) is
$$O\left\{\left(\frac{X}{d}\right)^\theta\prod_{p|\rho}\left(1+\frac{1}{p}\right)^{\frac{1}{2}}\int\limits_{-\infty}^{\infty}\frac{|t|^{\frac{1}{2}}\,dt}{\theta^2+t^2}\right\} = O\left\{\left(\frac{X}{d}\right)^\theta\prod_{p|\rho}\left(1+\frac{1}{p}\right)^{\frac{1}{2}}\frac{1}{\theta^{\frac{1}{2}}}\right\},$$
and the result stated follows.

If $\theta < \{\log(X/d)\}^{-1}$, we take the same contour as before modified by a detour round the right-hand side of the circle $|s| = 2\{\log(X/d)\}^{-1}$. On this circle
$$|(X/d)^s| \leqslant e^2,$$
the p-product goes as before, and
$$|\zeta(1-\theta+s)| > A\log(X/d).$$

Hence the integral round the circle is
$$O\left\{\log^{-\frac{1}{2}}\frac{X}{d}\prod_{p|\rho}\left(1+\frac{1}{p}\right)^{\frac{1}{2}}\int\left|\frac{ds}{s^2}\right|\right\} = O\left\{\log^{\frac{1}{2}}\frac{X}{d}\prod_{p|\rho}\left(1+\frac{1}{p}\right)^{\frac{1}{2}}\right\}.$$

The integral along the part of the line $\sigma = \theta$ above the circle is
$$O\left\{\left(\frac{X}{d}\right)^\theta\prod_{p|\rho}\left(1+\frac{1}{p}\right)^{\frac{1}{2}}\int\limits_{A(\log X/d)^{-1}}^{\infty}\frac{dt}{t^{\frac{3}{2}}}\right\} = O\left\{\left(\frac{X}{d}\right)^\theta\log^{\frac{1}{2}}\frac{X}{d}\prod_{p|\rho}\left(1+\frac{1}{p}\right)^{\frac{1}{2}}\right\}.$$

The lemma is thus proved in all cases.

10.13. LEMMA 10.13.

$$\sum_{\rho|dd_1} \frac{|\alpha_d \alpha_{d_1}|}{dd_1} = O\left\{\frac{1}{\rho} \prod_{p|\rho} \left(1+\frac{1}{p}\right)\right\}.$$

Defining α'_d as in § 10.9, we have

$$\sum_{\rho|dd_1} \frac{|\alpha_d \alpha_{d_1}|}{dd_1} \leqslant \sum_{\rho|dd_1} \frac{\alpha'_d \alpha'_{d_1}}{dd_1} = \sum_{\rho|D} \frac{1}{D},$$

where D is a number of the same class as d or d_1,

$$= \frac{1}{\rho} \prod_{p|\rho} \left(1-\frac{1}{p}\right)^{-1} = O\left\{\frac{1}{\rho} \prod_{p|\rho} \left(1+\frac{1}{p}\right)\right\}.$$

10.14. LEMMA 10.14.

$$S(\theta) = O\left(\frac{X^{2\theta}}{\log X}\right)$$

uniformly with respect to θ. In particular

$$S(0) = O\left(\frac{1}{\log X}\right).$$

By the formulae of § 10.11, and the above lemmas,

$$\sum_{\rho|\kappa\nu} \frac{\beta_\kappa \beta_\nu}{\kappa^{1-\theta}\nu} = O\left\{\frac{1}{\log^2 X} \sum_{\rho|dd_1} \frac{|\alpha_d \alpha_{d_1}|}{d^{1-\theta} d_1} \left(\frac{X}{d}\right)^\theta \log^{\frac{1}{2}} \frac{X}{d} \log^{\frac{1}{2}} \frac{X}{d_1} \prod_{p|\rho} \left(1+\frac{1}{p}\right)\right\}$$

$$= O\left\{\frac{X^\theta}{\log X} \prod_{p|\rho} \left(1+\frac{1}{p}\right) \sum_{\rho|dd_1} \frac{|\alpha_d \alpha_{d_1}|}{dd_1}\right\}$$

$$= O\left\{\frac{X^\theta}{\rho \log X} \prod_{p|\rho} \left(1+\frac{1}{p}\right)^2\right\}.$$

Hence

$$S(\theta) = O\left\{\frac{X^{2\theta}}{\log^2 X} \sum_{\rho \leqslant X^2} \frac{\phi_{-\theta}(\rho)}{\rho^2} \prod_{p|\rho} \left(1+\frac{1}{p}\right)^4\right\}$$

$$= O\left\{\frac{X^{2\theta}}{\log^2 X} \sum_{\rho \leqslant X^2} \frac{1}{\rho^{1+\theta}} \prod_{p|\rho} \left(1+\frac{1}{p}\right)^4\right\}$$

$$= O\left\{\frac{X^{2\theta}}{\log^2 X} \sum_{\rho \leqslant X^2} \frac{1}{\rho^{1+\theta}} \sum_{n|\rho} \frac{1}{n^{\frac{1}{2}}}\right\},$$

since

$$\prod_{p|\rho} \left(1+\frac{1}{p}\right)^4 = O\left\{\prod_{p|\rho} \left(1+\frac{4}{p}\right)\right\} = O\left\{\prod_{p|\rho} \left(1+\frac{1}{p^{\frac{1}{2}}}\right)\right\} = O\left(\sum_{n|\rho} \frac{1}{n^{\frac{1}{2}}}\right).$$

Hence

$$S(\theta) = O\left\{\frac{X^{2\theta}}{\log^2 X} \sum_{n \leqslant X^2} \sum_{\rho_1 \leqslant X^2/n} \frac{1}{(n\rho_1)^{1+\theta} n^{\frac{1}{2}}}\right\}$$

$$= O\left\{\frac{X^{2\theta}}{\log^2 X} \sum_{n=1}^{\infty} \frac{1}{n^{\frac{3}{2}+\theta}} \sum_{\rho_1 \leqslant X^2/n} \frac{1}{\rho_1^{1+\theta}}\right\}$$

$$= O\left\{\frac{X^{2\theta}}{\log^2 X} \sum_{n=1}^{\infty} \frac{1}{n^{\frac{3}{2}}} \sum_{\rho_1 \leqslant X^2} \frac{1}{\rho_1}\right\}$$

$$= O\left(\frac{X^{2\theta}}{\log X}\right).$$

10.15. Estimation of Σ_1. From (10.11.1), Lemma 10.14, and the inequality $|\beta_\nu| \leqslant 1$, we obtain

$$\Sigma_1 = O\left(\frac{1}{\delta^{\frac{1}{2}}\theta x^\theta \log X}\right) + O\left\{\frac{(\delta^{\frac{1}{2}} x X^2)^\theta}{\delta^{\frac{1}{2}}\theta x^\theta \log X}\right\} + O\left\{\frac{x^{1-\theta}\log(X/\delta)}{\theta} X^2 \log^2 X\right\}.$$

We shall ultimately take $X = \delta^{-c}$ and $h = (a \log X)^{-1}$, where a and c are suitable positive constants. Then $G = X^a = \delta^{-ac}$. If $x \leqslant G$, the last two terms can be omitted in comparison with the first if $GX^2 = O(\delta^{-\frac{1}{4}})$, i.e. if $(a+2)c \leqslant \frac{1}{4}$. We then have

$$\Sigma_1 = O\left(\frac{1}{\delta^{\frac{1}{2}}\theta x^\theta \log X}\right). \tag{10.15.1}$$

10.16. Estimation of Σ_2. If P and Q are positive, and $x \geqslant 1$,

$$\int_x^\infty e^{-Pu^2 + iQu^2} \frac{du}{u^\theta} = \frac{1}{2}\int_{x^2}^\infty \frac{e^{-Pv}}{v^{\frac{1}{2}\theta+\frac{1}{2}}} e^{iQv}\, dv = O\left(\frac{e^{-P}}{x^\theta Q}\right),$$

e.g. by applying the second mean-value theorem to the real and imaginary parts. Hence

$$\Sigma_2 = O\left[\frac{1}{x^\theta} \sum_{\kappa\lambda\mu\nu} \frac{1}{\lambda\nu} \sum_{mn}{}' \left|\frac{m^2\kappa^2}{\lambda^2} - \frac{n^2\mu^2}{\nu^2}\right|^{-1} \exp\left\{-\pi\left(\frac{m^2\kappa^2}{\lambda^2} + \frac{n^2\mu^2}{\nu^2}\right)\sin\delta\right\}\right].$$

The terms with $m\kappa/\lambda > n\mu/\nu$ contribute to the m, n sum

$$O\left\{\sum_{m=1}^\infty e^{-\pi m^2\kappa^2\lambda^{-2}\sin\delta} \sum_{n < m\kappa\nu/\lambda\mu} \left(\frac{m^2\kappa^2}{\lambda^2} - \frac{n^2\mu^2}{\nu^2}\right)^{-1}\right\}.$$

Now
$$\frac{m^2\kappa^2}{\lambda^2} - \frac{n^2\mu^2}{\nu^2} \geqslant \frac{m\kappa}{\lambda}\left(\frac{m\kappa}{\lambda} - \frac{n\mu}{\nu}\right) = \frac{m\kappa(m\kappa\nu - n\lambda\mu)}{\lambda^2\nu},$$

and
$$\sum_n \frac{1}{m\kappa\nu - n\lambda\mu} \leqslant 1 + \frac{1}{\lambda\mu} + \frac{1}{2\lambda\mu} + \dots = 1 + O\left(\frac{\log mX}{\lambda\mu}\right).$$

Hence the m, n sum is

$$O\left\{\frac{\lambda^2\nu}{\kappa}\sum_{m=1}^{\infty}\left(\frac{1}{m}+\frac{\log(mX)}{m\lambda\mu}\right)e^{-\pi m^2\kappa^2\lambda^{-2}\sin\delta}\right\}$$

$$= O\left\{\frac{\lambda^2\nu}{\kappa}\left(1+\frac{\log X}{\lambda\mu}\right)\log\frac{X^2}{\delta}+\frac{\lambda\nu}{\kappa\mu}\log^2\frac{X^2}{\delta}\right\}$$

$$= O\left(\frac{\lambda^2\nu}{\kappa}\log\frac{1}{\delta}\right)+O\left(\frac{\lambda\nu}{\kappa\mu}\log^2\frac{1}{\delta}\right),$$

since, as in §10.15, we have $X = \delta^{-c}$, with $0 < c \leqslant \frac{1}{8}$. The remaining terms may be treated similarly. Hence

$$\Sigma_2 = O\left\{\frac{1}{x^\theta}\sum_{\kappa\lambda\mu\nu}\left(\frac{\lambda}{\kappa}\log\frac{1}{\delta}+\frac{1}{\kappa\mu}\log^2\frac{1}{\delta}\right)\right\} = O\left(\frac{X^4}{x^\theta}\log^2\frac{1}{\delta}\right). \tag{10.16.1}$$

10.17. LEMMA 10.17. *Under the assumptions of* § 10.15

$$\int_{-\infty}^{\infty}\left|\int_t^{t+h}F(u)\,du\right|^2 dt = O\left(\frac{h}{\delta^{\frac{1}{2}}\log X}\right). \tag{10.17.1}$$

By (10.15.1) and (10.16.1),

$$J(x,\theta) = O\left(\frac{1}{\delta^{\frac{1}{2}}\theta x^\theta \log X}\right) \tag{10.17.2}$$

uniformly with respect to θ. Hence

$$\int_1^G |g(x)|^2\,dx = -\int_1^G x^\theta\frac{\partial J}{\partial x}\,dx = [-x^\theta J]_1^G+\theta\int_1^G x^{\theta-1}J\,dx$$

$$= O\left(\frac{1}{\delta^{\frac{1}{2}}\theta\log X}\right)+O\left(\theta\int_1^G\frac{dx}{\delta^{\frac{1}{2}}\theta x\log X}\right) = O\left(\frac{\log G}{\delta^{\frac{1}{2}}\log X}\right),$$

taking, for example, $\theta = \frac{1}{2}$. Also

$$\int_0^{\frac{1}{2}}\theta J(G,\theta)\,d\theta = \int_G^\infty |g(x)|^2\,dx\int_0^{\frac{1}{2}}\theta x^{-\theta}\,d\theta$$

$$= \int_G^\infty |g(x)|^2\left(\frac{1}{\log^2 x}-\frac{1}{2x^{\frac{1}{2}}\log x}-\frac{1}{x^{\frac{1}{2}}\log^2 x}\right)dx$$

$$\geqslant \int_G^\infty\frac{|g(x)|^2}{\log^2 x}\,dx-\frac{3}{2}\int_G^\infty\frac{|g(x)|^2}{x^{\frac{1}{2}}}\,dx$$

since $G = e^{1/h} \geqslant e$. Hence

$$\int_G^\infty \frac{|g(x)|^2}{\log^2 x}\,dx \leqslant \int_0^{\frac{1}{2}} \theta J(G,\theta)\,d\theta + \tfrac{3}{2}J(G,\tfrac{1}{2})$$

$$= O\left(\int_0^{\frac{1}{2}} \frac{d\theta}{\delta^{\frac{1}{2}}G^\theta \log X}\right) + O\left(\frac{1}{\delta^{\frac{1}{2}}G^{\frac{1}{2}}\log X}\right) = O\left(\frac{1}{\delta^{\frac{1}{2}}\log G\log X}\right).$$

Also $\phi(0) = O(X)$, $\phi(1) = O(\log X)$. The result therefore follows from the formulae of § 10.10.

10.18. So far the integrals considered have involved $F(t)$. We now turn to the integrals involving $|F(t)|$. The results about such integrals are expressed in the following lemmas.

LEMMA 10.18. $$\int_{-\infty}^{\infty} |F(t)|^2\,dt = O\left(\frac{\log 1/\delta}{\delta^{\frac{1}{2}}\log X}\right).$$

By the Fourier transform formulae, the left-hand side is equal to

$$2\int_0^\infty |f(y)|^2\,dy = 2\int_1^\infty \left|\frac{e^{-i(\frac{1}{4}\pi-\frac{1}{2}\delta)}}{2x}\phi(1)\phi(0) - g(x)\right|^2 dx$$

$$\leqslant 4\int_1^\infty |g(x)|^2\,dx + O(X^2\log^2 X).$$

Taking $x = 1$, $\theta = \{\log(1/\delta)\}^{-1}$ in (10.17.2), we have

$$\int_1^\infty |g(u)|^2 e^{-\log u/(\log 1/\delta)}\,du = O\left(\frac{\log 1/\delta}{\delta^{\frac{1}{2}}\log X}\right).$$

Hence $$\int_1^{\delta^{-2}} |g(u)|^2\,du = O\left(\frac{\log 1/\delta}{\delta^{\frac{1}{2}}\log X}\right).$$

We can estimate the integral over (δ^{-2},∞) in a comparatively trivial manner. As in § 10.11, this is less than

$$\sum_{m=1}^\infty \sum_{n=1}^\infty \sum_{\kappa\lambda\mu\nu} \frac{|\beta_\kappa \beta_\lambda \beta_\mu \beta_\nu|}{\lambda\nu} \int_{\delta^{-2}}^\infty \exp\left\{-\pi\left(\frac{m^2\kappa^2}{\lambda^2} + \frac{n^2\mu^2}{\nu^2}\right)u^2\sin\delta\right\}du.$$

Using, for example, $\kappa^2\lambda^{-2}\sin\delta > AX^{-2}\delta > A\delta^2$ (since $X = \delta^{-c}$ with $c < \frac{1}{2}$), and $|\beta_\nu| \leqslant 1$, this is

$$O\left\{X^2\log^2 X \sum_{m=1}^{\infty}\sum_{n=1}^{\infty}\int_{\delta^{-2}}^{\infty} e^{-A(m^2+n^2)\delta^2u^2}\,du\right\}$$

$$= O\left(X^2\log^2 X\int_{\delta^{-2}}^{\infty} e^{-A\delta^2u^2}\,du\right) = O(X^2\log^2 X\, e^{-A/\delta^2}),$$

which is of the required form.

10.19. LEMMA 10.19.

$$\int_{-\infty}^{\infty}\left\{\int_{t}^{t+h}|F(u)|\,du\right\}^2 dt = O\left(\frac{h^2\log 1/\delta}{\delta^{\frac{1}{2}}\log X}\right).$$

For the left-hand side does not exceed

$$\int_{-\infty}^{\infty}\left\{h\int_{t}^{t+h}|F(u)|^2\,du\right\} dt = h\int_{-\infty}^{\infty}|F(u)|^2\,du\int_{u-h}^{u}dt = h^2\int_{-\infty}^{\infty}|F(u)|^2\,du,$$

and the result follows from the previous lemma.

10.20. LEMMA 10.20. *If* $\delta = 1/T$,

$$\int_{0}^{T}|F(t)|\,dt > AT^{\frac{3}{4}}.$$

We have

$$\left(\int_{\frac{1}{2}+i}^{2+i} + \int_{2+i}^{2+iT} + \int_{2+iT}^{\frac{1}{2}+iT} + \int_{\frac{1}{2}+iT}^{\frac{1}{2}+i}\right)\zeta(s)\phi^2(s)\,ds = 0.$$

Since $\phi(s) = O(X^{\frac{1}{2}})$ for $\sigma \geqslant \frac{1}{2}$, the first term is $O(X)$, and the third is $O(XT^{\frac{1}{4}})$. Also

$$\zeta(s)\phi^2(s) = 1+\sum_{n=2}^{\infty}\frac{a_n}{n^s},$$

where $|a_n| \leqslant d_3(n)$. Hence

$$\int_{2+i}^{2+iT}\zeta(s)\phi^2(s)\,ds = i(T-1)+\sum_{n=2}^{\infty}a_n\int_{2+i}^{2+iT}\frac{ds}{n^s}$$

$$= i(T-1)+O\left(\sum_{n=2}^{\infty}\frac{d_3(n)}{n^2\log n}\right)$$

$$= iT+O(1).$$

It follows that $$\int_{0}^{T}\zeta(\tfrac{1}{2}+it)\phi^2(\tfrac{1}{2}+it)\,dt \sim T.$$

Hence
$$\int_0^T |F(t)|\,dt > A\int_0^T t^{-\frac{1}{4}}|\zeta(\tfrac{1}{2}+it)\phi^2(\tfrac{1}{2}+it)|\,dt$$
$$> AT^{-\frac{1}{4}}\int_{\frac{1}{2}T}^{T} |\zeta(\tfrac{1}{2}+it)\phi^2(\tfrac{1}{2}+it)|\,dt$$
$$> AT^{-\frac{1}{4}}\left|\int_{\frac{1}{2}T}^{T} \zeta(\tfrac{1}{2}+it)\phi^2(\tfrac{1}{2}+it)\,dt\right|$$
$$> AT^{\frac{3}{4}}.$$

10.21. LEMMA 10.21.
$$\int_0^T dt\int_t^{t+h} |F(u)|\,du > AhT^{\frac{3}{4}}.$$

The left-hand side is equal to
$$\int_0^{T+h} |F(u)|\,du \int_{\max(0,u-h)}^{\min(T,u)} dt \geqslant \int_h^T |F(u)|\,du\int_{u-h}^{u} dt = h\int_h^T |F(u)|\,du,$$
and the result follows from the previous lemma.

10.22. THEOREM 10.22.
$$N_0(T) > AT\log T.$$

Let E be the sub-set of $(0, T)$ where
$$\int_t^{t+h} |F(u)|\,du > \left|\int_t^{t+h} F(u)\,du\right|.$$
For such values of t, $F(u)$ must change sign in $(t, t+h)$, and hence so must $\Xi(u)$, and hence $\zeta(\frac{1}{2}+iu)$ must have a zero in this interval.

Since the two sides are equal except in E,
$$\int_E dt\int_t^{t+h} |F(u)|\,du \geqslant \int_E \left\{\int_t^{t+h} |F(u)|\,du - \left|\int_t^{t+h} F(u)\,du\right|\right\}dt$$
$$= \int_0^T \left\{\int_t^{t+h} |F(u)|\,du - \left|\int_t^{t+h} F(u)\,du\right|\right\}dt$$
$$> AhT^{\frac{3}{4}} - \int_0^T \left|\int_t^{t+h} F(u)\,du\right|\,dt.$$

The left-hand side is not greater than
$$\left\{\int_E dt\int_E \left(\int_t^{t+h} |F(u)|\,du\right)^2 dt\right\}^{\frac{1}{2}} \leqslant \left\{m(E)\int_{-\infty}^{\infty}\left(\int_t^{t+h} |F(u)|\,du\right)^2 dt\right\}^{\frac{1}{2}}$$
$$< A\{m(E)\}^{\frac{1}{2}} hT^{\frac{1}{4}}\left(\frac{\log T}{\log X}\right)^{\frac{1}{2}}$$

by Lemma 10.19 with $\delta = 1/T$. The second term on the right is not greater than

$$\left\{\int_0^T dt \int_0^T \left|\int_t^{t+h} F(u)\, du\right|^2 dt\right\}^{\frac{1}{2}} < \frac{Ah^{\frac{1}{2}}T^{\frac{3}{4}}}{\log^{\frac{1}{2}}X}$$

by Lemma 10.17. Hence

$$\{m(E)\}^{\frac{1}{2}} > A_1 T^{\frac{1}{2}}\left(\frac{\log X}{\log T}\right)^{\frac{1}{2}} - A_2 \frac{T^{\frac{1}{2}}}{h^{\frac{1}{2}}\log^{\frac{1}{2}}T},$$

where A_1 and A_2 denote the particular constants which occur. Since $X = T^c$ and $h = (a \log X)^{-1} = (ac \log T)^{-1}$,

$$\{m(E)\}^{\frac{1}{2}} > A_1 c^{\frac{1}{2}} T^{\frac{1}{2}} - A_2 (ac)^{\frac{1}{2}} T^{\frac{1}{2}}.$$

Taking a small enough, it follows that

$$m(E) > A_3 T.$$

Hence, of the intervals $(0, h)$, $(h, 2h)$,... contained in $(0, T)$, at least $[A_3 T/h]$ must contain points of E. If $(nh, (n+1)h)$ contains a point t of E, there must be a zero of $\zeta(\frac{1}{2}+iu)$ in $(t, t+h)$, and so in $(nh, (n+2)h)$. Allowing for the fact that each zero might be counted twice in this way, there must be at least

$$\tfrac{1}{2}[A_3 T/h] > AT \log T$$

zeros in $(0, T)$.

10.23. In this section we return to the function $\Xi^*(t)$ mentioned in § 10.1. In spite of its deficiencies as an approximation to $\Xi(t)$, it is of some interest to note that *all the zeros of* $\Xi^*(t)$ *are real.*†

A still better approximation to $\Phi(u)$ is

$$\Phi^{**}(u) = \pi(2\pi \cosh \tfrac{9}{2}u - 3 \cosh \tfrac{5}{2}u)e^{-2\pi \cosh 2u}.$$

This gives

$$\Xi^{**}(t) = 2\int_0^\infty \Phi^{**}(u) \cos ut\, du,$$

and we shall also prove that *all the zeros of* $\Xi^{**}(t)$ *are real.*

The function $K_z(a)$ is, for any value of a, an even integral function of z. We begin by proving that *if a is real all its zeros are purely imaginary.*

It is known that $w = K_z(a)$ satisfies the differential equation

$$\frac{d}{da}\left(a\frac{dw}{da}\right) = \left(a + \frac{z^2}{a}\right)w.$$

This is equivalent to the two equations

$$\frac{dw}{da} = \frac{W}{a}, \qquad \frac{dW}{da} = \left(a + \frac{z^2}{a}\right)w.$$

† Pólya (1), (2), (4).

These give
$$\frac{d}{da}(W\bar{w}) = \frac{1}{a}\{|W|^2+(a^2+z^2)|w|^2\}.$$

It is also easily verified that w and W tend to 0 as $a \to \infty$. It follows that, if w vanishes for a certain z and $a = a_0 > 0$, then
$$\int_{a_0}^{\infty} \{|W|^2+(a^2+z^2)|w|^2\}\frac{da}{a} = 0.$$

Taking imaginary parts,
$$2ixy\int_{a_0}^{\infty} \frac{|w|^2}{a}\,da = 0.$$

Here the integral is not 0, and $K_z(a)$ plainly does not vanish for z real, i.e. $y = 0$. Hence $x = 0$, the required result.

We also require the following lemma.

Let c *be a positive constant,* $F(z)$ *an integral function of genus* 0 *or* 1, *which takes real values for real* z, *and has no complex zeros and at least one real zero. Then all the zeros of*
$$F(z+ic)+F(z-ic) \tag{10.23.1}$$
are also real.

We have
$$F(z) = Cz^q e^{\alpha z}\prod_{n=1}^{\infty}\left(1-\frac{z}{\alpha_n}\right)e^{z/\alpha_n},$$
where C, α, α_1,... are real constants, $\alpha_n \neq 0$ for $n = 1, 2,..., \sum \alpha_n^{-2}$ is convergent, q a non-negative integer. Let z be a zero of (10.23.1). Then
$$|F(z+ic)| = |F(z-ic)|,$$
so that
$$1 = \left|\frac{F(z-ic)}{F(z+ic)}\right|^2 = \left\{\frac{x^2+(y-c)^2}{x^2+(y+c)^2}\right\}^q \prod_{n=1}^{\infty}\frac{(x-\alpha_n)^2+(y-c)^2}{(x-\alpha_n)^2+(y+c)^2}.$$

If $y > 0$, every factor on the right is < 1; if $y < 0$, every factor is > 1. Hence in fact $y = 0$.

The theorem that the zeros of $\Xi^*(t)$ are all real now follows on taking
$$F(z) = K_{\frac{1}{2}iz}(2\pi), \qquad c = \tfrac{9}{2}.$$

10.24. For the discussion of $\Xi^{**}(t)$ we require the following lemma.

Let $|f(t)| < Ke^{-|t|^{2+\delta}}$ *for some positive* δ, *so that*
$$F(z) = \frac{1}{\sqrt{(2\pi)}}\int_{-\infty}^{\infty} f(t)e^{izt}\,dt$$

is an integral function of z. Let all the zeros of $F(z)$ be real. Let $\phi(t)$ be an integral function of t of genus 0 or 1, real for real t. Then the zeros of

$$G(z) = \frac{1}{\sqrt{(2\pi)}} \int_{-\infty}^{\infty} f(t)\phi(it)e^{izt}\, dt$$

are also all real.

We have
$$\phi(t) = Ct^q e^{\alpha t} \prod_{m=1}^{\infty} \left(1 - \frac{t}{\alpha_m}\right) e^{t/\alpha_m},$$

where the constants are all real, and $\sum \alpha_m^{-2}$ is convergent. Let

$$\phi_n(t) = Ct^q e^{\alpha t} \prod_{m=1}^{n} \left(1 - \frac{t}{\alpha_m}\right) e^{t/\alpha_m}.$$

Then $\phi_n(t) \to \phi(t)$ uniformly in any finite interval, and (as in my *Theory of Functions*, § 8.25)
$$|\phi_n(t)| < Ke^{|t|^{2+\epsilon}}$$

uniformly with respect to n. Hence

$$G(z) = \lim_{n\to\infty} \frac{1}{\sqrt{(2\pi)}} \int_{-\infty}^{\infty} f(t)\phi_n(it)e^{izt}\, dt = \lim_{n\to\infty} G_n(z),$$

say. It is therefore sufficient to prove that, for every n, the zeros of $G_n(z)$ are real.

Now it is easily verified that $F(z)$ is an integral function of order less than 2. Hence, if its zeros are real, so are those of

$$(D-\alpha)F(z) = e^{\alpha z}\frac{d}{dz}\{e^{-\alpha z}F(z)\}$$

for any real α. Applying this principle repeatedly, we see that all the zeros of

$$H(z) = D^q(D-\alpha_1)...(D-\alpha_n)F(z) = \frac{1}{\sqrt{(2\pi)}} \int_{-\infty}^{\infty} f(t)(it)^q(it-\alpha_1)...(it-\alpha_n)e^{izt}\, dt$$

are real. Since

$$G_n(z) = \frac{(-1)^n C}{\alpha_1...\alpha_n} H\left(z+\alpha+\frac{1}{\alpha_1}+...+\frac{1}{\alpha_n}\right)$$

the result follows.

Taking
$$f(t) = 4\sqrt{(2\pi)}e^{-2\pi\cosh 2t}$$

we obtain
$$F(z) = K_{\frac{1}{2}iz}(2\pi),$$

all of whose zeros are real. If

$$\phi(t) = \tfrac{1}{2}\pi^2 \cos\tfrac{9}{2}t,$$

then $G(z) = \Xi^*(z)$, and it follows again that all the zeros of $\Xi^*(z)$ are real. If

$$\phi(t) = \tfrac{1}{2}\pi^2\left(\cos\frac{9}{2}t - \frac{3}{2\pi}\cos\frac{5}{2}t\right),$$

then $G(z) = \Xi^{**}(z)$. Hence all the zeros of $\Xi^{**}(z)$ are real.

10.25. By way of contrast to the Riemann zeta-function we shall now construct a function which has a similar functional equation, and for which the analogues of most of the theorems of this chapter are true; but which has no Euler product, and for which the analogue of the Riemann hypothesis is false.

We shall use the simplest properties of Dirichlet's L-functions (mod 5). These are defined for $\sigma > 1$ by

$$L_0(s) = \sum_{n=1}^{\infty} \frac{\chi_0(n)}{n^s} = \frac{1}{1^s} + \frac{1}{2^s} + \frac{1}{3^s} + \frac{1}{4^s} + \frac{1}{6^s} + \dots,$$

$$L_1(s) = \sum_{n=1}^{\infty} \frac{\chi_1(n)}{n^s} = \frac{1}{1^s} + \frac{i}{2^s} - \frac{i}{3^s} - \frac{1}{4^s} + \frac{1}{6^s} + \dots,$$

$$L_2(s) = \sum_{n=1}^{\infty} \frac{\chi_2(n)}{n^s} = \frac{1}{1^s} - \frac{i}{2^s} + \frac{i}{3^s} - \frac{1}{4^s} + \frac{1}{6^s} + \dots,$$

$$L_3(s) = \sum_{n=1}^{\infty} \frac{\chi_3(n)}{n^s} = \frac{1}{1^s} - \frac{1}{2^s} - \frac{1}{3^s} + \frac{1}{4^s} + \frac{1}{6^s} + \dots.$$

Each $\chi(n)$ has the period 5. It is easily verified that in each case

$$\chi(m)\chi(n) = \chi(mn)$$

if m is prime to n; and hence that

$$L(s) = \prod_p \left\{1 - \frac{\chi(p)}{p^s}\right\}^{-1} \quad (\sigma > 1).$$

It is also easily seen that

$$L_0(s) = \left(1 - \frac{1}{5^s}\right)\zeta(s),$$

so that $L_0(s)$ is regular except for a simple pole at $s = 1$. The other three series are convergent for any real positive s, and hence for $\sigma > 0$. Hence $L_1(s)$, $L_2(s)$, and $L_3(s)$ are regular for $\sigma > 0$.

Now consider the function

$$\begin{aligned} f(s) &= \tfrac{1}{2}\sec\theta\{e^{-i\theta}L_1(s) + e^{i\theta}L_2(s)\} \\ &= \frac{1}{1^s} + \frac{\tan\theta}{2^s} - \frac{\tan\theta}{3^s} - \frac{1}{4^s} + \frac{1}{6^s} + \dots \\ &= \frac{1}{5^s}\{\zeta(s, \tfrac{1}{5}) + \tan\theta\,\zeta(s, \tfrac{2}{5}) - \tan\theta\,\zeta(s, \tfrac{3}{5}) - \zeta(s, \tfrac{4}{5})\}, \end{aligned}$$

where $\zeta(s, a)$ is defined as in § 2.17.

By (2.17) $f(s)$ is an integral function of s, and for $\sigma < 0$ it is equal to

$$\frac{2\Gamma(1-s)}{5^s(2\pi)^{1-s}}\left\{\sin\tfrac{1}{2}\pi s \times\right.$$

$$\times \sum_{m=1}^{\infty}\left(\cos\frac{2m\pi}{5}+\tan\theta\cos\frac{4m\pi}{5}-\tan\theta\cos\frac{6m\pi}{5}-\cos\frac{8m\pi}{5}\right)\frac{1}{m^{1-s}}+$$

$$\left.+\cos\tfrac{1}{2}\pi s\sum_{m=1}^{\infty}\left(\sin\frac{2m\pi}{5}+\tan\theta\sin\frac{4m\pi}{5}-\tan\theta\sin\frac{6m\pi}{5}-\sin\frac{8m\pi}{5}\right)\frac{1}{m^{1-s}}\right\}$$

$$=\frac{4\Gamma(1-s)\cos\frac{1}{2}\pi s}{5^s(2\pi)^{1-s}}\sum_{m=1}^{\infty}\left(\sin\frac{2m\pi}{5}+\tan\theta\sin\frac{4m\pi}{5}\right)\frac{1}{m^{1-s}}.$$

If
$$\sin\frac{4\pi}{5}+\tan\theta\sin\frac{8\pi}{5}=\tan\theta\left(\sin\frac{2\pi}{5}+\tan\theta\sin\frac{4\pi}{5}\right), \tag{10.25.1}$$

this is equal to

$$\frac{4\Gamma(1-s)\cos\frac{1}{2}\pi s}{5^s(2\pi)^{1-s}}\left(\sin\frac{2\pi}{5}+\tan\theta\sin\frac{4\pi}{5}\right)f(1-s).$$

The equation (10.25.1) reduces to

$$\sin 2\theta = 2\cos\frac{2\pi}{5}=\frac{\sqrt{5}-1}{2},$$

and we take θ to be the root of this between 0 and $\frac{1}{4}\pi$. We obtain

$$\tan\theta=\frac{\sqrt{(10-2\sqrt{5})}-2}{\sqrt{5}-1},$$

$$\sin\frac{2\pi}{5}+\tan\theta\sin\frac{4\pi}{5}=\frac{\sqrt{5}}{2},$$

and $f(s)$ satisfies the functional equation

$$f(s)=\frac{2\Gamma(1-s)\cos\frac{1}{2}s\pi}{5^{s-\frac{1}{2}}(2\pi)^{1-s}}f(1-s).$$

There is now no difficulty in extending the theorems of this chapter to $f(s)$. We can write the above equation as

$$\left(\frac{5}{\pi}\right)^{\frac{1}{2}s}\Gamma(\tfrac{1}{2}+\tfrac{1}{2}s)f(s)=\left(\frac{5}{\pi}\right)^{\frac{1}{2}-\frac{1}{2}s}\Gamma(1-\tfrac{1}{2}s)f(1-s),$$

and putting $s=\frac{1}{2}+it$ we obtain an even integral function of t analogous to $\Xi(t)$.

We conclude that $f(s)$ has an infinity of zeros on the line $\sigma=\frac{1}{2}$, and that the number of such zeros between 0 and T is greater than AT.

On the other hand, we shall now prove that $f(s)$ *has an infinity of zeros in the half-plane* $\sigma > 1$.

If p is a prime, we define $\alpha(p)$ by

$$\alpha(p) = \tfrac{1}{2}(1+i)\chi_1(p)+\tfrac{1}{2}(1-i)\chi_2(p),$$

so that

$$\alpha(p) = \pm 1 \quad \text{or} \quad \pm i.$$

For composite n, we define $\alpha(n)$ by the equation

$$\alpha(n_1 n_2) = \alpha(n_1)\alpha(n_2).$$

Thus $|\alpha(n)|$ is always 0 or 1. Let

$$M(s,\chi) = \sum_{n=1}^{\infty} \frac{\alpha(n)\chi(n)}{n^s} = \prod_p \left(1-\frac{\alpha(p)\chi(p)}{p^s}\right)^{-1},$$

where χ denotes either χ_1 or χ_2. Let

$$N(s) = \tfrac{1}{2}\{M(s,\chi_1)+M(s,\chi_2)\}.$$

Now

$$\alpha(p)\chi_1(p) = \tfrac{1}{2}(1+i)\chi_1^2+\tfrac{1}{2}(1-i)\chi_1\chi_2,$$

$$\alpha(p)\chi_2(p) = \tfrac{1}{2}(1+i)\chi_1\chi_2+\tfrac{1}{2}(1-i)\chi_2^2,$$

and these are conjugate since $\chi_1^2 = \chi_2^2$ and χ_1^2 and $\chi_1\chi_2$ are real. Hence $M(s,\chi_1)$ and $M(s,\chi_2)$ are conjugate for real s, and $N(s)$ is real.

Let s be real, greater than 1, and $\rightarrow 1$. Then

$$\log M(s,\chi_1) = \sum_p \frac{\alpha(p)\chi_1(p)}{p^s}+O(1)$$

$$= \tfrac{1}{2}(1+i)\sum_p \frac{\chi_1^2(p)}{p^s}+\tfrac{1}{2}(1-i)\sum_p \frac{\chi_1(p)\chi_2(p)}{p^s}+O(1).$$

Now $\chi_1^2 = \chi_3$ and $\chi_1\chi_2 = \chi_0$. Hence

$$\sum_p \frac{\chi_1^2(p)}{p^s} = \sum_p \frac{\chi_3(p)}{p^s} = \log L_3(s)+O(1) = O(1),$$

$$\sum_p \frac{\chi_1(p)\chi_2(p)}{p^s} = \sum_p \frac{\chi_0(p)}{p^s} = \log L_0(s)+O(1) = \log\frac{1}{s-1}+O(1).$$

Hence

$$\log M(s,\chi_1) = \tfrac{1}{2}(1-i)\log\frac{1}{s-1}+O(1),$$

$$N(s) = \mathbf{R}M(s,\chi_1) = \frac{1}{\sqrt{(s-1)}}\cos\left(\tfrac{1}{2}\log\frac{1}{s-1}\right)e^{O(1)}.$$

It is clear from this formula that $N(s)$ *has a zero at each of the points* $s = 1+e^{-(2m+1)\pi}$ $(m = 1, 2,\ldots)$.

Now for $\sigma \geqslant 1+\delta$, and $\chi = \chi_1$ or χ_2,

$$\log L(s+i\tau,\chi)-\log M(s,\chi)$$
$$= \sum_{p\leqslant P}\left\{\log\left(1-\frac{\alpha(p)\chi(p)}{p^s}\right)-\log\left(\frac{1-p^{-i\tau}\chi(p)}{p^s}\right)\right\}+O\left(\frac{1}{P^\delta}\right)$$
$$= O\left\{\sum_{\substack{p\leqslant P\\ p\neq 5}}\frac{|\alpha(p)-p^{-i\tau}|}{p^\sigma}\right\}+O\left(\frac{1}{P^\delta}\right).$$

Let $\alpha(p) = e^{2\pi i\beta(p)}$. By Kronecker's theorem, given q, there is a number τ and integers x_p such that

$$\left|\tau\frac{\log p}{2\pi}+\beta(p)-x_p\right| \leqslant \frac{1}{q} \quad (p \leqslant P).$$

Then $$|\alpha(p)-p^{-i\tau}| = |e^{2\pi i\{\beta(p)+(\tau\log p)/2\pi\}}-1| \leqslant e^{2\pi/q}-1.$$

Hence $$\log L(s+i\tau,\chi)-\log M(s,\chi) = O\left(\frac{\log P}{q}\right)+O\left(\frac{1}{P^\delta}\right),$$

and we can make this as small as we please by choosing first P and then q. Using this with χ_1 and χ_2, it follows that, given $\epsilon > 0$ and $\delta > 0$, there is a τ such that

$$|f(s+i\tau)-N(s)| < \epsilon \quad (\sigma \geqslant 1+\delta).$$

Let $s_1 > 1$ be a zero of $N(s)$. For any $\eta > 0$ there exists an η_1 with $0 < \eta_1 < \eta$, $\eta_1 < s_1-1$, such that $N(s) \neq 0$ for $|s-s_1| = \eta_1$. Let

$$\epsilon = \min_{|s-s_1|=\eta_1} |N(s)|$$

and $\delta < s_1-\eta_1-1$. Then, by Rouché's theorem, $N(s)$ and

$$N(s)-\{N(s)-f(s+i\tau)\}$$

have the same number of zeros inside $|s-s_1| = \eta_1$, and so at least one. Hence $f(s)$ has at least one zero inside the circle $|s-s_1-i\tau| = \eta_1$.

A slight extension of the argument shows that the number of zeros of $f(s)$ in $\sigma > 1$, $0 < t \leqslant T$, exceeds AT as $T \to \infty$. For by the extension of Dirichlet's theorem (§ 8.2) the interval $(t_0, mq^P t_0)$ contains at least m values of t, differing by at least t_0, such that

$$\left|t\frac{\log p}{2\pi}-x_p'\right| \leqslant \frac{1}{q} \qquad (p \leqslant P).$$

The above argument then shows the existence of a zero in the neighbourhood of each point $s_1+i(\tau+t)$.

The method is due to Davenport and Heilbronn (1), (2); they proved that a class of functions, of which an example is

$$\sum_{m,n \neq 0,0} \sum \frac{1}{(m^2+5n^2)^s},$$

has an infinity of zeros for $\sigma > 1$. It has been shown by calculation† that this particular function has a zero in the critical strip, not on the critical line. The method throws no light on the general question of the occurrence of zeros of such functions in the critical strip, but not on the critical line.

NOTES FOR CHAPTER 10

10.26. In § 10.1 Titchmarsh's comment on Riemann's statement about the approximate formula for $N(T)$ is erroneous. It is clear that Riemann meant that the relative error $\{N(T)-L(T)\}/N(T)$ is $O(T^{-1})$.

10.27. Further work has been done on the problem mentioned at the end of § 10.25. Davenport and Heilbronn (1), (2) showed in general that if Q is any positive definite integral quadratic form of discriminant d, such that the class number $h(d)$ is greater than 1, then the Epstein Zeta-function

$$\zeta_Q(s) = \sum_{\substack{x,y=-\infty \\ (x,y)\neq(0,0)}}^{\infty} Q(x,y)^{-s} \qquad (\sigma > 1)$$

has zeros to the right of $\sigma = 1$. In fact they showed that the number of such zeros up to height T is at least of order T (and hence of exact order T). This result has been extended to the critical strip by Voronin [**3**], who proved that, for such functions $\zeta_Q(s)$, the number of zeros up to height T, for $\frac{1}{2} < \sigma_1 \leqslant \mathbf{I}(s) \leqslant \sigma_2 < 1$, is also of order at least T (and hence of exact order T). This answers the question raised by Titchmarsh at the end of § 10.25.

10.28. Much the most significant result on $N_0(T)$ is due to Levinson [**2**], who showed that

$$N_0(T) \geqslant \alpha N(T) \tag{10.28.1}$$

for large enough T, with $\alpha = 0{\cdot}342$. The underlying idea is to relate the distribution of zeros of $\zeta(s)$ to that of the zeros of $\zeta'(s)$. To put matters in

† Potter and Titchmarsh (1).

their proper perspective we first note that Berndt [1] has shown that

$$\#\{s=\sigma+it\colon 0<t\leqslant T, \zeta'(s)=0\} = \frac{T}{2\pi}\left(\log\frac{T}{4\pi}-1\right)+O(\log T),$$

and that Speiser (1) has proved that the Riemann Hypothesis is equivalent to the non-vanishing of $\zeta'(s)$ for $0<\sigma<\frac{1}{2}$. This latter result is related to the unconditional estimate

$$\begin{aligned}\#\{s=\sigma+it\colon -1<\sigma<\tfrac{1}{2}, T_1<t\leqslant T_2, \zeta'(s)=0\}\\ = \#\{s=\sigma+it\colon 0<\sigma<\tfrac{1}{2}, T_1<t\leqslant T_2, \zeta(s)=0\}\\ +O(\log T_2),\end{aligned} \qquad (10.28.2)$$

zeros being counted according to multiplicity. This is due to Levinson and Montgomery [1], who also gave a number of other interesting results on the distribution of the zeros of $\zeta'(s)$.

We sketch the proof of (10.28.2). We shall make frequent reference to the logarithmic derivative of the functional equation (2.6.4), which we write in the form

$$\begin{aligned}\frac{\zeta'(s)}{\zeta(s)}+\frac{\zeta'(1-s)}{\zeta(1-s)} &= \log\pi-\tfrac{1}{2}\left(\frac{\Gamma'(\frac{1}{2}s)}{\Gamma(\frac{1}{2}s)}+\frac{\Gamma'(\frac{1}{2}-\frac{1}{2}s)}{\Gamma(\frac{1}{2}-\frac{1}{2}s)}\right)\\ &= -F(s),\end{aligned} \qquad (10.28.3)$$

say. We note that $F(\frac{1}{2}+it)$ is always real, and that

$$F(s)=\log(t/2\pi)+O(1/t) \qquad (10.28.4)$$

uniformly for $t\geqslant 1$ and $|\sigma|\leqslant 2$. To prove (10.28.2) it suffices to consider the case in which the numbers T_j are chosen so that $\zeta(s)$ and $\zeta'(s)$ do not vanish for $t=T_j$, $-1\leqslant\sigma\leqslant\frac{1}{2}$. We examine the change in argument in $\zeta'(s)/\zeta(s)$ around the rectangle with vertices $\frac{1}{2}-\delta+iT_1$, $\frac{1}{2}-\delta+iT_2$, $-1+iT_2$, and $-1+iT_1$, where δ is a small positive number. Along the horizontal sides we apply the ideas of §9.4 to $\zeta(s)$ and $\zeta'(s)$ separately. We note that $\zeta(s)$ and $\zeta'(s)$ are each $O(t^A)$ for $-3\leqslant\sigma\leqslant 1$. Moreover we also have $|\zeta(-1+iT_j)|\gg T_j^{\frac{3}{2}}$, by the functional equation, and hence also

$$|\zeta'(-1+iT_j)|\gg T_j^{\frac{3}{2}}\left|\frac{\zeta'(-1+iT_j)}{\zeta(-1+iT_j)}\right|\gg T_j^{\frac{3}{2}}\log T_j,$$

by (10.28.3) and (10.28.4). The method of §9.4 therefore shows that $\arg\zeta(s)$ and $\arg\zeta'(s)$ both vary by $O(\log T_2)$ on the horizontal sides of the rectangle. On the vertical side $\sigma = -1$ we have

$$\frac{\zeta'(s)}{\zeta(s)} = \log\left(\frac{t}{2\pi}\right) + O(1),$$

by (10.28.3) and (10.28.4), so that the contribution to the total change in argument is $O(1)$. For the vertical side $\sigma = \frac{1}{2} - \delta$ we first observe from (10.28.3) and (10.28.4) that

$$\mathbf{R}\left(-\frac{\zeta'(\frac{1}{2}+it)}{\zeta(\frac{1}{2}+it)}\right) \geqslant 1 \tag{10.28.5}$$

if $t \geqslant T_1$ with T_1 sufficiently large. It follows that

$$\mathbf{R}\left(-\frac{\zeta'(\frac{1}{2}-\delta+it)}{\zeta(\frac{1}{2}-\delta+it)}\right) \geqslant \tfrac{1}{2} \tag{10.28.6}$$

for $T_1 \leqslant t \leqslant T_2$, if $\delta = \delta(T_2)$ is small enough. To see this, it suffices to examine a neighbourhood of a zero $\rho = \frac{1}{2} + i\gamma$ of $\zeta(s)$. Then

$$-\frac{\zeta'(s)}{\zeta(s)} = -\frac{m}{s-\rho} + m' + O(|s-\rho|),$$

where $m \geqslant 1$ is the multiplicity of ρ. The choice $s = \frac{1}{2} + it$ with $t \to \gamma$ therefore yields $\mathbf{R}(m') \geqslant 1$, by (10.28.5). Hence, on taking $s = \frac{1}{2} - \delta + it$, we find that

$$\mathbf{R}\left(-\frac{\zeta'(s)}{\zeta(s)}\right) = \frac{m\delta}{|s-\rho|^2} + \mathbf{R}(m') + O(|s-\rho|) \geqslant \tfrac{1}{2}$$

for $|s-\rho|$ small enough. The inequality (10.28.6) now follows. We therefore see that arg $\zeta'(s)/\zeta(s)$ varies by $O(1)$ on the vertical side $\mathbf{R}(s) = \frac{1}{2} - \delta$ of our rectangle, which completes the proof of (10.28.2).

If we write N for the quantity on the left of (10.28.2) it follows that

$$N_0(T_2) - N_0(T_1) = \{N(T_2) - N(T_1)\} - 2N + O(\log T_2), \tag{10.28.7}$$

so that we now require an upper bound for N. This is achieved by applying the 'mollifier method' of §§ 9.20–24 to $\zeta'(1-s)$. Let $\nu(\sigma, T_1, T_2)$ denote the number of zeros of $\zeta'(1-s)$ in the rectangle $\sigma \leqslant \mathbf{R}(s) \leqslant 2$, $T_1 < \mathbf{I}(s) < T_2$. The method produces an upper bound for

$$\int_u^2 \nu(\sigma, T_1, T_2)\, d\sigma, \tag{10.28.8}$$

which in turn yields an estimate $N \leqslant c\{N(T_2) - N(T_1)\}$ for large T_2. The constant c in this latter bound has to be calculated explicitly, and must be less than $\frac{1}{2}$ for (10.28.7) to be of use. This is in contrast to (9.20.5), in which the implied constant was not calculated explicitly, and would have been relatively large. It is difficult to have much feel in advance for how large the constant c produced by the method will be. The following very loose argument gives one some hope that c will turn out to be reasonably small, and so it transpires in practice.

In using (10.28.8) to obtain a bound for N we shall take

$$u = \tfrac{1}{2} - a/\log T_2,$$

where a is a positive constant to be chosen later. The zeros $\rho' = \beta' + i\gamma'$ of $\zeta'(1-s)$ have an asymmetrical distribution about the critical line. Indeed Levinson and Montgomery [**1**] showed that

$$\sum_{0<\gamma'\leqslant T} (\tfrac{1}{2} - \beta') \sim \frac{T}{2\pi} \operatorname{loglog} T,$$

whence β' is $\frac{1}{2} - (\operatorname{loglog} \gamma')/\log \gamma'$ on average. Thus one might reasonably hope that a fair proportion of such zeros have $\beta' < u$, thereby making the integral (10.28.8) rather small.

We now look in more detail at the method. In the first place, it is convenient to replace $\zeta'(1-s)$ by

$$\zeta(s) + \frac{\zeta'(s)}{F(s)} = G(s),$$

say. If we write $h(s) = \pi^{-\frac{1}{2}s}\,\Gamma(\frac{1}{2}s)$ then (10.28.3), together with the functional equation (2.6.4), yields

$$\zeta'(1-s) = -\frac{F(s)\,h(s)\,G(s)}{h(1-s)},$$

so that $G(s)$ and $\zeta'(1-s)$ have the same zeros for t large enough. Now let

$$\psi(s) = \sum_{n<y} b_n n^{-s} \tag{10.28.9}$$

be a suitable 'mollifier' for $G(s)$, and apply Littlewood's formula (9.9.1) to the function $G(s)\psi(s)$ and the rectangle with vertices $u+iT_1$, $2+iT_2$, $2+iT_1$, $u+iT_2$. Then, as in §9.16, we find that

$$N \leqslant \frac{\log T_2}{a} \int_u^2 \nu(\sigma, T_1, T_2)\,d\sigma$$

$$\leqslant \frac{\log T_2}{2\pi a} \int_{T_1}^{T_2} \log|G(u+it)\,\psi(u+it)|\,dt + O(\log T_2).$$

Moreover, as in §9.16 we have

$$\int_{T_1}^{T_2} \log|G(u+it)\,\psi(u+it)|\,dt$$

$$\leqslant \tfrac{1}{2}(T_2-T_1)\log\left(\frac{1}{T_2-T_1}\int_{T_1}^{T_2}|G(u+it)\,\psi(u+it)|^2\,dt\right).$$

Hence, if we can show that

$$\int_{T_1}^{T_2}|G(u+it)\,\psi(u+it)|^2\,dt \sim c(a)\,(T_2-T_1) \tag{10.28.10}$$

for suitable T_1, T_2, we will have

$$N \leqslant \left(\frac{\log c(a)}{2a}+o(1)\right)\{N(T_2)-N(T_1)\}, \tag{10.28.11}$$

whence

$$N_0(T_2)-N_0(T_1) \geqslant \left(1-\frac{\log c(a)}{a}+o(1)\right)\{N(T_2)-N(T_1)\}$$

by (10.28.7).

The computation of the mean value (10.28.10) is the most awkward part of Levinson's argument. In [**2**] he takes $y = T_2^{\frac{1}{2}-\varepsilon}$ and

$$b_n = \mu(n)\,n^{u-\frac{1}{2}}\frac{\log y/n}{\log y}.$$

This leads eventually to (10.28.10) with

$$c(a) = e^{2a}\left(\frac{1}{2a^3}+\frac{1}{24a}\right)-\frac{1}{2a^3}-\frac{1}{a^2}-\frac{25}{24a}+\frac{7}{12}-\frac{a}{12}.$$

The optimal choice of a is roughly $a = 1{\cdot}3$, which produces (10.28.1) with $= 0{\cdot}342$.

The method has been improved slightly by Levinson [**4**], [**5**], Lou [**1**] and Conrey [**1**] and the best constant thus far is $\alpha = 0{\cdot}3658$ (Conrey [**1**]). The principal restriction on the method is that on the size of y in (10.28.9). The above authors all take $y = T_2^{\frac{1}{2}-\varepsilon}$, but there is some scope for improvement via the ideas used in the mean-value theorems (7.24.5), (7.24.6), and (7.24.7).

10.29. An examination of the argument just given reveals that the right hand side of (10.28.11) gives an upper bound for $N+N^*$, where

$$N^* = \#\{s = \tfrac{1}{2}+it\colon T_1 < t \leqslant T_2, \zeta'(s) = 0\},$$

(zeros being counted according to multiplicities). However it is clear from (10.28.3) and (10.28.4) that $\zeta'(\frac{1}{2}+it)$ can only vanish if $\zeta(\frac{1}{2}+it)$ does. Consequently, if we write $N^{(r)}$ for the number of zeros of $\zeta(s)$ of multiplicity r, on the line segment $s = \frac{1}{2}+it$, $T_1 < t \leqslant T_2$, we will have

$$N^* = \sum_{r=2}^{\infty} (r-1)\, N^{(r)}.$$

Thus (10.28.7) may be replaced by

$$N^{(1)} - \sum_{r=3}^{\infty} (r-2)\, N^{(r)} = \{N(T_2) - N(T_1)\} - 2(N+N^*) + O(\log T_2).$$

If we now define $N^{(r)}(T)$ in analogy to $N^{(r)}$, but counting zeros $\frac{1}{2}+it$ with $0 < t \leqslant T$, we may deduce that

$$N^{(1)}(T) - \sum_{r=3}^{\infty} (r-2)\, N^{(r)}(T) \geqslant \alpha N(T), \tag{10.29.1}$$

for large enough T, and $\alpha = 0{\cdot}342$. In particular at least a third of the non-trivial zeros of $\zeta(s)$ not only lie on the critical line, but are simple. This observation is due independently to Heath-Brown [**5**] and Selberg (unpublished). The improved constants α mentioned above do not all allow this refinement. However it has been shown by Anderson [**1**] that (10.29.1) holds with $\alpha = 0{\cdot}3532$.

10.30. Levinson's method can be applied equally to the derivatives $\xi^{(m)}(s)$ of the function $\xi(s)$ given by (2.1.12). One can show that the zeros of these functions lie in the critical strip, and that the number of them, $N_m(T)$ say, for $0 < t \leqslant T$, is $N(T)+O_m(\log T)$. If the Riemann hypothesis holds then all these zeros must lie on the critical line. Thus it is of some interest to give unconditional estimates for

$$\liminf_{T\to\infty} N_m(T)^{-1}\, \#\{t\colon 0 < t \leqslant T, \xi^{(m)}(\tfrac{1}{2}+it) = 0\} = \alpha_m,$$

say. Levinson [**3**], [**5**] showed that $\alpha_1 \geqslant 0{\cdot}71$, and Conrey [**1**] improved and extended the method to give $\alpha_1 \geqslant 0{\cdot}8137$, $\alpha_2 \geqslant 0{\cdot}9584$ and in general $\alpha_m = 1+O(m^{-2})$.

XI

THE GENERAL DISTRIBUTION OF THE VALUES OF $\zeta(s)$

11.1. In the previous chapters we have been concerned almost entirely with the modulus of $\zeta(s)$, and the various values, particularly zero, which it takes. We now consider the problem of $\zeta(s)$ itself, and the values of s for which it takes any given value a.†

One method of dealing with this problem is to connect it with the famous theorem of Picard on functions which do not take certain values. We use the following theorem:‡

If $f(s)$ is regular and never 0 or 1 in $|s-s_0| \leqslant r$, and $|f(s_0)| \leqslant \alpha$, then $|f(s)| \leqslant A(\alpha, \theta)$ for $|s-s_0| \leqslant \theta r$, where $0 < \theta < 1$.

From this we deduce

Theorem 11.1. *$\zeta(s)$ takes every value, with one possible exception, an infinity of times in any strip $1-\delta < \sigma \leqslant 1+\delta$.*

Suppose, on the contrary, that $\zeta(s)$ takes the distinct values a and b only a finite number of times in the strip, and so *never* above $t = t_0$, say. Let $T > t_0+1$, and consider the function $f(s) = \{\zeta(s)-a\}/(b-a)$ in the circles C, C', of radii $\frac{1}{2}\delta$ and $\frac{1}{4}\delta$ $(0 < \delta < 1)$, and common centre $s_0 = 1+\frac{1}{4}\delta+iT$. Then

$$|f(s_0)| \leqslant \alpha = \{\zeta(1+\tfrac{1}{4}\delta)+|a|\}/|b-a|,$$

and $f(s)$ is never 0 or 1 in C. Hence

$$|f(s)| < A(\alpha)$$

in C', and so $|\zeta(\sigma+iT)| < A(a,b,\alpha)$ for $1 \leqslant \sigma \leqslant 1+\frac{1}{2}\delta$, $T > t_0+1$. Hence $\zeta(s)$ is bounded for $\sigma > 1$, which is false, by Theorem 8.4 (A). This proves the theorem.

We should, of course, expect the exceptional value to be 0.

If we assume the Riemann hypothesis, we can use a similar method inside the critical strip; but more detailed results independent of the Riemann hypothesis can be obtained by the method of Diophantine approximation. We devote the rest of the chapter to developments of this method.

† See Bohr (1)–(14), Bohr and Courant (1), Bohr and Jessen (1), (2), (5), Bohr and Landau (3), Borchsenius and Jessen (1), Jessen (1), van Kampen (1), van Kampen and Wintner (1), Kershner (1), Kershner and Wintner (1), (2), Wintner (1)–(4).

‡ See Landau's *Ergebnisse der Funktionentheorie*, § 24, or Valiron's *Integral Functions*, Ch. VI, § 3.

11.2. We restrict ourselves in the first place to the half-plane $\sigma > 1$; and we consider, not $\zeta(s)$ itself, but $\log \zeta(s)$, viz. the function defined for $\sigma > 1$ by the series

$$\log \zeta(s) = -\sum_p (p^{-s}+\tfrac{1}{2}p^{-2s}+...).$$

We consider at the same time the function

$$\frac{\zeta'(s)}{\zeta(s)} = \sum_p \log p(p^{-s}+p^{-2s}+...).$$

We observe that both functions are represented by Dirichlet series, absolutely convergent for $\sigma > 1$, and capable of being written in the form

$$F(s) = f_1(p_1^{-s})+f_2(p_2^{-s})+...,$$

where $f_n(z)$ is a power-series in z whose coefficients do not depend on s. In fact

$$f_n(z) = -\log(1-z), \qquad f_n(z) = z\log p_n/(1-z)$$

in the above two cases. In what follows $F(s)$ denotes either of the two functions.

11.3. We consider first the values which $F(s)$ takes on the line $\sigma = \sigma_0$, where σ_0 is an arbitrary number greater than 1. On this line

$$F(s) = \sum_{n=1}^{\infty} f_n(p_n^{-\sigma_0} e^{-it\log p_n}),$$

and, as t varies, the arguments $-t\log p_n$ are, of course, all related. But we shall see that there is an intimate connexion between the set U of values assumed by $F(s)$ on $\sigma = \sigma_0$ and the set V of values assumed by the function

$$\Phi(\sigma_0, \theta_1, \theta_2,...) = \sum_{n=1}^{\infty} f_n(p_n^{-\sigma_0} e^{2\pi i\theta_n})$$

of an infinite number of independent real variables θ_1, θ_2,... .

We shall in fact show that *the set U, which is obviously contained in V, is everywhere dense in V, i.e. that corresponding to every value v in V (i.e. to every given set of values θ_1, θ_2,...) and every positive ϵ, there exists a t such that*

$$|F(\sigma_0+it)-v| < \epsilon.$$

Since the Dirichlet series from which we start is absolutely convergent for $\sigma = \sigma_0$, it is obvious that we can find $N = N(\sigma_0, \epsilon)$ such that

$$\left|\sum_{n=N+1}^{\infty} f_n(p_n^{-\sigma_0} e^{2\pi i\mu_n})\right| < \tfrac{1}{3}\epsilon \tag{11.3.1}$$

for any values of the μ_n, and in particular for $\mu_n = \theta_n$, or for

$$\mu_n = -(t\log p_n)/2\pi.$$

Now since the numbers $\log p_n$ are linearly independent, we can, by Kronecker's theorem, find a number t and integers $g_1, g_2,..., g_N$ such that

$$|-t\log p_n - 2\pi\theta_n - 2\pi g_n| < \eta \quad (n = 1, 2,..., N),$$

η being an assigned positive number. Since $f_n(p_n^{-\sigma_0}e^{2\pi i\theta})$ is, for each n, a continuous function of θ, we can suppose η so small that

$$\left|\sum_{n=1}^{N} \{f_n(p_n^{-\sigma_0}e^{2\pi i\theta_n}) - f_n(p_n^{-\sigma_0}e^{-it\log p_n})\}\right| < \tfrac{1}{3}\epsilon. \tag{11.3.2}$$

The result now follows from (11.3.1) and (11.3.2).

11.4. We next consider the set W of values which $F(s)$ takes 'in the immediate neighbourhood' of the line $\sigma = \sigma_0$, i.e. the set of all values of w such that the equation $F(s) = w$ has, for every positive δ, a root in the strip $|\sigma-\sigma_0| < \delta$.

In the first place, it is evident that U is contained in W. Further, it is easy to see that U is everywhere dense in W. For, for sufficiently small δ (e.g. for $\delta < \frac{1}{2}(\sigma_0-1)$),

$$|F'(s)| < K(\sigma_0)$$

for all values of s in the strip $|\sigma-\sigma_0| < \delta$, so that

$$|F(\sigma_0+it) - F(\sigma_1+it)| < K(\sigma_0)|\sigma_1-\sigma_0| \quad (|\sigma_1-\sigma_0| < \delta). \tag{11.4.1}$$

Now each value w in W is assumed by $F(s)$ either on the line $\sigma = \sigma_0$, in which case it is a u, or at points σ_1+it arbitrarily near the line, in which case, in virtue of (11.4.1), we can find a u such that

$$|w-u| < K(\sigma_0)|\sigma_1-\sigma_0| < \epsilon.$$

We now proceed to prove that W *is identical with* V. Since U is contained in and is everywhere dense in both V and W, it follows that each of V and W is everywhere dense in the other. It is therefore obvious that W is contained in V, if V is closed.

We shall see presently that much more than this is true, viz. that V consists of all points of an area, including the boundary. The following direct proof that V is closed is, however, very instructive.

Let v^* be a limit-point of V, and let v_ν ($\nu = 1, 2,...$) be a sequence of v's tending to v^*. To each v_ν corresponds a point $P_\nu(\theta_{1,\nu}, \theta_{2,\nu},...)$ in the space of an infinite number of dimensions defined by $0 \leqslant \theta_{n,\nu} < 1$ ($n = 1, 2,...$), such that $\Phi(\sigma_0, \theta_{1,\nu},...) = v_\nu$.

Now since (P_ν) is a bounded set of points (i.e. all the coordinates are bounded), it has a limit-point P^* $(\theta_1^*, \theta_2^*,...)$, i.e. a point such that from (P_ν) we can choose a sequence (P_{ν_r}) such that each coordinate θ_{n,ν_r} of P_{ν_r} tends to the limit θ_n^* as $r \to \infty$.

It is now easy to prove that P^* corresponds to v^*, i.e. that

$$\Phi(\sigma_0, \theta_1^*,...) = v^*,$$

so that v^* is a point of V. For the series for v_{ν_r}, viz.

$$\sum_{n=1}^{\infty} f_n(p_n^{-\sigma_0} e^{2\pi i\theta_{n,\nu_r}}),$$

is uniformly convergent with respect to r, since (by Weierstrass's M-test) it is uniformly convergent with respect to all the θ's; further, the nth term tends to $f_n(p_n^{-\sigma_0} e^{2\pi i\theta_n^*})$ as $r \to \infty$. Hence

$$v^* = \lim_{r\to\infty} v_{\nu_r} = \lim_{r\to\infty} \sum_{n=1}^{\infty} f_n(p_n^{-\sigma_0} e^{2\pi i\theta_{n,\nu_r}}) = \Phi(\sigma_0, \theta_1^*,...),$$

which proves our result.

To establish the identity of V and W it remains to prove that V is contained in W. It is obviously sufficient (and also necessary) for this that W should be closed. But that W is closed does not follow, as might perhaps be supposed, from the mere fact that W is the set of values taken by a bounded analytic function in the immediate neighbourhood of a line. Thus e^{-z^2} is bounded and arbitrarily near to 0 in every strip including the real axis, but never actually assumes the value 0. The fact that W is closed (which we shall not prove directly) depends on the special nature of the function $F(s)$.

Let $v = \Phi(\sigma_0, \theta_1, \theta_2,...)$ be an arbitrary value contained in V. We have to show that v is a member of W, i.e. that, in every strip

$$|\sigma - \sigma_0| < \delta,$$

$F(s)$ assumes the value v.

Let
$$G(s) = \sum_{n=1}^{\infty} f_n(p_n^{-s} e^{2\pi i\theta_n}),$$

so that $G(\sigma_0) = v$. We choose a small circle C with centre σ_0 and radius less than δ such that $G(s) \neq v$ on the circumference. Let m be the minimum of $|G(s) - v|$ on C.

Kronecker's theorem enables us to choose t_0 such that, for every s in C,

$$|F(s+it_0) - G(s)| < m.$$

The proof is almost exactly the same as that used to show that U is everywhere dense in V. The series for $F(s)$ and $G(s)$ are uniformly convergent in the strip, and, for each fixed N, $\sum_1^N f_n(p_n^{-\sigma} e^{2\pi i\mu_n})$ is a continuous function of $\sigma, \mu_1,..., \mu_N$. It is therefore sufficient to show that we can choose t_0 so that the difference between the arguments of p_n^{-s} at $s = \sigma_0 + it_0$ and $p_n^{-s} e^{2\pi i\theta_n}$ at $s = \sigma_0$, and consequently that between the respective arguments at every pair of corresponding points of the two circles is (mod 2π) arbitrarily small for $n = 1, 2,..., N$. The possibility of this choice follows at once from Kronecker's theorem.

We now have

$$F(s+it_0)-v = \{G(s)-v\}+\{F(s+it_0)-G(s)\},$$

and on the circumference of C

$$|F(s+it_0)-G(s)| < m \leqslant |G(s)-v|.$$

Hence, by Rouché's theorem, $F(s+it_0)-v$ has in C the same number of zeros as $G(s)-v$, and so at least one. This proves the theorem.

11.5. We now proceed to the study of the set V. Let V_n be the set of values taken by $f_n(p_n^{-s})$ for $\sigma = \sigma_0$, i.e. the set taken by $f_n(z)$ for $|z| = p_n^{-\sigma_0}$. Then V is the 'sum' of the sets of points $V_1, V_2,...$, i.e. it is the set of all values $v_1+v_2+...$, where v_1 is any point of V_1, v_2 any point of V_2, and so on. For the function $\log\zeta(s)$, V_n consists of the points of the curve described by $-\log(1-z)$ as z describes the circle $|z| = p_n^{-\sigma_0}$; for $\zeta'(s)/\zeta(s)$ it consists of the points of the curve described by

$$-(z\log p_n)/(1-z).$$

We begin by considering the function $\zeta'(s)/\zeta(s)$. In this case we can find the set V explicitly. Let

$$w_n = -\frac{z_n\log p_n}{1-z_n}.$$

As z_n describes the circle $|z_n| = p_n^{-\sigma_0}$, w_n describes the circle with centre

$$c_n = -\frac{p_n^{-2\sigma_0}\log p_n}{1-p_n^{-2\sigma_0}}$$

and radius
$$\rho_n = \frac{p_n^{-\sigma_0}\log p_n}{1-p_n^{-2\sigma_0}}.$$

Let
$$w_n = c_n+w_n' = c_n+\rho_n e^{i\phi_n},$$

and let
$$c = \sum_{n=1}^{\infty} c_n = \frac{\zeta'(2\sigma_0)}{\zeta(2\sigma_0)}.$$

Then V is the set of all the values of

$$c+\sum_{n=1}^{\infty}\rho_n e^{i\phi_n}$$

for independent $\phi_1, \phi_2,...$. The set V' of the values of $\sum\rho_n e^{i\phi_n}$ is the 'sum' of an infinite number of circles with centre at the origin, whose radii $\rho_1, \rho_2,...$ form, as it is easy to see, a decreasing sequence. Let V_n' denote the nth circle.

Then $V_1'+V_2'$ is the area swept out by the circle of radius ρ_2 as its centre describes the circle with centre the origin and radius ρ_1. Hence, since $\rho_2 < \rho_1$, $V_1'+V_2'$ is the annulus with radii $\rho_1-\rho_2$ and $\rho_1+\rho_2$.

The argument clearly extends to any finite number of terms. Thus $V_1'+...+V_N'$ consists of all points of the annulus

$$\rho_1-\sum_{n=2}^{N}\rho_n \leqslant |w| \leqslant \sum_{n=1}^{N}\rho_n,$$

or, if the left-hand side is negative, of the circle

$$|w| \leqslant \sum_{n=1}^{N}\rho_n.$$

It is now easy to see that

(i) *if* $\rho_1 > \rho_2+\rho_3+...$, *the set* V' *consists of all points* w *of the annulus*

$$\rho_1-\sum_{n=2}^{\infty}\rho_n \leqslant |w| \leqslant \sum_{n=1}^{\infty}\rho_n;$$

(ii) *if* $\rho_1 \leqslant \rho_2+\rho_3+...$, V' *consists of all points* w *for which*

$$|w| \leqslant \sum_{n=1}^{\infty}\rho_n.$$

For example, in case (ii), let w_0 be an interior point of the circle. Then we can choose N so large that

$$\sum_{N+1}^{\infty}\rho_n < \sum_{n=1}^{N}\rho_n-|w_0|.$$

Hence
$$w_1 = w_0-\sum_{N+1}^{\infty}\rho_n e^{i\phi_n}$$

lies within the circle $V_1'+...+V_N'$ for any values of the ϕ_n, e.g. for $\phi_{N+1} = ... = 0$. Hence

$$w_1 = \sum_{n=1}^{N}\rho_n e^{i\phi_n}$$

for some values of $\phi_1,..., \phi_n$, and so

$$w_0 = \sum_{n=1}^{\infty}\rho_n e^{i\phi_n}$$

as required. That V' also includes the boundary in each case is clear on taking all the ϕ_n equal.

The complete result is that there is an absolute constant $D = 2{\cdot}57...$, determined as the root of the equation

$$\frac{2^{-D}\log 2}{1-2^{-2D}} = \sum_{n=2}^{\infty}\frac{p_n^{-D}\log p_n}{1-p_n^{-2D}},$$

such that for $\sigma_0 > D$ we are in case (i), and for $1 < \sigma_0 \leqslant D$ we are in case (ii). The radius of the outer boundary of V' is

$$R = \frac{\zeta'(2\sigma_0)}{\zeta(2\sigma_0)} - \frac{\zeta'(\sigma_0)}{\zeta(\sigma_0)}$$

in each case; the radius of the inner boundary in case (i) is

$$r = 2\rho_1 - R = 2^{1-\sigma_0}\log 2/(1-2^{-2\sigma_0}) - R.$$

Summing up, we have the following results for $\zeta'(s)/\zeta(s)$.

THEOREM 11.5 (A). *The values which $\zeta'(s)/\zeta(s)$ takes on the line $\sigma = \sigma_0 > 1$ form a set everywhere dense in a region $R(\sigma_0)$. If $\sigma_0 > D$, $R(\sigma_0)$ is the annulus (boundary included) with centre c and radii r and R; if $\sigma_0 \leqslant D$, $R(\sigma_0)$ is the circular area (boundary included) with centre c and radius R; c, r, and R are continuous functions of σ_0 defined by*

$$c = \zeta'(2\sigma_0)/\zeta(2\sigma_0), \quad R = c - \zeta'(\sigma_0)/\zeta(\sigma_0), \quad r = 2^{1-\sigma_0}\log 2/(1-2^{-2\sigma_0}) - R.$$

Further, as $\sigma_0 \to \infty$,

$$\lim c = \lim r = \lim R = 0, \qquad \lim c/R = \lim (R-r)/R = 0;$$

as $\sigma_0 \to D$, $\lim r = 0$; and as $\sigma_0 \to 1$, $\lim R = \infty$, $\lim c = \zeta'(2)/\zeta(2)$.

THEOREM 11.5 (B). *The set of values which $\zeta'(s)/\zeta(s)$ takes in the immediate neighbourhood of $\sigma = \sigma_0$ is identical with $R(\sigma_0)$. In particular, since c tends to a finite limit and R to infinity as $\sigma_0 \to 1$, $\zeta'(s)/\zeta(s)$ takes all values infinitely often in the strip $1 < \sigma < 1+\delta$, for an arbitrary positive δ.*

The above results evidently enable us to study the set of points at which $\zeta'(s)/\zeta(s)$ takes the assigned value a. We confine ourselves to giving the result for $a = 0$; this is the most interesting case, since the zeros of $\zeta'(s)/\zeta(s)$ are identical with those of $\zeta'(s)$.

THEOREM 11.5 (C). *There is an absolute constant E, between 2 and 3, such that $\zeta'(s) \neq 0$ for $\sigma > E$, while $\zeta'(s)$ has an infinity of zeros in every strip between $\sigma = 1$ and $\sigma = E$.*

In fact it is easily verified that the annulus $R(\sigma_0)$ includes the origin if $\sigma_0 = 2$, but not if $\sigma_0 = 3$.

11.6. We proceed now to the study of $\log \zeta(s)$. In this case the set V consists of the 'sum' of the curves V_n described by the points

$$w_n = -\log(1-z_n)$$

as z_n describes the circle $|z_n| = p_n^{-\sigma_0}$.

In the first place, V_n is a convex curve. For if

$$u+iv = w = f(z) = f(x+iy),$$

and z describes the circle $|z| = r$, then

$$\frac{du}{dx}+i\frac{dv}{dx} = f'(z)\left(1+i\frac{dy}{dx}\right) = f'(z)\frac{x+iy}{iy}.$$

Hence
$$\arctan\frac{dv}{du} = \arg\{zf'(z)\}-\tfrac{1}{2}\pi.$$

A sufficient condition that w should describe a convex curve as z describes $|z| = r$ is that the tangent to the path of w should rotate steadily through 2π as z describes the circle, i.e. that $\arg\{zf'(z)\}$ should increase steadily through 2π. This condition is satisfied in the case $f(z) = -\log(1-z)$; for $zf'(z) = z/(1-z)$ describes a circle enclosing the origin as z describes $|z| = r < 1$.

If $z = re^{i\theta}$, and $w = -\log(1-z)$, then

$$u = -\tfrac{1}{2}\log(1-2r\cos\theta+r^2), \qquad v = \arctan\frac{r\sin\theta}{1-r\cos\theta}.$$

The second equation leads to

$$r\cos\theta = \sin^2 v \pm \cos v(r^2-\sin^2 v)^{\frac{1}{2}}.$$

Hence, for real r and θ, $|v| \leqslant \arcsin r$. If $\cos\theta_1$ and $\cos\theta_2$ are the two values of $\cos\theta$ corresponding to a given v,

$$(1-2r\cos\theta_1+r^2)(1-2r\cos\theta_2+r^2) = (1-r^2)^2.$$

Hence if u_1 and u_2 are the corresponding values of u,

$$u_1+u_2 = -\log(1-r^2).$$

The curve V_n is therefore convex and symmetrical about the lines

$$u = -\tfrac{1}{2}\log(1-r^2) \quad \text{and} \quad v = 0.$$

Its diameters in the u and v directions are $\frac{1}{2}\log\{(1+r)/(1-r)\}$ and $\arcsin r$.

Let
$$c_n = -\tfrac{1}{2}\log(1-p_n^{-2\sigma_0})$$
and
$$w_n = c_n+w_n',$$
$$c = \sum_{n=1}^{\infty} c_n = \tfrac{1}{2}\log\zeta(2\sigma_0).$$

Then the points w_n' describe symmetrical convex figures with centre the origin. Let V' be the 'sum' of these figures.

It is now easy, by analogy with the previous case, to imagine the result. *The set V', which is plainly symmetrical about both axes, is either* (i) *the region bounded by two convex curves, one of which is entirely interior to the other, or* (ii) *the region bounded by a single convex curve. In each case the boundary is included as part of the region.*

This follows from a general theorem of Bohr on the 'summation' of a series of convex curves.

For our present purpose the following weaker but more obvious results will be sufficient. The set V' is included in the circle with centre the origin and radius

$$R = \sum_{n=1}^{\infty} \tfrac{1}{2}\log\frac{1+p_n^{-\sigma_0}}{1-p_n^{-\sigma_0}} = \tfrac{1}{2}\log\frac{\zeta^2(\sigma_0)}{\zeta(2\sigma_0)}.$$

If σ_0 is sufficiently large, V' lies entirely outside the circle of radius

$$\arcsin 2^{-\sigma_0} - \sum_{n=2}^{\infty} \tfrac{1}{2}\log\frac{1+p_n^{-\sigma_0}}{1-p_n^{-\sigma_0}} = \arcsin 2^{-\sigma_0} + \tfrac{1}{2}\log\frac{1+2^{-\sigma_0}}{1-2^{-\sigma_0}} - R.$$

If

$$\sum_{n=2}^{\infty} \arcsin p_n^{-\sigma_0} > \tfrac{1}{2}\log\frac{1+2^{-\sigma_0}}{1-2^{-\sigma_0}},$$

and so if σ_0 is sufficiently near to 1, V' includes all points inside the circle of radius

$$\sum_{n=1}^{\infty} \arcsin p_n^{-\sigma_0}.$$

In particular V' includes any given area, however large, if σ_0 is sufficiently near to 1.

We cannot, as in the case of circles, determine in all circumstances whether we are in case (i) or case (ii). It is not obvious, for example, whether there exists an absolute constant D' such that we are in case (i) or (ii) according as $\sigma_0 > D'$ or $1 < \sigma_0 \leqslant D'$. The discussion of this point demands a closer investigation of the geometry of the special curves with which we are dealing, and the question would appear to be one of considerable intricacy.

The relations between U, V, and W now give us the following analogues for $\log \zeta(s)$ of the results for $\zeta'(s)/\zeta(s)$.

THEOREM 11.6 (A). *On each line $\sigma = \sigma_0 > 1$ the values of* $\log\zeta(s)$ *are everywhere dense in a region $R(\sigma_0)$ which is either* (i) *the ring-shaped area bounded by two convex curves, or* (ii) *the area bounded by one convex curve. For sufficiently large values of σ_0 we are in case* (i), *and for values of σ_0 sufficiently near to* 1 *we are in case* (ii).

THEOREM 11.6 (B). *The set of values which* $\log\zeta(s)$ *takes in the immediate neighbourhood of $\sigma = \sigma_0$ is identical with $R(\sigma_0)$. In particular, since $R(\sigma_0)$ includes any given finite area when σ_0 is sufficiently near* 1, $\log\zeta(s)$ *takes every value an infinity of times in* $1 < \sigma < 1+\delta$.

As a consequence of the last result, we have

THEOREM 11.6 (C). *the function $\zeta(s)$ takes every value except* 0 *an infinity of times in the strip* $1 < \sigma < 1+\delta$.

This is a more precise form of Theorem 11.1.

11.7. We have seen above that $\log\zeta(s)$ takes any assigned value a an infinity of times in $\sigma > 1$. It is natural to raise the question *how often* the value a is taken, i.e. the question of the behaviour for large T of the number $M_a(T)$ of roots of $\log\zeta(s) = a$ in $\sigma > 1$, $0 < t < T$. This question is evidently closely related to the question as to how often, as $t \to \infty$, the point $(a_1 t, a_2 t,..., a_N t)$ of Kronecker's theorem, which, in virtue of the theorem, comes (mod 1) arbitrarily near every point in the N-dimensional unit cube, comes within a given distance of an assigned point $(b_1, b_2,..., b_N)$. The answer to this last question is given by the following theorem, which asserts that, roughly speaking, the point $(a_1 t,..., a_N t)$ comes near every point of the unit cube equally often, i.e. it does not give a preference to any particular region of the unit cube.

Let $a_1,..., a_N$ be linearly independent, and let γ be a region of the N-dimensional unit cube with volume Γ (in the Jordan sense). Let $I_\gamma(T)$ be the sum of the intervals between $t = 0$ and $t = T$ for which the point P $(a_1 t,..., a_N t)$ is (mod 1) *inside γ. Then*

$$\lim_{T\to\infty} I_\gamma(T)/T = \Gamma.$$

The region γ is said to have the volume Γ in the Jordan sense, if, given ϵ, we can find two sets of cubes with sides parallel to the axes, of volumes Γ_1 and Γ_2, included in and including γ respectively, such that

$$\Gamma_1 - \epsilon \leqslant \Gamma \leqslant \Gamma_2 + \epsilon.$$

If we call a point with coordinates of the form $(a_1 t,..., a_N t)$, mod 1, an 'accessible' point, Kronecker's theorem states that the accessible points are everywhere dense in the unit cube C. If now γ_1, γ_2 are two equal cubes with sides parallel to the axes, and with centres at accessible points P_1 and P_2, corresponding to t_1 and t_2, it is easily seen that

$$\lim I_{\gamma_1}(T)/I_{\gamma_2}(T) = 1.$$

For $(a_1 t,..., a_N t)$ will lie inside γ_2 when and only when $\{a_1(t+t_2-t_1),...\}$ lies inside γ_1.

Consider now a set of p non-overlapping cubes c, inside C, of side ϵ, each of which has its centre at an accessible point, and q of which lie inside γ; and a set of P overlapping cubes c', also centred on accessible points, whose union includes C and such that γ is included in a union of Q of them. Since the accessible points are everywhere dense, it is possible to choose the cubes such that q/P and Q/p are arbitrarily near to Γ. Now, denoting by $\sum_\gamma I_c(T)$ the sum of t-intervals in $(0, T)$ corresponding to the cubes c which lie in γ, and so on,

$$\sum_\gamma I_c(T) \Big/ \sum_C I_{c'}(T) \leqslant \frac{I_\gamma(T)}{T} \leqslant \sum_\gamma I_{c'}(T) \Big/ \sum_C I_c(T).$$

Making $T \to \infty$ we obtain

$$\frac{q}{P} \leqslant \varlimsup_{T\to\infty} \frac{I_\gamma(T)}{T} \leqslant \frac{Q}{p},$$

and the result follows.

11.8. We can now prove

THEOREM 11.8 (A). *If $\sigma = \sigma_0 > 1$ is a line on which* $\log \zeta(s)$ *comes arbitrarily near to a given number a, then in every strip $\sigma_0 - \delta < \sigma < \sigma_0 + \delta$ the value a is taken more than $K(a, \sigma_0, \delta)T$ times, for large T, in $0 < t < T$.*

To prove this we have to reconsider the argument of the previous sections, used to establish the existence of a root of $\log \zeta(s) = a$ in the strip, and use Kronecker's theorem in its generalized form. We saw that a sufficient condition that $\log \zeta(s) = a$ may have a root inside a circle with centre $\sigma_0 + it_0$ and radius 2δ is that, for a certain N and corresponding numbers $\theta_1,\ldots, \theta_N$, and a certain $\eta = \eta(\sigma_0, \delta, \theta_1,\ldots, \theta_N)$

$$|-t_0 \log p_n - 2\pi\theta_n - 2\pi g_n| < \eta \quad (n = 1, 2,\ldots, N).$$

From the generalized Kronecker's theorem it follows that the sum of the intervals between 0 and T in which t_0 satisfies this condition is asymptotically equal to $(\eta/2\pi)^N T$, and it is therefore greater than $\frac{1}{2}(\eta/2\pi)^N T$ for large T. Hence we can select more than $\frac{1}{8}(\eta/2\pi)^N T/\delta$ numbers t_0' in them, no two of which differ by less than 4δ. If now we describe circles with the points $\sigma_0 + it_0'$ as centres and radius 2δ, these circles will not overlap, and each of them will contain a zero of $\log \zeta(s) - a$. This gives the desired result.

We can also prove

THEOREM 11.8 (B). *There are positive constants $K_1(a)$ and $K_2(a)$ such that the number $M_a(T)$ of zeros of $\log \zeta(s) - a$ in $\sigma > 1$ satisfies the inequalities*

$$K_1(a)T < M_a(T) < K_2(a)T.$$

The lower bound follows at once from the above theorem. The upper bound follows from the more general result that *if b is any given constant, the number of zeros of $\zeta(s) - b$ in $\sigma > \frac{1}{2} + \delta$ $(\delta > 0)$, $0 < t < T$, is $O(T)$ as $T \to \infty$.*

The proof of this is substantially the same as that of Theorem 9.15 (A), the function $\zeta(s) - b$ playing the same part as $\zeta(s)$ did there. Finally the number of zeros of $\log \zeta(s) - a$ is not greater than the number of zeros of $\zeta(s) - e^a$, and so is $O(T)$.

11.9. We now turn to the more difficult question of the behaviour of $\zeta(s)$ in the critical strip. The difficulty, of course, is that $\zeta(s)$ is no longer represented by an absolutely convergent Dirichlet series. But by a device like that used in the proof of Theorem 9.17, we are able to obtain in the critical strip results analogous to those already obtained in the region of absolute convergence.

As before we consider $\log\zeta(s)$. For $\sigma \leqslant 1$, $\log\zeta(s)$ is defined, on each line $t =$ constant which does not pass through a singularity, by continuation along this line from $\sigma > 1$.

We require the following lemma.

LEMMA. *If $f(z)$ is regular for $|z-z_0| \leqslant R$, and*

$$\iint\limits_{|z-z_0|\leqslant R} |f(z)|^2\, dxdy = H,$$

then

$$|f(z)| \leqslant \frac{(H/\pi)^{\frac{1}{2}}}{R-R'} \quad (|z-z_0| \leqslant R' < R).$$

For if $|z'-z_0| \leqslant R'$,

$$\{f(z')\}^2 = \frac{1}{2\pi i}\int\limits_{|z-z'|=r} \frac{\{f(z)\}^2}{z-z'}\,dz = \frac{1}{2\pi}\int\limits_0^{2\pi} \{f(z'+re^{i\theta})\}^2\, d\theta.$$

Hence

$$|f(z')|^2 \int\limits_0^{R-R'} r\, dr \leqslant \frac{1}{2\pi}\int\limits_0^{R-R'}\int\limits_0^{2\pi} |f(z'+re^{i\theta})|^2 r\, drd\theta \leqslant \frac{H}{2\pi},$$

and the result follows.

THEOREM 11.9. *Let σ_0 be a fixed number in the range $\frac{1}{2} < \sigma \leqslant 1$. Then the values which $\log\zeta(s)$ takes on $\sigma = \sigma_0$, $t > 0$, are everywhere dense in the whole plane.*

Let

$$\zeta_N(s) = \zeta(s)\prod_{n=1}^{N}(1-p_n^{-s}).$$

This function is similar to the function $\zeta(s)M_X(s)$ of Chapter IX, but it happens to be more convenient here.

Let δ be a positive number less than $\frac{1}{2}(\sigma_0-\frac{1}{2})$. Then it is easily seen as in § 9.19 that for $N \geqslant N_0(\sigma_0, \epsilon)$, $T \geqslant T_0 = T_0(N)$,

$$\int\limits_1^T |\zeta_N(\sigma+it)-1|^2\, dt < \epsilon T$$

uniformly for $\sigma_0-\delta \leqslant \sigma \leqslant \sigma_1+\delta$ $(\sigma_1 > 1)$. Hence

$$\int\limits_1^T \int\limits_{\sigma_0-\delta}^{\sigma_1+\delta} |\zeta_N(\sigma+it)-1|^2\, d\sigma dt < (\sigma_1-\sigma_0+2\delta)\epsilon T.$$

Hence

$$\int\limits_{\nu-\frac{1}{2}}^{\nu+\frac{1}{2}} \int\limits_{\sigma_0-\delta}^{\sigma_1+\delta} |\zeta_N(\sigma+it)-1|^2\, d\sigma dt < (\sigma_1-\sigma_0+2\delta)\surd\epsilon$$

for more than $(1-\sqrt{\epsilon})T$ integer values of ν. Since this rectangle contains the circle with centre $s=\sigma+it$, where $\sigma_0 \leqslant \sigma \leqslant \sigma_1$, $\nu-\frac{1}{2}+\delta \leqslant t \leqslant \nu+\frac{1}{2}-\delta$, and radius δ, it is easily seen from the lemma that we can choose δ and ϵ so that *given* $0<\eta<1$, $0<\eta'<1$, *we have*

$$|\zeta_N(\sigma+it)-1| < \eta \quad (\sigma_0 \leqslant \sigma \leqslant \sigma_1) \tag{11.9.1}$$

for a set of values of t of measure greater than $(1-\eta')T$, *and for*

$$N \geqslant N_0(\sigma, \eta, \eta'), \qquad T \geqslant T_0(N).$$

Let
$$R_N(s) = -\sum_{N+1}^{\infty} \mathrm{Log}(1-p_n^{-s}) \quad (\sigma > 1),$$

where Log denotes the principal value of the logarithm. Then

$$\zeta_N(s) = \exp\{R_N(s)\}.$$

We want to show that $R_N(s) = \mathrm{Log}\,\zeta_N(s)$, i.e. that $|\mathbf{I}R_N(s)| < \frac{1}{2}\pi$, for $\sigma \geqslant \sigma_0$ and the values of t for which (11.9.1) holds. This is true for $\sigma=\sigma_1$ if σ_1 is sufficiently large, since $|R_N(s)| \to 0$ as $\sigma_1 \to \infty$. Also, by (11.9.1), $\mathbf{R}\zeta_N(s) > 0$ for $\sigma_0 \leqslant \sigma \leqslant \sigma_1$, so that $\mathbf{I}R_N(s)$ must remain between $-\frac{1}{2}\pi$ and $\frac{1}{2}\pi$ for all values of σ in this interval. This gives the desired result.

We have therefore

$$|R_N(s)| = |\mathrm{Log}[1+\{\zeta_N(s)-1\}]| < 2\,|\zeta_N(s)-1| < 2\eta$$

for $\sigma_0 \leqslant \sigma \leqslant \sigma_1$, $N \geqslant N_0(\sigma_0, \eta, \eta')$, $T \geqslant T_0(N)$, in a set of values of t of measure greater than $(1-\eta')T$.

Now consider the function

$$F_N(\sigma_0+it) = -\sum_{n=1}^{N} \log(1-p_n^{-\sigma_0-it}),$$

and in conjunction with it the function of N independent variables

$$\Phi_N(\theta_1,\ldots,\theta_N) = -\sum_{n=1}^{N} \log(1-p_n^{-\sigma_0}e^{2\pi i\theta_n}).$$

Since $\sum p_n^{-\sigma_0}$ is divergent, it is easily seen from our previous discussion of the values taken by $\log\zeta(s)$ that the set of values of Φ_N includes any given finite region of the complex plane if N is large enough. In particular, if a is any given number, we can find a number N and values of the θ's such that

$$\Phi_N(\theta_1,\ldots,\theta_N) = a.$$

We can then, by Kronecker's theorem, find a number t such that $|F_N(\sigma_0+it)-a|$ is arbitrarily small. But this in itself is not sufficient to prove the theorem, since this value of t does not necessarily make $|R_N(s)|$ small. An additional argument is therefore required.

Let

$$\Phi_{M,N} = -\sum_{n=M+1}^{N} \log(1-p_n^{-\sigma_0}e^{2\pi i\theta_n}) = \sum_{n=M+1}^{N} \sum_{m=1}^{\infty} \frac{p_n^{-m\sigma_0}e^{2\pi im\theta_n}}{m}.$$

Then, expressing the squared modulus of this as the product of conjugates, and integrating term by term, we obtain

$$\int_0^1\int_0^1 \cdots \int_0^1 |\Phi_{M,N}|^2\, d\theta_{M+1}\ldots d\theta_N = \sum_{n=M+1}^{N} \sum_{m=1}^{\infty} \frac{p_n^{-2m\sigma_0}}{m^2}$$

$$< \sum_{n=M+1}^{N} p_n^{-2\sigma_0} \sum_{m=1}^{\infty} \frac{1}{m^2} < A \sum_{n=M+1}^{\infty} p_n^{-2\sigma_0},$$

which can be made arbitrarily small, by choice of M, for all N. It therefore follows from the theory of Riemann integration of a continuous function that, given ϵ, we can divide up the $(N-M)$-dimensional unit cube into sub-cubes q_ν, each of volume λ, in such a way that

$$\lambda \sum_{\nu} \max_{q_\nu} |\Phi_{M,N}|^2 < \tfrac{1}{2}\epsilon^2.$$

Hence *for $M \geqslant M_0(\epsilon)$ and any $N > M$, we can find cubes of total volume greater than $\frac{1}{2}$ in which $|\Phi_{M,N}| < \epsilon$.*

We now choose our value of t as follows.

(i) Choose M so large, and give $\theta_1',\ldots, \theta_M'$ such values, that

$$\Phi_M(\theta_1',\ldots, \theta_M') = a.$$

It then follows from considerations of continuity that, given ϵ, we can find an M-dimensional cube with centre $\theta_1',\ldots,\theta_M'$ and side $d > 0$ throughout which

$$|\Phi_M(\theta_1,\ldots, \theta_M)-a| < \tfrac{1}{3}\epsilon.$$

(ii) We may also suppose that M has been chosen so large that, for any value of N, $|\Phi_{M,N}| < \frac{1}{3}\epsilon$ in certain $(N-M)$-dimensional cubes of total volume greater than $\frac{1}{2}$.

(iii) Having fixed M and d, we can choose N so large that, for $T > T_0(N)$, the inequality $|R_N(s)| < \frac{1}{3}\epsilon$ holds in a set of values of t of measure greater than $(1-\frac{1}{2}d^M)T$.

(iv) Let $I(T)$ be the sum of the intervals between 0 and T for which the point

$$\{-(t\log p_1)/2\pi,\ldots, -(t\log p_N)/2\pi\}$$

is (mod 1) inside one of the N-dimensional cubes, of total volume greater than $\frac{1}{2}d^M$, determined by the above construction. Then by the extended Kronecker's theorem, $I(T) > \frac{1}{2}d^MT$ if T is large enough. There are therefore values of t for which the point lies in one of these cubes, and for which at the same time $|R_N(s)| < \frac{1}{3}\epsilon$. For such a value of t

$$\begin{aligned}|\log\zeta(s)-a| &\leqslant |F_N(s)-a|+|R_N(s)|\\ &\leqslant |\Phi_M(\theta_1,\ldots,\theta_M)-a|+|\Phi_{M,N}|+|R_N(s)|\\ &< \tfrac{1}{3}\epsilon+\tfrac{1}{3}\epsilon+\tfrac{1}{3}\epsilon = \epsilon,\end{aligned}$$

and the result follows.

11.10. THEOREM 11.10. *Let $\frac{1}{2} < \alpha < \beta < 1$, and let a be any complex number. Let $M_{a,\alpha,\beta}(T)$ be the number of zeros of $\log\zeta(s)-a$ (defined as before) in the rectangle $\alpha < \sigma < \beta$, $0 < t < T$. Then there are positive constants $K_1(a,\alpha,\beta)$, $K_2(a,\alpha,\beta)$ such that*

$$K_1(a,\alpha,\beta)T < M_{a,\alpha,\beta}(T) < K_2(a,\alpha,\beta)T \quad (T > T_0).$$

We first observe that, for suitable values of the θ's, the series

$$-\sum_{n=1}^{\infty}\log(1-p_n^{-s}e^{2\pi i\theta_n})$$

is uniformly convergent in any finite region to the right of $\sigma = \frac{1}{2}$. This is true, for example, if $\theta_n = \frac{1}{2}n$ for sufficiently large values of n; for then

$$\sum_{n>n_0} p_n^{-s}e^{2\pi i\theta_n} = \sum_{n>n_0}(-1)^n p_n^{-s},$$

which is convergent for real $s > 0$, and hence uniformly convergent in any finite region to the right of the imaginary axis; and for any θ's $\sum |p_n^{-s}e^{2\pi i\theta_n}|^2 = \sum p_n^{-2\sigma}$ is uniformly convergent in any finite region to the right of $\sigma = \frac{1}{2}$.

If a is any given number, and the θ's have this property, we can choose n_1 so large that

$$\left|-\sum_{n=n_1+1}^{\infty}\log(1-p_n^{-s}e^{2\pi i\theta_n})\right| < \epsilon \quad (\sigma = \tfrac{1}{2}(\alpha+\beta)),$$

and at the same time so that the set of values of

$$-\sum_{n=1}^{n_1}\log(1-p_n^{-\frac{1}{2}\alpha-\frac{1}{2}\beta}e^{2\pi i\theta_n})$$

includes the circle with centre the origin and radius $|a|+|\epsilon|$. Hence by choosing first $\theta_{n_1+1},\ldots$, and then $\theta_1,\ldots,\theta_{n_1}$, we can find values of the θ's, say θ_1', $\theta_2',\ldots$, such that the series

$$G(s) = -\sum_{n=1}^{\infty}\log(1-p_n^{-s}e^{2\pi i\theta_n'})$$

is uniformly convergent in any finite region to the right of $\sigma = \frac{1}{2}$, and

$$G(\tfrac{1}{2}\alpha+\tfrac{1}{2}\beta) = a.$$

We can then choose a circle C of centre $\frac{1}{2}\alpha+\frac{1}{2}\beta$ and radius $\rho < \frac{1}{4}(\beta-\alpha)$ on which $G(s) \neq a$.

Let
$$m = \min_{s \text{ on } C} |G(s)-a|.$$

Now let
$$\Phi_{M,N}(s) = -\sum_{n=M+1}^{N} \log(1-p_n^{-s}e^{2\pi i\theta_n}).$$

Then, as in the previous proof,

$$\int_0^1 \dots \int_0^1 \iint_{|s-\frac{1}{2}\alpha-\frac{1}{2}\beta| \leqslant \frac{1}{2}(\beta-\alpha)} |\Phi_{M,N}(s)|^2 \, d\theta_{M+1} \dots d\theta_N \, d\sigma dt < A \sum_{M+1}^{\infty} p_n^{-2\alpha}.$$

Hence for $M \geqslant M_0(\epsilon)$ and any $N > M$ we can find cubes of total volume greater than $\frac{1}{2}$ in which

$$\iint_{|s-\frac{1}{2}\alpha-\frac{1}{2}\beta| \leqslant \frac{1}{2}(\beta-\alpha)} |\Phi_{M,N}(s)|^2 \, d\sigma dt < \epsilon$$

and so in which (by the lemma of § 11.9)

$$|\Phi_{M,N}(s)| < 2(\epsilon/\pi)^{\frac{1}{2}}(\beta-\alpha)^{-\frac{1}{2}} \quad (|s-\tfrac{1}{2}\alpha-\tfrac{1}{2}\beta| \leqslant \tfrac{1}{4}(\beta-\alpha)).$$

We also want a little more information about $R_N(s)$, viz. that $R_N(s)$ is regular, and $|R_N(s)| < \eta$, throughout the rectangle

$$|\sigma-\tfrac{1}{2}\alpha-\tfrac{1}{2}\beta| \leqslant \tfrac{1}{4}(\beta-\alpha), \quad t_0-\tfrac{1}{2} \leqslant t \leqslant t_0+\tfrac{1}{2},$$

for a set of values of t_0 of measure greater than $(1-\eta')T$. As before it is sufficient to prove this for $\zeta_N(s)-1$, and by the lemma it is sufficient to prove that

$$\phi(t_0) = \int_{\alpha}^{\beta} d\sigma \int_{t_0-1}^{t_0+1} |\zeta_N(s)-1|^2 \, dt < \epsilon$$

for such t_0, by choice of N. Now

$$\int_1^T \phi(t_0) \, dt = \int_{\alpha}^{\beta} d\sigma \int_1^T dt_0 \int_{t_0-1}^{t_0+1} |\zeta_N(s)-1|^2 \, dt$$

$$\leqslant \int_{\alpha}^{\beta} d\sigma \int_1^{T+1} |\zeta_N(s)-1|^2 \, dt \int_{t-1}^{t+1} dt_0 = 2\int_{\alpha}^{\beta} d\sigma \int_1^{T+1} |\zeta_N(s)-1|^2 \, dt < \epsilon T$$

by choice of N as before. Hence the measure of the set where $\phi(t_0) > \surd\epsilon$ is less than $\surd\epsilon T$, and the desired result follows.

It now follows as before that there is a set of values of t_0 in $(0, T)$, of measure greater than KT, such that for $|s-\frac{1}{2}\alpha-\frac{1}{2}\beta| \leqslant \frac{1}{4}(\beta-\alpha)$

$$\left|\sum_{n=1}^{M} \log(1-p_n^{-s}e^{2\pi i\theta'_n}) - \sum_{n=1}^{M} \log(1-p_n^{-s-it_0})\right| < \tfrac{1}{4}m,$$

$$|\Phi_{M,N}(s)| < \tfrac{1}{4}m,$$

and also
$$|R_N(s+it_0)| < \tfrac{1}{4}m.$$

At the same time we can suppose that M has been taken so large that

$$\left|G(s)+\sum_{n=1}^{M} \log(1-p_n^{-s}e^{2\pi i\theta'_n})\right| < \tfrac{1}{4}m \quad (\sigma \geqslant \alpha).$$

Then
$$|\log \zeta(s)-G(s)| < m$$

on the circle with centre $\frac{1}{2}\alpha+\frac{1}{2}\beta+it_0$ and radius ρ. Hence, as before, $\log\zeta(s)-a$ has at least one zero in such a circle. The number of such circles for $0 < t_0 < T$ which do not overlap is plainly greater than KT. The lower bound for $M_{a,\alpha,\beta}(T)$ therefore follows; the upper bound holds by the same argument as in the case $\sigma > 1$.

It has been proved by Bohr and Jensen, by a more detailed study of the situation, that there is a $K(a, \alpha, \beta)$ such that

$$M_{a,\alpha,\beta}(T) \sim K(a, \alpha, \beta)T.$$

An immediate corollary of Theorem 11.10 is that, *if $N_{a,\alpha,\beta}(T)$ is the number of points in the rectangle $\frac{1}{2} < \alpha < \sigma < \beta < 1$, $0 < t < T$ where $\zeta(s) = a$ $(a \neq 0)$, then*

$$N_{a,\alpha,\beta}(T) > K(a, \alpha, \beta)T \quad (T > T_0).$$

For $\zeta(s) = a$ if $\log\zeta(s) = \log a$, any one value of the right-hand side being taken. This result, in conjunction with Theorem 9.17, shows that the value 0 of $\zeta(s)$, if it occurs at all in $\sigma > \frac{1}{2}$, is at any rate quite exceptional, zeros being infinitely rarer than a-values for any value of a other than zero.

NOTES FOR CHAPTER 11

11.11. Theorem 11.9 has been generalized by Voronin [**1**], [**2**], who obtained the following 'universal' property for $\zeta(s)$. Let D_r be the closed disc of radius $r < \frac{1}{4}$, centred at $s = \frac{3}{4}$, and let $f(s)$ be any function continuous and non-vanishing on D_r, and holomorphic on the interior of D_r. Then for any $\varepsilon > 0$ there is a real number t such that

$$\max_{s\in D_r} |\zeta(s+it)-f(s)| < \varepsilon. \tag{11.11.1}$$

It follows that the curve

$$\gamma(t) = \big(\zeta(\sigma+it), \zeta'(\sigma+it), \ldots, \zeta^{(n-1)}(\sigma+it)\big)$$

is dense in $\mathbb{C}^n$, for any fixed σ in the range $\frac{1}{2} < \sigma < 1$. (In fact Voronin [**1**] establishes this for $\sigma = 1$ also.) To see this we choose a point $z = (z_0, z_1, \ldots, z_{n-1})$ with $z_0 \neq 0$, and take $f(s)$ to be a polynomial for which $f^{(m)}(\sigma) = z_m$ for $0 \leqslant m < n$. We then fix an R such that $0 < R < \frac{1}{4} - |\sigma - \frac{3}{4}|$, and such that $f(s)$ is nonvanishing on the closed disc $|s-\sigma| \leqslant R$. Thus, if $r = R + |\sigma - \frac{3}{4}|$, the disc D_r contains the circle $|s - \sigma| = R$, and hence (11.11.1) in conjunction with Cauchy's inequality

$$|g^{(m)}(z_0)| \leqslant \frac{m!}{R^m} \max_{|z-z_0|=R} |g(z)|,$$

yields

$$|\zeta^{(m)}(\sigma+it) - z_m| \leqslant \frac{m!}{R^m}\varepsilon \quad (0 \leqslant m < n).$$

Hence $\gamma(t)$ comes arbitrarily close to z. The required result then follows, since the available z are dense in $\mathbb{C}^n$.

Voronin's work has been extended by Bagchi [**1**] (see also Gonek [**1**]) so that D_r may be replaced by any compact subset D of the strip $\frac{1}{2} < \mathbf{R}(s) < 1$, whose complement in $\mathbb{C}$ is connected. The condition on f is then that it should be continuous and non-vanishing on D, and holomorphic on the interior (if any) of D. From this it follows that if Φ is any continuous function, and $h_1 < h_2 < \ldots < h_m$ are real constants, then $\zeta(s)$ cannot satisfy the differential–difference equation

$$\Phi\{\zeta(s+h_1), \zeta'(s+h_1), \ldots, \zeta^{(n_1)}(s+h_1), \zeta(s+h_2), \zeta'(s+h_2), \ldots, \zeta^{(n_2)}(s+h_2), \ldots\} = 0$$

unless Φ vanishes identically. This improves earlier results of Ostrowski [**1**] and Reich [**1**].

11.12. Levinson [**6**] has investigated further the distribution of the solutions $\rho_a = \beta_a + i\gamma_a$ of $\zeta(s) = a$. The principal results are that

$$\#\{\rho_a : 0 \leqslant \gamma_a \leqslant T\} = \frac{T}{2\pi}\log T + O(T)$$

and

$$\#\{\rho_a : 0 \leqslant \gamma_a \leqslant T, |\beta_a - \tfrac{1}{2}| \geqslant \delta\} = O_\delta(T) \quad (\delta > 0).$$

Thus (c.f. § 9.15) *all but an infinitesimal proportion of the zeros of* $\zeta(s) - a$ *lie in the strip* $\frac{1}{2} - \delta < \sigma < \frac{1}{2} + \delta$, *however small* δ *may be.*

In reviewing this work Montgomery (Math. Reviews **53** # 10737) quotes an unpublished result of Selberg, namely

$$\sum_{\substack{0 \leqslant \gamma_a \leqslant T \\ \beta_a \geqslant \frac{1}{2}}} (\beta_a - \tfrac{1}{2}) \sim \frac{1}{4\pi^{\frac{3}{2}}} T(\operatorname{loglog} T)^{\frac{1}{2}}. \tag{11.12.1}$$

This leads to a stronger version of the above principle, in which the infinite strip is replaced by the region

$$|\sigma - \tfrac{1}{2}| < \frac{\phi(t)(\operatorname{loglog} t)^{\frac{1}{2}}}{\log t},$$

where $\phi(t)$ is any positive function which tends to infinity with t. It should be noted for comparison with (11.12.1) that the estimate

$$\sum_{0 \leqslant \gamma_a \leqslant T} (\beta_a - \tfrac{1}{2}) = O(\log T)$$

is implicit in Levinson's work. It need hardly be emphasized that despite this result the numbers ρ_a are far from being symmetrically distributed about the critical line.

11.13. The problem of the distribution of values of $\zeta(\frac{1}{2}+it)$ is rather different from that of $\zeta(\sigma+it)$ with $\frac{1}{2}<\sigma<1$. In the first place it is not known whether the values of $\zeta(\frac{1}{2}+it)$ are everywhere dense, though one would conjecture so. Secondly there is a difference in the rates of growth with respect to t. Thus, for a fixed $\sigma > \frac{1}{2}$, Bohr and Jessen **(1)**, **(2)** have shown that there is a continuous function $F(z;\sigma)$ such that

$$\frac{1}{2T} m\{t \in [-T, T]: \log \zeta(\sigma+it) \in R\} \to \iint_R F(x+iy;\sigma)\, dx dy \quad (T \to \infty)$$

for any rectangle $R \subset \mathbb{C}$ whose sides are parallel to the real and imaginary axes. Here, as usual, m denotes Lebesgue measure, and $\log \zeta(s)$ is defined by continuous variation along lines parallel to the real axis, using (1.1.9) for $\sigma > 1$. By contrast, the corresponding result for $\sigma = \frac{1}{2}$ states that

$$\frac{1}{2T} m\left\{t \in [-T, T]: \frac{\log \zeta(\frac{1}{2}+it)}{\sqrt{[\frac{1}{2}\{\operatorname{loglog}(3+|t|)\}]}} \in R\right\} \to \frac{1}{2\pi} \iint_R e^{-(x^2+y^2)/2} dx dy$$
$$(T \to \infty).$$

(The right hand side gives a 2-dimensional normal distribution with mean 0 and variance 1.) This is an unpublished theorem of Selberg, which may be obtained via the method of Ghosh [**2**].

By using a different technique, based on the mean-value bounds of §7.23, Jutila [**4**] has obtained information on 'large deviations' of $\log|\zeta(\frac{1}{2}+it)|$. Specifically, he showed that there is a constant $A > 0$ such that

$$m\{t \in [0, T]: |\zeta(\tfrac{1}{2}+it)| \geqslant V\} \ll T \exp\left(-A \frac{\log^2 V}{\log\log T}\right),$$

uniformly for $1 \leqslant V \leqslant \log T$.

XII

DIVISOR PROBLEMS

12.1. The divisor problem of Dirichlet is that of determining the asymptotic behaviour as $x \to \infty$ of the sum

$$D(x) = \sum_{n \leqslant x} d(n),$$

where $d(n)$ denotes, as usual, the number of divisors of n. Dirichlet proved in an elementary way that

$$D(x) = x\log x + (2\gamma - 1)x + O(x^{\frac{1}{2}}). \tag{12.1.1}$$

In fact

$$\begin{aligned} D(x) &= \sum_{mn \leqslant x} \sum 1 = \sum_{m \leqslant \sqrt{x}} \sum_{n \leqslant \sqrt{x}} 1 + 2 \sum_{m \leqslant \sqrt{x}} \sum_{\sqrt{x} < n \leqslant x/m} 1 \\ &= [\sqrt{x}]^2 + 2 \sum_{m \leqslant \sqrt{x}} \left(\left[\frac{x}{m}\right] - [\sqrt{x}]\right) \\ &= 2 \sum_{m \leqslant \sqrt{x}} \left[\frac{x}{m}\right] - [\sqrt{x}]^2 \\ &= 2 \sum_{m \leqslant \sqrt{x}} \left\{\frac{x}{m} + O(1)\right\} - \{\sqrt{x} + O(1)\}^2 \\ &= 2x\{\log \sqrt{x} + \gamma + O(x^{-\frac{1}{2}})\} + O(\sqrt{x}) - \{x + O(\sqrt{x})\}, \end{aligned}$$

and (12.1.1) follows. Writing

$$D(x) = x\log x + (2\gamma - 1)x + \Delta(x)$$

we thus have

$$\Delta(x) = O(x^{\frac{1}{2}}). \tag{12.1.2}$$

Later researches have improved this result, but the exact order of $\Delta(x)$ is still undetermined.

The problem is closely related to that of the Riemann zeta-function. By (3.12.1) with $a_n = d(n)$, $s = 0$, $T \to \infty$, we have

$$D(x) = \frac{1}{2\pi i} \int_{c - i\infty}^{c + i\infty} \zeta^2(w) \frac{x^w}{w}\, dw \quad (c > 1),$$

provided that x is not an integer. On moving the line of integration to the left, we encounter a double pole at $w = 1$, the residue being $x\log x + (2\gamma - 1)x$, by (2.1.16). Thus

$$\Delta(x) = \frac{1}{2\pi i} \int_{c' - i\infty}^{c' + i\infty} \zeta^2(w) \frac{x^w}{w}\, dw \quad (0 < c' < 1).$$

The more general problem of

$$D_k(x) = \sum_{n \leqslant x} d_k(n),$$

where $d_k(n)$ is the number of ways of expressing n as a product of k factors, was also considered by Dirichlet. We have

$$D_k(x) = \frac{1}{2\pi i} \int\limits_{c-i\infty}^{c+i\infty} \zeta^k(w) \frac{x^w}{w}\, dw \quad (c > 1).$$

Here there is a pole of order k at $w = 1$, and the residue is of the form $xP_k(\log x)$, where P_k is a polynomial of degree $k-1$. We write

$$D_k(x) = xP_k(\log x) + \Delta_k(x), \tag{12.1.3}$$

so that $\Delta_2(x) = \Delta(x)$.

The classical elementary theorem† of the subject is

$$\Delta_k(x) = O(x^{1-1/k} \log^{k-2} x) \quad (k = 2, 3, \ldots,). \tag{12.1.4}$$

We have already proved this in the case $k = 2$. Now suppose that it is true in the case $k-1$. We have

$$\begin{aligned}
D_k(x) &= \sum_{n_1 n_2 \ldots n_k \leqslant x} 1 = \sum_{mn \leqslant x} d_{k-1}(n) \\
&= \sum_{m \leqslant x^{1/k}} \sum_{n \leqslant x/m} d_{k-1}(n) + \sum_{x^{1/k} < m \leqslant x} \sum_{n \leqslant x/m} d_{k-1}(n) \\
&= \sum_{m \leqslant x^{1/k}} \sum_{n \leqslant x/m} d_{k-1}(n) + \sum_{n \leqslant x^{1-1/k}} d_{k-1}(n) \sum_{x^{1/k} < m \leqslant x/n} 1 \\
&= \sum_{m \leqslant x^{1/k}} D_{k-1}\left(\frac{x}{m}\right) + \sum_{n \leqslant x^{1-1/k}} \left\{\frac{x}{n} - x^{1/k} + O(1)\right\} d_{k-1}(n) \\
&= \sum_{m \leqslant x^{1/k}} D_{k-1}\left(\frac{x}{m}\right) + x \sum_{n \leqslant x^{1-1/k}} \frac{d_{k-1}(n)}{n} - x^{1/k} D_{k-1}(x^{1-1/k}) + \\
&\qquad + O\{D_{k-1}(x^{1-1/k})\}.
\end{aligned}$$

Let us denote by $p_k(z)$ a polynomial in z, of degree $k-1$ at most, not always the same one. Then

$$\sum_{m \leqslant \xi} \frac{\log^{k-2} m}{m} = p_k(\log \xi) + O\left(\frac{\log^{k-2} \xi}{\xi}\right).$$

Hence
$$\sum_{m \leqslant x^{1/k}} \frac{x}{m} P_{k-1}\left(\frac{x}{m}\right) = xp_k(\log x) + O(x^{1-1/k} \log^{k-2} x).$$

Also

$$\begin{aligned}
\sum_{m \leqslant x^{1/k}} \Delta_{k-1}\left(\frac{x}{m}\right) &= O\left\{x^{1-1/(k-1)} \log^{k-3} x \sum_{m \leqslant x^{1/k}} \frac{1}{m^{1-1/(k-1)}}\right\} \\
&= O\{x^{1-1/(k-1)} \log^{k-3} x \,.\, x^{1/\{k(k-1)\}}\} = O(x^{1-1/k} \log^{k-3} x).
\end{aligned}$$

† See e.g. Landau (5).

The next term is

$$x\sum_{n\leqslant x^{1-1/k}}\frac{D_{k-1}(n)-D_{k-1}(n-1)}{n}=x\sum_{n\leqslant x^{1-1/k}}\frac{D_{k-1}(n)}{n(n+1)}+\frac{xD_{k-1}(N)}{N+1},$$

where $N=[x^{1-1/k}]$. Now

$$x\sum_{n\leqslant x^{1-1/k}}\frac{P_{k-1}(\log n)}{n+1}+x\frac{NP_{k-1}(\log N)}{N+1}=xp_k(\log x)+O(x^{1/k}\log^{k-2}x)$$

and

$$x\sum_{n\leqslant x^{1-1/k}}\frac{\Delta_{k-1}(n)}{n(n+1)}+\frac{x\Delta_{k-1}(N)}{N+1}=Cx-x\sum_{n>x^{1-1/k}}\frac{\Delta_{k-1}(n)}{n(n+1)}+\frac{x\Delta_{k-1}(N)}{N+1}$$

$$=Cx-x\sum_{n>x^{1-1/k}}O\left\{\frac{\log^{k-3}n}{n^{1+1/(k-1)}}\right\}+O(xN^{-1/(k-1)}\log^{k-3}N)$$

$$=Cx+O(x^{1-1/k}\log^{k-3}x).$$

Finally

$$x^{1/k}D_{k-1}(x^{1-1/k})=x^{1/k}\{x^{1-1/k}P_{k-1}(\log x^{1-1/k})+O(x^{(1-1/k)\{1-1/(k-1)\}}\log^{k-3}x)\}$$

$$=xp_{k-1}(\log x)+O(x^{1-1/k}\log^{k-3}x).$$

This proves (12.1.4).

We may define the order α_k of $\Delta_k(x)$ as the least number such that

$$\Delta_k(x)=O(x^{\alpha_k+\epsilon})$$

for every positive ϵ. Thus it follows from (12.1.4) that

$$\alpha_k\leqslant\frac{k-1}{k}\quad(k=2,3,\ldots). \tag{12.1.5}$$

The exact value of α_k has not been determined for any value of k.

12.2. The simplest theorem which goes beyond this elementary result is

THEOREM 12.2.†

$$\alpha_k\leqslant\frac{k-1}{k+1}\quad(k=2,3,4,\ldots).$$

Take $a_n=d_k(n)$, $\psi(n)=n^\epsilon$, $\alpha=k$, $s=0$, and let x be half an odd integer, in Lemma 3.12. Replacing w by s, this gives

$$D_k(x)=\frac{1}{2\pi i}\int_{c-iT}^{c+iT}\zeta^k(s)\frac{x^s}{s}\,ds+O\left(\frac{x^c}{T(c-1)^k}\right)+O\left(\frac{x^{1+\epsilon}}{T}\right)\quad(c>1).$$

† Voronoï (1), Landau (5).

Now take the integral round the rectangle $-a-iT$, $c-iT$, $c+iT$, $-a+iT$, where $a > 0$. We have, by (5.1.1) and the Phragmén–Lindelöf principle,

$$\zeta(s) = O(t^{(a+\frac{1}{2})(c-\sigma)/(a+c)})$$

in the rectangle. Hence

$$\int_{-a+iT}^{c+iT} \zeta^k(s)\frac{x^s}{s}\,ds = O\left(\int_{-a}^{c} T^{k(a+\frac{1}{2})(c-\sigma)/(a+c)-1}x^\sigma\,d\sigma\right)$$
$$= O(T^{k(a+\frac{1}{2})-1}x^{-a})+O(T^{-1}x^c),$$

since the integrand is a maximum at one end or the other of the range of integration. A similar result holds for the integral over

$$(-a-iT,\ c-iT).$$

The residue at $s = 1$ is $xP_k(\log x)$, and the residue at $s = 0$ is

$$\zeta^k(0) = O(1).$$

Finally

$$\int_{-a-iT}^{-a+iT} \zeta^k(s)\frac{x^s}{s}\,ds = \int_{-a-iT}^{-a+iT} \chi^k(s)\zeta^k(1-s)\frac{x^s}{s}\,ds$$
$$= \sum_{n=1}^{\infty} d_k(n) \int_{-a-iT}^{-a+iT} \frac{\chi^k(s)}{n^{1-s}}\frac{x^s}{s}\,ds$$
$$= ix^{-a}\sum_{n=1}^{\infty}\frac{d_k(n)}{n^{1+a}}\int_{-T}^{T}\frac{\chi^k(-a+it)}{-a+it}(nx)^{it}\,dt.$$

For $1 \leqslant t \leqslant T$,

$$\chi(-a+it) = Ce^{-it\log t+it\log 2\pi+it}t^{a+\frac{1}{2}}+O(t^{a-\frac{1}{2}})$$

and

$$\frac{1}{-a+it} = \frac{1}{it}+O\left(\frac{1}{t^2}\right).$$

The corresponding part of the integral is therefore

$$-iC^k\int_{1}^{T} e^{ikt(-\log t+\log 2\pi+1)}(nx)^{it}t^{(a+\frac{1}{2})k-1}\,dt+O(T^{(a+\frac{1}{2})k-1}),$$

provided that $(a+\frac{1}{2})k > 1$. This integral is of the form considered in Lemma 4.5, with

$$F(t) = kt(-\log t+\log 2\pi+1)+t\log nx.$$

Since

$$F''(t) = -\frac{k}{t} \leqslant -\frac{k}{T},$$

the integral is

$$O(T^{(a+\frac{1}{2})k-\frac{1}{2}}),$$

uniformly with respect to n and x. A similar result holds for the integral over $(-T,-1)$, while the integral over $(-1,1)$ is bounded. Hence

$$\Delta_k(x) = O\left(\frac{x^c}{T(c-1)^k}\right)+O\left(\frac{x^{1+\epsilon}}{T}\right)+O\left(\frac{T^{(a+\frac{1}{2})k-1}}{x^a}\right)+$$

$$+x^{-a}\sum_{n=1}^{\infty}\frac{d_k(n)}{n^{1+a}}O(T^{(a+\frac{1}{2})k-\frac{1}{2}})$$

$$= O\left(\frac{x^c}{T(c-1)^k}\right)+O\left(\frac{x^{1+\epsilon}}{T}\right)+O\left(\frac{T^{(a+\frac{1}{2})k-\frac{1}{2}}}{x^a}\right).$$

Taking $c = 1+\epsilon$, $a = \epsilon$, the terms are of the same order, apart from ϵ's, if

$$T = x^{2/(k+1)}.$$

Hence

$$\Delta_k(x) = O(x^{(k-1)/(k+1)+\epsilon}).$$

The restriction that x should be half an odd integer is clearly unnecessary to the result.

12.3. By using some of the deeper results on $\zeta(s)$ we can obtain a still better result for $k \geqslant 4$.

THEOREM 12.3.† $\alpha_k \leqslant \dfrac{k-1}{k+2} \quad (k = 4, 5,...).$

We start as in the previous theorem, but now take the rectangle as far as $\sigma = \frac{1}{2}$ only. Let us suppose that

$$\zeta(\tfrac{1}{2}+it) = O(t^\lambda).$$

Then

$$\zeta(s) = O(t^{\lambda(c-\sigma)/(c-\frac{1}{2})})$$

uniformly in the rectangle. The horizontal sides therefore give

$$O\left(\int_{\frac{1}{2}}^{c} T^{k\lambda(c-\sigma)/(c-\frac{1}{2})-1}x^\sigma\,d\sigma\right) = O(T^{k\lambda-1}x^{\frac{1}{2}})+O(T^{-1}x^c).$$

Also

$$\int_{\frac{1}{2}-iT}^{\frac{1}{2}+iT}\zeta^k(s)\frac{x^s}{s}\,ds = O(x^{\frac{1}{2}})+O\left(x^{\frac{1}{2}}\int_1^T|\zeta(\tfrac{1}{2}+it)|^k\frac{dt}{t}\right).$$

Now

$$\int_1^T|\zeta(\tfrac{1}{2}+it)|^k\frac{dt}{t} \leqslant \max_{1\leqslant t\leqslant T}|\zeta(\tfrac{1}{2}+it)|^{k-4}\int_1^T|\zeta(\tfrac{1}{2}+it)|^4\frac{dt}{t}$$

$$= O\left\{T^{(k-4)\lambda}\int_1^T|\zeta(\tfrac{1}{2}+it)|^4\frac{dt}{t}\right\}.$$

† Hardy and Littlewood (4).

Also
$$\phi(T) = \int_1^T |\zeta(\tfrac{1}{2}+it)|^4\,dt = O(T^{1+\epsilon}),$$
by (7.6.1), so that
$$\int_1^T |\zeta(\tfrac{1}{2}+it)|^4\frac{dt}{t} = \int_1^T \phi'(t)\frac{dt}{t} = \left[\frac{\phi(t)}{t}\right]_1^T + \int_1^T \frac{\phi(t)}{t^2}\,dt$$
$$= O(T^\epsilon)+O\left(\int_1^T \frac{1}{t^{1-\epsilon}}\,dt\right) = O(T^\epsilon).$$

Hence
$$\int_{\frac{1}{2}-iT}^{\frac{1}{2}+iT} \zeta^k(s)\frac{x^s}{s}\,ds = O(x^{\frac{1}{2}})+O(x^{\frac{1}{2}}T^{(k-4)\lambda+\epsilon}).$$

Altogether we obtain
$$\Delta_k(x) = O(T^{-1}x^c)+O(x^{\frac{1}{2}}T^{k\lambda-1})+O(x^{\frac{1}{2}}T^{(k-4)\lambda+\epsilon}).$$
The middle term is of smaller order than the last if $\lambda \leqslant \frac{1}{4}$. Taking $c = 1+\epsilon$, the other two terms are of the same order, apart from ϵ's, if
$$T = x^{1/\{2(k-4)\lambda+2\}}.$$
This gives
$$\Delta_k(x) = O(x^{[\{2(k-4)\lambda+1\}/\{2(k-4)\lambda+2\}]+\epsilon}).$$
Taking $\lambda = \frac{1}{6}+\epsilon$ (Theorems 5.5, 5.12) the result follows. Further slight improvements for $k \geqslant 5$ are obtained by using the results stated in § 5.18.

12.4. The above method does not give any new result for $k = 2$ or $k = 3$. For these values slight improvements on Theorem 12.2 have been made by special methods.

THEOREM 12.4.† $$\alpha_2 \leqslant \frac{27}{82}.$$

The argument of § 12.2 shows that
$$\Delta(x) = \frac{1}{2\pi i}\sum_{n=1}^\infty d(n)\int_{-a-iT}^{-a+iT} \frac{\chi^2(s)}{n^{1-s}}\frac{x^s}{s}\,ds+O\left(\frac{T^{2a}}{x^a}\right)+O\left(\frac{x^c}{T}\right) \quad (12.4.1)$$
where $a > 0$, $c > 1$. Let $T^2/(4\pi^2 x) = N+\frac{1}{2}$, where N is an integer, and consider the terms with $n > N$. As before, the integral over $1 \leqslant t \leqslant T$ is of the form
$$\frac{1}{x^a n^{1+a}}\int_1^T e^{iF(t)}\{t^{2a}+O(t^{2a-1})\}\,dt, \quad (12.4.2)$$

† van der Corput (4).

where
$$F(t) = 2t(-\log t+\log 2\pi+1)+t\log nx,$$
$$F'(t) = \log\frac{4\pi^2 nx}{t^2}.$$

Hence $F'(t) \geqslant \log\dfrac{n}{N+\frac{1}{2}}$, and (12.4.2) is
$$\frac{1}{x^a n^{1+a}}\left\{O\left(\frac{T^{2a}}{\log\{n/(N+\frac{1}{2})\}}\right)+O(T^{2a})\right\}.$$
For $n \geqslant 2N$ this contributes to (12.4.1)
$$O\left\{\frac{T^{2a}}{x^a}\sum_{n=2N}^{\infty}\frac{d(n)}{n^{1+a}}\right\} = O(N^{\epsilon}),$$
and for $N < n < 2N$ it contributes
$$O\left\{\frac{T^{2a}}{x^a}\sum_{n=N+1}^{2N}\frac{d(n)}{n^{1+a}\log\{n/(N+\frac{1}{2})\}}\right\} = O\left(N^{\epsilon}\sum_{m=1}^{N}\frac{1}{m}\right) = O(N^{\epsilon}).$$

Similarly for the integral over $-T \leqslant t \leqslant -1$; and the integral over $-1 < t < 1$ is clearly $O(x^{-a})$.

If $n \leqslant N$, we write
$$\int\limits_{-a-iT}^{-a+iT} = \int\limits_{-i\infty}^{i\infty} - \left(\int\limits_{iT}^{i\infty} + \int\limits_{-i\infty}^{-iT} + \int\limits_{-iT}^{-a-iT} + \int\limits_{-a+iT}^{iT}\right).$$
The first term is
$$\frac{1}{n}\int\limits_{-i\infty}^{i\infty} 2^{2s}\pi^{2s-2}\sin^2\tfrac{1}{2}s\pi\,\Gamma^2(1-s)\frac{(nx)^s}{s}\,ds$$
$$= -\frac{1}{n\pi^2}\int\limits_{1-i\infty}^{1+i\infty}\cos^2\tfrac{1}{2}w\pi\,\Gamma(w)\Gamma(w-1)\{2\pi\sqrt{(nx)}\}^{2-2w}\,dw$$
$$= -4i\sqrt{\left(\frac{x}{n}\right)}[K_1\{4\pi\sqrt{(nx)}\}+\tfrac{1}{2}\pi Y_1\{4\pi\sqrt{(nx)}\}]$$
in the usual notation of Bessel functions.†

The first integral in the bracket is
$$\int\limits_{T}^{\infty} e^{iF(t)}\left(A+\frac{A'}{t}+O(t^{-2})\right)dt = O\left\{\frac{1}{\log\{(N+\frac{1}{2})/n\}}\right\},$$
which gives
$$\sum_{n=1}^{N}\frac{d(n)}{n\log\{(N+\frac{1}{2})/n\}} = O(N^{\epsilon})$$

† See, e.g., Titchmarsh, *Fourier Integrals*, (7.9.8), (7.9.11).

as before; and similarly for the second integral. The last two give

$$O\left\{\sum_{n=1}^{N}\frac{d(n)}{n}\int_{-a}^{0}\left(\frac{nx}{T^2}\right)^{\sigma}d\sigma\right\} = O\left\{\sum_{n=1}^{N}\frac{d(n)}{n}\left(\frac{T^2}{nx}\right)^{a}\right\} = O\left\{\left(\frac{T^2}{x}\right)^{a}\right\}.$$

Altogether we have now proved that

$$\Delta(x) = -\frac{2\surd x}{\pi}\sum_{n=1}^{N}\frac{d(n)}{\surd n}[K_1\{4\pi\surd(nx)\}+\tfrac{1}{2}\pi Y_1\{4\pi\surd(nx)\}]+O\left(\frac{T^{2a}}{x^a}\right)+O\left(\frac{x^c}{T}\right). \tag{12.4.3}$$

By the usual asymptotic formulae† for Bessel functions, this may be replaced by

$$\Delta(x) = \frac{x^{\frac{1}{4}}}{\pi\surd 2}\sum_{n=1}^{N}\frac{d(n)}{n^{\frac{3}{4}}}\cos\{4\pi\surd(nx)-\tfrac{1}{4}\pi\}+O(x^{-\frac{1}{4}})+O\left(\frac{T^{2a}}{x^a}\right)+O\left(\frac{x^c}{T}\right). \tag{12.4.4}$$

Now

$$\sum_{n=1}^{N}d(n)e^{4\pi i\surd(nx)} = 2\sum_{m\leqslant\surd N}\;\sum_{n\leqslant N/m}e^{4\pi i\surd(mnx)}-\sum_{m\leqslant\surd N}\;\sum_{n\leqslant\surd N}e^{4\pi i\surd(mnx)}. \tag{12.4.5}$$

Consider the sum $\displaystyle\sum_{\frac{1}{2}N/m<n\leqslant N/m}e^{4\pi i\surd(mnx)}$.

We apply Theorem 5.13, with $k = 5$, and

$$f(n) = 2\surd(mnx), \qquad f^{(5)}(n) = A(mx)^{\frac{1}{2}}n^{-\frac{9}{2}}.$$

Hence the sum is

$$O\left\{\frac{N}{m}\left(\frac{(mx)^{\frac{1}{2}}}{(N/m)^{\frac{9}{2}}}\right)^{\frac{1}{30}}\right\}+O\left\{\left(\frac{N}{m}\right)^{\frac{7}{8}}\left(\frac{(N/m)^{\frac{9}{2}}}{(mx)^{\frac{1}{2}}}\right)^{\frac{1}{30}}\right\}$$
$$= O\{(N/m)^{\frac{17}{20}}(mx)^{\frac{1}{60}}\}+O\{(N/m)^{\frac{41}{40}}(mx)^{-\frac{1}{60}}\}.$$

Replacing N by $\frac{1}{2}N$, $\frac{1}{4}N$,..., and adding, the same result holds for the sum over $1 \leqslant n \leqslant N/m$. Hence the first term on the right of (12.4.5) is

$$O\Big(N^{\frac{17}{20}}x^{\frac{1}{60}}\sum_{m\leqslant\surd N}m^{-\frac{5}{6}}\Big)+O\Big(N^{\frac{41}{40}}x^{-\frac{1}{60}}\sum_{m\leqslant\surd N}m^{-\frac{25}{24}}\Big) = O(N^{\frac{14}{15}}x^{\frac{1}{60}})+O(N^{\frac{41}{40}}x^{-\frac{1}{60}}).$$

Similarly the second inner sum is

$$O\{(\surd N)^{\frac{17}{20}}(mx)^{\frac{1}{60}}\}+O\{(\surd N)^{\frac{41}{40}}(mx)^{-\frac{1}{60}}\},$$

and the whole sum is

$$O\Big(N^{\frac{17}{40}}x^{\frac{1}{60}}\sum_{m\leqslant\surd N}m^{\frac{1}{60}}\Big)+O\Big(N^{\frac{41}{80}}x^{-\frac{1}{60}}\sum_{m\leqslant\surd N}m^{-\frac{1}{60}}\Big)$$
$$= O(N^{\frac{14}{15}}x^{\frac{1}{60}})+O(N^{\frac{241}{240}}x^{-\frac{1}{60}}).$$

† Watson, *Theory of Bessel Functions*, §§ 7.21, 7.23.

Hence, multiplying by $e^{-\frac{1}{4}i\pi}$ and taking the real part,

$$\sum_{n=1}^{N} d(n)\cos\{4\pi\sqrt{(nx)}-\tfrac{1}{4}\pi\} = O(N^{\frac{14}{15}}x^{\frac{1}{60}})+O(N^{\frac{41}{40}}x^{-\frac{1}{60}}).$$

Using this and partial summation, (12.4.4) gives

$$\begin{aligned}\Delta(x) &= O(N^{\frac{14}{15}-\frac{3}{4}}x^{\frac{1}{4}+\frac{1}{60}})+O(N^{\frac{41}{40}-\frac{3}{4}}x^{\frac{1}{4}-\frac{1}{60}})+O(N^{a})+O(N^{-\frac{1}{2}}x^{c-\frac{1}{2}})\\ &= O(N^{\frac{11}{60}}x^{\frac{4}{15}})+O(N^{\frac{11}{40}}x^{\frac{7}{30}})+O(N^{a})+O(N^{-\frac{1}{2}}x^{c-\frac{1}{2}}).\end{aligned}$$

Taking $a = \epsilon$, $c = 1+\epsilon$, the first and last terms are of the same order, apart from ϵ's, if

$$N = [x^{\frac{14}{41}}].$$

Hence

$$\Delta(x) = O(x^{\frac{27}{82}+\epsilon}),$$

the result stated.

A similar argument may be applied to $\Delta_3(x)$. We obtain

$$\Delta_3(x) = \frac{x^{\frac{1}{3}}}{\pi\sqrt{3}} \sum_{n<T^3/(8\pi^3x)} \frac{d_3(n)}{n^{\frac{2}{3}}}\cos\{6\pi(nx)^{\frac{1}{3}}\}+O\left(\frac{x^{1+\epsilon}}{T}\right), \tag{12.4.6}$$

and deduce that

$$\alpha_3 \leqslant \tfrac{37}{75}.$$

The detailed argument is given by Atkinson (3).

If the series in (12.4.4) were absolutely convergent, or if the terms more or less cancelled each other, we should deduce that $\alpha_2 \leqslant \frac{1}{4}$; and it may reasonably be conjectured that this is the real truth. We shall see later that $\alpha_2 \geqslant \frac{1}{4}$, so that it would follow that $\alpha_2 = \frac{1}{4}$. Similarly from (12.4.6) we should obtain $\alpha_3 = \frac{1}{3}$; and so generally it may be conjectured that

$$\alpha_k = \frac{k-1}{2k}.$$

12.5. *The average order of* $\Delta_k(x)$. We may define β_k, the average order of $\Delta_k(x)$, to be the least number such that

$$\frac{1}{x}\int_0^x \Delta_k^2(y)\,dy = O(x^{2\beta_k+\epsilon})$$

for every positive ϵ. Since

$$\frac{1}{x}\int_0^x \Delta_k^2(y)\,dy = \frac{1}{x}\int_0^x O(y^{2\alpha_k+\epsilon})\,dy = O(x^{2\alpha_k+\epsilon}),$$

we have $\beta_k \leqslant \alpha_k$ for each k. In particular we obtain a set of upper bounds for the β_k from the above theorems.

As usual, the problem of average order is easier than that of order, and we can prove more about the β_k than about the α_k. We shall first prove the following theorem.†

† Titchmarsh (22).

THEOREM 12.5. *Let γ_k be the lower bound of positive numbers σ for which*

$$\int_{-\infty}^{\infty} \frac{|\zeta(\sigma+it)|^{2k}}{|\sigma+it|^2}\,dt < \infty. \tag{12.5.1}$$

Then $\beta_k = \gamma_k$; and

$$\frac{1}{2\pi}\int_{-\infty}^{\infty} \frac{|\zeta(\sigma+it)|^{2k}}{|\sigma+it|^2}\,dt = \int_0^{\infty} \Delta_k^2(x)x^{-2\sigma-1}\,dx \tag{12.5.2}$$

provided that $\sigma > \beta_k$.

We have $$D_k(x) = \frac{1}{2\pi i}\lim_{T\to\infty}\int_{c-iT}^{c+iT} \frac{\zeta^k(s)}{s}x^s\,ds \quad (c > 1).$$

Applying Cauchy's theorem to the rectangle $\gamma-iT$, $c-iT$, $c+iT$, $\gamma+iT$, where γ is less than, but sufficiently near to, 1, and allowing for the residue at $s = 1$, we obtain

$$\Delta_k(x) = \frac{1}{2\pi i}\lim_{T\to\infty}\int_{\gamma-iT}^{\gamma+iT} \frac{\zeta^k(s)}{s}x^s\,ds. \tag{12.5.3}$$

Actually (12.5.3) holds for $\gamma_k < \gamma < 1$. For† $\zeta^k(s)/s \to 0$ uniformly as $t \to \pm\infty$ in the strip. Hence if we integrate the integrand of (12.5.3) round the rectangle $\gamma'-iT$, $\gamma-iT$, $\gamma+iT$, $\gamma'+iT$, where

$$\gamma_k < \gamma' < \gamma < 1,$$

and make $T \to \infty$, we obtain the same result with γ' instead of γ.

If we replace x by $1/x$, (12.5.3) expresses the relation between the Mellin transforms

$$f(x) = \Delta_k(1/x), \qquad \mathfrak{F}(s) = \zeta^k(s)/s,$$

the relevant integrals holding also in the mean-square sense. Hence Parseval's formula for Mellin transforms‡ gives

$$\frac{1}{2\pi}\int_{-\infty}^{\infty} \frac{|\zeta(\gamma+it)|^{2k}}{|\gamma+it|^2}\,dt = \int_0^{\infty} \Delta_k^2\left(\frac{1}{x}\right)x^{2\gamma-1}\,dx = \int_0^{\infty} \Delta_k^2(x)x^{-2\gamma-1}\,dx \tag{12.5.4}$$

provided that $\gamma_k < \gamma < 1$.

It follows that, if $\gamma_k < \gamma < 1$,

$$\int_{\frac{1}{2}X}^{X} \Delta_k^2(x)x^{-2\gamma-1}\,dx < K = K(k,\gamma),$$

$$\int_{\frac{1}{2}X}^{X} \Delta_k^2(x)\,dx < KX^{2\gamma+1},$$

† By an application of the lemma of § 11.9.

‡ See Titchmarsh, *Theory of Fourier Integrals*, Theorem 71.

and, replacing X by $\frac{1}{2}X$, $\frac{1}{4}X$,..., and adding,

$$\int_1^X \Delta_k^2(x)\,dx < KX^{2\gamma+1}.$$

Hence $\beta_k \leqslant \gamma$, and so $\beta_k \leqslant \gamma_k$.

The inverse Mellin formula is

$$\frac{\zeta^k(s)}{s} = \int_0^\infty \Delta_k\left(\frac{1}{x}\right)x^{s-1}\,dx = \int_0^\infty \Delta_k(x)x^{-s-1}\,dx. \tag{12.5.5}$$

The right-hand side exists primarily in the mean-square sense, for $\gamma_k < \sigma < 1$. But actually the right-hand side is uniformly convergent in any region interior to the strip $\beta_k < \sigma < 1$; for

$$\int_{\frac{1}{2}X}^X |\Delta_k(x)|x^{-\sigma-1}\,dx \leqslant \left\{\int_{\frac{1}{2}X}^X \Delta_k^2(x)\,dx \int_{\frac{1}{2}X}^X x^{-2\sigma-2}\,dx\right\}^{\frac{1}{2}}$$

$$= \{O(X^{2\beta_k+1+\epsilon})O(X^{-2\sigma-1})\}^{\frac{1}{2}} = O(X^{\beta_k-\sigma+\epsilon}),$$

and on putting $X = 2, 4, 8,...$, and adding we obtain

$$\int_1^\infty |\Delta_k(x)|x^{-\sigma-1}\,dx < K.$$

It follows that the right-hand side of (12.5.5) represents an analytic function, regular for $\beta_k < \sigma < 1$. The formula therefore holds by analytic continuation throughout this strip. Also (by the argument just given) the right-hand side of (12.5.4) is finite for $\beta_k < \gamma < 1$. Hence so is the left-hand side, and the formula holds. Hence $\gamma_k \leqslant \beta_k$, and so, in fact, $\gamma_k = \beta_k$. This proves the theorem.

12.6. THEOREM 12.6 (A).†

$$\beta_k \geqslant \frac{k-1}{2k} \quad (k = 2, 3,...).$$

If $\frac{1}{2} < \sigma < 1$, by Theorem 7.2

$$C_\sigma T < \int_{\frac{1}{2}T}^T |\zeta(\sigma+it)|^2\,dt \leqslant \left\{\int_{\frac{1}{2}T}^T |\zeta(\sigma+it)|^{2k}\,dt\right\}^{1/k}\left(\int_{\frac{1}{2}T}^T dt\right)^{1-1/k}.$$

Hence
$$\int_{\frac{1}{2}T}^T |\zeta(\sigma+it)|^{2k}\,dt \geqslant 2^{k-1}C_\sigma^k T.$$

† Titchmarsh (22).

Hence, if $0 < \sigma < \frac{1}{2}$, $T > 1$,

$$\int_{-\infty}^{\infty} \frac{|\zeta(\sigma+it)|^{2k}}{|\sigma+it|^2}\,dt > \int_{\frac{1}{2}T}^{T} \frac{|\zeta(\sigma+it)|^{2k}}{|\sigma+it|^2}\,dt > \frac{C'}{T^2}\int_{\frac{1}{2}T}^{T} |\zeta(\sigma+it)|^{2k}\,dt$$

$$> C''T^{k(1-2\sigma)-2}\int_{\frac{1}{2}T}^{T} |\zeta(1-\sigma-it)|^{2k}\,dt \quad \text{(by the functional equation)}$$

$$\geqslant C''2^{k-1}C_{1-\sigma}^{k}\,T^{k(1-2\sigma)-1}.$$

This can be made as large as we please by choice of T if $\sigma < \frac{1}{2}(k-1)/k$.

Hence
$$\gamma_k \geqslant \frac{k-1}{2k}$$
and the theorem follows.

THEOREM 12.6 (B).†

$$\alpha_k \geqslant \frac{k-1}{2k} \quad (k = 2, 3, \ldots).$$

For $\alpha_k \geqslant \beta_k$.

Much more precise theorems of the same type are known. Hardy proved first that both

$$\Delta(x) > Kx^{\frac{1}{4}}, \qquad \Delta(x) < -Kx^{\frac{1}{4}}$$

hold for some arbitrarily large values of x, and then that $x^{\frac{1}{4}}$ may in each case be replaced by

$$(x\log x)^{\frac{1}{4}}\log\log x.$$

12.7. We recall that (§ 7.9) the numbers σ_k are defined as the lower bounds of σ such that

$$\frac{1}{T}\int_{1}^{T} |\zeta(\sigma+it)|^{2k}\,dt = O(1).$$

We shall next prove

THEOREM 12.7. *For each integer $k \geqslant 2$, a necessary and sufficient condition that*

$$\beta_k = \frac{k-1}{2k} \tag{12.7.1}$$

is that

$$\sigma_k \leqslant \frac{k+1}{2k}. \tag{12.7.2}$$

Suppose first that (12.7.2) holds. Then by the functional equation

$$\int_{1}^{T} |\zeta(\sigma+it)|^{2k}\,dt = O\left\{T^{k(1-2\sigma)}\int_{1}^{T} |\zeta(1-\sigma-it)|^{2k}\,dt\right\} = O(T^{k(1-2\sigma)+1})$$

† Hardy (2).

for $\sigma < \frac{1}{2}(k-1)/k$. It follows from the convexity of mean values that

$$\int_1^T |\zeta(\sigma+it)|^{2k}\,dt = O(T^{1+(\frac{1}{2}+1/2k+\epsilon/2k-\sigma)k})$$

for
$$\frac{k-1-\epsilon}{2k} < \sigma < \frac{k+1+\epsilon}{2k}.$$

The index of T is less than 2 if

$$\sigma > \frac{k-1+\epsilon}{2k}.$$

Then
$$\int_{\frac{1}{2}T}^{T} \frac{|\zeta(\sigma+it)|^{2k}}{|\sigma+it|^2}\,dt = O(T^{-\delta}) \quad (\delta > 0).$$

Hence (12.5.1) holds. Hence $\gamma_k \leqslant \frac{1}{2}(k-1)/k$. Hence $\beta_k \leqslant \frac{1}{2}(k-1)/k$, and so, by Theorem 12.6 (A), (12.7.1) holds.

On the other hand, if (12.7.1) holds, it follows from (12.5.2) that

$$\int_1^T |\zeta(\sigma+it)|^{2k}\,dt = O(T^2)$$

for $\sigma > \frac{1}{2}(k-1)/k$. Hence by the functional equation

$$\int_1^T |\zeta(\sigma+it)|^{2k}\,dt = O(T^{k(1-2\sigma)+2})$$

for $\sigma < \frac{1}{2}(k+1)/k$. Hence, by the convexity theorem, the left-hand side is $O(T^{1+\epsilon})$ for $\sigma = \frac{1}{2}(k+1)/k$; hence, in the notation of § 7.9, $\sigma'_k \leqslant \frac{1}{2}(k+1)/k$, and so (12.7.2) holds.

12.8. THEOREM 12.8.†

$$\beta_2 = \tfrac{1}{4}, \qquad \beta_3 = \tfrac{1}{3}, \qquad \beta_4 \leqslant \tfrac{3}{7}.$$

By Theorem 7.7, $\sigma_k \leqslant 1-1/k$. Since

$$1-\frac{1}{k} \leqslant \frac{k+1}{2k} \quad (k \leqslant 3)$$

it follows that $\beta_2 = \frac{1}{4}$, $\beta_3 = \frac{1}{3}$.

The available material is not quite sufficient to determine β_4. Theorem 12.6 (A) gives $\beta_4 \geqslant \frac{3}{8}$. To obtain an upper bound for it, we observe that, by Theorem 5.5. and (7.6.1),

$$\int_1^T |\zeta(\tfrac{1}{2}+it)|^8\,dt = O\left(T^{\frac{2}{3}+\epsilon}\int_1^T |\zeta(\tfrac{1}{2}+it)|^4\,dt\right) = O(T^{\frac{5}{3}+\epsilon}),$$

† The value of β_2 is due to Hardy (3), and that of β_3 to Cramér (4); for β_4 see Titchmarsh (22).

and, since $\sigma_4 \leqslant \frac{7}{10}$ by Theorem 7.10,

$$\int_1^T |\zeta(\tfrac{3}{10}+it)|^8\,dt = O\left(T^{\frac{8}{5}} \int_1^T |\zeta(\tfrac{7}{10}-it)|^8\,dt\right) = O(T^{\frac{13}{5}+\epsilon}).$$

Hence by the convexity theorem

$$\int_1^T |\zeta(\sigma+it)|^8\,dt = O(T^{4-\frac{14}{3}\sigma+\epsilon})$$

for $\frac{3}{10} < \sigma < \frac{1}{2}$. It easily follows that $\gamma_4 \leqslant \frac{3}{7}$, i.e. $\beta_4 \leqslant \frac{3}{7}$.

NOTES FOR CHAPTER 12

12.9. For large k the best available estimates for α_k are of the shape $\alpha_k \leqslant 1 - Ck^{-\frac{2}{3}}$, where C is a positive constant. The first such result is due to Richert [**2**]. (See also Karatsuba [**1**], Ivic [**3**; Theorem 13.3] and Fujii [**3**].) These results depend on bounds of the form (6.19.2).

For the range $4 \leqslant k \leqslant 8$ one has $\alpha_k \leqslant \frac{3}{4} - 1/k$ (Heath-Brown [**8**]) while for intermediate values of k a number of estimates are possible (see Ivic [**3**; Theorem 13.2]). In particular one has $\alpha_9 \leqslant \frac{35}{54}$, $\alpha_{10} \leqslant \frac{41}{60}$, $\alpha_{11} \leqslant \frac{7}{10}$, and $\alpha_{12} \leqslant \frac{5}{7}$.

12.10. The following bounds for α_2 have been obtained.

$\frac{33}{100} = 0{\cdot}330000\ldots$	van der Corput (2),
$\frac{27}{82} = 0{\cdot}329268\ldots$	van der Corput (4),
$\frac{15}{46} = 0{\cdot}326086\ldots$	Chih [**1**], Richert [**1**],
$\frac{12}{37} = 0{\cdot}324324\ldots$	Kolesnik [**1**],
$\frac{346}{1067} = 0{\cdot}324273\ldots$	Kolesnik [**2**],
$\frac{35}{108} = 0{\cdot}324074\ldots$	Kolesnik [**4**],
$\frac{139}{429} = 0{\cdot}324009\ldots$	Kolesnik [**5**].

In general the methods used to estimate α_2 and $\mu(\frac{1}{2})$ are very closely related. Suppose one has a bound

$$\sum_{M<m\leqslant M_1}\ \sum_{N<n\leqslant N_1} \exp\left[2\pi i\{x(mn)^{\frac{1}{2}} + cx^{-1}(mn)^{\frac{3}{2}}\}\right] \ll (MN)^{\frac{3}{4}} x^{2\vartheta-\frac{1}{2}}, \tag{12.10.1}$$

for any constant c, uniformly for $M < M_1 \leqslant 2M$, $N < N_1 \leqslant 2N$, and $MN \leqslant x^{2-4\vartheta}$. It then follows that $\mu(\frac{1}{2}) \leqslant \frac{1}{2}\vartheta$, $\alpha_2 \leqslant \vartheta$, and $E(T) \ll T^{\vartheta+\varepsilon}$ (for $E(T)$ as in §7.20). In practice those versions of the van der Corput method used to tackle $\mu(\frac{1}{2})$ and α_2 also apply to (12.10.1), which explains the similarity between the table of estimates given above and that presented in §5.21 for $\mu(\frac{1}{2})$. This is just one manifestation of the close similarity exhibited by the functions $E(T)$ and $\Delta(x)$, which has its origin in the formulae (7.20.6) and (12.4.4). The classical lattice-point problem for the circle falls within the same area of ideas. Thus, if the bound (12.10.1) holds, along with its analogue in which the summation condition $m \equiv 1 \pmod 4$ is imposed, then one has

$$\#\{(m, n) \in \mathbb{Z}^2 \colon m^2 + n^2 \leqslant x\} = \pi x + O(x^{\vartheta+\varepsilon}).$$

Jutila [3] has taken these ideas further by demonstrating a direct connection between the size of $\Delta(x)$ and that of $\zeta(\frac{1}{2}+it)$ and $E(T)$. In particular he has shown that if $\alpha_2 = \frac{1}{4}$ then $\mu(\frac{1}{2}) \leqslant \frac{3}{20}$ and $E(T) \ll T^{\frac{5}{16}+\varepsilon}$.

Further work has also been done on the problem of estimating α_3. The best result at present is $\alpha_3 \leqslant \frac{43}{96}$, due to Kolesnik [3]. For α_4, however, no sharpening of the bound $\alpha_4 \leqslant \frac{1}{2}$ given by Theorem 12.3 has yet been found. This result, dating from 1922, seems very resistant to any attempt at improvement.

12.11. The Ω-results attributed to Hardy in §12.6 may be found in Hardy [1]. However Hardy's argument appears to yield only

$$\Delta(x) = \Omega_+((x \log x)^{\frac{1}{4}} \log\log x), \tag{12.11.1}$$

and not the corresponding Ω_- result. The reason for this is that Dirichlet's Theorem is applicable for Ω_+, while Kronecker's Theorem is needed for the Ω_- result. By using a quantitative form of Kronecker's Theorem, Corrádi and Kátai [1] showed that

$$\Delta(x) = \Omega_-\left\{x^{\frac{1}{4}} \exp\left(c \frac{(\log\log x)^{\frac{1}{4}}}{(\log\log\log x)^{\frac{3}{4}}}\right)\right\},$$

for a certain positive constant c. This improved earlier work of Ingham [1] and Gangadharan [1]. Hardy's result (12.11.1) has also been sharpened by Hafner [1] who obtained

$$\Delta(x) = \Omega_+[(x \log x)^{\frac{1}{4}} (\log\log x)^{\frac{1}{4}(3+2\log 2)} \exp\{-c(\log\log\log x)^{\frac{1}{2}}\}]$$

for a certain positive constant c. For $k \geqslant 3$ he also showed [2] that, for a suitable positive constant c, one has

$$\Delta_k(x) = \Omega_*[(x \log x)^{(k-1)/2k} (\log\log x)^a \exp\{-c(\log\log\log x)^{\frac{1}{2}}\}],$$

where

$$a = \frac{k-1}{2k}(k\log k + k + 1)$$

and Ω_* is Ω_+ for $k = 3$ and $\Omega_\pm$ for $k \geqslant 4$.

12.12. As mentioned in §7.22 we now have $\sigma_4 \leqslant \frac{5}{8}$, whence $\beta_4 = \frac{3}{8}$, (Heath-Brown [**8**]). For $k = 2$ and 3 one can give asymptotic formulae for

$$\int_0^x \Delta_k(y)^2\,dy.$$

Thus Tong [**1**] showed that

$$\int_0^x \Delta_k(y)^2\,dy = \frac{x^{(2k-1)/k}}{(4k-2)\pi^2}\sum_{n=1}^{\infty} d_k(n)^2 n^{-(k+1)/k} + R_k(x)$$

with $R_2(x) \ll x(\log x)^5$ and

$$R_k(x) \ll x^{c_k+\varepsilon}, \qquad c_k = 2 - \frac{3-4\sigma_k}{2k(1-\sigma_k)-1}, \qquad (k \geqslant 3).$$

Taking $\sigma_3 \leqslant \frac{7}{12}$ (see §7.22) yields $c_3 \leqslant \frac{14}{9}$. However the available information concerning σ_k is as yet insufficient to give $c_k < (2k-1)/k$ for any $k \geqslant 4$. It is perhaps of interest to note that Hardy's result (12.11.1) implies $R_2(x) = \Omega\{x^{\frac{3}{4}}(\log x)^{-\frac{1}{4}}\}$, since any estimate $R_2(x) \ll F(x)$ easily leads to a bound $\Delta_2(x) \ll \{F(x)\log x\}^{\frac{1}{3}}$, by an argument analogous to that given for the proof of Lemma α in §14.13.

Ivić [**3**; Theorems 13.9 and 13.10] has estimated the higher moments of $\Delta_2(x)$ and $\Delta_3(x)$. In particular his results imply that

$$\int_0^x \Delta_2(y)^8\,dy \ll x^{3+\varepsilon}.$$

For $\Delta_3(x)$ his argument may be modified slightly to yield

$$\int_0^x |\Delta_3(y)|^3\,dy \ll x^{2+\varepsilon}.$$

These results are readily seen to contain the estimates $\alpha_2 \leqslant \frac{1}{3}$, $\beta_2 \leqslant \frac{1}{4}$ and $\alpha_3 \leqslant \frac{1}{2}$, $\beta_3 \leqslant \frac{1}{3}$ respectively.

XIII

THE LINDELÖF HYPOTHESIS

13.1. The Lindelöf hypothesis is that

$$\zeta(\tfrac{1}{2}+it) = O(t^\epsilon)$$

for every positive ϵ; or, what comes to the same thing, that

$$\zeta(\sigma+it) = O(t^\epsilon)$$

for every positive ϵ and every $\sigma \geqslant \frac{1}{2}$; for either statement is, by the theory of the function $\mu(\sigma)$, equivalent to the statement that $\mu(\sigma) = 0$ for $\sigma \geqslant \frac{1}{2}$. The hypothesis is suggested by various theorems in Chapters V and VII. It is also the simplest possible hypothesis on $\mu(\sigma)$, for on it the graph of $y = \mu(\sigma)$ consists simply of the two straight lines

$$y = \tfrac{1}{2}-\sigma \quad (\sigma \leqslant \tfrac{1}{2}), \qquad y = 0 \quad (\sigma \geqslant \tfrac{1}{2}).$$

We shall see later that the Lindelöf hypothesis is true if the Riemann hypothesis is true. The converse deduction, however, cannot be made —in fact (Theorem 13.5) the Lindelöf hypothesis is equivalent to a much less drastic, but still unproved, hypothesis about the distribution of the zeros.

In this chapter we investigate the consequences of the Lindelöf hypothesis. Most of our arguments are reversible, so that we obtain necessary and sufficient conditions for the truth of the hypothesis.

13.2. Theorem 13.2.† *Alternative necessary and sufficient conditions for the truth of the Lindelöf hypothesis are*

$$\frac{1}{T}\int_1^T |\zeta(\tfrac{1}{2}+it)|^{2k}\,dt = O(T^\epsilon) \quad (k = 1, 2,...); \tag{13.2.1}$$

$$\frac{1}{T}\int_1^T |\zeta(\sigma+it)|^{2k}\,dt = O(T^\epsilon) \quad (\sigma > \tfrac{1}{2},\ k = 1, 2,...); \tag{13.2.2}$$

$$\frac{1}{T}\int_1^T |\zeta(\sigma+it)|^{2k}\,dt \sim \sum_{n=1}^{\infty} \frac{d_k^2(n)}{n^{2\sigma}} \quad (\sigma > \tfrac{1}{2},\ k = 1, 2,...). \tag{13.2.3}$$

The equivalence of the first two conditions follows from the convexity theorem (§ 7.8), while that of the last two follows from the analysis of § 7.9. It is therefore sufficient to consider (13.2.1).

† Hardy and Littlewood (5).

The necessity of the condition is obvious. To prove that it is sufficient, suppose that $\zeta(\frac{1}{2}+it)$ is not $O(t^{\epsilon})$. Then there is a positive number λ, and a sequence of numbers $\frac{1}{2}+it_\nu$, such that $t_\nu \to \infty$ with ν, and

$$|\zeta(\tfrac{1}{2}+it_\nu)| > Ct_\nu^\lambda \quad (C > 0).$$

On the other hand, on differentiating (2.1.4) we obtain, for $t \geqslant 1$,

$$|\zeta'(\tfrac{1}{2}+it)| < Et,$$

E being a positive absolute constant. Hence

$$|\zeta(\tfrac{1}{2}+it)-\zeta(\tfrac{1}{2}+it_\nu)| = \left|\int_{t_\nu}^{t} \zeta'(\tfrac{1}{2}+iu)\,du\right| < 2E|t-t_\nu|t_\nu < \tfrac{1}{2}Ct_\nu^\lambda$$

if $|t-t_\nu| \leqslant t_\nu^{-1}$ and ν is sufficiently large. Hence

$$|\zeta(\tfrac{1}{2}+it)| > \tfrac{1}{2}Ct_\nu^\lambda \quad (|t-t_\nu| \leqslant t_\nu^{-1}).$$

Take $T = \frac{2}{3}t_\nu$, so that the interval $(t_\nu-t_\nu^{-1}, t_\nu+t_\nu^{-1})$ is included in $(T, 2T)$ if ν is sufficiently large. Then

$$\int_T^{2T} |\zeta(\tfrac{1}{2}+it)|^{2k}\,dt > \int_{t_\nu-t_\nu^{-1}}^{t_\nu+t_\nu^{-1}} (\tfrac{1}{2}Ct_\nu^\lambda)^{2k}\,dt = 2(\tfrac{1}{2}C)^{2k}t_\nu^{2k\lambda-1},$$

which is contrary to hypothesis if k is large enough. This proves the theorem.

We could plainly replace the right-hand side of (13.2.1) by $O(T^A)$ without altering the theorem or the proof.

13.3. THEOREM 13.3. *A necessary and sufficient condition for the truth of the Lindelöf hypothesis is that, for every positive integer k and $\sigma > \frac{1}{2}$,*

$$\zeta^k(s) = \sum_{n \leqslant t^\delta} \frac{d_k(n)}{n^s} + O(t^{-\lambda}) \quad (t > 0), \tag{13.3.1}$$

where δ is any given positive number less than 1, and $\lambda = \lambda(k, \delta, \sigma) > 0$.

We may express this roughly by saying that, on the Lindelöf hypothesis, the behaviour of $\zeta(s)$, or of any of its positive integral powers, is dominated, throughout the right-hand half of the critical strip, by a section of the associated Dirichlet series whose length is less than any positive power of t, however small. The result may be contrasted with what we can deduce, without unproved hypothesis, from the approximate functional equation.

Taking $a_n = d_k(n)$ in Lemma 3.12, we have (if x is half an odd integer)

$$\sum_{n<x} \frac{d_k(n)}{n^s} = \frac{1}{2\pi i}\int_{c-iT}^{c+iT} \zeta^k(s+w)\frac{x^w}{w}\,dw + O\left(\frac{x^c}{T(\sigma+c-1)^k}\right)$$

where $c > 1-\sigma+\epsilon$. Now let $0 < t < T-1$, and integrate round the rectangle $\frac{1}{2}-\sigma-iT$, $c-iT$, $c+iT$, $\frac{1}{2}-\sigma+iT$. We have

$$\frac{1}{2\pi i} \int\limits_{\text{rectangle}} \zeta^k(s+w)\frac{x^w}{w}\,dw = \zeta^k(s)+\frac{x^{1-s}}{1-s}P\left(\frac{1}{1-s}, \log x\right)$$
$$= \zeta^k(s)+O(x^{1-\sigma+\epsilon}t^{-1+\epsilon}),$$

P being a polynomial in its arguments. Also

$$\left(\int\limits_{\frac{1}{2}-\sigma-iT}^{c-iT} + \int\limits_{c+iT}^{\frac{1}{2}-\sigma+iT}\right) \zeta^k(s+w)\frac{x^w}{w}\,dw = O(x^c T^{-1+\epsilon})$$

by the Lindelöf hypothesis; and

$$\int\limits_{\frac{1}{2}-\sigma-iT}^{\frac{1}{2}-\sigma+iT} \zeta^k(s+w)\frac{x^w}{w}\,dw = O\left\{x^{\frac{1}{2}-\sigma} \int\limits_{-T}^{T} \frac{|\zeta^k(\frac{1}{2}+it+iv)|}{|\frac{1}{2}+iv|}\,dv\right\}$$
$$= O(x^{\frac{1}{2}-\sigma}T^{\epsilon})$$

by the Lindelöf hypothesis. Hence

$$\zeta^k(s) = \sum_{n<x}\frac{d_k(n)}{n^s}+O\left\{\frac{x^c}{T(\sigma+c-1)^k}\right\}+O(x^{1-\sigma+\epsilon}t^{\epsilon-1})+$$
$$+O(x^c T^{-1+\epsilon})+O(x^{\frac{1}{2}-\sigma}T^{\epsilon}),$$

and (13.3.1) follows on taking $x = [t^{\delta}]+\frac{1}{2}$, $c = 2$, $T = t^3$.

Conversely, the condition is clearly sufficient, since it gives

$$\zeta^k(s) = O\Big(\sum_{n\leqslant t^{\delta}} n^{\epsilon-\sigma}\Big)+O(t^{-\lambda}) = O(t^{\delta(1-\epsilon-\sigma)}),$$

where δ is arbitrarily small.

The result may be used to prove the equivalence of the conditions of the previous section, without using the general theorems quoted.

13.4. Another set of conditions may be stated in terms of the numbers α_k and β_k of the previous chapter.

THEOREM 13.4. *Alternative necessary and sufficient conditions for the truth of the Lindelöf hypothesis are*

$$\alpha_k \leqslant \tfrac{1}{2} \qquad (k = 2, 3,...), \tag{13.4.1}$$

$$\beta_k \leqslant \tfrac{1}{2} \qquad (k = 2, 3,...), \tag{13.4.2}$$

$$\beta_k = \frac{k-1}{2k} \quad (k = 2, 3,...). \tag{13.4.3}$$

As regards sufficiency, we need only consider (13.4.2), since the other conditions are formally more stringent. Now (13.4.2) gives $\gamma_k \leqslant \frac{1}{2}$, and so

$$\int_{\frac{1}{2}T}^{T} \frac{|\zeta(\sigma+it)|^{2k}}{|\sigma+it|^2}\,dt = O(1) \quad (\sigma > \tfrac{1}{2}),$$

$$\int_{\frac{1}{2}T}^{T} |\zeta(\sigma+it)|^{2k}\,dt = O(T^2) \quad (\sigma > \tfrac{1}{2}).$$

The truth of the Lindelöf hypothesis follows from this, as in § 13.2.

Now suppose that the Lindelöf hypothesis is true. We have, as in § 12.2,

$$D_k(x) = \frac{1}{2\pi i}\int_{2-iT}^{2+iT} \zeta^k(s)\frac{x^s}{s}\,ds + O\left(\frac{x^2}{T}\right).$$

Now integrate round the rectangle with vertices at $\frac{1}{2}-iT$, $2-iT$, $2+iT$, $\frac{1}{2}+iT$. We have

$$\int_{\frac{1}{2}\pm iT}^{2\pm iT} \zeta^k(s)\frac{x^s}{s}\,ds = O(x^2T^{\epsilon-1}),$$

$$\int_{\frac{1}{2}-iT}^{\frac{1}{2}+iT} \zeta^k(s)\frac{x^s}{s}\,ds = O\left\{x^{\frac{1}{2}}\int_{-T}^{T} |\tfrac{1}{2}+it|^{\epsilon-1}\,dt\right\} = O(x^{\frac{1}{2}}T^{\epsilon}).$$

The residue at $s = 1$ accounts for the difference between $D_k(x)$ and $\Delta_k(x)$. Hence

$$\Delta_k(x) = O(x^{\frac{1}{2}}T^{\epsilon}) + O(x^2T^{\epsilon-1}).$$

Taking $T = x^2$, it follows that $\alpha_k \leqslant \frac{1}{2}$. Hence also $\beta_k \leqslant \frac{1}{2}$. But in fact $\sigma_k \leqslant \frac{1}{2}$ on the Lindelöf hypothesis, so that, by Theorem 12.7, (13.4.3) also follows.

13.5. *The Lindelöf hypothesis and the zeros.*

THEOREM 13.5.† *A necessary and sufficient condition for the truth of the Lindelöf hypothesis is that, for every $\sigma > \frac{1}{2}$,*

$$N(\sigma, T+1) - N(\sigma, T) = o(\log T).$$

The necessity of the condition is easily proved. We apply Jensen's formula

$$\log\frac{r^n}{r_1 \ldots r_n} = \frac{1}{2\pi}\int_0^{2\pi} \log|f(re^{i\theta})|\,d\theta - \log|f(0)|,$$

where $r_1,\ldots$ are the moduli of the zeros of $f(s)$ in $|s| \leqslant r$, to the circle with centre $2+it$ and radius $\frac{3}{2}-\frac{1}{4}\delta$, $f(s)$ being $\zeta(s)$. On the Lindelöf

† Backlund (4).

hypothesis the right-hand side is less than $o(\log t)$; and, if there are N zeros in the concentric circle of radius $\frac{3}{2}-\frac{1}{2}\delta$, the left-hand side is greater than

$$N\log\{(\tfrac{3}{2}-\tfrac{1}{4}\delta)/(\tfrac{3}{2}-\tfrac{1}{2}\delta)\}.$$

Hence the number of zeros in the circle of radius $\frac{3}{2}-\frac{1}{2}\delta$ is $o(\log t)$; and the result stated, with $\sigma = \frac{1}{2}+\delta$, clearly follows by superposing a number (depending on δ only) of such circles.

To prove the converse,† let C_1 be the circle with centre $2+iT$ and radius $\frac{3}{2}-\delta$ $(\delta > 0)$, and let Σ_1 denote a summation over zeros of $\zeta(s)$ in C_1. Let C_2 be the concentric circle of radius $\frac{3}{2}-2\delta$. Then for s in C_2

$$\psi(s) = \frac{\zeta'(s)}{\zeta(s)} - \sum\nolimits_1 \frac{1}{s-\rho} = O\left(\frac{\log T}{\delta}\right).$$

This follows from Theorem 9.6 (A), since for each term which is in one of the sums

$$\sum\nolimits_1 \frac{1}{s-\rho}, \qquad \sum_{|t-\gamma|<1} \frac{1}{s-\rho},$$

but not in the other, $|s-\rho| \geqslant \delta$; and the number of such terms is $O(\log T)$.

Let C_3 be the concentric circle of radius $\frac{3}{2}-3\delta$, C the concentric circle of radius $\frac{1}{2}$. Then $\psi(s) = o(\log T)$ for s in C, since each term is $O(1)$, and by hypothesis the number of terms is $o(\log T)$. Hence Hadamard's three-circles theorem gives, for s in C_3,

$$|\psi(s)| < \{o(\log T)\}^{\alpha}\{O(\delta^{-1}\log T)\}^{\beta}$$

where $\alpha+\beta = 1$, $0 < \beta < 1$, α and β depending on δ only. Thus in C_3

$$\psi(s) = o(\log T),$$

for any given δ.

Now

$$\begin{aligned}\int\limits_{\frac{1}{2}+3\delta}^{2} \psi(s)\, d\sigma &= \log\zeta(2+it) - \log\zeta(\tfrac{1}{2}+3\delta+it) - \\ &\qquad - \sum\nolimits_1 \{\log(2+it-\rho) - \log(\tfrac{1}{2}+3\delta+it-\rho)\} \\ &= O(1) - \log\zeta(\tfrac{1}{2}+3\delta+it) + o(\log T) + \\ &\qquad + \sum\nolimits_1 \log(\tfrac{1}{2}+3\delta+it-\rho),\end{aligned}$$

since $\sum_1$ has $o(\log T)$ terms. Also, if $t = T$, the left-hand side is $o(\log T)$. Hence, putting $t = T$ and taking real parts,

$$\log|\zeta(\tfrac{1}{2}+3\delta+iT)| = o(\log T) + \sum\nolimits_1 \log|\tfrac{1}{2}+3\delta+iT-\rho|.$$

Since $|\frac{1}{2}+3\delta+iT-\rho| < A$ in C_1, it follows that

$$\log|\zeta(\tfrac{1}{2}+3\delta+iT)| < o(\log T),$$

i.e. the Lindelöf hypothesis is true.

† Littlewood (4).

13.6. THEOREM 13.6 (A).† *On the Lindelöf hypothesis*

$$S(t) = o(\log t).$$

The proof is the same as Backlund's proof (§ 9.4) that, without any hypothesis, $S(t) = O(\log t)$, except that we now use $\zeta(s) = O(t^\epsilon)$ where we previously used $\zeta(s) = O(t^A)$.

THEOREM 13.6 (B).‡ *On the Lindelöf hypothesis*

$$S_1(t) = o(\log t).$$

Integrating the real part of (9.6.3) from $\frac{1}{2}$ to $\frac{1}{2}+3\delta$,

$$\int\limits_{\frac{1}{2}}^{\frac{1}{2}+3\delta} \log|\zeta(s)|\,d\sigma = \sum_{|\gamma-t|<1} \int\limits_{\frac{1}{2}}^{\frac{1}{2}+3\delta} \log|s-\rho|\,d\sigma + O(\delta\log t),$$

where $\rho = \beta+i\gamma$ runs through zeros of $\zeta(s)$. Now

$$\int\limits_{\frac{1}{2}}^{\frac{1}{2}+3\delta} \log|s-\rho|\,d\sigma = \frac{1}{2}\int\limits_{\frac{1}{2}}^{\frac{1}{2}+3\delta} \log\{(\sigma-\beta)^2+(\gamma-t)^2\}\,d\sigma \leqslant \frac{3\delta}{2}\log 2$$

and

$$\geqslant \int\limits_{\frac{1}{2}}^{\frac{1}{2}+3\delta} \log|\sigma-\beta|\,d\sigma \geqslant \int\limits_{\frac{1}{2}}^{\frac{1}{2}+3\delta} \log|\sigma-\tfrac{1}{2}-\tfrac{3}{2}\delta|\,d\sigma = 3\delta(\log\tfrac{3}{2}\delta-1).$$

Hence

$$\int\limits_{\frac{1}{2}}^{\frac{1}{2}+3\delta} \log|\zeta(s)|\,d\sigma = \sum_{|\gamma-t|<1} O\left(\delta\log\frac{1}{\delta}\right) + O(\delta\log t)$$

$$= O(\delta\log 1/\delta\,.\log t).$$

Also, as in the proof of Theorem 13.5,

$$\log\zeta(s) = \textstyle\sum_1 \log(s-\rho)+o(\log t) \quad (\tfrac{1}{2}+3\delta \leqslant \sigma \leqslant 2).$$

Hence

$$\int\limits_{\frac{1}{2}+3\delta}^{2} \log|\zeta(s)|\,d\sigma = \sum\nolimits_1 \int\limits_{\frac{1}{2}+3\delta}^{2} \log|s-\rho|\,d\sigma + o(\log t)$$

$$= \textstyle\sum_1 O(1)+o(\log t)$$

$$= o(\log t).$$

Hence, by Theorem 9.9,

$$S_1(t) = \frac{1}{\pi}\int\limits_{\frac{1}{2}}^{2} \log|\zeta(s)|\,d\sigma + O(1)$$

$$= O(\delta\log 1/\delta\,.\log t)+o(\log t)+O(1),$$

and the result follows on choosing first δ and then t.

† Cramér (1), Littlewood (4). ‡ Littlewood (4).

NOTES FOR CHAPTER 13

13.7. Since the proof of Theorem 13.6(A) is not quite straightforward we give the details. Let

$$g(z) = \tfrac{1}{2}\{\zeta(z+2+iT)+\zeta(z+2-iT)\}$$

and define $n(r)$ to be the number of zeros of $g(z)$ in the disc $|z| \leqslant r$. As in § 9.4 one finds that $S(T) \ll n(\tfrac{3}{2})+1$. Moreover, by Jensen's Thorem, one has

$$\int_0^R \frac{n(r)}{r}\,dr = \frac{1}{2\pi}\int_0^{2\pi} \log|g(Re^{i\vartheta})|\,d\vartheta - \log|g(0)|. \tag{13.7.1}$$

With our choice of $g(z)$ we have $\log|g(0)| = \log|\mathbf{R}\zeta(2+iT)| = O(1)$. We shall take $R = \tfrac{3}{2}+\delta$. Then, on the Lindelöf Hypothesis, one finds that

$$|\zeta(Re^{i\vartheta}+2\pm iT)| \leqslant T^{\varepsilon}$$

for $\cos\vartheta \geqslant -3/(2R)$ and T sufficiently large. The remaining range for ϑ is an interval of length $O(\delta^{\frac{1}{2}})$. Here we write $\mathbf{R}(Re^{i\vartheta}+2) = \sigma$, so that $\tfrac{1}{2}-\delta \leqslant \sigma \leqslant \tfrac{1}{2}$. Then, using the convexity of the μ function, together with the facts that $\mu(0) = \tfrac{1}{2}$ and, on the Lindelöf Hypothesis, that $\mu(\tfrac{1}{2}) = 0$, we have $\mu(\sigma) \leqslant \delta$. It follows that

$$|\zeta(Re^{i\vartheta}+2\pm iT)| \leqslant T^{\delta+\varepsilon}$$

for $\cos\vartheta \leqslant -3/2R$, and large enough T. We now see that the right-hand side of (13.7.1) is at most

$$O(\varepsilon\log T) + O\{\delta^{\frac{1}{2}}(\delta+\varepsilon)\log T\}.$$

Since

$$\frac{\delta}{R}\,n(\tfrac{3}{2}) \leqslant \int_0^R \frac{n(r)}{r}\,dr$$

we conclude that

$$n(\tfrac{3}{2}) = O\left\{\left(\frac{\varepsilon}{\delta}+\delta^{-\frac{1}{2}}(\delta+\varepsilon)\right)\log T\right\},$$

and on taking $\delta = \varepsilon^{\frac{2}{3}}$ we obtain $n(\tfrac{3}{2}) = O(\varepsilon^{\frac{1}{3}}\log T)$, from which the result follows.

13.8. It has been observed by Ghosh and Goldston (in unpublished work) that the converse of Theorem 13.6(B) follows from Lemma 21 of Selberg (**5**).

THEOREM 13.8. *If $S_1(t) = o(\log t)$, then the Lindelöf hypothesis holds.*

We reproduce the arguments used by Selberg and by Ghosh and Goldston here. Let $\frac{1}{2} \leqslant \sigma \leqslant 2$, and consider the integral

$$\frac{1}{2\pi i}\int_{5-i\infty}^{5+i\infty} \frac{\log\zeta(s+iT)}{4-(s-\sigma)^2}\,ds.$$

Since $\log\zeta(s+iT) \ll 2^{-\mathbf{R}(s)}$ the integral is easily seen to vanish, by moving the line of integration to the right. We now move the line of integration to the left, to $\mathbf{R}(s) = \sigma$, passing a pole at $s = 2+\sigma$, with residue $-\frac{1}{4}\log\zeta(2+\sigma+iT) = O(1)$. We must make detours around $s = 1-iT$, if $\sigma < 1$, and around $s = \rho - iT$, if $\sigma < \beta$. The former, if present, will produce an integral contributing $O(T^{-2})$, and the latter, if present, will be

$$-\int_0^{\beta-\sigma} \frac{du}{4-\{u+i(\gamma-T)\}^2}.$$

It follows that

$$\frac{1}{2\pi}\int_{-\infty}^{\infty} \frac{\log\zeta(\sigma+it+iT)}{4+t^2}\,dt - \sum_{\beta>\sigma}\int_0^{\beta-\sigma} \frac{du}{4-\{u+i(\gamma-T)\}^2} = O(1),$$

for $T \geqslant 1$. We now take real parts and integrate for $\frac{1}{2} \leqslant \sigma \leqslant 2$. Then by Theorem 9.9 we have

$$\frac{1}{2}\int_{-\infty}^{\infty} \frac{S_1(t+T)}{4+t^2}\,dt = \sum_{\beta>\frac{1}{2}}\int_0^{\beta-\frac{1}{2}} (\beta-\tfrac{1}{2}-u)\mathbf{R}\left(\frac{1}{4-\{u+i(\gamma-T)\}^2}\right)du + O(1). \tag{13.8.1}$$

By our hypothesis the integral on the left is $o(\log T)$. Moreover

$$\mathbf{R}\left(\frac{1}{4-\{u+i(\gamma-T)\}^2}\right) \geqslant \begin{cases} A\ (>0) & \text{if } |\gamma-T| \leqslant 1, \\ 0, & \text{otherwise.} \end{cases}$$

If $\sigma > \frac{1}{2}$ is given, then each zero counted by $N(\sigma, T+1) - N(\sigma, T)$ contributes at least $\frac{1}{2}(\sigma-\frac{1}{2})^2 A$ to the sum on the right of (13.8.1), whence $N(\sigma, T+1) - N(\sigma, T) = o(\log T)$. Theorem 13.8 therefore follows from Theorem 13.5.

XIV

CONSEQUENCES OF THE RIEMANN HYPOTHESIS

14.1. In this chapter we assume the truth of the unproved Riemann hypothesis, that all the complex zeros of $\zeta(s)$ lie on the line $\sigma = \frac{1}{2}$. It will be seen that a perfectly coherent theory can be constructed on this basis, which perhaps gives some support to the view that the hypothesis is true. A proof of the hypothesis would make the 'theorems' of this chapter essential parts of the theory, and would make unnecessary much of the tentative analysis of the previous chapters.

The Riemann hypothesis, of course, leaves nothing more to be said about the 'horizontal' distribution of the zeros. From it we can also deduce interesting consequences both about the 'vertical' distribution of the zeros and about the order problems. In most cases we obtain much more precise results with the hypothesis than without it. But even a proof of the Riemann hypothesis would not by any means complete the theory. The finer shades in the behaviour of $\zeta(s)$ would still not be completely determined.

On the Riemann hypothesis, the function $\log\zeta(s)$, as well as $\zeta(s)$, is regular for $\sigma > \frac{1}{2}$ (except at $s = 1$). This is the basis of most of the analysis of this chapter.

We shall not repeat the words 'on the Riemann hypothesis', which apply throughout the chapter.

14.2. THEOREM 14.2.† *We have*

$$\log\zeta(s) = O\{(\log t)^{2-2\sigma+\epsilon}\} \tag{14.2.1}$$

uniformly for $\frac{1}{2} < \sigma_0 \leqslant \sigma \leqslant 1$.

Apply the Borel–Carathéodory theorem to the function $\log\zeta(z)$ and the circles with centre $2+it$ and radii $\frac{3}{2}-\frac{1}{2}\delta$ and $\frac{3}{2}-\delta$ $(0 < \delta < \frac{1}{2})$. On the larger circle

$$\mathbf{R}\{\log\zeta(z)\} = \log|\zeta(z)| < A\log t.$$

Hence, on the smaller circle,

$$\begin{aligned}|\log\zeta(z)| &\leqslant \frac{3-2\delta}{\frac{1}{2}\delta}A\log t + \frac{3-\frac{3}{2}\delta}{\frac{1}{2}\delta}|\log|\zeta(2+it)|| \\ &< A\delta^{-1}\log t.\end{aligned} \tag{14.2.2}$$

† Littlewood (1).

Now apply Hadamard's three-circles theorem to the circles C_1, C_2, C_3 with centre σ_1+it $(1<\sigma_1\leqslant t)$, passing through the points $1+\eta+it$, $\sigma+it$, $\frac{1}{2}+\delta+it$. The radii are thus

$$r_1=\sigma_1-1-\eta, \qquad r_2=\sigma_1-\sigma, \qquad r_3=\sigma_1-\tfrac{1}{2}-\delta.$$

If the maxima of $|\log\zeta(z)|$ on the circles are M_1, M_2, M_3, we obtain

$$M_2\leqslant M_1^{1-a}M_3^a,$$

where

$$a=\log\frac{r_2}{r_1}\Big/\log\frac{r_3}{r_1}=\log\left(1+\frac{1+\eta-\sigma}{\sigma_1-1-\eta}\right)\Big/\log\left(1+\frac{\frac{1}{2}+\eta-\delta}{\sigma_1-1-\eta}\right)$$

$$=\frac{1+\eta-\sigma}{\frac{1}{2}+\eta-\delta}+O\left(\frac{1}{\sigma_1}\right)=2-2\sigma+O(\delta)+O(\eta)+O\left(\frac{1}{\sigma_1}\right).$$

By (14.2.2), $M_3<A\delta^{-1}\log t$; and, since

$$\log\zeta(s)=\sum_{n=2}^{\infty}\frac{\Lambda_1(n)}{n^s} \quad (\Lambda_1(n)\leqslant 1), \tag{14.2.3}$$

$$M_1\leqslant\max_{x\geqslant 1+\eta}\left|\sum_{n=2}^{\infty}\frac{\Lambda_1(n)}{n^z}\right|\leqslant\sum_{n=2}^{\infty}\frac{1}{n^{1+\eta}}<\frac{A}{\eta}.$$

Hence

$$|\log\zeta(\sigma+it)|<\left(\frac{A}{\eta}\right)^{1-a}\left(\frac{A\log t}{\delta}\right)^a<\frac{A}{\eta^{1-a}\delta^a}(\log t)^{2-2\sigma+O(\delta)+O(\eta)+O(1/\sigma_1)}.$$

The result stated follows on taking δ and η small enough and σ_1 large enough. More precisely, we can take

$$\sigma_1=\frac{1}{\delta}=\frac{1}{\eta}=\log\log t;$$

since

$$(\log t)^{O(\delta)}=e^{O(\delta\log\log t)}=e^{O(1)}=O(1),$$

etc., we obtain

$$\log\zeta(s)=O\{\log\log t(\log t)^{2-2\sigma}\} \quad \left(\frac{1}{2}+\frac{1}{\log\log t}\leqslant\sigma\leqslant 1\right). \tag{14.2.4}$$

Since the index of $\log t$ in (14.2.1) is less than unity if ϵ is small enough, it follows that (with a new ϵ)

$$-\epsilon\log t<\log|\zeta(s)|<\epsilon\log t \quad (t>t_0(\epsilon)),$$

i.e. we have both

$$\zeta(s)=O(t^\epsilon), \tag{14.2.5}$$

$$\frac{1}{\zeta(s)}=O(t^\epsilon) \tag{14.2.6}$$

for every $\sigma>\frac{1}{2}$. In particular, *the truth of the Lindelöf hypothesis follows from that of the Riemann hypothesis.*

It also follows that *for every fixed* $\sigma > \frac{1}{2}$, *as* $T \to \infty$

$$\int_1^T \frac{dt}{|\zeta(\sigma+it)|^2} \sim \frac{\zeta(2\sigma)}{\zeta(4\sigma)} T.$$

For $\sigma > 1$ this follows from (7.1.2) and (1.2.7). For $\frac{1}{2} < \sigma \leqslant 1$ it follows from (14.2.6) and the analysis of § 7.9, applied to $1/\zeta(s)$ instead of to $\zeta^k(s)$.

14.3. The function† $\nu(\sigma)$. For each $\sigma > \frac{1}{2}$ we define $\nu(\sigma)$ as the lower bound of numbers a such that

$$\log \zeta(s) = O(\log^a t).$$

It is clear from (14.2.3) that $\nu(\sigma) \leqslant 0$ for $\sigma > 1$; and from (14.2.2) that $\nu(\sigma) \leqslant 1$ for $\frac{1}{2} < \sigma \leqslant 1$; and in fact from (14.2.1) that $\nu(\sigma) \leqslant 2-2\sigma$ for $\frac{1}{2} < \sigma \leqslant 1$.

On the other hand, since $\Lambda_1(2) = 1$, (14.2.3) gives

$$|\log \zeta(s)| \geqslant \frac{1}{2^\sigma} - \sum_{n=3}^{\infty} \frac{\Lambda_1(n)}{n^\sigma},$$

and hence $\nu(\sigma) \geqslant 0$ if σ is so large that the right-hand side is positive. Since

$$\sum_{n=3}^{\infty} \frac{\Lambda_1(n)}{n^\sigma} \leqslant \sum_{n=3}^{\infty} \frac{1}{n^\sigma} < \int_2^{\infty} \frac{dx}{x^\sigma} = \frac{2^{1-\sigma}}{\sigma-1}$$

this is certainly true for $\sigma \geqslant 3$. Hence $\nu(\sigma) = 0$ for $\sigma \geqslant 3$.

Now let $\frac{1}{2} < \sigma_1 < \sigma < \sigma_2 \leqslant 4$, and suppose that

$$\log \zeta(\sigma_1+it) = O(\log^a t), \qquad \log \zeta(\sigma_2+it) = O(\log^b t).$$

Let
$$g(s) = \log \zeta(s)\{\log(-is)\}^{-k(s)},$$

where $k(s)$ is the linear function of s such that $k(\sigma_1) = a$, $k(\sigma_2) = b$, viz.

$$k(s) = \frac{(s-\sigma_1)b+(\sigma_2-s)a}{\sigma_2-\sigma_1}.$$

Here
$$\{\log(-is)\}^{-k(s)} = e^{-k(s)\log\log(-is)},$$

where
$$\log(-is) = \log(t-i\sigma), \qquad \log\log(-is) \;\; (t > e)$$

denote the branches which are real for $\sigma = 0$. Thus

$$\log(-is) = \log t + \log\left(1-\frac{i\sigma}{t}\right) = \log t + O\left(\frac{1}{t}\right),$$

$$\log\log(-is) = \log\log t + \log\left\{1+O\left(\frac{1}{t\log t}\right)\right\}$$
$$= \log\log t + O(1/t).$$

† Bohr and Landau (3), Littlewood (5).

Hence

$$|\{\log(-is)\}^{-k(s)}| = e^{-\mathbf{R}\{k(s)\log\log(-is)\}} = e^{-k(\sigma)\log\log t+O(1/t)}$$
$$= (\log t)^{-k(\sigma)}\{1+O(1/t)\}.$$

Hence $g(s)$ is bounded on the lines $\sigma = \sigma_1$ and $\sigma = \sigma_2$; and it is $O(\log^K t)$ for some K uniformly in the strip. Hence, by the theorem of Phragmén and Lindelöf, it is bounded in the strip. Hence

$$\log \zeta(s) = O\{(\log t)^{k(\sigma)}\},$$

i.e.

$$\nu(\sigma) \leqslant k(\sigma) = \frac{(\sigma-\sigma_1)b+(\sigma_2-\sigma)a}{\sigma_2-\sigma_1}. \tag{14.3.1}$$

Taking $\sigma = 3$, $\sigma_2 = 4$, $\nu(3) = 0$, $b = 0$, we obtain $a \geqslant 0$. Hence $\nu(\sigma) \geqslant 0$ for $\sigma > \frac{1}{2}$. Hence $\nu(\sigma) = 0$ for $\sigma > 1$.

Since $\nu(\sigma)$ is finite for every $\sigma > \frac{1}{2}$, we can take $a = \nu(\sigma_1)+\epsilon$, $b = \nu(\sigma_2)+\epsilon$ in (14.3.1). Making $\epsilon \to 0$, we obtain

$$\nu(\sigma) \leqslant \frac{(\sigma-\sigma_1)\nu(\sigma_2)+(\sigma_2-\sigma)\nu(\sigma_1)}{\sigma_2-\sigma_1},$$

i.e. $\nu(\sigma)$ *is a convex function of* σ. Hence it is continuous, and it is non-increasing since it is ultimately zero.

We can also show that $\zeta'(s)/\zeta(s)$ *has the same* ν*-function as* $\log \zeta(s)$. Let $\nu_1(s)$ be the ν-function of $\zeta'(s)/\zeta(s)$. Since

$$\frac{\zeta'(s)}{\zeta(s)} = \frac{1}{2\pi i} \int\limits_{|s-z|=\delta} \frac{\log \zeta(z)}{(s-z)^2}\, dz = O\left\{\frac{1}{\delta}(\log t)^{\nu(\sigma-\delta)+\epsilon}\right\},$$

we have

$$\nu_1(\sigma) \leqslant \nu(\sigma-\delta)$$

for every positive δ; and since $\nu(\sigma)$ is continuous it follows that

$$\nu_1(\sigma) \leqslant \nu(\sigma).$$

We can show, as in the case of $\nu(\sigma)$, that $\nu_1(\sigma)$ is non-increasing, and is zero for $\sigma \geqslant 3$. Hence for $\sigma < 3$

$$\log \zeta(s) = -\int\limits_{\sigma}^{3} \frac{\zeta'(x+it)}{\zeta(x+it)}\, dx - \log \zeta(3+it)$$
$$= O\left\{\int\limits_{\sigma}^{3} (\log t)^{\nu_1(x)+\epsilon}\, dx\right\} + O(1)$$
$$= O\{(\log t)^{\nu_1(\sigma)+\epsilon}\},$$

i.e.

$$\nu(\sigma) \leqslant \nu_1(\sigma).$$

The exact value of $\nu(\sigma)$ is not known for any value of σ less than 1. All we know is

THEOREM 14.3. *For* $\frac{1}{2} < \sigma < 1$,

$$1-\sigma \leqslant \nu(\sigma) \leqslant 2(1-\sigma).$$

The upper bound follows from Theorem 14.2 and the lower bound from Theorem 8.12. The same lower bound can, however, be obtained in another and in some respects simpler way, though this proof, unlike the former, depends essentially on the Riemann hypothesis. For the proof we require some new formulae.

14.4. THEOREM 14.4.† *As* $t \to \infty$,

$$-\frac{\zeta'(s)}{\zeta(s)} = \sum_{n=1}^{\infty} \frac{\Lambda(n)}{n^s} e^{-\delta n} + \sum_{\rho} \delta^{s-\rho}\Gamma(\rho-s) + O(\delta^{\sigma-\frac{1}{4}}\log t), \quad (14.4.1)$$

uniformly for $\frac{1}{2} \leqslant \sigma \leqslant \frac{9}{8}$, $e^{-\sqrt{t}} \leqslant \delta \leqslant 1$.

Taking $a_n = \Lambda(n)$, $f(s) = -\zeta'(s)/\zeta(s)$ in the lemma of § 7.9, we have

$$\sum_{n=1}^{\infty} \frac{\Lambda(n)}{n^s} e^{-\delta n} = -\frac{1}{2\pi i} \int_{2-i\infty}^{2+i\infty} \Gamma(z-s)\frac{\zeta'(z)}{\zeta(z)} \delta^{s-z}\, dz. \quad (14.4.2)$$

Now, by Theorem 9.6 (A),

$$\frac{\zeta'(s)}{\zeta(s)} = \sum_{|t-\gamma|<1} \frac{1}{s-\frac{1}{2}-i\gamma} + O(\log t),$$

and there are $O(\log t)$ terms in the sum. Hence

$$\frac{\zeta'(s)}{\zeta(s)} = O(\log t)$$

on any line $\sigma \neq \frac{1}{2}$. Also

$$\frac{\zeta'(s)}{\zeta(s)} = O\left(\frac{\log t}{\min|t-\gamma|}\right) + O(\log t)$$

uniformly for $-1 \leqslant \sigma \leqslant 2$. Since each interval $(n, n+1)$ contains values of t whose distance from the ordinate of any zero exceeds $A/\log n$, there is a t_n in any such interval for which

$$\frac{\zeta'(s)}{\zeta(s)} = O(\log^2 t) \quad (-1 \leqslant \sigma \leqslant 2,\ t = t_n).$$

† Littlewood (5), to the end of § 14.8.

By the theorem of residues,

$$\frac{1}{2\pi i}\left(\int\limits_{2-it_n}^{2+it_n}+\int\limits_{2+it_n}^{\frac{1}{4}+it_n}+\int\limits_{\frac{1}{4}+it_n}^{\frac{1}{4}-it_n}+\int\limits_{\frac{1}{4}-it_n}^{2-it_n}\right)\Gamma(z-s)\frac{\zeta'(z)}{\zeta(z)}\delta^{s-z}\,dz$$
$$=\frac{\zeta'(s)}{\zeta(s)}+\sum_{-t_n<\gamma<t_n}\Gamma(\rho-s)\delta^{s-\rho}-\Gamma(1-s)\delta^{s-1}.$$

The integrals along the horizontal sides tend to zero as $n\to\infty$, so that

$$\sum_{n=1}^{\infty}\frac{\Lambda(n)}{n^s}e^{-\delta n}=-\frac{1}{2\pi i}\int\limits_{\frac{1}{4}-i\infty}^{\frac{1}{4}+i\infty}\Gamma(z-s)\frac{\zeta'(z)}{\zeta(z)}\delta^{s-z}\,dz-$$
$$-\frac{\zeta'(s)}{\zeta(s)}-\sum_{\rho}\Gamma(\rho-s)\delta^{s-\rho}+\Gamma(1-s)\delta^{s-1}.$$

Since $\Gamma(z-s)=O(e^{-A|y-t|})$, the integral is

$$O\left\{\int\limits_{-\infty}^{\infty}e^{-A|y-t|}\log(|y|+2)\delta^{\sigma-\frac{1}{4}}\,dy\right\}$$
$$=O\left\{\int\limits_{0}^{2t}e^{-A|y-t|}\log(|2t|+2)\delta^{\sigma-\frac{1}{4}}\,dy\right\}+$$
$$+O\left\{\left(\int\limits_{-\infty}^{0}+\int\limits_{2t}^{\infty}\right)e^{-\frac{1}{2}A|y|}\log(|y|+2)\delta^{\sigma-\frac{1}{4}}\,dy\right\}$$
$$=O(\delta^{\sigma-\frac{1}{4}}\log t)+O(\delta^{\sigma-\frac{1}{4}})=O(\delta^{\sigma-\frac{1}{4}}\log t).$$

Also

$$\Gamma(1-s)\delta^{s-1}=O(e^{-At}\delta^{\sigma-1})=O(e^{-At}\delta^{-\frac{1}{2}})$$
$$=O(e^{-At+\frac{1}{2}\sqrt{t}})=O(e^{-At})=O(\delta^{\sigma-\frac{1}{4}}\log t).$$

This proves the theorem.

14.5. We can now prove more precise results about $\zeta'(s)/\zeta(s)$ and $\log\zeta(s)$ than those expressed by the inequality $\nu(\sigma)\leqslant 2-2\sigma$.

THEOREM 14.5. *We have*

$$\frac{\zeta'(s)}{\zeta(s)}=O\{(\log t)^{2-2\sigma}\},\tag{14.5.1}$$

$$\log\zeta(s)=O\left\{\frac{(\log t)^{2-2\sigma}}{\log\log t}\right\},\tag{14.5.2}$$

uniformly for $\frac{1}{2}<\sigma_0\leqslant\sigma\leqslant\sigma_1<1$.

We have

$$\left|\frac{\zeta'(s)}{\zeta(s)}\right|\leqslant\sum_{n=1}^{\infty}\frac{\Lambda(n)}{n^{\sigma}}e^{-\delta n}+\delta^{\sigma-\frac{1}{2}}\sum_{\rho}|\Gamma(\rho-s)|+O(\delta^{\sigma-\frac{1}{4}}\log t).$$

Now

$$\sum_{n=1}^{\infty}\frac{\Lambda(n)}{n^{\sigma}}e^{-\delta n} = -\frac{1}{2\pi i}\int\limits_{2-i\infty}^{2+i\infty}\Gamma(z-\sigma)\frac{\zeta'(z)}{\zeta(z)}\delta^{\sigma-z}\,dz = O(\delta^{\sigma-1}),$$

since we may move the line of integration to $\mathbf{R}(z) = \frac{3}{4}$, and the leading term is the residue at $z = 1$. Also

$$|\Gamma(\rho-s)| < Ae^{-A|\gamma-t|}$$

uniformly for σ in the above range. Hence

$$\sum|\Gamma(\rho-s)| < A\sum_{\gamma}e^{-A|t-\gamma|} = A\sum_{n=1}^{\infty}\sum_{n-1\leqslant|t-\gamma|<n}e^{-A|t-\gamma|}.$$

The number of terms in the inner sum is

$$O\{\log(t+n)\} = O(\log t)+O\{\log(n+1)\}.$$

Hence we obtain

$$O\Big[\sum_{n=1}^{\infty}e^{-An}\{\log t+\log(n+1)\}\Big] = O(\log t).$$

Hence
$$\frac{\zeta'(s)}{\zeta(s)} = O(\delta^{\sigma-1})+O(\delta^{\sigma-\frac{1}{2}}\log t)+O(\delta^{\sigma-\frac{1}{4}}\log t),$$

and taking $\delta = (\log t)^{-2}$ we obtain the first result.

Again for $\sigma_0 \leqslant \sigma \leqslant \sigma_1$

$$\log\zeta(s) = \log\zeta(\sigma_1+it)-\int\limits_{\sigma}^{\sigma_1}\frac{\zeta'(x+it)}{\zeta(x+it)}\,dx$$

$$= O\{(\log t)^{2-2\sigma_1+\epsilon}\}+O\left\{\int\limits_{\sigma}^{\sigma_1}(\log t)^{2-2x}\,dx\right\}$$

$$= O\{(\log t)^{2-2\sigma_1+\epsilon}\}+O\left\{\frac{(\log t)^{2-2\sigma}}{\log\log t}\right\}.$$

If $\sigma \leqslant \sigma_2 < \sigma_1$ and $\epsilon < 2(\sigma_1-\sigma_2)$, this is of the required form; and since σ_1 and so σ_2 may be as near to 1 as we please, the second result (with σ_2 for σ_1) follows.

14.6. To obtain the alternative proof of the inequality $\nu(\sigma) \geqslant 1-\sigma$ we require an approximate formula for $\log\zeta(s)$.

THEOREM 14.6. *For fixed* α *and* σ *such that* $\frac{1}{2} < \alpha < \sigma \leqslant 1$, *and* $e^{-\sqrt{t}} \leqslant \delta \leqslant 1$,

$$\log\zeta(s) = \sum_{n=1}^{\infty}\frac{\Lambda_1(n)}{n^s}e^{-\delta n}+O\{\delta^{\sigma-\alpha}(\log t)^{\nu(\alpha)+\epsilon}\}+O(1).$$

Moving the line of integration in (14.4.2) to $\mathbf{R}(w) = \alpha$, we have

$$\sum_{n=1}^{\infty} \frac{\Lambda(n)}{n^s} e^{-\delta n} = -\frac{\zeta'(s)}{\zeta(s)} - \Gamma(1-s)\,\delta^{s-1} - \frac{1}{2\pi i}\int_{\alpha-i\infty}^{\alpha+i\infty} \Gamma(z-s)\frac{\zeta'(z)}{\zeta(z)}\,\delta^{s-z}\,dz.$$

Since $\zeta'(s)/\zeta(s)$ has the ν-function $\nu(\sigma)$, the integral is of the form

$$O\left\{\delta^{\sigma-\alpha}\int_{-\infty}^{\infty} e^{-A|y-t|}\{\log(|y|+2)\}^{\nu(\alpha)+\epsilon}\,dy\right\} = O\{\delta^{\sigma-\alpha}(\log t)^{\nu(\alpha)+\epsilon}\};$$

and $\Gamma(1-s)\delta^{s-1}$ is also of this form, as in § 14.4. Hence

$$-\frac{\zeta'(s)}{\zeta(s)} = \sum_{n=1}^{\infty} \frac{\Lambda(n)}{n^s} e^{-\delta n} + O\{\delta^{\sigma-\alpha}(\log t)^{\nu(\alpha)+\epsilon}\}.$$

This result holds uniformly in the range $[\sigma, \frac{9}{8}]$, and so we may integrate over this interval. We obtain

$$\log\zeta(s) - \sum_{n=1}^{\infty} \frac{\Lambda_1(n)}{n^s} e^{-\delta n} + O\{\delta^{\sigma-\alpha}(\log t)^{\nu(\alpha)+\epsilon}\}$$
$$= \log\zeta(\tfrac{9}{8}+it) - \sum_{n=1}^{\infty} \frac{\Lambda_1(n)}{n^{\frac{9}{8}+it}} e^{-\delta n} = O(1),$$

as required.

14.7. Proof that $\nu(\sigma) \geqslant 1-\sigma$. Theorem 14.6 enables us to extend the method of Diophantine approximation, already used for $\sigma > 1$, to values of σ between $\frac{1}{2}$ and 1. It gives

$$\log|\zeta(s)| = \sum_{n=1}^{\infty} \frac{\Lambda_1(n)}{n^\sigma} \cos(t\log n) e^{-\delta n} + O\{\delta^{\sigma-\alpha}(\log t)^{\nu(\alpha)+\epsilon}\} + O(1),$$

$$= \sum_{n=1}^{N} \frac{\Lambda_1(n)}{n^\sigma} \cos(t\log n) e^{-\delta n} + O\Big(\sum_{n=N+1}^{\infty} e^{-\delta n}\Big) + O\{\delta^{\sigma-\alpha}(\log t)^{\nu(\alpha)+\epsilon}\} + O(1)$$

for all values of N. Now by Dirichlet's theorem (§ 8.2) there is a number t in the range $2\pi \leqslant t \leqslant 2\pi q^N$, and integers $x_1,\ldots, x_N$, such that

$$\left|t\frac{\log n}{2\pi} - x_n\right| \leqslant \frac{1}{q} \quad (n = 1, 2,\ldots, N).$$

Let us assume for the moment that this number t satisfies the condition of Theorem 14.6 that $e^{-\sqrt{t}} \leqslant \delta$. It gives

$$\sum_{n=1}^{N} \frac{\Lambda_1(n)}{n^\sigma} \cos(t\log n) e^{-\delta n} \geqslant \sum_{n=1}^{N} \frac{\Lambda_1(n)}{n^\sigma} \cos\frac{2\pi}{q} e^{-\delta n}$$
$$= \sum_{n=1}^{N} \frac{\Lambda_1(n)}{n^\sigma} e^{-\delta n} + O\Big(\frac{1}{q}\Big) \sum_{n=1}^{N} \frac{1}{n^\sigma}.$$

Now

$$\sum_{n=1}^{N} \frac{\Lambda_1(n)}{n^\sigma} e^{-\delta n} \geqslant \frac{1}{\log N} \sum_{n=1}^{N} \frac{\Lambda(n)}{n^\sigma} e^{-\delta n}$$

$$\geqslant \frac{1}{\log N} \sum_{n=1}^{\infty} \frac{\Lambda(n)}{n^\sigma} e^{-\delta n} + O\Big(\sum_{n=N+1}^{\infty} e^{-\delta n}\Big)$$

$$> \frac{K(\sigma)\delta^{\sigma-1}}{\log N} + O\left(\frac{e^{-\delta N}}{\delta}\right)$$

as in § 14.5. Hence

$$\log|\zeta(s)| > \frac{K(\sigma)\delta^{\sigma-1}}{\log N} + O\left(\frac{e^{-\delta N}}{\delta}\right) + O\left(\frac{N^{1-\sigma}}{q}\right) + O\{\delta^{\sigma-\alpha}(\log t)^{\nu(\alpha)+\epsilon}\} + O(1).$$

Take $q = N = [\delta^{-a}]$, where $a > 1$. The second and third terms on the right are then bounded. Also

$$\log t \leqslant N \log q + \log 2\pi \leqslant \frac{a}{\delta^a} \log\frac{1}{\delta} + \log 2\pi,$$

so that

$$\delta \leqslant K(\log t)^{-1/a+\epsilon}.$$

Hence

$$\log|\zeta(s)| > K(\log t)^{1-\sigma-\eta} + O\{(\log t)^{\alpha-\sigma+\nu(\alpha)+\eta'}\},$$

where η and η' are functions of a which tend to zero as $a \to 1$.

If the first term on the right is of larger order than the second, it follows at once that $\nu(\sigma) \geqslant 1-\sigma$. Otherwise

$$\alpha - \sigma + \nu(\alpha) \geqslant 1 - \sigma,$$

and making $\alpha \to \sigma$ the result again follows.

We have still to show that the t of the above argument satisfies $e^{-\sqrt{t}} \leqslant \delta$. Suppose on the contrary that $\delta < e^{-\sqrt{t}}$ for some arbitrarily small values of δ. Now, by (8.4.4),

$$|\zeta(s)| \geqslant \left(\cos\frac{2\pi}{q} - 2N^{1-\sigma}\right)\zeta(\sigma) > \frac{A}{\sigma-1}(\tfrac{1}{2} - 2N^{1-\sigma})$$

for $\sigma > 1$, $q \geqslant 6$. Taking $\sigma = 1 + \log 8/\log N$,

$$|\zeta(s)| > \frac{A}{\sigma-1} = A \log N > A \log\frac{1}{\delta} > At^{\frac{1}{2}}.$$

Since $|\zeta(s)| \to \infty$ and $t \geqslant 2\pi$, $t \to \infty$, and the above result contradicts Theorem 3.5. This completes the proof.

14.8. The function $\zeta(1+it)$. We are now in a position to obtain fairly precise information about this function. We shall first prove

THEOREM 14.8. *We have*

$$|\log \zeta(1+it)| \leqslant \log\log\log t + A. \tag{14.8.1}$$

In particular

$$\zeta(1+it) = O(\log\log t), \tag{14.8.2}$$

$$\frac{1}{\zeta(1+it)} = O(\log\log t). \tag{14.8.3}$$

Taking $\sigma = 1$, $\alpha = \frac{3}{4}$ in Theorem 14.6, we have

$$|\log\zeta(1+it)| \leqslant \sum_{n=1}^{\infty} \frac{\Lambda_1(n)}{n} e^{-\delta n} + O(\delta^{\frac{1}{4}}\log t) + O(1)$$

$$\leqslant \sum_{n=1}^{N} \frac{\Lambda_1(n)}{n} + \sum_{n=N+1}^{\infty} e^{-\delta n} + O(\delta^{\frac{1}{4}}\log t) + O(1)$$

$$\leqslant \log\log N + O(e^{-\delta N}/\delta) + O(\delta^{\frac{1}{4}}\log t) + O(1)$$

by (3.14.4). Taking $\delta = \log^{-4}t$, $N = 1+[\log^5 t]$, the result follows.

Comparing this result with Theorems 8.5 and 8.8, we see that, as far as the order of the functions $\zeta(1+it)$ and $1/\zeta(1+it)$ is concerned, the result is final. It remains to consider the values of the constants involved in the inequalities.

14.9. We define a function $\beta(\sigma)$ as

$$\beta(\sigma) = \frac{\nu(\sigma)}{2-2\sigma}.$$

By the convexity of $\nu(\sigma)$ we have, for $\frac{1}{2} < \sigma < \sigma' < 1$,

$$\nu(\sigma') \leqslant \frac{(1-\sigma')\nu(\sigma)+(\sigma'-\sigma)\nu(1)}{1-\sigma} = \frac{1-\sigma'}{1-\sigma}\nu(\sigma),$$

i.e.

$$\beta(\sigma') \leqslant \beta(\sigma).$$

Thus $\beta(\sigma)$ is non-increasing in $(\frac{1}{2}, 1)$. We write

$$\beta(\tfrac{1}{2}) = \lim_{\sigma\to\frac{1}{2}+0} \beta(\sigma), \qquad \beta(1) = \lim_{\sigma\to 1-0} \beta(\sigma).$$

Then by Theorem 14.3, for $\frac{1}{2} < \sigma < 1$,

$$\tfrac{1}{2} \leqslant \beta(1) \leqslant \beta(\sigma) \leqslant \beta(\tfrac{1}{2}) \leqslant 1.$$

We shall now prove†

THEOREM 14.9. *As* $t \to \infty$

$$|\zeta(1+it)| \leqslant 2\beta(1)e^{\gamma}\{1+o(1)\}\log\log t, \tag{14.9.1}$$

$$\frac{1}{|\zeta(1+it)|} \leqslant 2\beta(1)\frac{6e^{\gamma}}{\pi^2}\{1+o(1)\}\log\log t. \tag{14.9.2}$$

† Littlewood (6).

We observe that the $O(1)$ in Theorem 14.6 is actually $o(1)$ if $\delta \to 0$. Also, taking $\sigma = 1$,

$$\delta^{1-\alpha}(\log t)^{\nu(\alpha)+\epsilon} = o(1)$$

if

$$\delta = (\log t)^{-2\beta(\alpha)-\eta} \quad (\eta > 0).$$

Hence, for such δ,

$$\log \zeta(1+it) = \sum_{n=1}^{\infty} \frac{\Lambda_1(n)}{n^{1+it}} e^{-\delta n} + o(1)$$

$$= \sum_{p,m} \frac{e^{-\delta p^m}}{mp^{m(1+it)}} + o(1)$$

$$= \sum_{p,m} \frac{e^{-\delta mp}}{mp^{m(1+it)}} + \sum_p \sum_{m>1} \frac{e^{-\delta p^m} - e^{-\delta mp}}{mp^{m(1+it)}} + o(1).$$

Now the modulus of the second double sum does not exceed

$$\sum_p \sum_{m>1} \frac{e^{-\delta p^m} - e^{-\delta mp}}{p^m}.$$

This is evidently uniformly convergent for $\delta \geqslant 0$, the summand being less than p^{-m}. Since each term tends to zero with δ the sum is $o(1)$. Hence

$$\log \zeta(1+it) = \sum_{p,m} \frac{e^{-\delta mp}}{mp^{m(1+it)}} + o(1)$$

$$= -\sum_p \log\left(1 - \frac{e^{-\delta p}}{p^{1+it}}\right) + o(1)$$

$$= -\sum_{p \leqslant \varpi} \log\left(1 - \frac{e^{-\delta p}}{p^{1+it}}\right) + O\left(\sum_{\varpi+1}^{\infty} e^{-\delta n}\right) + o(1).$$

The second term is $O(e^{-\delta\varpi}/\delta) = o(1)$ if $\varpi = [\delta^{-1-\epsilon}]$. Also

$$1 - \frac{1}{p} \leqslant \left|1 - \frac{e^{-\delta p}}{p^{1+it}}\right| \leqslant 1 + \frac{1}{p}.$$

Hence, by (3.15.2),

$$\log |\zeta(1+it)| \leqslant -\sum_{p \leqslant \varpi} \log\left(1 - \frac{1}{p}\right) + o(1)$$

$$= \log\log \varpi + \gamma + o(1),$$

or

$$|\zeta(1+it)| \leqslant e^{\gamma + o(1)} \log \varpi.$$

Now

$$\log \varpi \leqslant (1+\epsilon) \log \frac{1}{\delta} = (1+\epsilon)\{2\beta(\alpha)+\eta\} \log\log t,$$

and taking α arbitrarily near to 1, we obtain (14.9.1). Similarly, by (3.15.3),

$$\log\frac{1}{|\zeta(1+it)|} \leqslant \sum_{p\leqslant\varpi} \log\left(1+\frac{1}{p}\right)+o(1)$$
$$= \log\log\varpi+\log\frac{6e^{\gamma}}{\pi^2}+o(1),$$

and (14.9.2) follows from this.

Comparing Theorem 14.9 with Theorems 8.9 (A) and (B), we see that, since we know only that $\beta(1) \leqslant 1$, in each problem a factor 2 remains in doubt. It is possible that $\beta(1) = \frac{1}{2}$, and if this were so each constant would be determined exactly.

14.10. The function $S(t)$. We shall next discuss the behaviour of this function on the Riemann hypothesis.

If $\frac{1}{2} < \alpha < \sigma < \beta$, $T < t < T'$, we have

$$\log\zeta(s) = \frac{1}{2\pi i}\left(\int\limits_{\beta+iT}^{\beta+iT'} + \int\limits_{\beta+iT'}^{\alpha+iT'} + \int\limits_{\alpha+iT'}^{\alpha+iT} + \int\limits_{\alpha+iT}^{\beta+iT}\right)\frac{\log\zeta(z)}{z-s}\,dz.$$

Let $\beta > 2$. By (14.2.2),

$$\int\limits_{\alpha+iT}^{2+iT} \frac{\log\zeta(z)}{z-s}\,dz = O\left\{\frac{1}{t-T}\int\limits_{\alpha}^{2} |\log\zeta(x+iT)|\,dx\right\} = O\left(\frac{\log T}{t-T}\right).$$

Also
$$\int\limits_{2+iT}^{\beta+iT} \frac{\log\zeta(z)}{z-s}\,dz = \sum_{n=2}^{\infty} \Lambda_1(n) \int\limits_{2+iT}^{\beta+iT} \frac{n^{-z}}{z-s}\,dz.$$

Now

$$\int\limits_{2+iT}^{\beta+iT} \frac{n^{-z}}{z-s}\,dz = \left[\frac{-n^{-z}}{(z-s)\log n}\right]_{2+iT}^{\beta+iT} - \frac{1}{\log n}\int\limits_{2+iT}^{\beta+iT} \frac{n^{-z}}{(z-s)^2}\,dz$$
$$= O\left\{\frac{1}{n^2(t-T)}\right\}+O\left\{\frac{1}{n^2}\int\limits_{-\infty}^{\infty} \frac{dx}{(x-\sigma)^2+(t-T)^2}\right\} = O\left\{\frac{1}{n^2(t-T)}\right\}.$$

Hence
$$\int\limits_{2+iT}^{\beta+iT} \frac{\log\zeta(z)}{z-s}\,dz = O\left(\frac{1}{t-T}\right),$$

and hence
$$\int\limits_{\alpha+iT}^{\beta+iT} \frac{\log\zeta(z)}{z-s}\,dz = O\left(\frac{\log T}{t-T}\right)$$

uniformly with respect to β. Similarly for the integral over

$$(\beta+iT',\ \alpha+iT').$$

Also
$$\int\limits_{\beta+iT}^{\beta+iT'} \frac{\log\zeta(z)}{z-s}\,dz = O\left(\frac{T'-T}{\beta-\sigma}\right).$$

Making $\beta \to \infty$, it follows that

$$\log\zeta(s) = \frac{1}{2\pi i}\int\limits_{\alpha+iT}^{\alpha+iT'} \frac{\log\zeta(z)}{s-z}\,dz + O\left(\frac{\log T}{t-T}\right) + O\left(\frac{\log T'}{T'-t}\right). \tag{14.10.1}$$

A similar argument shows that, if $\mathbf{R}(s') < \frac{1}{2}$,

$$0 = \frac{1}{2\pi i}\int\limits_{\alpha+iT}^{\alpha+iT'} \frac{\log\zeta(z)}{s'-z}\,dz + O\left(\frac{\log T}{t-T}\right) + O\left(\frac{\log T'}{T'-t}\right). \tag{14.10.2}$$

Taking $s' = 2\alpha-\sigma+it$, so that

$$s'-z = 2\alpha-\sigma+it-(\alpha+iy) = \alpha-iy-(\sigma-it),$$

and replacing (14.10.2) by its conjugate, we have

$$0 = \frac{1}{2\pi i}\int\limits_{\alpha+iT}^{\alpha+iT'} \frac{\log|\zeta(z)|-i\arg\zeta(z)}{z-s}\,dz + O\left(\frac{\log T}{t-T}\right) + O\left(\frac{\log T'}{T'-t}\right). \tag{14.10.3}$$

From (14.10.1) and (14.10.3) it follows that

$$\log\zeta(s) = \frac{1}{\pi i}\int\limits_{\alpha+iT}^{\alpha+iT'} \frac{\log|\zeta(z)|}{s-z}\,dz + O\left(\frac{\log T}{t-T}\right) + O\left(\frac{\log T'}{T'-t}\right) \tag{14.10.4}$$

and
$$\log\zeta(s) = \frac{1}{\pi}\int\limits_{\alpha+iT}^{\alpha+iT'} \frac{\arg\zeta(z)}{s-z}\,dz + O\left(\frac{\log T}{t-T}\right) + O\left(\frac{\log T'}{T'-t}\right). \tag{14.10.5}$$

14.11. We can now show that each of the functions

$$\max\{\log|\zeta(s)|, 0\}, \qquad \max\{-\log|\zeta(s)|, 0\},$$
$$\max\{\arg\zeta(s), 0\}, \qquad \max\{-\arg\zeta(s), 0\}$$

has the same ν-function as $\log\zeta(s)$. Consider, for example,

$$\max\{\arg\zeta(s), 0\},$$

and let its ν-function be $\nu_1(\sigma)$. Since

$$|\arg\zeta(s)| \leqslant |\log\zeta(s)|$$

we have at once
$$\nu_1(\sigma) \leqslant \nu(\sigma).$$

Also (14.10.5) gives

$$\arg\zeta(s) = \frac{1}{\pi}\int_{T}^{T'} \frac{\sigma-\alpha}{(\sigma-\alpha)^2+(t-y)^2}\arg\zeta(\alpha+iy)\,dy+O\left(\frac{\log T}{t-T}\right)+O\left(\frac{\log T'}{T'-t}\right) \tag{14.11.1}$$

$$< A(\log T')^{\nu_1(\alpha)+\epsilon}\int_{T}^{T'} \frac{\sigma-\alpha}{(\sigma-\alpha)^2+(t-y)^2}\,dy+O\left(\frac{\log T}{t-T}\right)+O\left(\frac{\log T'}{T'-t}\right)$$

$$< A(\log t)^{\nu_1(\alpha)+\epsilon}+O(t^{-1}\log t),$$

taking, for example, $T = \frac{1}{2}t$, $T' = 2t$.

It is clear from this that $\nu_1(\sigma)$ is non-increasing. Also the Borel–Carathéodory inequality, applied to circles with centre $2+it$ and radii $2-\alpha-\delta$, $2-\alpha-2\delta$, gives

$$|\log\zeta(\alpha+\delta+it)| < \frac{A}{\delta}\left\{(\log t)^{\nu_1(\alpha)+\epsilon}+\frac{\log t}{t}\right\}+\frac{A}{\delta}|\log|\zeta(2+it)||.$$

If $\alpha+\delta < 1$, so that $\nu(\alpha+\delta) > 0$, it follows that

$$\nu(\alpha+\delta) \leqslant \nu_1(\alpha)+\epsilon.$$

Since ϵ and δ may be as small as we please, and $\nu(\sigma)$ is continuous, it follows that

$$\nu(\alpha) \leqslant \nu_1(\alpha).$$

Hence
$$\nu_1(\sigma) = \nu(\sigma) \quad (\tfrac{1}{2} < \sigma < 1).$$

Similarly all the ν-functions are equal.

14.12. *Ω-results† for $S(t)$ and $S_1(t)$.*

THEOREM 14.12 (A). *Each of the inequalities*

$$S(t) > (\log t)^{\frac{1}{2}-\epsilon}, \tag{14.12.1}$$

$$S(t) < -(\log t)^{\frac{1}{2}-\epsilon} \tag{14.12.2}$$

has solutions for arbitrarily large values of t.

Making $\alpha \to \frac{1}{2}$ in (14.11.1), by bounded convergence

$$\arg\zeta(s) = \int_{\frac{1}{2}t}^{2t} \frac{\sigma-\frac{1}{2}}{(\sigma-\frac{1}{2})^2+(t-y)^2}S(y)\,dy+O\left(\frac{\log t}{t}\right) \quad (\sigma > \tfrac{1}{2}). \tag{14.12.3}$$

If $S(t) < \log^a t$ for all large t, this gives

$$\arg\zeta(s) < A\log^a t\int_{\frac{1}{2}t}^{2t} \frac{\sigma-\frac{1}{2}}{(\sigma-\frac{1}{2})^2+(t-y)^2}\,dy+O\left(\frac{\log t}{t}\right)$$

$$< A\log^a t+O(t^{-1}\log t).$$

† Landau (1), Bohr and Landau (3), Littlewood (5).

The above analysis shows that this is false if $a < \nu(\sigma)$, which is satisfied if $a < \frac{1}{2}$ and σ is near enough to $\frac{1}{2}$. This proves the first result, and the other may be proved similarly.

THEOREM 14.12 (B).

$$S_1(t) = \Omega\{(\log t)^{\frac{1}{2}-\epsilon}\}.$$

From (14.10.5) with $\alpha \to \frac{1}{2}$ we have

$$\log \zeta(s) = i \int_{\frac{1}{2}t}^{2t} \frac{S(y)}{s-\frac{1}{2}-iy}\, dy + O(1)$$

$$= i\left[\frac{S_1(y)}{s-\frac{1}{2}-iy}\right]_{\frac{1}{2}t}^{2t} + \int_{\frac{1}{2}t}^{2t} \frac{S_1(y)}{(s-\frac{1}{2}-iy)^2}\, dy + O(1)$$

$$= \int_{\frac{1}{2}t}^{2t} \frac{S_1(y)}{(s-\frac{1}{2}-iy)^2}\, dy + O(1) \tag{14.12.4}$$

since $S_1(y) = O(\log y)$. The result now follows as before.

In view of the result of Selberg stated in § 9.9, this theorem is true independently of the Riemann hypothesis. In the case of $S(t)$, Selberg's method gives only an index $\frac{1}{3}$ instead of the index $\frac{1}{2}$ obtained on the Riemann hypothesis.

14.13. We now turn to results of the opposite kind.† We know that without any hypothesis

$$S(t) = O(\log t), \qquad S_1(t) = O(\log t),$$

and that on the Lindelöf hypothesis, and *a fortiori* on the Riemann hypothesis, each O can be replaced by o. On the Riemann hypothesis we should expect something more precise. The result actually obtained is

THEOREM 14.13.

$$S(t) = O\left(\frac{\log t}{\log\log t}\right), \tag{14.13.1}$$

$$S_1(t) = O\left(\frac{\log t}{(\log\log t)^2}\right). \tag{14.13.2}$$

We first prove three lemmas.

LEMMA α. *Let*

$$\phi(t) = \max_{1 \leqslant u \leqslant t} |S_1(u)|,$$

so that $\phi(t)$ is non-decreasing, and $\phi(t) = O(\log t)$. Then

$$S(t) = O[\{\phi(2t)\log t\}^{\frac{1}{2}}].$$

† Landau (11), Cramér (1), Littlewood (4), Titchmarsh (3).

This is independent of the Riemann hypothesis. We have

$$N(t) = L(t)+R(t),$$

where $L(t)$ is defined by (9.3.1), and $R(t) = S(t)+O(1/t)$. Now

$$N(T+x)-N(T) \geqslant 0 \quad (0 < x < T).$$

Hence

$$R(T+x)-R(T) \geqslant -\{L(T+x)-L(T)\} > -Ax\log T.$$

Hence

$$\int\limits_T^{T+x} R(t)\,dt = xR(T)+\int\limits_0^x \{R(T+u)-R(T)\}\,du$$

$$> xR(T)-A\int\limits_0^x u\log T\,du$$

$$> xR(T)-Ax^2\log T.$$

Hence

$$R(T) < \frac{1}{x}\int\limits_T^{T+x} R(t)\,dt+Ax\log T$$

$$= \frac{S_1(T+x)-S_1(T)}{x}+O\left(\frac{1}{T}\right)+Ax\log T$$

$$= O\left\{\frac{\phi(2T)}{x}\right\}+O\left(\frac{1}{T}\right)+Ax\log T.$$

Taking $x = \{\phi(2T)/\log T\}^{\frac{1}{2}}$, the upper bound for $S(T)$ follows. Similarly by considering integrals over $(T-x, T)$ we obtain the lower bound.

Lemma β. *Let* $\sigma \leqslant 1$, *and let*

$$F(T) = \max|\log\zeta(s)|+\log^{\frac{1}{4}}T \quad \left(\sigma-\tfrac{1}{2} \geqslant \frac{1}{\log\log T},\quad 4 \leqslant t \leqslant T\right).$$

Then

$$\log\zeta(s) = O\{F(T+1)e^{-A(\sigma-\frac{1}{2})\log\log T}\}$$

$$\left(\frac{1}{2}+\frac{1}{\log\log T} \leqslant \sigma \leqslant 2,\quad 4 \leqslant t \leqslant T\right).$$

We apply Hadamard's three-circles theorem as in § 14.2, but now take

$$\sigma_1 = \frac{3}{2}+\frac{1}{\log\log T},\quad \eta = \tfrac{1}{4},\quad \delta = \frac{1}{\log\log T},\quad \sigma \leqslant \tfrac{5}{4}.$$

We obtain

$$M_2 < AM_3^a = AM_3(1/M_3)^{1-a},$$

where

$$M_3 \leqslant F(T+1),$$

and

$$1-a = \log\frac{r_3}{r_2}\Big/\log\frac{r_3}{r_1} = \log\left(1+\frac{\sigma-\frac{1}{2}-\delta}{\sigma_1-\sigma}\right)\Big/\log\left(\frac{\sigma_1-\frac{1}{2}-\delta}{\sigma_1-1-\eta}\right)$$

$$> A(\sigma-\tfrac{1}{2}-\delta).$$

Hence

$$M_2 < AF(T+1)^{1-A(\sigma-\frac{1}{2}-\delta)} \leqslant AF(T+1)(\log^{\frac{1}{4}} T)^{-A(\sigma-\frac{1}{2}-\delta)}.$$

This gives the required result if $\sigma \leqslant \frac{5}{4}$, and for $\frac{5}{4} \leqslant \sigma \leqslant 2$ it is trivial, if the A is small enough.

LEMMA γ. *For* $\sigma > \frac{1}{2}$, $0 < \xi < \frac{1}{2}t$,

$$\log \zeta(s) = i \int_{t-\xi}^{t+\xi} \frac{S(y)}{s-\frac{1}{2}-iy}\, dy + O\left\{\frac{\phi(2t)}{\xi}\right\} + O(1). \qquad (14.13.3)$$

We have

$$\int_{t+\xi}^{2t} \frac{S(y)}{s-\frac{1}{2}-iy}\, dy = \left[\frac{S_1(y)}{s-\frac{1}{2}-iy}\right]_{t+\xi}^{2t} - i \int_{t+\xi}^{2t} \frac{S_1(y)}{(s-\frac{1}{2}-iy)^2}\, dy$$

$$= O\left\{\frac{\phi(2t)}{\xi}\right\} + O\left\{\phi(2t) \int_{t+\xi}^{2t} \frac{dy}{(\sigma-\frac{1}{2})^2+(y-t)^2}\right\} = O\left\{\frac{\phi(2t)}{\xi}\right\},$$

and similarly for the integral over $(\frac{1}{2}t, t-\xi)$. The result therefore follows from (14.12.4).

Proof of Theorem 14.13. By Lemmas α and γ,

$$\log \zeta(s) = O\{\phi(4t)\log t\}^{\frac{1}{2}} \int_{t-\xi}^{t+\xi} \frac{dy}{\{(\sigma-\frac{1}{2})^2+(y-t)^2\}^{\frac{1}{2}}} + O\left\{\frac{\phi(2t)}{\xi}\right\} + O(1)$$

$$= O\left[\{\phi(4t)\log t\}^{\frac{1}{2}} \frac{\xi}{\sigma-\frac{1}{2}}\right] + O\left\{\frac{\phi(4t)}{\xi}\right\} + O(1)$$

for $\sigma-\frac{1}{2} \geqslant 1/\log\log T$, $4 \leqslant t \leqslant T$. Taking

$$\xi = A\left\{\frac{\phi(4t)}{\log t}\right\}^{\frac{1}{4}} \frac{1}{(\log\log T)^{\frac{1}{2}}},$$

we obtain
$$\log \zeta(s) = O[(\log T)^{\frac{1}{4}}(\log\log T)^{\frac{1}{2}}\{\phi(4T)\}^{\frac{3}{4}}].$$

Hence by Lemma β, for $\sigma-\frac{1}{2} \geqslant 1/\log\log T$,

$$\log \zeta(s) = O[(\log T)^{\frac{1}{4}}(\log\log T)^{\frac{1}{2}}\{\phi(4T+4)\}^{\frac{3}{4}} e^{-A(\sigma-\frac{1}{2})\log\log T}].$$

Hence

$$\int_{\frac{1}{2}+1/\log\log T}^{2} \log|\zeta(s)|\, d\sigma = O[(\log T)^{\frac{1}{4}}(\log\log T)^{-\frac{1}{2}}\{\phi(4T+4)\}^{\frac{3}{4}}]. \qquad (14.13.4)$$

Again, the real part of (14.13.3) may be written

$$\log|\zeta(s)| = \int_0^{\xi} \frac{x}{(\sigma-\frac{1}{2})^2+x^2}\{S(t-x)-S(t+x)\}\,dx+O\left\{\frac{\phi(2t)}{\xi}\right\}+O(1). \tag{14.13.5}$$

Hence

$$\int_{\frac{1}{2}}^{\frac{1}{2}+\mu} \log|\zeta(s)|\,d\sigma = \int_0^{\xi} \arctan\frac{\mu}{x}\{S(t-x)-S(t+x)\}\,dx+ +O\{\mu\,\phi(2t)/\xi\}+O(\mu)$$

$$= O[\xi\{\phi(4t)\log t\}^{\frac{1}{2}}]+O\{\mu\,\phi(2t)/\xi\}+O(\mu).$$

Taking $\mu = 1/\log\log T$, and ξ as before,

$$\int_{\frac{1}{2}}^{\frac{1}{2}+1/\log\log T} \log|\zeta(s)|\,d\sigma = O[(\log T)^{\frac{1}{4}}(\log\log T)^{-\frac{1}{2}}\{\phi(4T)\}^{\frac{3}{4}}]. \tag{14.13.6}$$

Now (14.13.4), (14.13.6), and Theorem 9.9 give

$$S_1(t) = O[(\log T)^{\frac{1}{4}}(\log\log T)^{-\frac{1}{2}}\{\phi(5T)\}^{\frac{3}{4}}] \quad (4 \leqslant t \leqslant T). \tag{14.13.7}$$

Varying t and taking the maximum,

$$\phi(T) = O[(\log T)^{\frac{1}{4}}(\log\log T)^{-\frac{1}{2}}\{\phi(5T)\}^{\frac{3}{4}}].$$

Let

$$\psi(T) = \max_{4\leqslant t\leqslant T} \frac{(\log\log t)^2\phi(t)}{\log t},$$

so that $\psi(T)$ is non-decreasing and

$$\phi(T) \leqslant \frac{\log T}{(\log\log T)^2}\psi(T).$$

Then (14.13.7) gives

$$\phi(T) = O\left[\frac{\log T}{(\log\log T)^2}\{\psi(5T)\}^{\frac{3}{4}}\right]$$

or

$$\frac{\phi(T)(\log\log T)^2}{\log T} = O[\{\psi(5T)\}^{\frac{3}{4}}] = O[\{\psi(5T_1)\}^{\frac{3}{4}}] \quad (T \leqslant T_1).$$

Varying T and taking the maximum,

$$\psi(T_1) = O[\{\psi(5T_1)\}^{\frac{3}{4}}].$$

But $\psi(5T_1) < 5\psi(T_1)$ for some arbitrarily large T_1; for otherwise

$$\psi(5^n t_0) \geqslant 5^n\psi(t_0),$$

i.e. $\psi(T) > AT$ for some arbitrarily large T, which is not so, since in fact $\phi(T) = O(\log T)$, $\psi(T) = O\{(\log\log T)^2\}$. Hence

$$\psi(T_1) < A\{\psi(T_1)\}^{\frac{3}{4}}, \qquad \psi(T_1) < A,$$

for some arbitrarily large T_1, and so for all T_1, since ψ is non-decreasing.

Hence $$\phi(T) = O\left\{\frac{\log T}{(\log\log T)^2}\right\}.$$

This proves (14.13.2), and (14.13.1) then follows from Lemma α.

The argument can be extended to show that, if $S_n(t)$ is the nth integral of $S(t)$, then

$$S_n(t) = O\left\{\frac{\log t}{(\log\log t)^{n+1}}\right\}. \tag{14.13.8}$$

14.14. Theorem 14.13 also enables us to prove inequalities for $\zeta(s)$ in the immediate neighbourhood of $\sigma = \frac{1}{2}$, a region not touched by previous arguments. We obtain first

THEOREM 14.14 (A).

$$\zeta(\tfrac{1}{2}+it) = O\left\{\exp\left(A\frac{\log t}{\log\log t}\right)\right\}. \tag{14.14.1}$$

We have

$$S(t+x)-S(t) = \{N(t+x)-N(t)\}-\{L(t+x)-L(t)\}-\{f(t+x)-f(t)\},$$

where $f(t)$ is the $O(1/t)$ of (9.3.2), and arises from the asymptotic formula for $\log\Gamma(s)$. Thus $f'(t) = O(1/t^2)$, and since $N(t+x) \geqslant N(t)$

$$S(t+x)-S(t) > -Ax\log t+O(x/t^2) > -Ax\log t.$$

Hence, by (14.13.5),

$$\log|\zeta(s)| < A\int_0^\xi \frac{x^2\log t}{(\sigma-\frac{1}{2})^2+x^2}\,dx+O\left\{\frac{\log t}{\xi(\log\log t)^2}\right\}+O(1)$$

$$< A\xi\log t+O\left\{\frac{\log t}{\xi(\log\log t)^2}\right\}+O(1)$$

uniformly for $\sigma > \frac{1}{2}$, and so by continuity for $\sigma = \frac{1}{2}$. Taking

$$\xi = 1/\log\log t$$

the result follows.

THEOREM 14.14 (B). *We have*

$$-\frac{A\log t}{\log\log t}\log\left\{\frac{2}{(\sigma-\frac{1}{2})\log\log t}\right\} < \log|\zeta(s)| < \frac{A\log t}{\log\log t}$$

$$(\tfrac{1}{2} < \sigma \leqslant \tfrac{1}{2}+A/\log\log t), \tag{14.14.2}$$

$$\arg\zeta(s) = O\left(\frac{\log t}{\log\log t}\right) \quad (\tfrac{1}{2} \leqslant \sigma \leqslant \tfrac{1}{2}+A/\log\log t). \tag{14.14.3}$$

By (14.13.1) and (14.13.3),

$$\log\zeta(s) = O\left\{\frac{\log t}{\log\log t}\int_0^\xi \frac{dx}{\sqrt{\{(\sigma-\frac{1}{2})^2+x^2\}}}\right\}+O\left\{\frac{\log t}{\xi(\log\log t)^2}\right\}+O(1).$$

Now
$$\int_0^{\xi} \frac{du}{\sqrt{\{(\sigma-\frac{1}{2})^2+x^2\}}} = \int_0^{\xi/(\sigma-\frac{1}{2})} \frac{dx'}{\sqrt{(1+x'^2)}},$$
which is less than 1 if $\xi \leqslant \sigma-\frac{1}{2}$, and otherwise is less than
$$1+\int_1^{\xi/(\sigma-\frac{1}{2})} \frac{dx'}{x'} = 1+\log\frac{\xi}{\sigma-\frac{1}{2}}.$$
Taking $\xi = 1/\log\log t$, the lower bound in (14.14.2) follows. The upper bound follows from the argument of the previous section. Lastly, taking imaginary parts in (14.13.3),
$$\arg\zeta(s) = \int_0^{\xi} \frac{\sigma-\frac{1}{2}}{x^2+(\sigma-\frac{1}{2})^2}\{S(t+x)-S(t-x)\}\,dx+$$
$$+O\left\{\frac{\log t}{\xi(\log\log t)^2}\right\}+O(1)$$
$$= O\left\{\frac{\log t}{\log\log t}\int_0^{\xi} \frac{\sigma-\frac{1}{2}}{x^2+(\sigma-\frac{1}{2})^2}\,dx\right\}+O\left\{\frac{\log t}{\xi(\log\log t)^2}\right\}+O(1).$$
Now
$$\int_0^{\xi} \frac{\sigma-\frac{1}{2}}{x^2+(\sigma-\frac{1}{2})^2}\,dx < \int_0^{\infty} \frac{\sigma-\frac{1}{2}}{x^2+(\sigma-\frac{1}{2})^2}\,dx = \tfrac{1}{2}\pi.$$
Hence, taking $\xi = 1$, (14.14.3) follows uniformly for $\sigma > \frac{1}{2}$, and so by continuity for $\sigma = \frac{1}{2}$.

In particular
$$\log\zeta(s) = O\left(\frac{\log t}{\log\log t}\right) \quad \left(\sigma = \frac{1}{2}+\frac{A}{\log\log t}\right). \tag{14.14.4}$$

From (14.14.4), (14.5.2), and a Phragmén–Lindelöf argument it follows that
$$\log\zeta(s) = O\left\{\frac{(\log t)^{2-2\sigma}}{\log\log t}\right\} \tag{14.14.5}$$
uniformly for
$$\frac{1}{2}+\frac{A}{\log\log t} \leqslant \sigma \leqslant 1-\delta.$$

14.15. Another result in the same order of ideas is an approximate formula for $\log\zeta(s)$, which should be compared with Theorem 9.6 (B).

THEOREM 14.15. *For $\frac{1}{2} \leqslant \sigma \leqslant 2$,*
$$\log\zeta(s) = \sum_{|t-\gamma|<1/\log\log t} \log(s-\rho)+O\left(\frac{\log t\log\log\log t}{\log\log t}\right), \tag{14.15.1}$$
where $\rho = \frac{1}{2}+i\gamma$ runs through zeros of $\zeta(s)$.

In Lemma α of § 3.9, let

$$f(s) = \zeta(s), \qquad s_0 = \frac{1}{2}+\frac{1}{\sqrt{3}}\delta+iT, \qquad r = \frac{4}{\sqrt{3}}\delta,$$

where $\delta = 1/\log\log T$. By (14.14.4)

$$\left|\frac{1}{\zeta(s_0)}\right| \leqslant \exp\left(\frac{A\log T}{\log\log T}\right).$$

The upper bound in (14.14.2) gives

$$|\zeta(s)| < \exp\left(\frac{A\log T}{\log\log T}\right)$$

for $|s-s_0| \leqslant r$, $\sigma \geqslant \frac{1}{2}$; and for $|s-s_0| \leqslant r$, $\sigma < \frac{1}{2}$, the functional equation gives

$$|\zeta(s)| < At^{\frac{1}{2}-\sigma}|\zeta(1-s)| < At^{\sqrt{3}\delta}\exp\left(\frac{A\log T}{\log\log T}\right) < \exp\left(\frac{A\log T}{\log\log T}\right).$$

It therefore follows from (3.9.1) that

$$\log\zeta(s)-\log\zeta(s_0)-\sum_{|s_0-\rho|\leqslant 2\delta/\sqrt{3}}\log(s-\rho)+ \\ +\sum_{|s_0-\rho|\leqslant 2\delta/\sqrt{3}}\log(s_0-\rho) = O\left(\frac{\log T}{\log\log T}\right)$$

for $|s-s_0| \leqslant \frac{3}{8}r$, and so in particular for $\frac{1}{2} \leqslant \sigma \leqslant \frac{1}{2}+\delta$, $t = T$.

Now
$$\log\zeta(s_0) = O\left(\frac{\log T}{\log\log T}\right).$$

Also
$$s_0-\rho = \frac{1}{\sqrt{3}}\delta+i(T-\gamma).$$

Hence
$$\frac{1}{\sqrt{3}}\delta \leqslant |s_0-\rho| < A,$$

and so, if the logarithm has its principal value,

$$\log(s_0-\rho) = O\left(\log\frac{1}{\delta}\right) = O(\log\log\log T).$$

Also the number of values of ρ in the above sums does not exceed

$$N\left(T+\frac{2\delta}{\sqrt{3}}\right)-N\left(T-\frac{2\delta}{\sqrt{3}}\right) = O(\delta\log T)+O\left(\frac{\log T}{\log\log T}\right) = O\left(\frac{\log T}{\log\log T}\right),$$

by Theorem 14.13. Hence

$$\sum_{|s_0-\rho|\leqslant 2\delta/\sqrt{3}}\log(s_0-\rho) = O\left(\frac{\log T\log\log\log T}{\log\log T}\right).$$

Since $|T-\gamma| \leqslant \delta$ if $|s_0-\rho| \leqslant 2\delta/\sqrt{3}$, the result follows, with T for t and $\frac{1}{2} \leqslant \sigma \leqslant \frac{1}{2}+\delta$. It is also true for $\frac{1}{2}+\delta < \sigma \leqslant 2$, since in this region

$$\log \zeta(s) = O\left(\frac{\log T}{\log\log T}\right),$$

and, as in the case of the other sum,

$$\sum_{|s_0-\rho| \leqslant 2\delta/\sqrt{3}} \log(s-\rho) = O\left(\frac{\log T \log\log\log T}{\log\log T}\right).$$

This proves the theorem.

For $\zeta'(s)/\zeta(s)$ we obtain similarly from Lemma α of § 3.9

$$\frac{\zeta'(s)}{\zeta(s)} = \sum_{|t-\gamma| \leqslant 1/\log\log t} \frac{1}{s-\rho} + O(\log t). \tag{14.15.2}$$

14.16. THEOREM 14.16. *Each interval $[T, T+1]$ contains a value of t such that*

$$|\zeta(s)| > \exp\left(-A\frac{\log t}{\log\log t}\right) \quad (\tfrac{1}{2} \leqslant \sigma \leqslant 2). \tag{14.16.1}$$

Let $\delta = 1/\log\log T$. Then the lower bound (14.16.1) holds automatically for $\sigma \geqslant \frac{1}{2}+\delta$, by (14.14.4). We therefore assume that $\frac{1}{2} \leqslant \sigma \leqslant \frac{1}{2}+\delta$. If $s = \sigma+it$ and $s_0 = \frac{1}{2}+\delta+it$ then, on integrating (14.15.2), we find

$$\log\frac{\zeta(s)}{\zeta(s_0)} = \sum_{|t-\gamma| \leqslant \delta} \log\left(\frac{s-\rho}{s_0-\rho}\right) + O\left(\frac{\log T}{\log\log T}\right).$$

Moreover $\log\zeta(s_0) = O\left(\dfrac{\log T}{\log\log T}\right)$ by (14.14.4) so that, on taking real parts

$$\log|\zeta(s)| = \sum_{|t-\gamma| \leqslant \delta} \log\left|\frac{s-\rho}{s_0-\rho}\right| + O\left(\frac{\log T}{\log\log T}\right)$$

$$\geqslant \sum_{|t-\gamma| \leqslant \delta} \log\frac{|t-\gamma|}{2\delta} + O\left(\frac{\log T}{\log\log T}\right),$$

since $|s-\rho| \geqslant |t-\gamma|$ and $|s_0-\rho| \leqslant 2\delta$. We now observe that

$$\int_T^{T+1} \sum_{|t-\gamma| \leqslant \delta} \log\frac{|t-\gamma|}{2\delta}\,dt = \sum_{T-\delta \leqslant \gamma \leqslant T+1+\delta} \int_{\max(\gamma-\delta, T)}^{\min(\gamma+\delta, T+1)} \log\frac{|t-\gamma|}{2\delta}\,dt$$

$$\geqslant \sum_{T-\delta \leqslant \gamma \leqslant T+1+\delta} \int_{\gamma-\delta}^{\gamma+\delta} \log\frac{|t-\gamma|}{2\delta}\,dt$$

$$= \sum_{T-\delta \leqslant \gamma \leqslant T+1+\delta} (-2\delta - 2\delta\log 2)$$

$$\geqslant -A\delta\log T,$$

as there are $O(\log T)$ terms in the sum. Hence there is a t for which

$$\sum_{|t-\gamma|\leqslant\delta} \log\frac{|t-\gamma|}{2\delta} \geqslant -A\delta\log T$$

and the result follows.

In particular, if ϵ is any positive number, each $(T, T+1)$ contains a t such that

$$\frac{1}{\zeta(s)} = O(t^{\epsilon}) \quad (\tfrac{1}{2}\leqslant\sigma\leqslant 2). \tag{14.16.2}$$

14.17. Mean-value theorems† for $S(t)$ and $S_1(t)$. We consider first $S_1(t)$. We begin by proving

THEOREM 14.17. *For $\frac{1}{2}T\leqslant t\leqslant T$, $\delta = T^{-\frac{1}{2}}$,*

$$\pi S_1(t) = C-\sum_{n=2}^{\infty}\frac{\Lambda_1(n)\cos(t\log n)}{n^{\frac{1}{2}}\log n}e^{-\delta n}+O\left(\frac{1}{\log\log T}\right) \tag{14.17.1}$$

where

$$C = \int_{\frac{1}{2}}^{\infty}\log|\zeta(\sigma)|\,d\sigma.$$

Making $\beta\to\infty$ in (9.9.4), we have

$$\pi S_1(t) = C-\int_{\frac{1}{2}}^{\infty}\log|\zeta(\sigma+it)|\,d\sigma. \tag{14.17.2}$$

Now, integrating (14.4.1) from s to $\frac{9}{8}+it$,

$$\log\zeta(s)-\log\zeta(\tfrac{9}{8}+it) = \sum_{n=2}^{\infty}\frac{\Lambda_1(n)}{n^s}e^{-\delta n}-\sum_{n=2}^{\infty}\frac{\Lambda_1(n)}{n^{\frac{9}{8}+it}}e^{-\delta n}+$$

$$+\sum_{\rho}\int_{\sigma}^{\frac{9}{8}}\delta^{s_1-\rho}\Gamma(\rho-s_1)\,d\sigma_1+O(\delta^{\sigma-\frac{1}{4}}\log t) \quad (\tfrac{1}{2}\leqslant\sigma\leqslant\tfrac{9}{8}).$$

Also, if $\sigma\geqslant\frac{9}{8}$,

$$\log\zeta(\sigma+it)-\sum_{n=2}^{\infty}\frac{\Lambda_1(n)}{n^s}e^{-\delta n} = \sum_{n=2}^{\infty}\frac{\Lambda_1(n)}{n^s}(1-e^{-\delta n})$$

$$= O\left(\sum_{n=2}^{\infty}n^{-\sigma}(1-e^{-\delta n})\right) = O\left(\sum_{2\leqslant n\leqslant 1/\delta}n^{1-\sigma}\delta+\sum_{n>1/\delta}n^{-\sigma}\right)$$

$$= O\{(\delta^{\sigma-1}+2^{-\sigma}\delta)\log t\}. \tag{14.17.3}$$

Hence, for $\frac{1}{2}\leqslant\sigma\leqslant\frac{9}{8}$,

$$\log\zeta(s) = \sum_{n=2}^{\infty}\frac{\Lambda_1(n)}{n^s}e^{-\delta n}+\sum_{\rho}\int_{\sigma}^{\frac{9}{8}}\delta^{s_1-\rho}\Gamma(\rho-s_1)\,d\sigma_1+O(\delta^{\frac{1}{8}})+O(\delta^{\sigma-\frac{1}{4}}\log t),$$

† Littlewood (5), Titchmarsh (2).

and integrating over $\frac{1}{2} \leqslant \sigma \leqslant \frac{9}{8}$,

$$\int_{\frac{1}{2}}^{\frac{9}{8}} \log \zeta(s)\, d\sigma = \int_{\frac{1}{2}}^{\frac{9}{8}} \left(\sum_{n=2}^{\infty} \frac{\Lambda_1(n)}{n^s} e^{-\delta n}\right) d\sigma + \sum_{\rho} \int_{\frac{1}{2}}^{\frac{9}{8}} (\sigma_1 - \tfrac{1}{2})\delta^{s_1-\rho}\Gamma(\rho - s_1)\, d\sigma_1 + O(\delta^{\frac{1}{9}}) + O(\delta^{\frac{1}{4}} \log t).$$

Also, by (14.17.3),

$$\int_{\frac{9}{8}}^{\infty} \log \zeta(s)\, d\sigma = \int_{\frac{9}{8}}^{\infty} \left(\sum_{n=2}^{\infty} \frac{\Lambda_1(n)}{n^s} e^{-\delta n}\right) d\sigma + O(\delta^{\frac{1}{9}}),$$

and
$$\int_{\frac{1}{2}}^{\infty} \sum_{n=2}^{\infty} \frac{\Lambda_1(n)}{n^s} e^{-\delta n}\, d\sigma = \sum_{n=2}^{\infty} \frac{\Lambda_1(n)}{n^{\frac{1}{2}+it} \log n} e^{-\delta n},$$

the inversion being justified by absolute convergence. Hence

$$\pi S_1(t) = C - \sum_{n=2}^{\infty} \frac{\Lambda_1(n) \cos(t \log n)}{n^{\frac{1}{2}} \log n} e^{-\delta n} + O\left\{\sum_{\rho} \int_{\frac{1}{2}}^{\frac{9}{8}} (\sigma_1 - \tfrac{1}{2})\delta^{\sigma_1 - \frac{1}{2}} |\Gamma(\rho - s_1)|\, d\sigma_1\right\} + O(\delta^{\frac{1}{9}}) + O(\delta^{\frac{1}{4}} \log t).$$

Now
$$\Gamma(\rho - s_1) = O(e^{-A|\gamma - t|}) \quad (|\gamma - t| \geqslant 1),$$

$$\Gamma(\rho - s_1) = O\left(\frac{1}{|\rho - s_1|}\right) = O\left\{\frac{1}{\{(\sigma_1 - \frac{1}{2})^2 + |\gamma - t|^2\}^{\frac{1}{2}}}\right\} \quad (|\gamma - t| < 1).$$

Hence

$$\sum_{\rho} = \sum_{|\gamma - t| < 1/\log\log t} + \sum_{1/\log\log t \leqslant |\gamma - t| < 1} + \sum_{|\gamma - t| \geqslant 1}$$

$$= O\left(\sum_{|\gamma - t| < 1/\log\log t} \int_{\frac{1}{2}}^{\frac{9}{8}} \delta^{\sigma_1 - \frac{1}{2}}\, d\sigma_1\right) + O\left(\sum_{1/\log\log t \leqslant |\gamma - t| < 1} \frac{1}{|\gamma - t|} \int_{\frac{1}{2}}^{\frac{9}{8}} (\sigma_1 - \tfrac{1}{2})\delta^{\sigma_1 - \frac{1}{2}}\, d\sigma_1\right) + O\left(\sum_{|\gamma - t| \geqslant 1} e^{-A|\gamma - t|} \int_{\frac{1}{2}}^{\frac{9}{8}} (\sigma_1 - \tfrac{1}{2})\delta^{\sigma_1 - \frac{1}{2}}\, d\sigma_1\right).$$

Now
$$\int_{\frac{1}{2}}^{\frac{9}{8}} \delta^{\sigma_1-\frac{1}{2}}\,d\sigma_1 < \int_0^\infty e^{-x\log 1/\delta}\,dx = \log^{-1}(1/\delta),$$

$$\int_{\frac{1}{2}}^{\frac{9}{8}} (\sigma_1-\tfrac{1}{2})\delta^{\sigma_1-\frac{1}{2}}\,d\sigma_1 < \int_0^\infty xe^{-x\log 1/\delta}\,dx = \log^{-2}(1/\delta).$$

As in § 14.5,
$$\sum_{|\gamma-t|\geqslant 1} e^{-A|\gamma-t|} = O(\log t).$$

Also, by (14.13.1), for $t-1 \leqslant t' \leqslant t+1$
$$N\left(t'+\frac{1}{\log\log t}\right)-N(t') = O\left(\frac{\log t}{\log\log t}\right).$$

Hence
$$\sum_{|\gamma-t|\leqslant 1/\log\log t} 1 = O\left(\frac{\log t}{\log\log t}\right),$$
and
$$\sum_{t+1/\log\log t<\gamma\leqslant t+1} \frac{1}{\gamma-t} = \sum_{m<\log\log t}\ \sum_{t+m/\log\log t\leqslant\gamma<t+(m+1)/\log\log t} \frac{1}{\gamma-t}$$
$$= \sum_{m<\log\log t} O\left(\frac{1}{m/\log\log t}\,\frac{\log t}{\log\log t}\right) = O(\log t\log\log\log t).$$

Hence
$$\sum_\rho = O\left(\frac{\log t}{\log 1/\delta\,\log\log t}\right)+O\left(\frac{\log t\log\log\log t}{\log^2 1/\delta}\right)+O\left(\frac{\log t}{\log^2 1/\delta}\right) = O\left(\frac{1}{\log\log t}\right)$$
for the given δ and t. This proves the theorem.

14.18. Lemma 14.18. *If $a_n = O(1)$, $\delta < \frac{1}{2}$, then*
$$\frac{2}{T}\int_{\frac{1}{2}T}^{T} \left|\sum_{n=1}^\infty \frac{a_n}{n^s}e^{-\delta n}\right|^2 dt = \sum_{n=1}^\infty \frac{|a_n|^2}{n^{2\sigma}}e^{-2\delta n}+O\left(\frac{1}{T\delta}\log\frac{1}{\delta}\right)$$
uniformly for $\sigma \geqslant \frac{1}{2}$. Similarly, if $a_n = O(\log n)$, the formula holds with a remainder term
$$O\left(\frac{1}{T\delta}\log^3\frac{1}{\delta}\right).$$

The left-hand side is
$$\sum_{m=1}^\infty\sum_{n=1}^\infty \frac{2}{T}\int_{\frac{1}{2}T}^{T} \frac{a_m\bar{a}_n}{(mn)^\sigma}e^{-(m+n)\delta}\left(\frac{n}{m}\right)^{it}dt = \sum_{m=n}+\sum_{m\neq n}.$$

Clearly
$$\sum_{m=n} = \sum_{n=1}^\infty \frac{|a_n|^2}{n^{2\sigma}}e^{-2\delta n}.$$

Also
$$\sum_{m\neq n} = O\left(\frac{1}{T}\right)\sum_{m<n}\frac{e^{-n\delta}}{(mn)^{\frac{1}{2}}\log n/m}.$$

Now

$$\sum_{n=m+1}^{2m} \frac{e^{-n\delta}}{(mn)^{\frac{1}{2}}\log n/m} = O\left(\sum_{n=m+1}^{2m} \frac{e^{-m\delta}}{m\log\{1+(n-m)/m\}}\right)$$

$$= O\left(e^{-m\delta}\sum_{n=m+1}^{2m}\frac{1}{n-m}\right) = O(e^{-m\delta}\log m),$$

$$\sum_{n=2m+1}^{\infty} \frac{e^{-n\delta}}{(mn)^{\frac{1}{2}}\log n/m} = O\left(\frac{1}{m}\sum_{n=2m+1}^{\infty} e^{-n\delta}\right) = O\left(\frac{e^{-m\delta}}{m\delta}\right).$$

Hence $$\sum_{m\neq n} = O\left(\frac{1}{T}\right)\sum_{m=1}^{\infty}\left(\log m+\frac{1}{m\delta}\right)e^{-m\delta} = O\left(\frac{1}{T\delta}\log\frac{1}{\delta}\right).$$

This proves the first part. In the second part we have a pair of logarithms running throughout the remainder terms, and this is easily seen to produce the extra $\log^2 1/\delta$ in the result.

14.19. THEOREM 14.19. *As* $T\to\infty$,

$$\frac{1}{T}\int_0^T \{S_1(t)\}^2\,dt \sim \frac{C^2}{\pi^2}+\frac{1}{2\pi^2}\sum_{n=2}^{\infty}\frac{\Lambda_1^2(n)}{n\log^2 n}.$$

Let $$f(t) = Ce^{-\delta}-\sum_{n=2}^{\infty}\frac{\Lambda_1(n)\cos(t\log n)}{n^{\frac{1}{2}}\log n}e^{-\delta n}.$$

Then, as in the lemma,

$$\frac{2}{T}\int_{\frac{1}{2}T}^{T}\{f(t)\}^2\,dt = C^2e^{-2\delta}+\frac{1}{2}\sum_{n=2}^{\infty}\frac{\Lambda_1^2(n)}{n\log^2 n}e^{-2\delta n}+O\left(\frac{\log 1/\delta}{T\delta}\right),$$

and we can replace δ by 0 in the first two terms on the right with error

$$O(\delta)+O\left(\sum_{n=2}^{\infty}\frac{1-e^{-2\delta n}}{n\log^2 n}\right) = O(\delta)+O\left(\sum_{n\leqslant 1/\delta}\frac{\delta}{\log^2 n}\right)+O\left(\sum_{n>1/\delta}\frac{1}{n\log^2 n}\right)$$

$$= O\{1/\log(1/\delta)\}.$$

Hence, taking $\delta = T^{-\frac{1}{2}}$,

$$\frac{2}{T}\int_{\frac{1}{2}T}^{T}\{f(t)\}^2\,dt = C^2+\frac{1}{2}\sum_{n=2}^{\infty}\frac{\Lambda_1^2(n)}{n\log^2 n}+O\left(\frac{1}{\log T}\right).$$

Hence

$$\frac{2\pi^2}{T}\int_{\frac{1}{2}T}^{T}\{S_1(t)\}^2\,dt = \frac{2}{T}\int_{\frac{1}{2}T}^{T}\left\{f(t)+O\left(\frac{1}{\operatorname{loglog} T}\right)\right\}^2 dt$$

$$= \frac{2}{T}\int_{\frac{1}{2}T}^{T}\{f(t)\}^2\,dt+O\left\{\frac{1}{T\operatorname{loglog} T}\int_{\frac{1}{2}T}^{T}|f(t)|\,dt\right\}+O\left\{\frac{1}{(\operatorname{loglog} T)^2}\right\},$$

and, since $$\int_{\frac{1}{2}T}^{T} |f(t)|\,dt \leqslant \left[\int_{\frac{1}{2}T}^{T} dt \int_{\frac{1}{2}T}^{T} \{f(t)\}^2\,dt\right]^{\frac{1}{2}} = O(T),$$
it follows that
$$\frac{2}{T}\int_{\frac{1}{2}T}^{T} \{S_1(t)\}^2\,dt = \frac{C^2}{\pi^2}+\frac{1}{2\pi^2}\sum_{n=2}^{\infty}\frac{\Lambda_1^2(n)}{n\log^2 n}+O\left(\frac{1}{\log\log T}\right).$$
Replacing T by $\frac{1}{2}T$, $\frac{1}{4}T$,... and adding, we obtain the result.

14.20. The corresponding problem involving $S(t)$ is naturally more difficult, but it has recently been solved by A. Selberg (4). The solution depends on the following formula for $\zeta'(s)/\zeta(s)$

THEOREM 14.20. *Without any hypothesis*
$$\frac{\zeta'(s)}{\zeta(s)} = -\sum_{n<x^2}\frac{\Lambda_x(n)}{n^s}+\frac{x^{2(1-s)}-x^{1-s}}{(1-s)^2\log x}+$$
$$+\frac{1}{\log x}\sum_{q=1}^{\infty}\frac{x^{-2q-s}-x^{-2(2q+s)}}{(2q+s)^2}+\frac{1}{\log x}\sum_{\rho}\frac{x^{\rho-s}-x^{2(\rho-s)}}{(s-\rho)^2},$$
where
$$\Lambda_x(n) = \Lambda(n)\quad (1 \leqslant n \leqslant x),\qquad \frac{\Lambda(n)\log(x^2/n)}{\log x}\quad (x \leqslant n \leqslant x^2).$$
Let $\alpha = \max(2, 1+\sigma)$. Then
$$\frac{1}{2\pi i}\int_{\alpha-i\infty}^{\alpha+i\infty}\frac{x^{z-s}-x^{2(z-s)}}{(z-s)^2}\frac{\zeta'(z)}{\zeta(z)}\,dz$$
$$= -\frac{1}{2\pi i}\sum_{n=1}^{\infty}\Lambda(n)\int_{\alpha-i\infty}^{\alpha+i\infty}\frac{x^{z-s}-x^{2(z-s)}}{(z-s)^2 n^z}\,dz$$
$$= -\frac{1}{2\pi i}\sum_{n=1}^{\infty}\frac{\Lambda(n)}{n^s}\int_{\alpha-\sigma-i\infty}^{\alpha-\sigma+i\infty}\frac{x^w-x^{2w}}{w^2 n^w}\,dw$$
$$= -\frac{1}{2\pi i}\sum_{n=1}^{\infty}\frac{\Lambda(n)}{n^s}\left\{\int_{\alpha-\sigma-i\infty}^{\alpha-\sigma+i\infty}\left(\frac{x}{n}\right)^w\frac{dw}{w^2}-\int_{\alpha-\sigma-i\infty}^{\alpha-\sigma+i\infty}\left(\frac{x^2}{n}\right)^w\frac{dw}{w^2}\right\}$$
$$= -\sum_{n\leqslant x}\frac{\Lambda(n)}{n^s}\left(\log\frac{x}{n}-\log\frac{x^2}{n}\right)-\sum_{x<n\leqslant x^2}\frac{\Lambda(n)}{n^s}\left(-\log\frac{x^2}{n}\right)$$
$$= \log x\sum_{n\leqslant x^2}\frac{\Lambda_x(n)}{n^s}.$$

Now consider the residues obtained by moving the line of integration to the left. The residue at $z = s$ is $-\log x\,\zeta'(s)/\zeta(s)$; that at $z = 1$ is

$$-\frac{x^{1-s}-x^{2(1-s)}}{(1-s)^2};$$

those at $z = -2q$ and $z = \rho$ are

$$\frac{x^{-2q-s}-x^{2(-2q-s)}}{(-2q-s)^2}, \qquad \frac{x^{\rho-s}-x^{2(\rho-s)}}{(\rho-s)^2},$$

respectively. The result now easily follows.

14.21. THEOREM 14.21. *For $t > 2$, $4 \leqslant x \leqslant t^2$,*

$$\sigma_1 = \frac{1}{2}+\frac{1}{\log x},$$

we have

$$S(t) = -\frac{1}{\pi}\sum_{n<x^3}\frac{\Lambda_x(n)}{n^{\sigma_1}}\frac{\sin(t\log n)}{\log n}+O\left\{\frac{1}{\log x}\left|\sum_{n<x^2}\frac{\Lambda_x(n)}{n^{\sigma_1+it}}\right|\right\}+O\left(\frac{\log t}{\log x}\right).$$

By the previous theorem,

$$\frac{\zeta'(\sigma+it)}{\zeta(\sigma+it)} = -\sum_{n<x^2}\frac{\Lambda_x(n)}{n^{\sigma+it}}+O\left\{\frac{x^{2(1-\sigma)}+x^{1-\sigma}}{t^2\log x}\right\}+\frac{2\omega x^{\frac{1}{2}-\sigma}}{\log x}\sum_{\gamma}\frac{1}{(\sigma_1-\frac{1}{2})^2+(t-\gamma)^2} \tag{14.21.1}$$

for $\sigma \geqslant \sigma_1$, where $|\omega| < 1$. Now

$$\frac{x^{2(1-\sigma)}+x^{1-\sigma}}{t^2\log x} \leqslant \frac{x^{1-2\sigma}+x^{-\sigma}}{\log x} < 2x^{\frac{1}{2}-\sigma}.$$

Hence

$$\frac{\zeta'(\sigma+it)}{\zeta(\sigma+it)} = -\sum_{n<x^2}\frac{\Lambda_x(n)}{n^{\sigma+it}}+2\omega x^{\frac{1}{2}-\sigma}\sum_{\gamma}\frac{\sigma_1-\frac{1}{2}}{(\sigma_1-\frac{1}{2})^2+(t-\gamma)^2}+O(x^{\frac{1}{2}-\sigma}). \tag{14.21.2}$$

Now by (2.12.7)

$$\frac{\zeta'(s)}{\zeta(s)} = \sum_{\rho}\left(\frac{1}{s-\rho}+\frac{1}{\rho}\right)+O(\log t).$$

Hence

$$\mathbf{R}\frac{\zeta'(\sigma+it)}{\zeta(\sigma+it)} = \sum_{\gamma}\left\{\frac{\sigma-\frac{1}{2}}{(\sigma-\frac{1}{2})^2+(t-\gamma)^2}+\frac{\frac{1}{2}}{\frac{1}{4}+\gamma^2}\right\}+O(\log t)$$

$$= \sum_{\gamma}\frac{\sigma-\frac{1}{2}}{(\sigma-\frac{1}{2})^2+(t-\gamma)^2}+O(\log t).$$

Taking real parts in (14.21.2), substituting this on the left, and taking $\sigma = \sigma_1$,

$$\sum_{\gamma} \frac{\sigma_1 - \frac{1}{2}}{(\sigma_1 - \frac{1}{2})^2 + (t-\gamma)^2} + O(\log t)$$

$$= -\mathbf{R} \sum_{n<x^2} \frac{\Lambda_x(n)}{n^{\sigma_1+it}} + \frac{2\omega'}{e} \sum_{\gamma} \frac{\sigma_1 - \frac{1}{2}}{(\sigma_1 - \frac{1}{2})^2 + (t-\gamma)^2} \qquad (|\omega'| < 1).$$

Hence

$$\left(1 - \frac{2\omega'}{e}\right) \sum_{\gamma} \frac{\sigma_1 - \frac{1}{2}}{(\sigma_1 - \frac{1}{2})^2 + (t-\gamma)^2} = -\mathbf{R} \sum_{n<x^2} \frac{\Lambda_x(n)}{n^{\sigma_1+it}} + O(\log t).$$

Here

$$1 - \frac{2\omega'}{e} > 1 - \frac{2}{e} > \frac{1}{4}.$$

Hence

$$\sum_{\gamma} \frac{\sigma_1 - \frac{1}{2}}{(\sigma_1 - \frac{1}{2})^2 + (t-\gamma)^2} = O\left| \sum_{n<x^2} \frac{\Lambda_x(n)}{n^{\sigma_1+it}} \right| + O(\log t). \qquad (14.21.3)$$

Inserting this in (14.21.2), we get

$$\frac{\zeta'(\sigma+it)}{\zeta(\sigma+it)} = -\sum_{n<x^2} \frac{\Lambda_x(n)}{n^{\sigma+it}} + O\left\{ x^{\frac{1}{2}-\sigma} \left| \sum_{n<x^2} \frac{\Lambda_x(n)}{n^{\sigma_1+it}} \right| \right\} + O(x^{\frac{1}{2}-\sigma} \log t). \qquad (14.21.4)$$

Now

$$\arg \zeta(\tfrac{1}{2}+it) = -\int_{\frac{1}{2}}^{\infty} \mathbf{I} \frac{\zeta'(\sigma+it)}{\zeta(\sigma+it)}\, d\sigma$$

$$= -\int_{\sigma_1}^{\infty} \mathbf{I} \frac{\zeta'(\sigma+it)}{\zeta(\sigma+it)}\, d\sigma - (\sigma_1 - \tfrac{1}{2}) \mathbf{I} \frac{\zeta'(\sigma_1+it)}{\zeta(\sigma_1+it)} +$$

$$+ \int_{\frac{1}{2}}^{\sigma_1} \mathbf{I} \left\{ \frac{\zeta'(\sigma_1+it)}{\zeta(\sigma_1+it)} - \frac{\zeta'(\sigma+it)}{\zeta(\sigma+it)} \right\} d\sigma$$

$$= J_1 + J_2 + J_3.$$

By (14.21.4)

$$J_1 = \mathbf{I} \int_{\sigma_1}^{\infty} \sum_{n<x^2} \frac{\Lambda_x(n)}{n^{\sigma+it}}\, d\sigma + O\left\{ \left| \sum_{n<x^2} \frac{\Lambda_x(n)}{n^{\sigma_1+it}} \right| \int_{\sigma_1}^{\infty} x^{\frac{1}{2}-\sigma}\, d\sigma \right\} + O\left\{ \log t \int_{\sigma_1}^{\infty} x^{\frac{1}{2}-\sigma}\, d\sigma \right\}$$

$$= \mathbf{I} \sum_{n<x^2} \frac{\Lambda_x(n)}{n^{\sigma_1+it} \log n} + O\left\{ \frac{1}{\log x} \left| \sum_{n<x^2} \frac{\Lambda_x(n)}{n^{\sigma_1+it}} \right| \right\} + O\left(\frac{\log t}{\log x} \right).$$

Also, by (14.21.4) with $\sigma = \sigma_1$,

$$|J_2| \leqslant (\sigma_1-\tfrac{1}{2})\left|\frac{\zeta'(\sigma_1+it)}{\zeta(\sigma_1+it)}\right|$$

$$= O\left\{(\sigma_1-\tfrac{1}{2})\left|\sum_{n<x^2}\frac{\Lambda_x(n)}{n^{\sigma_1+it}}\right|\right\}+O\{(\sigma_1-\tfrac{1}{2})\log t\}$$

$$= O\left\{\frac{1}{\log x}\left|\sum_{n<x^2}\frac{\Lambda_x(n)}{n^{\sigma_1+it}}\right|\right\}+O\left(\frac{\log t}{\log x}\right).$$

It remains to estimate J_3. For $\frac{1}{2} < \sigma \leqslant \sigma_1$,

$$\mathbf{I}\left\{\frac{\zeta'(\sigma_1+it)}{\zeta(\sigma_1+it)}-\frac{\zeta'(\sigma+it)}{\zeta(\sigma+it)}\right\} = \sum_\rho \mathbf{I}\left(\frac{1}{\sigma_1+it-\rho}-\frac{1}{\sigma+it-\rho}\right)+O(\log t)$$

$$= \sum_\gamma \frac{(t-\gamma)\{(\sigma-\frac{1}{2})^2-(\sigma_1-\frac{1}{2})^2\}}{\{(\sigma-\frac{1}{2})^2+(t-\gamma)^2\}\{(\sigma_1-\frac{1}{2})^2+(t-\gamma)^2\}}+O(\log t).$$

Hence

$$\left|\mathbf{I}\left\{\frac{\zeta'(\sigma_1+it)}{\zeta(\sigma_1+it)}-\frac{\zeta'(\sigma+it)}{\zeta(\sigma+it)}\right\}\right|$$

$$\leqslant \sum_\gamma \frac{|t-\gamma|(\sigma_1-\frac{1}{2})^2}{\{(\sigma-\frac{1}{2})^2+(t-\gamma)^2\}\{(\sigma_1-\frac{1}{2})^2+(t-\gamma)^2\}}+O(\log t).$$

Hence

$$|J_3| \leqslant \sum_\gamma \frac{(\sigma_1-\frac{1}{2})^2}{(\sigma_1-\frac{1}{2})^2+(t-\gamma)^2}\int_{\frac{1}{2}}^{\infty}\frac{|t-\gamma|\,d\sigma}{(\sigma-\frac{1}{2})^2+(t-\gamma)^2}+O\{(\sigma_1-\tfrac{1}{2})\log t\}$$

$$\leqslant \tfrac{1}{2}\pi(\sigma_1-\tfrac{1}{2})\sum_\gamma \frac{(\sigma_1-\frac{1}{2})}{(\sigma_1-\frac{1}{2})^2+(t-\gamma)^2}+O\{(\sigma_1-\tfrac{1}{2})\log t\}$$

$$= O\left\{\frac{1}{\log x}\left|\sum_{n<x^2}\frac{\Lambda_x(n)}{n^{\sigma_1+it}}\right|\right\}+O\left(\frac{\log t}{\log x}\right),$$

by (14.21.3). The theorem follows from these results.

Theorem 14.21 leads to an alternative proof of Theorem 14.13; for taking $x = \sqrt{(\log t)}$ we obtain

$$S(t) = O\left(\sum_{n\leqslant x^2}\frac{1}{n^{\frac{1}{2}}}\right)+O\left(\frac{1}{\log x}\sum_{n\leqslant x^2}\frac{\log n}{n^{\frac{1}{2}}}\right)+O\left(\frac{\log t}{\log x}\right)$$

$$= O(x)+O(x)+O\left(\frac{\log t}{\log x}\right)$$

$$= O\left(\frac{\log t}{\log\log t}\right).$$

14.22. THEOREM 14.22. *For*

$$T^a \leqslant x \leqslant T^{\frac{1}{2}} \quad (0 < a \leqslant \tfrac{1}{2})$$

$$\int\limits_{\frac{1}{2}T}^{T} \left\{S(t)+\frac{1}{\pi}\sum_{p<x^2}\frac{\sin(t\log p)}{\sqrt{p}}\right\}^2 dt = O(T).$$

We have

$$S(t)+\frac{1}{\pi}\sum_{p<x^2}\frac{\sin(t\log p)}{\sqrt{p}} = \frac{1}{\pi}\sum_{p<x^2}\frac{\Lambda(p)-\Lambda_x(p)p^{\frac{1}{2}-\sigma_1}}{p^{\frac{1}{2}}\log p}\sin(t\log p)+$$

$$+O\left\{\frac{1}{\log x}\left|\sum_{p<x^2}\frac{\Lambda_x(p)}{p^{\sigma_1+it}}\right|\right\}+O\left\{\left|\sum_{p^2<x^2}\frac{\Lambda_x(p^2)}{p^{2(\sigma_1+it)}\log p}\right|\right\}+$$

$$+O\left\{\frac{1}{\log x}\left|\sum_{p^2<x^2}\frac{\Lambda_x(p^2)}{p^{2(\sigma_1+it)}}\right|\right\}+O\left(\sum_{r>2}\sum_{p}\frac{1}{p^{\frac{1}{2}r}}\right)+O\left(\frac{\log t}{\log x}\right).$$

The last term is bounded if $\frac{1}{2}T \leqslant t \leqslant T$, $x \geqslant T^a$, where a is a fixed positive constant. The last term but one is

$$O\left(\sum_{p}\sum_{r=3}^{\infty}\frac{1}{p^{\frac{1}{2}r}}\right) = O\left(\sum_{p}\frac{p^{-\frac{3}{2}}}{1-p^{-\frac{1}{2}}}\right) = O(1).$$

Now consider the first term on the right. If $p \leqslant x$,

$$\Lambda(p)-\Lambda_x(p)p^{\frac{1}{2}-\sigma_1} = (1-p^{\frac{1}{2}-\sigma_1})\log p$$
$$= (1-p^{-1/\log x})\log p = (1-e^{-\log p/\log x})\log p = O\left(\frac{\log^2 p}{\log x}\right),$$

and, if $x < p \leqslant x^2$, it is

$$O\{\Lambda_x(p)\} = O\left\{\log p\frac{\log x^2/p}{\log x}\right\} = O\left(\frac{\log^2 p}{\log x}\right).$$

Hence the first term is the imaginary part of

$$\sum_{p<x^2}\frac{\alpha_p}{p^{\frac{1}{2}+it}},$$

where
$$\alpha_p = \frac{\Lambda(p)-\Lambda_x(p)p^{\frac{1}{2}-\sigma_1}}{\log p} = O(\log p/\log x).$$

Now

$$\int\limits_{\frac{1}{2}T}^{T}\left|\sum_{p<x^2}\frac{\alpha_p}{p^{\frac{1}{2}+it}}\right|^2 dt = \sum_{p<x^2}\sum_{q<x^2}\frac{\alpha_p\,\alpha_q}{p^{\frac{1}{2}}q^{\frac{1}{2}}}\int\limits_{\frac{1}{2}T}^{T}\left(\frac{q}{p}\right)^{it}dt$$

$$= O\left(T\sum_{p<x^2}\frac{\alpha_p^2}{p}\right)+O\left(\sum_{p\neq q}\sum\frac{|\alpha_p\,\alpha_q|}{p^{\frac{1}{2}}q^{\frac{1}{2}}}\frac{1}{|\log p/q|}\right).$$

Since
$$\alpha_p^2 = O\left(\frac{\log^2 p}{\log^2 x}\right) = O\left(\frac{\log p}{\log x}\right),$$
the first term is
$$O\left(T\frac{1}{\log x}\sum_{p<x^2}\frac{\log p}{p}\right) = O(T),$$
by (3.14.3). The second term is
$$O\left(\sum_{p<x^2}\sum_{q\leqslant\frac{1}{2}p}\frac{|\alpha_p\,\alpha_q|}{p^{\frac{1}{2}}q^{\frac{1}{2}}}\right)+O\left(\sum_{p<x^2}\sum_{\frac{1}{2}p<q<p}\frac{|\alpha_p\,\alpha_q|}{p.(p-q)/p}\right)$$
$$= O\left(\sum_{p<x^2}\frac{\log p}{p^{\frac{1}{2}}\log x}\right)^2+O\left(\sum_{p<x^2}\frac{\log^2 p}{\log^2 x}.\log p\right)$$
$$= O\left(\sum_{p<x^2}\frac{1}{p^{\frac{1}{2}}}\right)^2+O\left(\sum_{p<x^2}\log p\right)$$
$$= O(x^2)+O(x^2) = O(T)$$
if $x \leqslant \surd T$.

A similar argument clearly applies to the second term. In the third term, the sum is of the form
$$\sum_{p<x}\frac{\alpha_p'}{p^{1+2it}},$$
where $\alpha_p' = O(1)$; and
$$\int_{\frac{1}{2}T}^{T}\left|\sum_{p<x}\frac{\alpha_p'}{p^{1+2it}}\right|^2 dt = \sum_{p<x}\sum_{q<x}\frac{\alpha_p'\,\alpha_q'}{pq}\int_{\frac{1}{2}T}^{T}\left(\frac{q}{p}\right)^{it}dt$$
$$= O\left(T\sum_{p<x}\frac{1}{p^2}\right)+O\left(\sum\sum_{p\neq q}\frac{1}{pq|\log p/q|}\right)$$
$$= O(T)+O\left(\sum_{p<x}\sum_{q\leqslant\frac{1}{2}p}\frac{1}{pq}\right)+O\left(\sum_{p<x}\sum_{\frac{1}{2}p<q<p}\frac{1}{p^2.(p-q)/p}\right)$$
$$= O(T)+O(\log^2 x)+O(\log^2 x) = O(T).$$
Similarly for the fourth term, and the result follows.

14.23. THEOREM 14.23. *If* $T^a \leqslant x \leqslant T^{\frac{1}{4}}$,
$$\int_{\frac{1}{2}T}^{T}\left\{\sum_{p<x^2}\frac{\sin(t\log p)}{\surd p}\right\}^2 dt = \tfrac{1}{4}T\log\log T+O(T).$$

This is

$$\sum_{p<x^2}\sum_{q<x^2}\frac{1}{p^{\frac{1}{2}}q^{\frac{1}{2}}}\int\limits_{\frac{1}{2}T}^{T}\sin(t\log p)\sin(t\log q)\,dt$$

$$=\sum_{p<x^2}\frac{1}{p}\left\{\tfrac{1}{4}T+O\left(\frac{1}{\log p}\right)\right\}+O\left(\sum_{p\neq q}\sum\frac{1}{p^{\frac{1}{2}}q^{\frac{1}{2}}|\log p/q|}\right).$$

Now, by (3.14.5),

$$\sum_{p<x^2}\frac{1}{p}=\log\log x^2+O(1)=\log\log T+O(1)$$

and (since $p_n > An\log n$)

$$\sum\frac{1}{p\log p}=O(1).$$

Hence the first term is

$$\tfrac{1}{4}T\log\log T+O(T).$$

Also $$|\log p/q|>A/p>A/x^2.$$

Hence the remainder is

$$O\left\{x^2\left(\sum_{p<x^2}\frac{1}{p^{\frac{1}{2}}}\right)^2\right\}=O(x^2.x^2)=O(x^4),$$

and the result follows if $x\leqslant T^{\frac{1}{4}}$.

14.24. THEOREM 14.24.

$$\int\limits_0^T\{S(t)\}^2\,dt\sim\frac{1}{2\pi^2}T\log\log T.$$

For

$$\int\limits_{\frac{1}{2}T}^{T}\{S(t)\}^2\,dt=\int\limits_{\frac{1}{2}T}^{T}\left\{S(t)+\frac{1}{\pi}\sum_{p<x^2}\frac{\sin(t\log p)}{\sqrt{p}}-\frac{1}{\pi}\sum_{p<x^2}\frac{\sin(t\log p)}{\sqrt{p}}\right\}^2dt$$

$$=\int\limits_{\frac{1}{2}T}^{T}\left\{S(t)+\frac{1}{\pi}\sum_{p<x^2}\frac{\sin(t\log p)}{\sqrt{p}}\right\}^2dt-$$

$$-\frac{2}{\pi}\int\limits_{\frac{1}{2}T}^{T}\left\{S(t)+\frac{1}{\pi}\sum_{p<x^2}\frac{\sin(t\log p)}{\sqrt{p}}\right\}\sum_{p<x^2}\frac{\sin(t\log p)}{\sqrt{p}}\,dt+$$

$$+\frac{1}{\pi^2}\int\limits_{\frac{1}{2}T}^{T}\left\{\sum_{p<x^2}\frac{\sin(t\log p)}{\sqrt{p}}\right\}^2dt$$

$$=O(T)+O\{T(\log\log T)^{\frac{1}{2}}\}+\frac{1}{4\pi^2}T\log\log T+O(T)$$

(using Schwarz's inequality on the middle term). The result then follows by addition.

It can be proved in a similar way that

$$\int_0^T \{S(t)\}^{2k}\,dt \sim \frac{(2k)!}{k!(2\pi)^{2k}}T(\log\log T)^k \tag{14.24.1}$$

for every positive integer k.

14.25. The Dirichlet series for $1/\zeta(s)$.

It was proved in § 3.13 that the formula

$$\frac{1}{\zeta(s)} = \sum_{n=1}^{\infty}\frac{\mu(n)}{n^s},$$

which is elementary for $\sigma > 1$, holds also for $\sigma = 1$. On the Riemann hypothesis we can go much farther than this.†

THEOREM 14.25 (A). *The series*

$$\sum_{n=1}^{\infty}\frac{\mu(n)}{n^s} \tag{14.25.1}$$

is convergent, and its sum is $1/\zeta(s)$, *for every s with* $\sigma > \frac{1}{2}$.

In the lemma of § 3.12, take $a_n = \mu(n)$, $f(s) = 1/\zeta(s)$, $c = 2$, and x half an odd integer. We obtain

$$\sum_{n<x}\frac{\mu(n)}{n^s} = \frac{1}{2\pi i}\int_{2-iT}^{2+iT}\frac{1}{\zeta(s+w)}\frac{x^w}{w}\,dw + O\left(\frac{x^2}{T}\right)$$

$$= \frac{1}{2\pi i}\left(\int_{2-iT}^{\frac{1}{2}-\sigma+\delta-iT} + \int_{\frac{1}{2}-\sigma+\delta-iT}^{\frac{1}{2}-\sigma+\delta+iT} + \int_{\frac{1}{2}-\sigma+\delta+iT}^{2+iT}\right)\frac{1}{\zeta(s+w)}\frac{x^w}{w}\,dw + \frac{1}{\zeta(s)} + O\left(\frac{x^2}{T}\right),$$

where $0 < \delta < \sigma - \frac{1}{2}$.

By (14.2.5), the first and third integrals are

$$O\left(T^{-1+\epsilon}\int_{\frac{1}{2}-\sigma+\delta}^{2} x^u\,du\right) = O(T^{-1+\epsilon}x^2),$$

and the second integral is

$$O\left\{x^{\frac{1}{2}-\sigma+\delta}\int_{-T}^{T}(1+|t|)^{-1+\epsilon}\,dt\right\} = O(x^{\frac{1}{2}-\sigma+\delta}T^{\epsilon}).$$

† Littlewood (1).

Hence
$$\sum_{n<x} \frac{\mu(n)}{n^s} = \frac{1}{\zeta(s)} + O(T^{-1+\epsilon}x^2) + O(T^{\epsilon}x^{\frac{1}{2}-\sigma+\delta}).$$

Taking, for example, $T = x^3$, the O-terms tend to zero as $x \to \infty$, and the result follows.

Conversely, if (14.25.1) is convergent for $\sigma > \frac{1}{2}$, it is uniformly convergent for $\sigma \geqslant \sigma_0 > \frac{1}{2}$, and so in this region represents an analytic function, which is $1/\zeta(s)$ for $\sigma > 1$ and so throughout the region. Hence the Riemann hypothesis is true. We have in fact

THEOREM 14.25 (B). *The convergence of* (14.25.1) *for* $\sigma > \frac{1}{2}$ *is a necessary and sufficient condition for the truth of the Riemann hypothesis.*

We shall write
$$M(x) = \sum_{n \leqslant x} \mu(n).$$

Then we also have

THEOREM 14.25 (C). *A necessary and sufficient condition for the Riemann hypothesis is*
$$M(x) = O(x^{\frac{1}{2}+\epsilon}). \tag{14.25.2}$$

The lemma of § 3.12 with $s = 0$, x half an odd integer, gives

$$\begin{aligned} M(x) &= \frac{1}{2\pi i} \int_{2-iT}^{2+iT} \frac{1}{\zeta(w)} \frac{x^w}{w}\,dw + O\left(\frac{x^2}{T}\right) \\ &= \frac{1}{2\pi i}\left(\int_{2-iT}^{\frac{1}{2}+\delta-iT} + \int_{\frac{1}{2}+\delta-iT}^{\frac{1}{2}+\delta+iT} + \int_{\frac{1}{2}+\delta+iT}^{2+iT}\right) \frac{1}{\zeta(w)} \frac{x^w}{w}\,dw + O\left(\frac{x^2}{T}\right) \\ &= O\left(\int_{-T}^{T} (1+|v|)^{-1+\epsilon} x^{\frac{1}{2}+\delta}\,dv\right) + O(T^{\epsilon-1}x^2) + O\left(\frac{x^2}{T}\right) \\ &= O(T^{\epsilon}x^{\frac{1}{2}+\delta}) + O(x^2T^{\epsilon-1}), \end{aligned} \tag{14.25.3}$$

by (14.2.6). Taking $T = x^2$, (14.25.2) follows if x is half an odd integer, and so generally.

Conversely, if (14.25.2) holds, then by partial summation (14.25.1) converges for $\sigma > \frac{1}{2}$, and the Riemann hypothesis follows.

14.26. The finer theory of $M(x)$ is extremely obscure, and the results are not nearly so precise as the corresponding ones in the prime-number problem. The best O-result known is

THEOREM 14.26 (A).†

$$M(x) = O\left\{x^{\frac{1}{2}}\exp\left(A\frac{\log x}{\log\log x}\right)\right\}. \tag{14.26.1}$$

To prove this, take

$$\delta = \tfrac{1}{2}+\frac{1}{\log\log T}, \qquad T = x^2,$$

in the formula (14.25.3). By (14.14.2),

$$\left|\frac{1}{\zeta(w)}\right| \leqslant \exp\left(A\frac{\log T}{\log\log T}\right)$$

on the horizontal sides of the contour. The contribution of these is therefore

$$O\left\{x^2\frac{1}{T}\exp\left(A\frac{\log T}{\log\log T}\right)\right\} = O\left\{\exp\left(A\frac{\log x}{\log\log x}\right)\right\}.$$

On the vertical side, (14.14.2) gives

$$\left|\frac{1}{\zeta(\frac{1}{2}+\delta+iv)}\right| \leqslant \exp\left(A\frac{\log v}{\log\log v}\log\frac{2\log\log T}{\log\log v}\right)$$

for $v_0 \leqslant v \leqslant T$. Now it is easily seen that the right-hand side is a steadily increasing function of v in this interval. Hence

$$\left|\frac{1}{\zeta(w)}\right| \leqslant \exp\left(A\frac{\log T}{\log\log T}\right) \quad (v_0 \leqslant v \leqslant T).$$

Hence the integral along the vertical side is of the form

$$O(x^{\frac{1}{2}+\delta})+O\left\{x^{\frac{1}{2}+\delta}\exp\left(A\frac{\log T}{\log\log T}\right)\int\limits_{v_0}^{T}\frac{dv}{v}\right\}$$

$$= O\left\{x^{\frac{1}{2}+\delta}\exp\left(A\frac{\log T}{\log\log T}\right)\log T\right\} = O\left\{x^{\frac{1}{2}}\exp\left(A\frac{\log x}{\log\log x}\right)\right\}.$$

This proves the theorem.

THEOREM 14.26 (B). $$M(x) = \Omega(x^{\frac{1}{2}}). \tag{14.26.2}$$

This is true without any hypothesis. For if the Riemann hypothesis is false, Theorem 14.25 (C) shows that

$$M(x) = \Omega(x^{a})$$

† Landau (13), Titchmarsh (3).

with some a greater than $\frac{1}{2}$. On the other hand, if the Riemann hypothesis is true, then for $\sigma > \frac{1}{2}$

$$\frac{1}{\zeta(s)} = \sum_{n=1}^{\infty} \frac{\mu(n)}{n^s} = \sum_{n=1}^{\infty} M(n)\left\{\frac{1}{n^s} - \frac{1}{(n+1)^s}\right\} = s\int_1^{\infty} \frac{M(x)}{x^{s+1}}\,dx. \tag{14.26.3}$$

Suppose that

$$|M(x)| \leqslant M_0 \ (1 \leqslant x < x_0), \quad \leqslant \delta x^{\frac{1}{2}} \ (x \geqslant x_0).$$

Then

$$\left|\frac{1}{\zeta(s)}\right| \leqslant |s|M_0 \int_1^{x_0} \frac{dx}{x^{\frac{3}{2}}} + |s|\delta \int_{x_0}^{\infty} \frac{dx}{x^{\sigma+\frac{1}{2}}}$$

$$< |s|M_0 \int_1^{\infty} \frac{dx}{x^{\frac{3}{2}}} + |s|\delta \int_1^{\infty} \frac{dx}{x^{\sigma+\frac{1}{2}}}$$

$$= 2|s|M_0 + \frac{|s|\delta}{\sigma - \frac{1}{2}}. \tag{14.26.4}$$

But if $\rho = \frac{1}{2}+i\gamma$ is a simple zero of $\zeta(s)$, and $s = \sigma+i\gamma$, $\sigma \to \frac{1}{2}$, then

$$\frac{1}{\zeta(s)} = \frac{1}{\zeta(s)-\zeta(\rho)} \sim \frac{1}{(\sigma-\frac{1}{2})\zeta'(\rho)}.$$

We therefore obtain a contradiction if

$$\delta < \frac{1}{|\rho\,\zeta'(\rho)|}.$$

This proves the theorem.

14.27. Formulae connecting the functions of prime-number theory with series of the form

$$\sum x^{\rho}, \quad \sum \frac{x^{\rho}}{\rho},$$

etc., are well known, and are discussed in the books of Landau and Ingham. Here we prove a similar formula for the function $M(x)$.

THEOREM 14.27. *There is a sequence T_ν, $\nu \leqslant T_\nu \leqslant \nu+1$, such that*

$$M(x) = \lim_{\nu\to\infty} \sum_{|\gamma|<T_\nu} \frac{x^{\rho}}{\rho\zeta'(\rho)} - 2 + \sum_{n=1}^{\infty} \frac{(-1)^{n-1}(2\pi/x)^{2n}}{(2n)!\,n\,\zeta(2n+1)} \tag{14.27.1}$$

if x is not an integer. If x is an integer, $M(x)$ is to be replaced by

$$M(x) - \tfrac{1}{2}\mu(x).$$

In writing the series we have supposed for simplicity that all the zeros of $\zeta(s)$ are simple; obvious modifications are required if this is not so.

For a fixed non-integral x, (3.12.1), with $a_n = \mu(n)$, $s = 0$, $c = 2$, and w replaced by s, gives

$$M(x) = \frac{1}{2\pi i}\int\limits_{2-iT}^{2+iT} \frac{x^s}{s}\frac{1}{\zeta(s)}\,ds + O\left(\frac{1}{T}\right).$$

If x is an integer, $\frac{1}{2}\mu(x)$ is to be subtracted from the left-hand side. By the calculus of residues, the first term on the right is equal to

$$\sum_{|\gamma|<T} \frac{x^\rho}{\rho\zeta'(\rho)} - 2 + \sum_{n=1}^{N} \frac{(-1)^{n-1}(2\pi/x)^{2n}}{(2n)!\,n\zeta(2n+1)} +$$
$$+\frac{1}{2\pi i}\left(\int\limits_{2-iT}^{-2N-1-iT} + \int\limits_{-2N-1-iT}^{-2N-1+iT} + \int\limits_{-2N-1+iT}^{2+iT}\right)\frac{x^s}{s\zeta(s)}\,ds,$$

where T is not the ordinate of a zero. Now

$$\int\limits_{-2N-1-iT}^{-2N-1+iT} \frac{x^s}{s\,\zeta(s)}\,ds = \int\limits_{2N+2-iT}^{2N+2+iT} \frac{x^{1-s}}{(1-s)\zeta(1-s)}\,ds$$
$$= \int\limits_{2N+2-iT}^{2N+2+iT} \frac{x^{1-s}}{1-s}\,\frac{2^{s-1}\pi^s}{\cos\frac{1}{2}s\pi\Gamma(s)}\,\frac{1}{\zeta(s)}\,ds.$$

Here

$$\frac{1}{\Gamma(s)} = O(|e^{s-(s-\frac{1}{2})\log s}|) = O(e^{\sigma-(\sigma-\frac{1}{2})\log|s|+\frac{1}{2}\pi|t|})$$
$$= O(e^{\sigma-(\sigma-\frac{1}{2})\log\sigma+\frac{1}{2}\pi|t|}).$$

Hence the integral is

$$O\left\{\int\limits_{-T}^{T} \frac{1}{T}\left(\frac{2\pi}{x}\right)^{2N+2} e^{2N+2-(2N+\frac{3}{2})\log(2N+2)}\,dt\right\},$$

which tends to zero as $N \to \infty$, for a fixed T. Hence we obtain

$$\sum_{|\gamma|<T} \frac{x^\rho}{\rho\zeta'(\rho)} - 2 + \sum_{n=1}^{\infty} \frac{(-1)^{n-1}(2\pi/x)^{2n}}{(2n)!\,n\zeta(2n+1)} + \frac{1}{2\pi i}\left(\int\limits_{2-iT}^{-\infty-iT} + \int\limits_{-\infty+iT}^{2+iT}\right)\frac{x^s}{s\zeta(s)}\,ds.$$

Also

$$\int\limits_{-\infty+iT}^{-1+iT} \frac{x^s}{s\zeta(s)}\,ds = \int\limits_{2+iT}^{\infty+iT} \frac{x^{1-s}}{(1-s)\zeta(1-s)}\,ds = \int\limits_{2+iT}^{\infty+iT} \frac{x^{1-s}}{1-s}\,\frac{2^{s-1}\pi^s}{\cos\frac{1}{2}s\pi\Gamma(s)}\,\frac{1}{\zeta(s)}\,ds$$
$$= O\left\{\int\limits_{2}^{\infty} \frac{1}{T}\left(\frac{2\pi}{x}\right)^{\sigma} e^{\sigma-(\sigma-\frac{1}{2})\log\sigma}\,d\sigma\right\} = O\left(\frac{1}{T}\right).$$

Also by (14.16.2) we can choose $T = T_\nu$ ($\nu \leqslant T_\nu \leqslant \nu+1$) such that

$$\frac{1}{\zeta(s)} = O(t^\epsilon) \quad (\tfrac{1}{2} \leqslant \sigma \leqslant 2,\ t = T_\nu).$$

Hence for $-1 \leqslant \sigma \leqslant \frac{1}{2}$, $t = T_\nu$

$$\frac{1}{\zeta(s)} = O\left\{\frac{|t|^{\sigma-\frac{1}{2}}}{|\zeta(1-s)|}\right\} = O(t^\epsilon).$$

Hence
$$\int\limits_{-1+iT_\nu}^{2+iT_\nu} \frac{x^s}{s\,\zeta(s)}\,ds = O(T_\nu^{\epsilon-1}).$$

Similarly for the integral over $(2-iT, -\infty-iT)$, and the result stated follows.

It follows from the above theorem that

$$\sum \frac{1}{|\rho\zeta'(\rho)|}$$

is divergent; if it were convergent,

$$\sum \frac{x^\rho}{\rho\zeta'(\rho)}$$

would be uniformly convergent over any finite interval, and $M(x)$ would be continuous.

14.28. The Mertens hypothesis.† It was conjectured by Mertens, from numerical evidence, that

$$|M(n)| < \surd n \quad (n > 1). \tag{14.28.1}$$

This has not been proved or disproved. It implies the Riemann hypothesis, but is not apparently a consequence of it. A slightly less precise hypothesis would be

$$M(x) = O(x^{\frac{1}{2}}). \tag{14.28.2}$$

The problem has a certain similarity to that of the function $\psi(x)-x$ in prime-number theory, where

$$\psi(x) = \sum_{n \leqslant x} \Lambda(n).$$

On the Riemann hypothesis, $\psi(x)-x = O(x^{\frac{1}{2}+\epsilon})$, but it is not of the form $O(x^{\frac{1}{2}})$, and in fact

$$\psi(x)-x = \Omega(x^{\frac{1}{2}}\log\log\log x). \tag{14.28.3}$$

The influence of the factor $\log\log\log x$ is quite inappreciable as far as

† See references in Landau's *Handbuch*, and von Sterneck (1).

the calculations go, and it might be conjectured that (14.28.2) could be disproved similarly. We shall show, however, that there is an essential difference between the two problems, and that the proof of (14.28.3) cannot be extended to the other case, at any rate in any obvious way.

The proof of (14.28.3) depends on the fact that the real part of

$$\sum_{\gamma>0} \frac{e^{i\rho z}}{\rho}$$

is unbounded in the neighbourhood of $z = 0$. To deal with $M(x)$ in the same way, we should have to prove that the real part of

$$f(z) = \sum_{\gamma>0} \frac{e^{i\rho z}}{\rho\,\zeta'(\rho)} \quad (\mathbf{R}(z) > 0)$$

is unbounded in the neighbourhood of $z = 0$. This, however, is not the case. For consider the integral

$$\frac{1}{2\pi i} \int \frac{e^{isz}}{s\,\zeta(s)}\,ds$$

taken round the rectangle $(-1,\ 2,\ 2+iT_n,\ -1+iT_n)$, where the T_n are those of the previous section, and an indentation is made above $s = 0$. The integral along the upper side of the contour tends to 0 as $n \to \infty$, and we calculate that

$$f(z) = \frac{1}{2\pi i} \int_{2}^{2+i\infty} \frac{e^{isz}}{s\,\zeta(s)}\,ds - \frac{1}{2\pi i} \int_{-1}^{-1+i\infty} \frac{e^{isz}}{s\,\zeta(s)}\,ds + \frac{1}{2\pi i} \int_{-1}^{2} \frac{e^{isz}}{s\,\zeta(s)}\,ds.$$

The last term tends to a finite limit as $z \to 0$. Also

$$|e^{isz}| = e^{y-xt} \leqslant e^{y} \quad (s = -1+it,\ z = x+it,\ x > 0)$$

and $1/\zeta(-1+it) = O(t^{-\frac{1}{2}})$. The second term is therefore bounded for $\mathbf{R}(z) > 0$.

The first term is equal to

$$\frac{1}{2\pi i} \sum_{n=1}^{\infty} \mu(n) \int_{2}^{2+i\infty} \frac{e^{isz}}{sn^{s}}\,ds.$$

Now, if $n > 1$,

$$\int_{2}^{2+i\infty} \frac{e^{s(iz-\log n)}}{s}\,ds = \left[\frac{e^{s(iz-\log n)}}{s(iz-\log n)}\right]_{2}^{2+i\infty} + \frac{1}{iz-\log n} \int_{2}^{2+i\infty} \frac{e^{s(iz-\log n)}}{s^{2}}\,ds$$

and $$|e^{(2+it)(iz-\log n)}| = e^{-2y-2\log n-tx} \leqslant n^{-2}e^{-2y}.$$

Hence $$\int_{2}^{2+i\infty} \frac{e^{s(iz-\log n)}}{s}\,ds = O\left(\frac{1}{n^{2}\log n}\right)$$

uniformly in the neighbourhood of $z = 0$. Hence

$$\sum_{n=2}^{\infty} \mu(n) \int_{2}^{2+i\infty} \frac{e^{isz}}{sn^s}\, ds = O(1).$$

If $z = re^{i\theta}$, we have

$$\int_{2}^{2+i\infty} \frac{e^{isz}}{s}\, ds = \int_{2}^{e^{i(\frac{1}{2}\pi-\theta)}} + \int_{e^{i(\frac{1}{2}\pi-\theta)}}^{\infty e^{i(\frac{1}{2}\pi-\theta)}}$$

$$= O(1) + \int_{1}^{\infty} \frac{e^{-r\lambda}}{\lambda}\, d\lambda$$

$$= O(1) + \int_{r}^{\infty} \frac{e^{-x}}{x}\, dx$$

$$= O(1) + \int_{r}^{1} \frac{dx}{x} + \int_{r}^{1} \frac{e^{x}-1}{x}\, dx + \int_{1}^{\infty} \frac{e^{-x}}{x}\, dx$$

$$= \log\frac{1}{r} + O(1).$$

Hence
$$f(z) = \frac{1}{2\pi i} \log\frac{1}{r} + O(1),$$

and consequently $\mathbf{R}f(z)$ is bounded.

14.29. In this section we shall investigate the consequences of the hypothesis that

$$\int_{1}^{X} \left\{\frac{M(x)}{x}\right\}^2 dx = O(\log X). \tag{14.29.1}$$

This is less drastic than the Mertens hypothesis, since it clearly follows from (14.28.2). The corresponding formula with $M(x)$ replaced by $\psi(x)-x$ is a consequence of the Riemann hypothesis.†

THEOREM 14.29 (A). *If* (14.29.1) *is true, all the zeros of* $\zeta(s)$ *on the critical line are simple.*

By (14.26.3),

$$\left|\frac{1}{\zeta(s)}\right| \leqslant |s| \int_{1}^{\infty} \frac{|M(x)|}{x^{\sigma+1}}\, dx = |s| \int_{1}^{\infty} \frac{|M(x)|}{x^{\frac{1}{2}\sigma+\frac{3}{4}}} \frac{1}{x^{\frac{1}{2}\sigma+\frac{1}{4}}}\, dx$$

$$\leqslant |s| \left\{\int_{1}^{\infty} \frac{M^2(x)}{x^{\sigma+\frac{3}{2}}}\, dx \int_{1}^{\infty} \frac{dx}{x^{\sigma+\frac{1}{2}}}\right\}^{\frac{1}{2}} = \frac{|s|}{(\sigma-\frac{1}{2})^{\frac{1}{2}}} \left\{\int_{1}^{\infty} \frac{M^2(x)}{x^{\sigma+\frac{3}{2}}}\, dx\right\}^{\frac{1}{2}}.$$

† Cramér (5).

Let
$$f(X) = \int_1^X \left\{\frac{M(x)}{x}\right\}^2 dx.$$

Then
$$\int_1^\infty \frac{M^2(x)}{x^{\sigma+\frac{3}{2}}}\,dx = \int_1^\infty \frac{f'(x)}{x^{\sigma-\frac{1}{2}}}\,dx = (\sigma-\tfrac{1}{2})\int_1^\infty \frac{f(x)}{x^{\sigma+\frac{1}{2}}}\,dx$$

$$= O\left\{(\sigma-\tfrac{1}{2})\int_1^\infty \frac{\log x}{x^{\sigma+\frac{1}{2}}}\,dx\right\} = O\left(\int_1^\infty \frac{dx}{x^{\sigma+\frac{1}{2}}}\right) = O\left(\frac{1}{\sigma-\frac{1}{2}}\right).$$

Hence
$$\frac{1}{\zeta(s)} = O\left(\frac{|s|}{\sigma-\frac{1}{2}}\right).$$

Let ρ be a zero and $s = \rho+h$, where $h > 0$. Then $\sigma = \frac{1}{2}+h$, and hence
$$\frac{1}{\zeta(\rho+h)} = O\left(\frac{|\rho+h|}{h}\right). \tag{14.29.2}$$

This would be false for $h \to 0$ if ρ were a zero of order higher than the first, so that the result follows.

Multiplying each side of (14.29.2) by h, and making $h \to 0$, we obtain
$$\frac{1}{\zeta'(\rho)} = O(|\rho|). \tag{14.29.3}$$

We can, however, prove more than this.

THEOREM 14.29 (B). *If* (14.29.1) *is true,*
$$\sum \frac{1}{|\rho\zeta'(\rho)|^2} \tag{14.29.4}$$
is convergent.

This follows from an argument of the 'Bessel's inequality' type. We have
$$0 \leqslant \int_1^X \left\{\frac{M(x)}{x} - \sum_{|\gamma|<T} \frac{x^{\rho-1}}{\rho\zeta'(\rho)}\right\}^2 dx$$

$$= \int_1^X \left\{\frac{M(x)}{x}\right\}^2 dx + \sum_{|\gamma|<T}\sum_{|\gamma'|<T} \frac{1}{\rho\rho'\zeta'(\rho)\zeta'(\rho')}\int_1^X x^{\rho+\rho'-2}\,dx -$$

$$-2\sum_{|\gamma|<T} \frac{1}{\rho\zeta'(\rho)}\int_1^X M(x)x^{\rho-2}\,dx.$$

In the first sum, the terms with $\rho' = 1-\rho$ are

$$\sum_{|\gamma|<T} \frac{1}{\rho(1-\rho)\zeta'(\rho)\zeta'(1-\rho)} \int_1^X \frac{dx}{x} = \log X \sum_{|\gamma|<T} \frac{1}{|\rho\zeta'(\rho)|^2},$$

since $1-\rho$ is the conjugate of ρ. In the remaining terms, $\rho = \frac{1}{2}+i\gamma$. $\rho' = \frac{1}{2}+i\gamma'$, where $\gamma' \neq -\gamma$. Hence

$$\int_1^X x^{\rho+\rho'-2}\,dx = \frac{X^{\rho+\rho'-1}-1}{\rho+\rho'-1} = O\left(\frac{1}{|\gamma+\gamma'|}\right).$$

Hence the sum of these terms is less than $K_1 = K_1(T)$.

In the last sum we write

$$\int_1^X M(x)x^{\rho-2}\,dx = \int_1^X M(x)x^{\rho-2}\left(1-\frac{x}{X}\right)dx + \frac{1}{X}\int_1^X M(x)x^{\rho-1}\,dx.$$

The last term is

$$O\left\{\frac{1}{X}\int_1^X |M(x)|x^{-\frac{1}{2}}\,dx\right\} = O\left[\frac{1}{X}\left\{\int_1^X \frac{M^2(x)}{x^2}\,dx \int_1^X x\,dx\right\}^{\frac{1}{2}}\right]$$

$$= O\left\{\int_1^X \frac{M^2(x)}{x^2}\right\}^{\frac{1}{2}} = O(\log^{\frac{1}{2}}X),$$

by (14.29.1). Also

$$\int_1^X M(x)x^{\rho-2}\left(1-\frac{x}{X}\right)dx = \frac{1}{2\pi i}\int_{2-i\infty}^{2+i\infty} \frac{X^{w+\rho-1}-1}{\zeta(w)w(w+\rho)(w+\rho-1)}\,dw. \qquad (14.29.5)$$

To prove this, insert the Dirichlet series for $1/\zeta(w)$ on the right-hand side and integrate term by term. This is justified by absolute convergence. We obtain

$$\sum_{n=1}^{\infty} \frac{\mu(n)}{2\pi i}\int_{2-i\infty}^{2+i\infty} \frac{1}{n^w}\,\frac{X^{w+\rho-1}-1}{w(w+\rho)(w+\rho-1)}\,dw.$$

Evaluating the integral in the usual way by the calculus of residues, we obtain

$$\sum_{n\leqslant X} \mu(n)\left\{\frac{X^{\rho-1}-n^{\rho-1}}{\rho-1} - \frac{X^\rho - n^\rho}{X\rho}\right\} = \sum_{n\leqslant X}\mu(n)\int_n^X \left(x^{\rho-2}-\frac{x^{\rho-1}}{X}\right)dx$$

$$= \int_1^X \sum_{n\leqslant X}\mu(n)x^{\rho-2}\left(1-\frac{x}{X}\right)dx = \int_1^X M(x)x^{\rho-2}\left(1-\frac{x}{X}\right)dx,$$

and (14.29.5) follows.

Let U be not the ordinate of a zero. Then the right-hand side of (14.29.5) is equal to

$$\frac{1}{2\pi i}\left(\int_{2-i\infty}^{2-iU}+\int_{2-iU}^{\frac{1}{4}-iU}+\int_{\frac{1}{4}-iU}^{\frac{1}{4}+iU}+\int_{\frac{1}{4}+iU}^{2+iU}+\int_{2+iU}^{2+i\infty}\right)+$$

$$+\text{sum of residues in } -U<\mathbf{I}(w)<U.$$

Let ρ'' run through zeros of $\zeta(s)$ with imaginary parts between $-U$ and U. Let $U>T$. Then there is a pole at $w=1-\rho$, with residue

$$\frac{\log X}{(1-\rho)\zeta'(1-\rho)}.$$

At the other ρ'' the residues are

$$\frac{X^{\rho''+\rho-1}-1}{\zeta'(\rho'')\rho''(\rho''+\rho-1)(\rho''+\rho)}=O\left(\frac{1}{|(\rho''+\rho-1)(\rho''+\rho)|}\right)=O\left(\frac{1}{|\gamma''+\gamma|^2}\right),$$

by (14.29.3), and

$$\sum_{\substack{-U<\gamma''<U\\ \gamma''\neq-\gamma}}\frac{1}{|\gamma''+\gamma|^2}\leqslant\sum_{\gamma''\neq-\gamma}\frac{1}{|\gamma''+\gamma|^2}<K_2,$$

where K_2 depends on T, if $|\gamma|<T$, but not on U.

Again

$$\int_{2+iU}^{2+i\infty}\frac{X^{w+\rho-1}-1}{\zeta(w)w(w+\rho)(w+\rho-1)}\,dw=O\left(X^{\frac{3}{2}}\int_{U}^{\infty}\frac{dv}{v(v+\gamma)^2}\right)$$

$$=O\left(\frac{X^{\frac{3}{2}}}{U(U+\gamma)}\right)=O\left(\frac{X^{\frac{3}{2}}}{U(U-T)}\right),$$

and similarly for the integral over $(2-i\infty,\ 2-iU)$. Also by (14.2.6) and the functional equation

$$\frac{1}{\zeta(\frac{1}{4}+it)}=O\left\{\frac{|t|^{-\frac{1}{4}}}{|\zeta(\frac{3}{4}-it)|}\right\}=O(|t|^{\epsilon-\frac{1}{4}}).$$

Hence, since $|w+\rho|\geqslant\frac{3}{4}$, $|w+\rho-1|\geqslant\frac{1}{4}$,

$$\int_{\frac{1}{4}-iU}^{\frac{1}{4}+iU}\frac{X^{w+\rho-1}-1}{\zeta(w)w(w+\rho)(w+\rho-1)}\,dw=O\left\{\int_{-U}^{U}\frac{|v|^{\epsilon-\frac{1}{4}}}{(\frac{1}{16}+v^2)^{\frac{1}{2}}}\,dv\right\}=O(1).$$

Finally, by Theorem 14.16, we can choose a sequence of values of U such that

$$\frac{1}{\zeta(w)}=O(|w|)\quad (t=U,\ \tfrac{1}{2}\leqslant\sigma\leqslant 2).$$

By the functional equation the same result then holds for $\frac{1}{4} \leqslant \sigma \leqslant \frac{1}{2}$ also. Hence

$$\int\limits_{\frac{1}{4}+iU}^{2+iU} \frac{X^{w+\rho-1}-1}{\zeta(w)w(w+\rho)(w+\rho-1)}\,dw = O\left(\frac{X^{\frac{3}{2}}}{U^{1-\epsilon}(U+\gamma)^2}\right) = O\left\{\frac{X^{\frac{3}{2}}}{U^{1-\epsilon}(U-T)^2}\right\},$$

and similarly for the integral over $(2-iU, \frac{1}{4}-iU)$. Making $U \to \infty$, it follows that

$$\int\limits_1^X M(x)x^{\rho-2}\left(1-\frac{x}{X}\right)dx = \frac{\log X}{(1-\rho)\zeta'(1-\rho)}+R,$$

where $|R| < K_3 = K_3(T)$ if $|\gamma| < T$.

Hence we obtain

$$0 \leqslant A\log X+\log X \sum_{|\gamma|<T} \frac{1}{|\rho\zeta'(\rho)|^2} - 2\log X \sum_{|\gamma|<T} \frac{1}{|\rho\zeta'(\rho)|^2} + A\log^{\frac{1}{2}}X+K_4(T),$$

$$\sum_{|\gamma|<T} \frac{1}{|\rho\zeta'(\rho)|^2} \leqslant A+\frac{A}{\log^{\frac{1}{2}}X}+\frac{K_4(T)}{\log X}.$$

Making $X \to \infty$,
$$\sum_{|\gamma|<T} \frac{1}{|\rho\zeta'(\rho)|^2} \leqslant A.$$

Since the right-hand side is now independent of T, the result follows.

In particular
$$\frac{1}{\zeta'(\rho)} = o(|\rho|).$$

14.30. *If* (14.29.1) *is true,*†

$$\frac{1}{\zeta'(\frac{1}{2}+it)} = O\left\{\exp\left(\frac{A\log^2 t}{\log\log t}\right)\right\}.$$

Suppose that the interval $(t-t^{-3}, t+t^{-3})$ contains γ, the ordinate of a zero. By differentiating (2.1.4) twice,

$$\zeta''(\tfrac{1}{2}+it) = O(t).$$

Using this and (14.29.3), we obtain

$$|\zeta'(\tfrac{1}{2}+it)| = \left|\zeta'(\tfrac{1}{2}+i\gamma)+\int\limits_{\frac{1}{2}+i\gamma}^{\frac{1}{2}+it} \zeta''(s)\,ds\right|$$

$$> \frac{A}{\gamma}-At|t-\gamma|$$

$$> \frac{A}{t}-\frac{A}{t^2} > \frac{A}{t}.$$

† Cramér and Landau (1).

Suppose on the contrary that $(t-t^{-3}, t+t^{-3})$ is free from ordinates of zeros. Theorem 14.15 gives

$$\log|\zeta(\tfrac{1}{2}+it)| = \sum_{|t-\gamma|\leqslant 1/\log\log t} \log|t-\gamma|+O\left(\frac{\log t\log\log\log t}{\log\log t}\right).$$

There are $O(\log t/\log\log t)$ terms in the sum, each being now $O(\log t)$. Hence

$$\log|\zeta(\tfrac{1}{2}+it)| = O\left(\frac{\log^2 t}{\log\log t}\right), \qquad \frac{1}{\zeta(\frac{1}{2}+it)} = O\left\{\exp\left(\frac{A\log^2 t}{\log\log t}\right)\right\}.$$

Now $\pi^{-\frac{1}{2}s}\Gamma(\frac{1}{2}s)\zeta(s)$ is real on $\sigma = \frac{1}{2}$. Hence

$$-\tfrac{1}{2}\log\pi+\frac{1}{2}\frac{\Gamma'(\frac{1}{2}s)}{\Gamma(\frac{1}{2}s)}+\frac{\zeta'(s)}{\zeta(s)}$$

is purely imaginary on $\sigma = \frac{1}{2}$. Hence, on $\sigma = \frac{1}{2}$,

$$\left|\frac{\zeta'(s)}{\zeta(s)}\right| \geqslant -\mathbf{R}\frac{\zeta'(s)}{\zeta(s)} = -\tfrac{1}{2}\log\pi+\tfrac{1}{2}\mathbf{R}\frac{\Gamma'(\frac{1}{2}s)}{\Gamma(\frac{1}{2}s)} = \tfrac{1}{2}\log t+O(1) \to \infty.$$

Hence (without any hypothesis)

$$|\zeta'(\tfrac{1}{2}+it)| \geqslant |\zeta(\tfrac{1}{2}+it)| \quad (t > t_0).$$

This proves the theorem.

14.31. *Let* $\frac{1}{2}+i\gamma$, $\frac{1}{2}+i\gamma'$ *be consecutive complex zeros of* $\zeta(s)$. *If* (14.29.1) *is true*

$$\gamma'-\gamma > \frac{A}{\gamma}\exp\left(-A\frac{\log\gamma}{\log\log\gamma}\right).$$

We have

$$0 = \int_{\gamma}^{\gamma'} \zeta'(\tfrac{1}{2}+it)\,dt = (\gamma'-\gamma)\zeta'(\tfrac{1}{2}+i\gamma)+\int_{\gamma}^{\gamma'}(\gamma'-t)\zeta''(\tfrac{1}{2}+it)\,dt.$$

Hence by (14.29.3)

$$\frac{\gamma'-\gamma}{\gamma} < A\left|\int_{\gamma}^{\gamma'}(\gamma'-t)\zeta''(\tfrac{1}{2}+it)\,dt\right|$$

$$< A\max_{\gamma\leqslant t\leqslant\gamma'}|\zeta''(\tfrac{1}{2}+it)|\int_{\gamma}^{\gamma'}(\gamma'-t)\,dt = A(\gamma'-\gamma)^2\max_{\gamma\leqslant t\leqslant\gamma'}|\zeta''(\tfrac{1}{2}+it)|.$$

Now

$$\zeta''(\tfrac{1}{2}+it) = \frac{1}{\pi i}\int_0^{2\pi}\frac{\zeta(\frac{1}{2}+it+re^{i\theta})}{(re^{i\theta})^3}ire^{i\theta}\,d\theta = O\left\{\frac{1}{r^2}\int_0^{2\pi}|\zeta(\tfrac{1}{2}+it+re^{i\theta})|\,d\theta\right\}$$

$$= O\left\{\frac{1}{r^2}\exp\left(A\frac{\log t}{\log\log t}\right)(1+t^{Ar})\right\}$$

by (14.14.1) and the functional equation. Taking $r = 1/\log t$,

$$\zeta''(\tfrac{1}{2}+it) = O\left\{\exp\left(A\frac{\log t}{\log\log t}\right)\right\}$$

and the result follows.

14.32. *Necessary and sufficient conditions for the Riemann hypothesis.* Two such conditions have been given in § 14.25. Other similar conditions occur in the prime-number problem.†

A different kind of condition was stated by M. Riesz.‡ Let

$$F(x) = \sum_{k=1}^{\infty} \frac{(-1)^{k+1}x^k}{(k-1)!\,\zeta(2k)}. \tag{14.32.1}$$

Then a simple application of the calculus of residues gives

$$F(x) = \frac{i}{2}\int\limits_{a-i\infty}^{a+i\infty} \frac{x^s}{\Gamma(s)\zeta(2s)\sin\pi s}\,ds = \frac{i}{2\pi}\int\limits_{a-i\infty}^{a+i\infty} \frac{\Gamma(1-s)}{\zeta(2s)}x^s\,ds,$$

where $\frac{1}{2} < a < 1$. Taking a just greater than $\frac{1}{2}$, it clearly follows that

$$F(x) = O(x^{\frac{1}{2}+\epsilon}).$$

On the Riemann hypothesis we could move the line of integration to $a = \frac{1}{4}+\epsilon$ (using (14.2.5)) and obtain similarly

$$F(x) = O(x^{\frac{1}{4}+\epsilon}). \tag{14.32.2}$$

Conversely, by Mellin's inversion formula,

$$\frac{\Gamma(1-s)}{\zeta(2s)} = -\int\limits_0^{\infty} F(x)x^{-1-s}\,ds.$$

If (14.32.2) holds, the integral converges uniformly for $\sigma \geqslant \sigma_0 > \frac{1}{4}$; the analytic function represented is therefore regular for $\sigma > \frac{1}{4}$, and the truth of the Riemann hypothesis follows. Hence (14.32.2) is a necessary and sufficient condition for the Riemann hypothesis.

A similar condition stated by Hardy and Littlewood§ is

$$\sum_{k=1}^{\infty} \frac{(-x)^k}{k!\,\zeta(2k+1)} = O(x^{-\frac{1}{4}}). \tag{14.32.3}$$

These conditions have a superficial attractiveness since they depend explicitly only on values taken by $\zeta(s)$ at points in $\sigma > 1$; but actually no use has ever been made of them.

† Landau, *Vorlesungen*, ii. 108–56. ‡ M. Riesz (1).
§ Hardy and Littlewood (2).

Conditions for the Riemann hypothesis also occur in the theory of Farey series. Let the fractions h/k with $0 < h \leqslant k$, $(h,k) = 1$, $k \leqslant N$, arranged in order of magnitude, be denoted by r_ν ($\nu = 1, 2,..., \Phi(N)$, where $\Phi(N) = \phi(1)+...+\phi(N)$). Let

$$\delta_\nu = r_\nu - \nu/\Phi(N)$$

be the distance between r_ν and the corresponding fraction obtained by dividing up the interval (0, 1) into $\Phi(N)$ equal parts. Then a necessary and sufficient condition for the Riemann hypothesis is‡

$$\sum_{\nu=1}^{\Phi(N)} \delta_r^2 = O\left(\frac{1}{N^{1-\epsilon}}\right). \tag{14.32.4}$$

An alternative necessary and sufficient condition is§

$$\sum_{\nu=1}^{\Phi(N)} |\delta_\nu| = O(N^{\frac{1}{2}+\epsilon}). \tag{14.32.5}$$

Details are given in Landau's *Vorlesungen*, ii. 167–77.

Still another condition|| can be expressed in terms of the formulae of § 10.1. If $\Xi(t)$ and $\Phi(u)$ are related by (10.1.3), a necessary and sufficient condition that all the zeros of $\Xi(t)$ should be real is that

$$\int_{-\infty}^{\infty}\int_{-\infty}^{\infty} \Phi(\alpha)\Phi(\beta)e^{i(\alpha+\beta)x}e^{(\alpha-\beta)y}(\alpha-\beta)^2\,d\alpha d\beta \geqslant 0 \tag{14.32.6}$$

for all real values of x and y. But no method has been suggested of showing whether such criteria are satisfied or not.

A sufficient condition†† for the Riemann hypothesis is that the partial sums $\sum_{\nu=1}^{n} \nu^{-s}$ of the series for $\zeta(s)$ should have no zeros in $\sigma > 1$.

NOTES FOR CHAPTER 14

14.33. The argument of §14.5 may be extended to the strip $\frac{1}{2} \leqslant \sigma_0 \leqslant \sigma \leqslant \frac{9}{8}$, giving

$$\frac{\zeta'(s)}{\zeta(s)} = O\left(\frac{\delta^{\sigma-1}-1}{1-\sigma}\right) + O(\delta^{\sigma-\frac{1}{2}}\log t).$$

The choice $\delta = (\log t)^{-2}$ then yields

$$\frac{\zeta'(s)}{\zeta(s)} \ll \frac{(\log t)^{2-2\sigma}-1}{1-\sigma}$$

‡ Franel (1). § Landau (16). || See Pólya (3), § 7. †† Turán (3).

uniformly for $\sigma_0 \leqslant \sigma \leqslant \frac{9}{8}$ and $t \geqslant 2$, and hence

$$\log \zeta(s) \ll \begin{cases} \log \dfrac{1}{\sigma - 1} & \text{if} \quad 1 + \dfrac{1}{\operatorname{loglog} t} \leqslant \sigma \leqslant \frac{9}{8}, \\ \dfrac{(\log t)^{2-2\sigma} - 1}{(1-\sigma)\operatorname{loglog} t} + \operatorname{logloglog} t & \text{if} \quad \sigma_0 \leqslant \sigma \leqslant 1 + \dfrac{1}{\operatorname{loglog} t}. \end{cases}$$

These results, together with those of § 14.14 are the sharpest conditional order-estimates available at present.

14.34. The Ω-result given by Theorem 14.12(A) has been sharpened by Montgomery [**3**], to give

$$S(t) = \Omega_{\pm}\left(\frac{(\log t)^{\frac{1}{2}}}{(\operatorname{loglog} t)^{\frac{1}{2}}}\right)$$

on the Riemann hypothesis. A minor modification of his method also yields

$$S_1(t) = \Omega_{\pm}\left(\frac{(\log t)^{\frac{1}{2}}}{(\operatorname{loglog} t)^{\frac{3}{2}}}\right).$$

It may be conjectured that these are best possible.

Mueller [**2**] has shown, on the Riemann hypothesis, that if c is a suitable constant, then $S(t)$ changes sign in any interval $[T, T + c \operatorname{loglog} T]$.

Further results and conjectures on the vertical distribution of the zeros are given by Montgomery [**2**], who investigated the pair correlation function

$$F(\alpha, T) = \frac{1}{N(T)} \sum_{0 < \gamma, \gamma' \leqslant T} T^{i\alpha(\gamma - \gamma')} w(\gamma - \gamma'),$$

where $w(u) = 4/(4 + u^2)$. This is a real-valued, even, non-negative function of α, and satisfies

$$F(\alpha, T) = \alpha + T^{-2\alpha} \log T + O\left(\frac{1}{\log T}\right) + O(\alpha T^{\alpha - 1}) + O(T^{-\frac{3}{2}\alpha}) \tag{14.34.1}$$

for $\alpha \geqslant 0$, whence $F(\alpha, T) \to \alpha$ as $T \to \infty$, uniformly for $0 < \delta \leqslant \alpha \leqslant 1 - \delta$. Montgomery conjectured that in general

$$F(\alpha, T) \to \min(\alpha, 1) \tag{14.34.2}$$

uniformly for $0 \leqslant \delta \leqslant \alpha \leqslant A$. This is related to a number of conjectures on the distribution of prime numbers. (See Gallagher and Mueller [**1**], Heath-Brown [**10**], and joint work of Goldston and Montgomery in the course of publication.) From (14.34.2) one may deduce that

$$\#\left\{\gamma, \gamma' \in [0, T]: \frac{2\pi\alpha}{\log T} \leqslant \gamma - \gamma' \leqslant \frac{2\pi\beta}{\log T}\right\}$$

$$\sim N(T)\left\{\delta(\alpha, \beta) + \int_\alpha^\beta 1 - \left(\frac{\sin \pi u}{u}\right)^2 du\right\}$$

for fixed α, β, as $T \to \infty$. Here $\delta(\alpha, \beta) = 1$ or 0 according as $\alpha \leqslant 0 \leqslant \beta$ or not.

Using (14.34.1), Montgomery showed that

$$\sum_{0<\gamma\leqslant T}{}' \, m(\rho)^2 \leqslant \{\tfrac{4}{3} + o(1)\} N(T),$$

where $m(\rho)$ is the multiplicity of ρ, and Σ' counts zeros without regard to multiplicity. One may deduce, in the notation of §10.29, that

$$N^{(1)}(T) \geqslant \{\tfrac{2}{3} + o(1)\} N(T), \tag{14.34.3}$$

on the Riemann hypothesis. The conjecture (14.34.2) would indeed yield $N^{(1)}(T) \sim N(T)$, i.e. 'almost all' the zeros would be simple. Montgomery also used (14.34.1) to show that

$$\liminf_{n\to\infty} \frac{\gamma_{n+1} - \gamma_n}{(2\pi/\log \gamma_n)} \leqslant 0{\cdot}68; \tag{14.34.4}$$

here $2\pi/\log \gamma_n$ is the average spacing between zeros.

By using a different method, Conrey, Ghosh, and Gonek (in work in the course of publication) have improved (14.34.3). Their starting point is the observation that

$$\left|\sum_{0<\gamma\leqslant T} M(\tfrac{1}{2} + i\gamma)\zeta'(\tfrac{1}{2} + i\gamma)\right|^2 \leqslant N^{(1)}(T) \sum_{0<\gamma\leqslant T} |M(\tfrac{1}{2} + i\gamma)\zeta'(\tfrac{1}{2} + i\gamma)|^2, \tag{14.34.5}$$

by Cauchy's inequality. The function $M(s)$ is taken to be a mollifier

$$M(s) = \sum_{n\leqslant y} \mu(n) P\left(\frac{\log y/n}{\log y}\right) n^{-s}, \qquad y = T^{\frac{1}{2}-\varepsilon},$$

where the polynomial $P(x)$ is chosen optimally as $\frac{3}{2}x - \frac{1}{2}x^2$. One may write the sums occurring in (14.34.5) as integrals

$$\frac{1}{2\pi i}\int_P \frac{\zeta'(s)}{\zeta(s)} M(s)\zeta'(s)\,ds$$

and

$$\frac{1}{2\pi i}\int_P \frac{\zeta'(s)}{\zeta(s)} M(s)M(1-s)\zeta'(s)\zeta'(1-s)\,ds,$$

taken around an appropriate rectangular path P. The estimation of these is long and complicated, but leads ultimately to the lower bound

$$N^{(1)}(T) \geqslant \{\tfrac{19}{27}+o(1)\}N(T).$$

The estimate (14.34.4) has also been improved, firstly by Montgomery and Odlyzko [**1**], and then by Conrey, Ghosh and Gonek [**1**]. The latter work produces the constant 0·5172. The corresponding lower bound

$$\limsup_{n\to\infty} \frac{\gamma_{n+1}-\gamma_n}{2\pi/\log\gamma_n} \geqslant \lambda > 1 \tag{14.34.6}$$

has been considered by Mueller [**1**], as well as in the two papers just cited. Here the best result known is that of Conrey, Ghosh, and Gonek [**1**], which has $\lambda = 2{\cdot}337$. Indeed, further work by Conrey, Ghosh, and Gonek, which is in the course of publication at the time of writing, yields $\lambda = 2{\cdot}68$ subject to the generalized Riemann hypothesis (i.e. a Riemann hypothesis for $\zeta(s)$ and all Dirichlet L-functions $L(s,\chi)$.) Moreover it seems likely that this condition may be relaxed to the ordinary Riemann hypothesis with further work.

If one asks for bounds of the form (14.34.4) and (14.34.6) which are satisfied by a positive proportion of zeros (as in § 9.25) then one may take constants 0·77 and 1·33 (Conrey, Ghosh, Goldston, Gonek, and Heath-Brown [**1**]).

14.35. It should be remarked in connection with § 14.24 that Selberg (4) proved Theorem 14.24 with error term $O(T)$, while the method here yields only $O\{T(\log\log T)^{\frac{1}{2}}\}$. Moreover he obtained the error term $O\{T(\log\log T)^{k-1}\}$ for (14.24.1).

14.36. The argument of the final paragraph of § 14.27 may be quantified, and then yields

$$\sum_{|\gamma|\leqslant T} |\zeta'(\tfrac{1}{2}+i\gamma)|^{-1} \gg T,$$

uniformly for $T \geqslant T_0$, assuming the Riemann hypothesis and that all the zeros are simple. However a slightly better result comes from combining the asymptotic formula

$$\sum_{0<\gamma\leqslant T} |\zeta'(\tfrac{1}{2}+i\gamma)|^2 \sim \tfrac{1}{12}N(T)(\log T)^2$$

of Gonek [2] with the bound (14.34.3). Using Hölder's inequality one may then derive the estimate

$$\sum_{cT<\gamma<T}^{*} \frac{1}{|\zeta'(\frac{1}{2}+i\gamma)|} \gg T,$$

where Σ^* counts simple zero only, and $c>0$ is a suitable numerical constant.

14.37. The Mertens hypothesis has been disproved by Odlyzko and te Riele [1], who showed that

$$\limsup_{x\to\infty} \frac{M(x)}{\sqrt{x}} > 1{\cdot}06$$

and

$$\liminf_{x\to\infty} \frac{M(x)}{\sqrt{x}} < -1{\cdot}009.$$

Their treatment is indirect, and produces no specific x for which $|M(x)|>x^{\frac{1}{2}}$. The method used is computational, and depends on solving numerically the inequalities occurring in Kronecker's theorem, so as to make the first few terms of (14.27.1) pull in the same direction. To this extent Odlyzko and te Riele follow the earlier work of Jurkat and Peyerimhoff [1], but they use a much more efficient algorithm for solving the Diophantine approximation problem.

14.38. Turán (3) conjectured that

$$\sum_{n\leqslant x} \frac{\lambda(n)}{n} \geqslant 0 \qquad (14.38.1)$$

for all $x>0$, where $\lambda(n)$ is the Liouville function, given by (1.2.11). He showed that his condition, given in §14.32, implies the above conjecture, which in turn implies the Riemann hypothesis. However Haselgrove [2] proved that (14.38.1) is false in general, thereby showing that Turán's condition does not hold. Later Spira [1] found by calculation that

$$\sum_{n=1}^{19} n^{-s}$$

has a zero in the region $\sigma>1$.

XV

CALCULATIONS RELATING TO THE ZEROS

15.1. It is possible to verify by means of calculation that all the complex zeros of $\zeta(s)$ up to a certain point lie exactly (not merely approximately) on the critical line. As a simple example we shall find roughly the position of the first complex zero in the upper half-plane, and show that it lies on the critical line.

We consider the function $Z(t) = e^{i\vartheta}\zeta(\frac{1}{2}+it)$ defined in § 4.17. This is real for real values of t, so that, if $Z(t_1)$ and $Z(t_2)$ have opposite signs, $Z(t)$ vanishes between t_1 and t_2, and so $\zeta(s)$ has a zero on the critical line between $\frac{1}{2}+it_1$ and $\frac{1}{2}+it_2$.

It follows from (2.2.1) that $\zeta(\frac{1}{2}) < 0$, then from (2.1.12) that $\xi(\frac{1}{2}) > 0$, i.e. that $\Xi(0) > 0$; and then from (4.17.3) that $Z(0) < 0$.

We shall next consider the value $t = 6\pi$. Now the argument of § 4.14 shows that, if x is half an odd integer,

$$\left|\zeta(s) - \sum_{n<x} \frac{1}{n^s}\right| \leqslant \frac{x^{1-\sigma}}{|1-s|} + \frac{2x^{-\sigma}}{2\pi - |t|/x}. \tag{15.1.1}$$

Hence, taking $t > 0$,

$$\left|Z(t) - \sum_{n<x} \frac{\cos(t\log n - \vartheta)}{n^{\frac{1}{2}}}\right| \leqslant \frac{x^{\frac{1}{2}}}{t} + \frac{2x^{\frac{1}{2}}}{2\pi x - t}. \tag{15.1.2}$$

For $x = \frac{9}{2}$, $t = 6\pi$, the right-hand side is about 0·6.

We next require an approximation to ϑ. We have

$$e^{-2i\vartheta} = \chi(\tfrac{1}{2}+it) = \pi^{it}\frac{\Gamma(\frac{1}{4}-\frac{1}{2}it)}{\Gamma(\frac{1}{4}+\frac{1}{2}it)},$$

so that

$$\vartheta = -\tfrac{1}{2}t\log\pi + \mathbf{I}\log\Gamma(\tfrac{1}{4}+\tfrac{1}{2}it)$$

$$= \tfrac{1}{2}t\log\frac{t}{2\pi} - \tfrac{1}{2}t - \tfrac{1}{8}\pi + O\left(\frac{1}{t}\right).$$

It may be verified that the term $O(1/t)$ is negligible in the calculations.

Writing $\vartheta = 2\pi K$, and using the values

$$\log 2 = 0{\cdot}6931, \qquad \log 3 = 1{\cdot}0986,$$

it is found that

$$K = 0{\cdot}1166, \qquad 3\log 3 - K = 3{\cdot}179$$

$$3\log 2 - K = 1{\cdot}963, \qquad 3\log 4 - K = 4{\cdot}042,$$

approximately. Hence the cosines in (15.1.2) are all positive, and $\cos 2\pi K = 0{\cdot}74...$. Hence $Z(6\pi) > 0$.

There is therefore one zero at least on the critical line between $t = 0$ and $t = 6\pi$.

Again, the formulae of § 9.3 give

$$N(T) = 1+2K+\frac{1}{\pi}\Delta \arg \zeta(s),$$

where Δ denotes variation along $(2, 2+iT, \frac{1}{2}+iT)$. Now $\mathbf{R}\,\zeta(s) > 0$ on $\sigma = 2$, and an argument similar to that already used, but depending on (15.1.1), shows that $\mathbf{R}\,\zeta(s) > 0$ on $(2+iT, \frac{1}{2}+iT)$, if $T = 6\pi$. Hence $|\Delta \arg \zeta(s)| < \frac{1}{2}\pi$, and

$$N(6\pi) < \tfrac{3}{2}+2K < 2.$$

Hence there is at most one complex zero with imaginary part less than 6π, and so in fact just one, namely the one on the critical line.

15.2. It is plain that the above process can be continued as long as the appropriate changes of sign of the function $Z(t)$ occur. Defining $K = K(t)$, as before, let t_ν be such that

$$K(t_\nu) = \tfrac{1}{2}\nu-1 \quad (\nu = 1, 2,...). \tag{15.2.1}$$

Then (15.1.2) gives

$$Z(t_\nu) \sim (-1)^\nu \sum_{n<x} \frac{\cos(t_\nu \log n)}{n^{\frac{1}{2}}}.$$

If the sum is dominated by its first term, it is positive, and so $Z(t_\nu)$ has the sign of $(-1)^\nu$. If this is true for ν and $\nu+1$, $Z(t)$ has a zero in the interval $(t_\nu, t_{\nu+1})$.

The value $t = 6\pi$ in the above argument is a rough approximation to t_2.

The ordinates of the first six zeros are

$$14{\cdot}13, \quad 21{\cdot}02, \quad 25{\cdot}01, \quad 30{\cdot}42, \quad 32{\cdot}93, \quad 37{\cdot}58$$

to two decimal places.† Some of these have been calculated with great accuracy.

15.3. The calculations which the above process requires are very laborious if t is at all large. A much better method is to use the formula (4.17.5) arising from the approximate functional equation. Let us write $t = 2\pi u$,

$$\alpha_n = \alpha_n(u) = n^{-\frac{1}{2}} \cos 2\pi(K-u\log n),$$

and

$$h(\xi) = \frac{\cos 2\pi(\xi^2-\xi-\frac{1}{16})}{\cos 2\pi\xi}.$$

† See the references Gram (6), Lindelöf (3), in Landau's *Handbuch*.

Then (4.17.5) gives

$$Z(2\pi u) = 2\sum_{n=1}^{m} \alpha_n(u) + (-1)^{m-1}u^{-\frac{1}{4}}h(\surd u - m) + R(u),$$

where $m = [\surd u]$, and $R(u) = O(u^{-\frac{3}{4}})$. The $\alpha_n(u)$ can be found, for given values of u, from a table of the function $\cos 2\pi x$. In the interval $0 \leqslant \xi \leqslant \frac{1}{2}$, $h(\xi)$ decreases steadily from 0·92388 to 0·38268, and $h(1-\xi) = h(\xi)$.

For the purpose of calculation we require a numerical upper bound for $R(u)$. A rather complicated formula of this kind is obtained in Titchmarsh (17), Theorem 2. For values of u which are not too small it can be much simplified, and in fact it is easy to deduce that

$$|R(u)| < \frac{3}{2u^{\frac{3}{4}}} \quad (u > 125).$$

This inequality is sufficient for most purposes.

Occasionally, when $Z(2\pi u)$ is too small, a second term of the Riemann–Siegel asymptotic formula has to be used.

The values of u for which the calculations are performed are the solutions of (15.2.1), since they make α_1 alternately 1 and -1. In the calculations described in Titchmarsh (17), I began with

$$u = 1{\cdot}6, \qquad K = -0{\cdot}04865$$

and went as far as

$$u = 62{\cdot}785, \qquad K = 98{\cdot}5010.$$

The values of u were obtained in succession, and are rather rough approximations to the u_ν, so that the K's are not quite integers or integers and a half.

It was shown in this way that the first 198 zeros of $\zeta(s)$ above the real axis all lie on the line $\sigma = \frac{1}{2}$.

The calculations were carried a great deal farther by Dr. Comrie.† Proceeding on the same lines, it was shown that the first 1,041 zeros of $\zeta(s)$ above the real axis all lie on the critical line, in the interval $0 < t < 1{,}468$.

One interesting point which emerges from these calculations is that $Z(t_\nu)$ does not always have the same sign as $(-1)^\nu$. A considerable number of exceptional cases were found; but in each of these cases there is a neighbouring point t'_ν such that $Z(t'_\nu)$ has the sign of $(-1)^\nu$, and the succession of changes of sign of $Z(t)$ is therefore not interrupted.

15.4. As far as they go, these calculations are all in favour of the truth of the Riemann hypothesis. Nevertheless, it may be that they do

† See Titchmarsh (18).

not go far enough to reveal the real state of affairs. At the end of the table constructed by Dr. Comrie there are only fifteen terms in the series for $Z(t)$, and this is a very small number when we are dealing with oscillating series of this kind. Indeed there is one feature of the table which may suggest a change in its character farther on. In the main, the result is dominated by the first term α_1, and later terms more or less cancel out. Occasionally (e.g. at $K = 435$) all, or nearly all, the numbers α_n have the same sign, and $Z(t)$ has a large maximum or minimum. As we pass from this to neighbouring values of t, the first few α_n undergo violent changes, while the later ones vary comparatively slowly. The term α_n appears when $u = n^2$, and here

$$\cos 2\pi(K - u\log n) = \cos \pi\{u\log(u/n^2) - u - \tfrac{1}{8} + ...\}$$
$$= \cos \pi(n^2 + \tfrac{1}{8} + ...) = (-1)^n \cos \tfrac{1}{8}\pi + ...,$$

and

$$\frac{d}{du}(K - u\log n) = \tfrac{1}{2}\log u - \log n - \frac{1}{192\pi^2 u^2} + ... \sim -\frac{1}{192\pi^2 u^2}.$$

At its first appearance in the table α_n will therefore be approximately $(-1)^n n^{-\frac{1}{2}} \cos \frac{1}{8}\pi$, and it will vary slowly for some time after its appearance.

It is conceivable that if t, and so the number of terms, were large enough, there might be places where the smaller slowly varying terms would combine to overpower the few quickly varying ones, and so prevent the graph of $Z(t)$ from crossing the zero line between successive maxima. There are too few terms in the table already constructed to test this possibility.

There are, of course, relations between the numbers α_n which destroy any too simple argument of this kind. If the Riemann hypothesis is true, there must be some relation, at present hidden, which prevents the suggested possibility from ever occurring at all.

No doubt the whole matter will soon be put to the test of modern methods of calculation. Naturally the Riemann hypothesis cannot be proved by calculation, but, if it is false, it could be disproved by the discovery of exceptions in this way.

NOTES FOR CHAPTER 15

15.5. A number of workers have checked the Riemann hypothesis over increasingly large ranges. At the time of writing the most extensive calculation is that of van de Lune and te Riele (as reported in Odlyzko and te Riele [**1**]), who have found that the first $1{\cdot}5 \times 10^9$ non-trivial zeros are simple and lie on the critical line.

ORIGINAL PAPERS

[*This list includes that given in my Cambridge tract; it does not include papers referred to in Landau's* Handbuch der Lehre von der Verteilung der Primzahlen, *1909*.]

ABBREVIATIONS

A.M.	*Acta Mathematica.*
C.R.	*Comptes rendus de l'Académie des sciences* (Paris).
J.L.M.S.	*Journal of the London Mathematical Society.*
J.M.	*Journal für die reine und angewandte Mathematik.*
M.A.	*Mathematische Annalen.*
M.Z.	*Mathematische Zeitschrift.*
P.C.P.S.	*Proceedings of the Cambridge Philosophical Society.*
P.L.M.S.	*Proceedings of the London Mathematical Society.*
Q.J.O.	*Quarterly Journal of Mathematics* (Oxford Series).

Ananda-Rau, K.

(1) The infinite product for $(s-1)\zeta(s)$, *M.Z.* 20 (1924), 156–64.

Arwin, A.

(1) A functional equation from the theory of the Riemann $\zeta(s)$-function, *Annals of Math.* (2), 24 (1923), 359–66.

Atkinson, F. V.

(1) The mean value of the zeta-function on the critical line, *Q.J.O.* 10 (1939) 122–8.

(2) The mean value of the zeta-function on the critical line, *P.L.M.S.* (2), 47 (1941), 174–200.

(3) A divisor problem, *Q.J.O.* 12 (1941), 193–200.

(4) A mean value property of the Riemann zeta-function, *J.L.M.S.* 23 (1948), 128–35.

(5) The Abel summation of certain Dirichlet series, *Q.J.O.* 19 (1948), 59–64.

(6) The Riemann zeta-function, *Duke Math. J.* 17 (1950), 63–8.

Babini, J.

(1) Über einige Eigenschaften der Riemannschen $\zeta(z)$-Funktion, *An. Soc. Ci. Argent.* 118 (1934), 209–15.

Backlund, R.

(1) Einige numerische Rechnungen, die Nullpunkte der Riemannschen ζ-Funktion betreffend, *Öfversigt Finska Vetensk. Soc.* (A), 54 (1911–12), No. 3.

(2) Sur les zéros de la fonction $\zeta(s)$ de Riemann, *C.R.* 158 (1914), 1979–81.

(3) Über die Nullstellen der Riemannschen Zetafunktion, *A.M.* 41 (1918), 345–75.

(4) Über die Beziehung zwischen Anwachsen und Nullstellen der Zetafunktion *Öfversigt Finska Vetensk. Soc.* 61 (1918–19), No. 9.

Beaupain, J.

(1) Sur la fonction $\zeta(s, w)$ et la fonction $\zeta(s)$ de Riemann, *Acad. Royale de Belgique* (2), 3 (1909), No. 1.

Bellman, R.

(1) The Dirichlet divisor problem, *Duke Math. J.* 14 (1947), 411–17.

(2) An analog of an identity due to Wilton, *Duke Math. J.* 16 (1949), 539–45.

(3) Wigert's approximate functional equation and the Riemann zeta-function, *Duke Math. J.* 16 (1949), 547–52.

BOHR, H.

(1) En Saetning om ζ-Functionen, *Nyt. Tidss. for Math.* (B), 21 (1910), 61–6.

(2) Über das Verhalten von $\zeta(s)$ in der Halbebene $\sigma > 1$, *Göttinger Nachrichten* (1911), 409–28.

(3) Sur l'existence de valeurs arbitrairement petites de la fonction $\zeta(s) = \zeta(\sigma+it)$ de Riemann pour $\sigma > 1$, *Oversigt Vidensk. Selsk. København* (1911), 201–8.

(4) Sur la fonction $\zeta(s)$ dans le demi-plan $\sigma > 1$, *C.R.* 154 (1912), 1078–81.

(5) Über die Funktion $\zeta'(s)/\zeta(s)$, *J.M.* 141 (1912), 217–34.

(6) En nyt Bevis for, at den Riemann'ske Zetafunktion $\zeta(s) = \zeta(\sigma+it)$ har uendelij mange Nulpunkten indenfor Parallel-strimlen $0 \leqslant \sigma \leqslant 1$, *Nyt. Tidss. for Math.* (B) 23 (1912), 81–5.

(7) Om de Vaerdier, den Riemann'ske Funktion $\zeta(\sigma+it)$ antager i Halvplanen $\sigma > 1$, *2. Skand. Math. Kongr.* (1912), 113–21.

(8) Note sur la fonction zéta de Riemann $\zeta(s) = \zeta(\sigma+it)$ sur la droite $\sigma = 1$, *Oversigt Vidensk. Selsk. København* (1913), 3–11.

(9) Lösung des absoluten Konvergenzproblems einer allgemeinen Klasse Dirichletscher Reihen, *A.M.* 36 (1913), 197–240.

(10) Sur la fonction $\zeta(s)$ de Riemann, *C.R.* 158 (1914), 1986–8.

(11) Zur Theorie der Riemannschen Zetafunktion im kritischen Streifen, *A.M.* 40 (1915), 67–100.

(12) Die Riemannsche Zetafunktion, *Deutsche Math. Ver.* 24 (1915), 1–17.

(13) Über eine quasi-periodische Eigenschaft Dirichletscher Reihen mit Anwendung auf die Dirichletschen L-Funktionen, *M.A.* 85 (1922), 115–22.

(14) Über diophantische Approximationen und ihre Anwendungen auf Dirichletsche Reihen, besonders auf die Riemannsche Zetafunktion, *5. Skand. Math. Kongr.* (1923), 131–54.

(15) Another proof of Kronecker's theorem, *P.L.M.S.* (2), 21 (1922), 315–16.

(16) Again the Kronecker theorem, *J.L.M.S.* 9 (1934), 5–6.

BOHR, H., and COURANT, R.

(1) Neue Anwendungen der Theorie der Diophantischen Approximationen auf die Riemannsche Zetafunktion, *J.M.* 144 (1914), 249–74.

BOHR, H., and JESSEN, B.

(1), (2) Über die Werteverteilung der Riemannschen Zetafunktion, *A.M.* 54 (1930), 1–35 and ibid. 58 (1932), 1–55.

(3) One more proof of Kronecker's theorem, *J.L.M.S.* 7 (1932), 274–5.

(4) Mean-value theorems for the Riemann zeta-function, *Q.J.O.* 5 (1934), 43–7.

(5) On the distribution of the values of the Riemann zeta-function, *Amer. J. Math.* 58 (1936), 35–44.

BOHR, H., and LANDAU, E.

(1) Über das Verhalten von $\zeta(s)$ und $\zeta_{\mathfrak{K}}(s)$ in der Nähe der Geraden $\sigma = 1$, *Göttinger Nachrichten* (1910), 303–30.

(2) Über die Zetafunktion, *Rend. di Palermo*, 32 (1911), 278–85.

(3) Beiträge zur Theorie der Riemannschen Zetafunktion, *M.A.* 74 (1913), 3–30.

(4) Ein Satz über Dirichletsche Reihen mit Anwendung auf die ζ-Funktion und die L-Funktionen, *Rend. di Palermo*, 37 (1914), 269–72.

(5) Sur les zéros de la fonction $\zeta(s)$ de Riemann, *C.R.* 158 (1914), 106–10.

(6) Über das Verhalten von $1/\zeta(s)$ auf der Geraden $\sigma = 1$, *Göttinger Nachrichten* (1923), 71–80.

(7) Nachtrag zu unseren Abhandlungen aus den Jahrgängen 1910 und 1923, *Göttinger Nachrichten* (1924), 168–72.

Bohr, H., Landau, E., and Littlewood, J. E.

(1) Sur la fonction $\zeta(s)$ dans le voisinage de la droite $\sigma = \frac{1}{2}$, *Bull. Acad. Belgique*, 15 (1913), 1144–75.

Borchsenius, V., and Jessen, B.

(1) Mean motions and values of the Riemann zeta-function, *A.M.* 80 (1948), 97–166.

Bouwkamp, C. J.

(1) Über die Riemannsche Zetafunktion für positive, gerade Werte des Argumentes, *Nieuw Arch. Wisk.* 19 (1936), 50–8.

Brika, M.

(1) Über eine Gestalt der Riemannschen Reihe $\zeta(s)$ für s = gerade ganze Zahl, *Bull. Soc. Math. Grèce*, 14 (1933), 36–8.

Brun, V.

(1) On the function $[x]$, *P.C.P.S.* 20 (1920), 299–303.

(2) Deux transformations élémentaires de la fonction zéta de Riemann, *Revista Ci. Lima*, 41 (1939), 517–25.

Burrau, C.

(1) Numerische Lösung der Gleichung $\frac{2^{-D}\log 2}{1-2^{-2D}} = \sum_{n=2}^{\infty} \frac{p_n^{-D}\log p_n}{1-p_n^{-2D}}$ wo p_n die Reihe der Primzahlen von 3 an durchläuft, *J.M.* 142 (1912), 51–3.

Carlson, F.

(1) Über die Nullstellen der Dirichletschen Reihen und der Riemannschen ζ-Funktion, *Arkiv för Mat. Astr. och Fysik*, 15 (1920), No. 20.

(2), (3) Contributions à la théorie des séries de Dirichlet, *Arkiv för Mat. Astr. och Fysik*, 16 (1922), No. 18, and 19 (1926), No. 25.

Chowla, S. D.

(1) On some identities involving zeta-functions, *Journal Indian Math. Soc.* 17 (1928), 153–63.

Corput, J. G. van der

(1) Zahlentheoretische Abschätzungen, *M.A.* 84 (1921), 53–79.

(2) Verschärfung der Abschätzung beim Teilerproblem, *M.A.* 87 (1922), 39–65.

(3) Neue zahlentheoretische Abschätzungen, erste Mitteilung, *M.A.* 89 (1923), 215–54.

(4) Zum Teilerproblem, *M.A.* 98 (1928), 697–716.

(5) Zahlentheoretische Abschätzungen, mit Anwendung auf Gitterpunktprobleme, *M.Z.* 28 (1928), 301–10.

(6) Neue zahlentheoretische Abschätzungen, zweite Mitteilung, *M.Z.* 29 (1929), 397–426.

(7) Über Weylsche Summen, *Mathematica B* (1936–7), 1–30.

Corput, J. G. van der, and Koksma, J. F.

(1) Sur l'ordre de grandeur de la fonction $\zeta(s)$ de Riemann dans la bande critique, *Annales de Toulouse* (3), 22 (1930), 1–39.

Craig, C. F.

(1) On the Riemann ζ-Function, *Bull. Amer. Math. Soc.* 29 (1923), 337–40.

Cramér, H.

(1) Über die Nullstellen der Zetafunktion, *M.Z.* 2 (1918), 237–41.

(2) Studien über die Nullstellen der Riemannschen Zetafunktion, *M.Z.* 4 (1919), 104–30.

(3) Bemerkung zu der vorstehenden Arbeit des Herrn E. Landau, *M.Z.* 6 (1920), 155–7.

(4) Über das Teilerproblem von Piltz, *Arkiv för Mat. Astr. och. Fysik*, 16 (1922), No. 21.

(5) Ein Mittelwertsatz in der Primzahltheorie, *M.Z.* 12 (1922), 147–53.

CRAMÉR, H., and LANDAU, E.

(1) Über die Zetafunktion auf der Mittellinie des kritischen Streifens, *Arkiv för Mat. Astr. och Fysik*, 15 (1920), No. 28.

CRUM, M. M.

(1) On some Dirichlet series, *J.L.M.S.* 15 (1940), 10–15.

DAVENPORT, H.

(1) Note on mean-value theorems for the Riemann zeta-function, *J.L.M.S.* 10 (1935), 136–8.

DAVENPORT, H., and HEILBRONN, H.

(1), (2) On the zeros of certain Dirichlet series I, II, *J.L.M.S.* 11 (1936), 181–5 and 307–12.

DENJOY, A.

(1) L'hypothèse de Riemann sur la distribution des zéros de $\zeta(s)$, reliée à la théorie des probabilités, *C.R.* 192 (1931), 656–8.

DEURING, M.

(1) Imaginäre quadratische Zahlkörper mit der Klassenzahl 1, *M.Z.* 37 (1933), 405–15.

(2) On Epstein's zeta-function, *Annals of Math.* (2) 38 (1937), 584–93.

ESTERMANN, T.

(1) On certain functions represented by Dirichlet series, *P.L.M.S.* (2), 27 (1928), 435–48.

(2) On a problem of analytic continuation, *P.L.M.S.* 27 (1928), 471–82.

(3) A proof of Kronecker's theorem by induction, *J.L.M.S.* 8 (1933), 18–20.

FAVARD, J.

(1) Sur la répartition des points où une fonction presque périodique prend une valeur donnée, *C.R.* 194 (1932), 1714–16.

FEJÉR, L.

(1) Nombre de changements de signe d'une fonction dans un intervalle et ses moments, *C.R.* 158 (1914), 1328–31.

FEKETE, M.

(1) Sur une limite inférieure des changements de signe d'une fonction dans un intervalle, *C.R.* 158 (1914), 1256–8.

(2) The zeros of Riemann's zeta-function on the critical line, *J.L.M.S.* 1 (1926), 15–19.

FLETT, T. M.

(1) On the function $\sum_{n=1}^{\infty} \frac{1}{n} \sin \frac{t}{n}$, *J.L.M.S.* 25 (1950), 5–19.

(2) On a coefficient problem of Littlewood and some trigonometrical sums, *Q.J.O.* (2) 2 (1951), 26–52.

FRANEL, J.

(1) Les suites de Farey et le problème des nombres premiers, *Göttinger Nachrichten* (1924), 198–201.

GABRIEL, R. M.

(1) Some results concerning the integrals of moduli of regular functions along certain curves, *J.L.M.S.* 2 (1927), 112–17.

GRAM, J. P.

(1) Tafeln für die Riemannsche Zetafunktion, *Skriften København* (8), 9 (1925), 311–25.

GRONWALL, T. H.

(1) Sur la fonction $\zeta(s)$ de Riemann au voisinage de $\sigma = 1$, *Rend. di Palermo*, 35 (1913), 95–102.

(2) Über das Verhalten der Riemannschen Zeta-funktion auf der Geraden $\sigma = 1$, *Arch. der Math. u. Phys.* (3) 21 (1913), 231–8.

GROSSMAN, J.

(1) Über die Nullstellen der Riemannschen Zeta-funktion und der Dirichletschen L-Funktionen, *Dissertation*, Göttingen (1913).

GUINAND, A. P.

(1) A formula for $\zeta(s)$ in the critical strip, *J.L.M.S.* 14 (1939), 97–100.

(2) Some Fourier transforms in prime-number theory, *Q.J.O.* 18 (1947), 53–64.

(3) Some formulae for the Riemann zeta-function, *J.L.M.S.* 22 (1947), 14–18.

(4) Fourier reciprocities and the Riemann zeta-function, *P.L.M.S.* 51 (1949), 401–14.

HADAMARD, J.

(1) Une application d'une formule intégrale relative aux séries de Dirichlet, *Bull. Soc. Math. de France*, 56, ii (1927), 43–4.

HAMBURGER, H.

(1), **(2)**, **(3)** Über die Riemannsche Funktionalgleichung der ζ-Funktion, *M.Z.* 10 (1921), 240–54; 11 (1922), 224–45; 13 (1922), 283–311.

(4) Über einige Beziehungen, die mit der Funktionalgleichung der Riemannschen ζ-Funktion äquivalent sind, *M.A.* 85 (1922), 129–40.

HARDY, G. H.

(1) Sur les zéros de la fonction $\zeta(s)$ de Riemann, *C.R.* 158 (1914), 1012–14.

(2) On Dirichlet's divisor problem, *P.L.M.S.* (2), 15 (1915), 1–25.

(3) On the average order of the arithmetical functions $\mathrm{P}(n)$ and $\Delta(n)$, *P.L.M.S.* (2), 15 (1915), 192–213.

(4) On some definite integrals considered by Mellin, *Messenger of Math.* 49 (1919), 85–91.

(5) Ramanujan's trigonometrical function $c_q(n)$, *P.C.P.S.* 20 (1920), 263–71.

(6) On the integration of Fourier series, *Messenger of Math.* 51 (1922), 186–92.

(7) A new proof of the functional equation for the zeta-function, *Mat. Tidsskrift*, B (1922), 71–3.

(8) Note on a theorem of Mertens, *J.L.M.S.* 2 (1926), 70–2.

HARDY, G. H., INGHAM, A. E., and PÓLYA, G.

(1) Theorems concerning mean values of analytic functions, *Proc. Royal Soc.* (A), 113 (1936), 542–69.

HARDY, G. H., and LITTLEWOOD, J. E.

(1) Some problems of Diophantine approximation, *Internat. Congress of Math., Cambridge* (1912), 1, 223–9.

(2) Contributions to the theory of the Riemann zeta-function and the theory of the distribution of primes, *A.M.* 41 (1918), 119–96.

(3) The zeros of Riemann's zeta-function on the critical line, *M.Z.* 10 (1921), 283–317.
(4) The approximate functional equation in the theory of the zeta-function, with applications to the divisor problems of Dirichlet and Piltz, *P.L.M.S.* (2), 21 (1922), 39–74.
(5) On Lindelöf's hypothesis concerning the Riemann zeta-function, *Proc. Royal Soc.* (A), 103 (1923), 403–12.
(6) The approximate functional equations for $\zeta(s)$ and $\zeta^2(s)$, *P.L.M.S.* (2), 29 (1929), 81–97.

Hartman, P.
(1) Mean motions and almost periodic functions, *Trans. Amer. Math. Soc.* 46 (1939), 66–81.

Haselgrove, C. B.
(1) A connexion between the zeros and the mean values of $\zeta(s)$, *J.L.M.S.* 24 (1949), 215–22.

Hasse, H.
(1) Beweis des Analogons der Riemannschen Vermutung für die Artinschen und F. K. Schmidtschen Kongruenzzetafunktionen in gewissen elliptischen Fällen, *Göttinger Nachrichten*, 42 (1933), 253–62.

Haviland, E. K.
(1) On the asymptotic behavior of the Riemann ξ-function, *Amer. J. Math.* 67 (1945), 411–16.

Hecke, E.
(1) Über die Lösungen der Riemannschen Funktionalgleichung, *M.Z.* 16 (1923), 301–7.
(2) Über die Bestimmung Dirichletscher Reihen durch ihre Funktionalgleichung, *M.A.* 112 (1936), 664–99.
(3) Herleitung des Euler-Produktes der Zetafunktion und einiger L-Reihen aus ihrer Funktionalgleichung, *M.A.* 119 (1944), 266–87.

Heilbronn, H.
(1) Über den Primzahlsatz von Herrn Hoheisel, *M.Z.* 36 (1933), 394–423.

Hille, E.
(1) A problem in 'Factorisatio Numerorum', *Acta Arith.* 2 (1936), 134–44.

Hoheisel, G.
(1) Normalfolgen und Zetafunktion, *Jahresber. Schles. Gesell.* 100 (1927), 1–7.
(2) Eine Illustration zur Riemannschen Vermutung, *M.A.* 99 (1928), 150–61.
(3) Über das Verhalten des reziproken Wertes der Riemannschen Zeta-Funktion, *Sitzungsber. Preuss. Akad. Wiss.* (1929), 219–23.
(4) Nullstellenanzahl und Mittelwerte der Zetafunktion, *Sitzungsber. Preuss. Akad. Wiss.* (1930), 72–82.
(5) Primzahlprobleme in der Analysis, *Sitzungsber. Preuss. Akad. Wiss.* (1930), 580–8.

Hölder, O.
(1) Über gewisse der Möbiusschen Funktion $\mu(n)$ verwandte zahlentheoretische Funktionen, der Dirichletschen Multiplikation und eine Verallgemeinerung der Umkehrungsformeln, *Ber. Verh. sächs. Akad. Leipzig*, 85 (1933), 3–28.

Hua, L. K.
(1) An improvement of Vinogradov's mean-value theorem and several applications, *Q.J.O.* 20 (1949), 48–61.

HUTCHINSON, J. I.

(1) On the roots of the Riemann zeta-function, *Trans. Amer. Math. Soc.* 27 (1925), 49–60.

INGHAM, A. E.

(1) Mean-value theorems in the theory of the Riemann zeta-function, *P.L.M.S.* (2), 27 (1926), 273–300.

(2) Some asymptotic formulae in the theory of numbers, *J.L.M.S.* 2 (1927), 202–8.

(3) Notes on Riemann's ζ-function and Dirichlet's L-functions, *J.L.M.S.* 5 (1930), 107–12.

(4) Mean-value theorems and the Riemann zeta-function, *Q.J.O.* 4 (1933), 278–90.

(5) On the difference between consecutive primes, *Q.J.O.* 8 (1937), 255–66.

(6) On the estimation of $N(\sigma, T)$, *Q.J.O.* 11 (1940), 291–2.

(7) On two conjectures in the theory of numbers, *Amer. J. Math.* 64 (1942), 313–19.

JARNÍK, V., and LANDAU, E.

(1) Untersuchungen über einen van der Corputschen Satz, *M.Z.* 39 (1935), 745–67.

JESSEN, B.

(1) Eine Integrationstheorie für Funktionen unendlich vieler Veränderlichen, mit Anwendung auf das Werteverteilungsproblem für fastperiodische Funktionen, insbesondere für die Riemannsche Zetafunktion, *Mat. Tidsskrift*, B (1932), 59–65.

(2) Mouvement moyen et distribution des valeurs des fonctions presque-périodiques, *10. Skand. Math. Kongr.* (1946), 301–12.

JESSEN, B., and WINTNER, A.

(1) Distribution functions and the Riemann zeta-function, *Trans. Amer. Math. Soc.* 38 (1935), 48–88.

KAC, M., and STEINHAUS, H.

(1) Sur les fonctions indépendantes (*N*), *Studia Math.* 7 (1938), 1–15.

KAMPEN, E. R. VAN

(1) On the addition of convex curves and the densities of certain infinite convolutions, *Amer. J. Math.* 59 (1937), 679–95.

KAMPEN, E. R. VAN, and WINTNER, A.

(1) Convolutions of distributions on convex curves and the Riemann zeta-function, *Amer. J. Math.* 59 (1937), 175–204.

KERSHNER, R.

(1) On the values of the Riemann ζ-function on fixed lines $\sigma > 1$, *Amer. J. Math.* 59 (1937), 167–74.

KERSHNER, R., and WINTNER, A.

(1) On the boundary of the range of values of $\zeta(s)$, *Amer. J. Math.* 58 (1936), 421–5.

(2) On the asymptotic distribution of $\zeta'/\zeta(s)$ in the critical strip, *Amer. J. Math.* 59 (1937), 673–8.

KIENAST, A.

(1) Über die Dirichletschen Reihen für $\zeta^{\rho}(s)$, $L^{\rho}(s)$, *Comment. Math. Helv.* 8 (1936), 359–70.

KLOOSTERMAN, H. D.

(1) Een integraal voor de ζ-functie van Riemann, *Christiaan Huygens Math. Tijdschrift*, 2 (1922), 172–7.

KLUYVER, J. C.

(1) On certain series of Mr. Hardy, *Quart. J. of Math.* 50 (1924), 185–92.

KOBER, H.

(1) Transformationen einer bestimmten Besselschen Reihe sowie von Potenzen der Riemannschen ζ-Funktion und von verwandten Funktionen, *J.M.* 173 (1935), 65–78.

(2) Eine der Riemannschen verwandte Funktionalgleichung, *M.Z.* 39 (1935), 630–3.

(3) Funktionen, die den Potenzen der Riemannschen Zetafunktion verwandt sind, und Potenzreihen, die über den Einheitskreis nicht fortsetzbar sind, *J.M.* 174 (1936), 206–25.

(4) Eine Mittelwertformel der Riemannschen Zetafunktion, *Compositio Math.* 3 (1936), 174–89.

KOCH, H. VON

(1) Contribution à la théorie des nombres premiers, *A.M.* 33 (1910), 293–320.

KOSLIAKOV, N.

(1) Some integral representations of the square of Riemann's function $\Xi(t)$, *C.R. Acad. Sci. U.R.S.S.* 2 (1934), 401–4.

(2) Integral for the square of Riemann's function, *C.R. Acad. Sci. U.R.S.S.* N.S. 2 (1936), 87–90.

(3) Some formulae for the functions $\zeta(s)$ and $\zeta_r(s)$, *C.R. Acad. Sci. U.R.S.S.* (2), 25 (1939), 567–9.

KRAMASCHKE, L.

(1) Nullstellen der Zetafunktion, *Deutsche Math.* 2 (1937), 107–10.

KUSMIN, R.

(1) Sur les zéro de la fonction $\zeta(s)$ de Riemann, *C.R. Acad. Sci. U.R.S.S.* 2 (1934), 398–400.

LANDAU, E.

(1) Zur Theorie der Riemannschen Zetafunktion, *Vierteljahrsschr. Naturf. Ges. Zürich.* 56 (1911), 125–48.

(2) Über die Nullstellen der Zetafunktion, *M.A.* 71 (1911), 548–64.

(3) Ein Satz über die ζ-Funktion, *Nyt. Tidss.* 22 (B), (1911), 1–7.

(4) Über einige Summen, die von den Nullstellen der Riemannschen Zetafunktion abhangen, *A.M.* 35 (1912), 271–94.

(5) Über die Anzahl der Gitterpunkte in gewissen Bereichen, *Göttinger Nachrichten* (1912), 687–771.

(6) Gelöste and ungelöste Probleme aus der Theorie der Primzahlverteilung und der Riemannschen Zetafunktion, *Jahresber. der Deutschen Math. Ver.* 21 (1912), 208–28.

(7) Gelöste and ungelöste Probleme aus der Theorie der Primzahlverteilung und der Riemannschen Zetafunktion, *Proc. 5 Internat. Math. Congr.* (1913), 1, 93–108.

(8) Über die Hardysche Entdeckung unendlich vieler Nullstellen der Zetafunktion mit reellem Teil $\frac{1}{2}$, *M.A.* 76 (1915), 212–43.

(9) Über die Wigertsche asymptotische Funktionalgleichung für die Lambertsche Reihe, *Arch. d. Math. u. Phys.* (3), 27 (1916), 144–6.

(10) Neuer Beweis eines Satzes von Herrn Valiron, *Jahresber. der Deutschen Math. Ver.* 29 (1920), 239.

(11) Über die Nullstellen der Zetafunktion, *M.Z.* 6 (1920), 151–4.

(**12**) Über die Nullstellen der Dirichletschen Reihen und der Riemannschen ζ-Funktion, *Arkiv för Mat. Astr. och Fysik*, 16 (1921), No. 7.

(**13**) Über die Möbiussche Funktion, *Rend. di Palermo*, 48 (1924), 277–80.

(**14**) Über die Wurzeln der Zetafunktion, *M.Z.* 20 (1924), 98–104.

(**15**) Über die ζ-Funktion und die L-Funktionen, *M.Z.* 20 (1924), 105–25.

(**16**) Bemerkung zu der vorstehenden Arbeit von Herrn Franel, *Göttinger Nachrichten* (1924), 202–6.

(**17**) Über die Riemannsche Zetafunktion in der Nähe von $\sigma = 1$, *Rend. di Palermo*, 50 (1926), 423–7.

(**18**) Über die Zetafunktion und die Hadamardsche Theorie der ganzen Funktionen, *M.Z.* 26 (1927), 170–5.

(**19**) Über das Konvergenzgebiet einer mit der Riemannschen Zetafunktion zusammenhängenden Reihe, *M.A.* 97 (1927), 251–90.

(**20**) Bemerkung zu einer Arbeit von Hrn. Hoheisel über die Zetafunktion, *Sitzungsber. Preuss. Akad. Wiss.* (1929), 271–5.

(**21**) Über die Fareyreihe und die Riemannsche Vermutung, *Göttinger Nachrichten* (1932), 347–52.

(**22**) Über den Wertevorrat von $\zeta(s)$ in der Halbebene $\sigma > 1$, *Göttinger Nachrichten* (1933), 81–91.

LANDAU, E., and WALFISZ, A.

(**1**) Über die Nichtfortsetzbarkeit einiger durch Dirichletsche Reihen definierter Funktionen, *Rend. di Palermo* 44 (1919), 82–6.

LERCH, M.

(**1**) Über die Bestimmung der Koeffizienten in der Potenzreihe für die Funktion $\zeta(s)$, *Časopis*, 43 (1914), 511–22.

LETTENMEYER, F.

(**1**) Neuer Beweis des allgemeinen Kroneckerschen Approximationssatzes, *P.L.M.S.* (2) 21 (1923), 306–14.

LEVINSON, N.

(**1**) On Hardy's theorem on the zeros of the zeta function, *J. Math. Phys. Mass. Inst. Tech.* 19 (1940), 159–60.

LITTLEWOOD, J. E.

(**1**) Quelques conséquences de l'hypothèse que la fonction $\zeta(s)$ de Riemann n'a pas de zéros dans le demi-plan $\mathbf{R}(s) > \frac{1}{2}$, *C.R.* 154 (1912), 263–6.

(**2**) Researches in the theory of the Riemann ζ-function, *P.L.M.S.* (2) 20 (1922), Records xxii–xxviii.

(**3**) Two notes on the Riemann zeta-function, *P.C.P.S.* 22 (1924), 234–42.

(**4**) On the zeros of the Riemann zeta-function, *P.C.P.S.* 22 (1924), 295–318.

(**5**) On the Riemann zeta-function, *P.L.M.S.* (2), 24 (1925), 175–201.

(**6**) On the function $1/\zeta(1+ti)$, *P.L.M.S.* (2), 27 (1928), 349–57.

MAIER, W.

(**1**) Gitterfunktionen der Zahlebene, *M.A.* 113 (1936), 363–79.

MALURKAR, S. L.

(**1**) On the application of Herr Mellin's integrals to some series, *Journal Indian Math. Soc.* 16 (1925), 130–8.

MATTSON, R.

(**1**) Eine neue Darstellung der Riemann'schen Zetafunktion, *Arkiv för Mat. Astr. och Fysik*, 19 (1926), No. 26.

MELLIN, H.

(1) Über die Nullstellen der Zetafunktion, *Annales Acad. Scientiarium Fennicae* (A), 10 (1917), No. 11.

MEULENBELD, B.

(1) Een approximatieve Functionaalbetrekking van de Zetafunctie van Riemann, *Dissertation, Groningen*, 1936.

MIKOLÁS, M.

(1) Sur l'hypothèse de Riemann, *C.R.* 228 (1949), 633–6.

MIN, S. H.

(1) On the order of $\zeta(\frac{1}{2}+it)$, *Trans. Amer. Math. Soc.* 65 (1949), 448–72.

MIYATAKE, O.

(1) On Riemann's ξ-function, *Tôhoku Math. Journal*, 46 (1939), 160–72.

MORDELL, L. J.

(1) Some applications of Fourier series in the analytic theory of numbers, *P.C.P.S.* 34 (1928), 585–96.

(2) Poisson's summation formula and the Riemann zeta-function, *J.L.M.S.* 4 (1929), 285–91.

(3) On the Riemann hypothesis and imaginary quadratic fields with a given class number, *J.L.M.S.* 9 (1934), 289–98.

MÜNTZ, C. H.

(1) Beziehungen der Riemannschen ζ-Funktion zu willkürlichen reellen Funktionen, *Mat. Tidsskrift*, B (1922), 39–47.

MUTATKER, V. L.

(1) On some formulae in the theory of the zeta-function, *Journal Indian Math. Soc.* 19 (1932), 220–4.

NEVANLINNA, F. and R.

(1), (2) Über die Nullstellen der Riemannschen Zetafunktion, *M.Z.* 20 (1924), 253–63, and 23 (1925), 159–60.

OSTROWSKI, A.

(1) Notiz über den Wertevorrat der Riemannschen ζ-Funktion am Rande des kritischen Streifens, *Jahresbericht Deutsch. Math. Verein.* 43 (1933), 58–64.

PALEY, R. E. A. C., and WIENER, N.

(1) Notes on the theory and application of Fourier transforms V, *Trans. Amer. Math. Soc.* 35 (1933), 768–81.

PHILLIPS, ERIC

(1) The zeta-function of Riemann; further developments of van der Corput's method, *Q.J.O.* 4 (1933), 209–25.

(2) A note on the zeros of $\zeta(s)$, *Q.J.O.* 6 (1935), 137–45.

POL, B. VAN DER,

(1) An electro-mechanical investigation of the Riemann zeta-function in the critical strip, *Bull. Amer. Math. Soc.* 53 (1947), 976–81.

PÓLYA, G.

(1) Bemerkung über die Integraldarstellung der Riemannschen ξ-Funktion, *A.M.* 48 (1926), 305–17.

(2) On the zeros of certain trigonometric integrals, *J.L.M.S.* 1 (1926), 98–9.

(3) Über die algebraisch-funktiontheoretischen Untersuchungen von J. L. W. V. Jensen, *Kgl. Danske Videnskabernes Selskab.* 7 (1927), No. 17.

(4) Über trigonometrische Integrale mit nur reellen Nullstellen, *J.M.* 158 (1927), 6–18.

Popov, A. I.

(1) Several series containing primes and roots of $\zeta(s)$, *C.R. Acad. Sci. U.R.S.S.*, n.s. 41 (1943), 362–3.

Potter, H. S. A., and Titchmarsh, E. C.

(1) The zeros of Epstein's zeta-functions, *P.L.M.S.* (2) 39 (1935), 372–84.

Rademacher, H.

(1) Ein neuer Beweis für die Funktionalgleichung der ζ-Funktion. *M.Z.* 31 (1930), 39–44.

Ramanujan, S.

(1) New expressions for Riemann's functions $\xi(s)$ and $\Xi(t)$, *Quart. J. of Math.* 46 (1915), 253–61.

(2) Some formulae in the analytic theory of numbers, *Messenger of Math.* 45 (1915), 81–4.

(3) On certain trigonometrical sums and their applications in the theory of numbers, *Trans. Camb. Phil. Soc.* 22 (1918), 259–76.

Ramaswami, V.

(1) Notes on Riemann's ζ-function, *J.L.M.S.* 9 (1934), 165–9.

Riesz, M.

(1) Sur l'hypothèse de Riemann, *A.M.* 40 (1916), 185–90.

Schnee, W.

(1) Die Funktionalgleichung der Zetafunktion und der Dirichletschen Reihen mit periodischen Koeffizienten, *M.Z.* 31 (1930), 378–90.

Selberg, A.

(1) On the zeros of Riemann's zeta-function on the critical line, *Arch. for Math. og Naturv.* B, 45 (1942), 101–14.

(2) On the zeros of Riemann's zeta-function, *Skr. Norske Vid. Akad. Oslo* (1942), no. 10.

(3) On the normal density of primes in small intervals, and the difference between consecutive primes, *Arch. for Math. og Naturv.* B, 47 (1943), No. 6.

(4) On the remainder in the formula for $N(T)$, the number of zeros of $\zeta(s)$ in the strip $0 < t < T$, *Avhandlinger Norske Vid. Akad. Oslo* (1944), no. 1.

(5) Contributions to the theory of the Riemann zeta-function, *Arch. for Math. og Naturv.* B, 48 (1946), no. 5.

(6) The zeta-function and the Riemann hypothesis, *10. Skand. Math. Kongr.* (1946), 187–200.

(7) An elementary proof of the prime-number theorem, *Ann. of Math.* (2) 50 (1949), 305–13.

Selberg, S.

(1) Bemerkung zu einer Arbeit von Viggo Brun über die Riemannsche Zetafunktion, *Norske Vid. Selsk. Forh.* 13 (1940), 17–19.

Siegel, C. L.

(1) Bemerkung zu einem Satz von Hamburger über die Funktionalgleichung der Riemannschen Zetafunktion, *M.A.* 86 (1922), 276–9.

(2) Über Riemanns Nachlass zur analytischen Zahlentheorie, *Quellen und Studien zur Geschichte der Math. Astr. und Physik, Abt. B: Studien*, 2 (1932), 45–80.

(3) Contributions to the theory of the Dirichlet L-series and the Epstein Zeta-functions, *Annals of Math.* 44 (1943), 143–72.

Speiser, A.

(1) Geometrisches zur Riemannschen Zetafunktion, *M.A.* 110 (1934), 514–21.

STEEN, S. W. P.

(1) A linear transformation connected with the Riemann zeta-function, *P.L.M.S.* (2), 41 (1936), 151–75.

STERNECK, R. VON

(1) Neue empirische Daten über die zahlentheoretische Funktion $\sigma(n)$, *Internat. Congress of Math., Cambridge* (1912), 1, 341–3.

SZÁSZ, O.

(1) Introduction to the theory of divergent series, Cincinnati, 1944 (*Math. Rev.* 6, 45).

TAYLOR, P. R.

(1) On the Riemann zeta-function, *Q.J.O.* 16 (1945), 1–21.

TCHUDAKOFF, N. G.

(1) Sur les zéros de la fonction $\zeta(s)$, *C.R.* 202 (1936), 191–3.

(2) On zeros of the function $\zeta(s)$, *C.R. Acad. Sci. U.R.S.S.* 1 (x) (1936), 201–4.

(3) On zeros of Dirichlet's L-functions, *Mat. Sbornik* (1) 43 (1936), 591–602.

(4) On Weyl's sums, *Mat. Sbornik* (2) 44 (1937), 17–35.

(5) On the functions $\zeta(s)$ and $\pi(x)$, *C.R. Acad. Sci. U.R.S.S.* 21 (1938), 421–2.

THIRUVENKATACHARYA, V.

(1) On some properties of the zeta-function, *Journal Indian Math. Soc.* 19 (1931), 92–6.

TITCHMARSH, E. C.

(1) The mean-value of the zeta-function on the critical line, *P.L.M.S.* (2) 27 (1928), 137–50.

(2) On the remainder in the formula for $N(T)$, the number of zeros of $\zeta(s)$ in the strip $0 < t < T$, *P.L.M.S.* (2), 27 (1928), 449–58.

(3) A consequence of the Riemann hypothesis, *J.L.M.S.* 2 (1927), 247–54.

(4) On an inequality satisfied by the zeta-function of Riemann, *P.L.M.S.* (2), 28 (1928), 70–80.

(5) On the zeros of the Riemann zeta-function, *P.L.M.S.* (2) 30 (1929), 319–21.

(6) Mean value theorems in the theory of the Riemann zeta-function, *Messenger of Math.* 58 (1929), 125–9.

(7) The zeros of Dirichlet's L-functions, *P.L.M.S.* (2) 32 (1931), 488–500.

(8)–(12) On van der Corput's method and the zeta-function of Riemann, *Q.J.O.* 2 (1931), 161–73; 2 (1931), 313–20; 3 (1932), 133–41; 5 (1934), 98–105; 5 (1934), 195–210.

(13) On the Riemann zeta-function, *P.C.P.S.* 28 (1932), 273–4.

(14) On the function $1/\zeta(1+it)$, *Q.J.O.* 4 (1933), 64–70.

(15) On Epstein's zeta-function, *P.L.M.S.* (2) 36 (1934), 485–500.

(16) The lattice-points in a circle *P.L.M.S.* (2) 38 (1935), 96–115.

(17), (18) The zeros of the Riemann zeta-function, *Proc. Royal Soc.* (A), 151 (1935), 234–55, and ibid. 157 (1936), 261–3.

(19) The mean value of $|\zeta(\frac{1}{2}+it)|^4$, *Q.J.O.* 8 (1937), 107–12.

(20) On $\zeta(s)$ and $\pi(x)$, *Q.J.O.* 9 (1938), 97–108.

(21) The approximate functional equation for $\zeta^2(s)$, *Q.J.O.* 9 (1938), 109–14.

(22) On divisor problems, *Q.J.O.* 9 (1938), 216–20.

(23) A convexity theorem, *J.L.M.S.* 13 (1938), 196–7.

(24) On the order of $\zeta(\frac{1}{2}+it)$, *Q.J.O.* 13 (1942), 11–17.

(25) Some properties of the Riemann zeta-function, *Q.J.O.* 14 (1943), 16–26.

(26) On the zeros of the Riemann zeta function, *Q.J.O.* 18 (1947), 4–16.

TORELLI, G.

(1) Studio sulla funzione $\zeta(s)$ di Riemann, *Napoli Rend.* (3) 19 (1913), 212–16.

TSUJI, M.

(1) On the zeros of the Riemann zeta-function, *Proc. Imp. Acad. Tokyo*, 18 (1942), 631–4.

TURÁN, P.

(1) Über die Verteilung der Primzahlen I, *Acta Szeged*, 10 (1941), 81–104.

(2) On Riemann's hypothesis, *Bull. Acad. Sci. U.R.S.S.* 11 (1947), 197–262.

(3) On some approximative Dirichlet-polynomials in the theory of the zeta-function of Riemann, *Danske Vidensk. Selskab*, 24 (1948), Nr. 17.

TURING, A. M.

(1) A method for the calculation of the zeta-function, *P.L.M.S.* (2), 48 (1943), 180–97.

UTZINGER, A. A.

(1) Die reellen Züge der Riemannschen Zetafunktion, *Dissertation, Zürich*, 1934 (see *Zentralblatt für Math.* 10, 163).

VALIRON, G.

(1) Sur les fonctions entières d'ordre nul et d'ordre fini, *Annales de Toulouse* (3), 5 (1914), 117–257.

VALLÉE POUSSIN, C. DE LA

(1), (2) Sur les zéros de $\zeta(s)$ de Riemann, *C.R.* 163 (1916), 418–21 and 471–3.

VINOGRADOV, I. M.

(1) On Weyl's sums, *Mat. Sbornik*, 42 (1935), 521–30.

(2) A new method of resolving of certain general questions of the theory of numbers, *Mat. Sbornik* (1), 43 (1936), 9–19.

(3) A new method of estimation of trigonometrical sums, *Mat. Sbornik* (1), 43 (1936), 175–188.

(4) The method of trigonometrical sums in the theory of numbers, *Trav. Inst. Math. Stekloff* 23 (1947).

VORONOÏ, G.

(1) Sur un problème du calcul des fonctions asymptotiques, *J.M.* 126 (1903), 241–82.

(2) Sur une fonction transcendante et ses applications à la sommation de quelques séries, *Annales de l'École Normale* (3) 21 (1904), 207–68 and 459–534.

WALFISZ, A.

(1) Zur Abschätzung von $\zeta(\frac{1}{2}+it)$, *Göttinger Nachrichten* (1924), 155–8.

(2) Über Gitterpunkte in mehrdimensionalen Ellipsoiden VIII, *Travaux de l'Institut Math. de Tbilissi*, 5 (1938), 181–96.

WALTHER, A.

(1) Über die Extrema der Riemannschen Zetafunktion bei reellem Argument, *Jahresbericht Deutsch. Math. Verein.* 34 (1925), 171–7.

(2) Anschauliches zur Riemannschen Zetafunktion, *A.M.* 48 (1926), 393–400.

WANG, F. T.

(1) A remark on the mean-value theorem of Riemann's zeta-function, *Science Reports Tôhoku Imperial Univ.* (1) 25 (1936), 381–91.

(2) On the mean-value theorem of Riemann's zeta-function, *Science Reports Tôhoku Imperial Univ.* (1) 25 (1936), 392–414.

(3) A note on zeros of Riemann zeta-function, *Proc. Imp. Acad. Tokyo*, 12, (1937), 305–6.

(4) A formula on Riemann zeta-function, *Ann. of Math.* (2) 46 (1945), 88–92.

(5) A note on the Riemann zeta-function, *Bull. Amer. Math. Soc.* 52 (1946), 319–21.

(6) A mean-value theorem of the Riemann zeta function, *Q.J.O.* 18 (1947), 1–3.

WATSON, G. N.

(1) Some properties of the extended zeta-function, *P.L.M.S.* (2) 12 (1913), 288–96.

WENNBERG, S.

(1) Zur Theorie der Dirichletschen Reihen, *Dissertation, Upsala,* 1920.

WEYL, H.

(1) Über die Gleichverteilung von Zahlen mod. Eins, *M.A.* 77 (1916), 313–52.

(2) Zur Abschätzung von $\zeta(1+ti)$, *M.Z.* 10 (1921), 88–101.

WHITTAKER, J. M.

(1) An inequality for the Riemann zeta-function, *P.L.M.S.* (2), 41 (1936), 544–52.

(2) A mean-value theorem for analytic functions, *P.L.M.S.* (2), 42 (1936), 186–95.

WIGERT, S.

(1) Sur la série de Lambert et son application à la théorie des nombres, *A.M.* 41 (1916), 197–218.

(2) Sur la théorie de la fonction $\zeta(s)$ de Riemann, *Arkiv for Mat. Astr. och Fysik* (1919), No. 12.

(3) On a problem concerning the Riemann ζ-function, *P.C.P.S.* 21 (1921), 17–21.

WILSON, B. M.

(1) Proofs of some formulae enunciated by Ramanujan, *P.L.M.S.* (2) 21 (1922), 235–55.

WILTON, J. R.

(1) Note on the zeros of Riemann's ζ-function, *Messenger of Math.* 45 (1915), 180–3.

(2) A proof of Burnside's formula for $\log \Gamma(x+1)$ and certain allied properties of Riemann's ζ-function, *Messenger of Math.* 52 (1922), 90–3.

(3) A note on the coefficients in the expansion of $\zeta(s, x)$ in powers of $s-1$, *Quart. J. of Math.* 50 (1927), 329–32.

(4) An approximate functional equation for the product of two ζ-functions, *P.L.M.S.* (2), 31 (1930), 11–17.

(5) The mean value of the zeta-function on the critical line, *J.L.M.S.* 5 (1930), 28–32.

WINTNER, A.

(1) A note on the distribution of the zeros of the zeta-function, *Amer. J. Math.* 57 (1935), 101–2.

(2) A note on the Riemann ξ-function, *J.L.M.S.* 10 (1935), 82–3.

(3) The almost periodic behavior of the function $1/\zeta(1+it)$, *Duke Math. J.* 2 (1936), 443–6.

(4) Riemann's hypothesis and almost periodic behavior, *Revista Ci. Lima,* 41 (1939), 575–85.

(5) On the asymptotic behavior of the Riemann zeta-function on the line $\sigma = 1$, *Amer. J. Math.* 63 (1941), 575–80.

(6) Riemann's hypothesis and harmonic analysis, *Duke Math. J.* 10 (1943), 99–105.

(7) The behavior of Euler's product on the boundary of convergence, *Duke Math. J.* 10 (1943), 429–40.

(8) Random factorizations and Riemann's hypothesis, *Duke Math. J.* 11 (1944), 267–75.

FURTHER REFERENCES

ANDERSON, R. J.

[1] Simple zeros of the Riemann zeta-function, *J. Number Theory*, 17 (1983), 176–182.

ATKINSON, F. V.

[1] The mean value of the Riemann zeta-function, *Acta Math.* 81 (1949), 353–376.

BAGCHI, B.

[1] A joint universality theorem for Dirichlet L-functions, *Math. Zeit.* 181 (1982), 319–335.

BALASUBRAMANIAN, R.

[1] An improvement of a theorem of Titchmarsh on the mean square of $|\zeta(\frac{1}{2}+it)|$, *Proc. London Math. Soc.* (3), 36 (1978), 540–576.

BALASUBRAMANIAN, R., CONREY, J. B., and HEATH-BROWN, D. R.

[1] Asymptotic mean square of the product of the Riemann zeta-function and a Dirichlet polynomial, *J. Reine Angew. Math.* 357 (1985), 161–181.

BALASUBRAMANIAN, R., and RAMACHANDRA, K.

[1] The place of an identity of Ramanujan in prime number theory, *Proc. Indian Acad. Sci.* 83 A (1976), 156–165.

[2] On the frequency of Titchmarsh's phenomenon for $\zeta(s)$. III, *Proc. Indian Acad. Sci.* 86 A (1977), 341–351.

[3] On the zeros of the Riemann Zeta function and L-series -II, *Hardy-Ramanujan J.* 5 (1982), 1–30.

BERNDT, B. C.

[1] The number of zeros for $\zeta^{(k)}(s)$, *J. London Math. Soc.* (2), 2 (1970), 577–580.

BURGESS, D. A.

[1] On character sums and L-series. II, *Proc. London Math. Soc.* (3), 13 (1963), 524–536.

CHEN, J.-R.

[1] On the order of $\zeta(\frac{1}{2}+it)$, *Chinese Math. Acta*, 6 (1965), 463–478.

CHIH, T.-T.

[1] A divisor problem, *Acad. Sinica Sci. Record*, 3 (1950), 177–182.

CONREY, J. B.

[1] Zeros of derivatives of Riemann's xi-function on the critical line, *J. Number Theory*, 16 (1983), 49–74.

CONREY, J. B., and GHOSH, A.

[1] On mean values of the zeta-function, *Mathematika*, 31 (1984), 159–161.

CONREY, J. B., GHOSH, A., GOLDSTON, D., GONEK, S. M., and HEATH-BROWN, D. R.

[1] On the distribution of gaps between zeros of the zeta-function, *Quart. J. Math. Oxford* (2), 36 (1985), 43–51.

CONREY, J. B., GHOSH, A., and GONEK, S. M.

[1] A note on gaps between zeros of the zeta function, *Bull. London Math. Soc.* 16 (1984), 421–424.

CORRÁDI, K., and KÁTAI, I.

[1] A comment on K. S. Gangadharan's paper entitled "Two classical lattice point problems", *Magyar Tud. Akad. Mat. Fiz. Oszt. Közl.* 17 (1967), 89–97.

DESHOUILLERS, J.-M., and IWANIEC, H.
[1] Kloosterman sums and Fourier coefficients of cusp forms, *Invent. Math.* 70 (1982), 219–288.
[2] Power mean values of the Riemann zeta-function, *Mathematatika*, 29 (1982), 202–212.
[3] Power mean values of the Riemann zeta-function II, *Acta Arith.* 48 (1984), 305–312.
DIAMOND, H.
[1] Elementary methods in the study of the distribution of prime numbers, *Bull. Amer. Math. Soc.* 7 (1982), 553–589.
ERDŐS, P.
[1] On a new method in elementary number theory which leads to an elementary proof of the prime number theorem, *Proc. Nat. Acad. Sci. USA*, 35 (1949), 374–384.
ESTERMANN, T.
[1] On the representation of a number as the sum of two products, *Proc. London Math. Soc.* (2), 31 (1930), 123–133.
[2] On Kloosterman's sum, *Mathematika*, 8 (1961), 83–86.
FUJII, A.
[1] On the distribution of the zeros of the Riemann Zeta function in short intervals, *Bull. Amer. Math. Soc.* 81 (1975), 139–142.
[2] On the difference between r consecutive ordinates of the Riemann Zeta function, *Proc. Japan Acad.* 51 (1975), 741–743.
[3] On the problem of divisors, *Acta Arith.* 31 (1976), 355–360.
GALLAGHER, P. X., and MUELLER, J. H.
[1] Primes and zeros in short intervals, *J. Reine Angew. Math.* 303/304 (1978) 205–220.
GANGADHARAN, K. S.
[1] Two classical lattice point problems, *Proc. Camb. Phil. Soc.* 57 (1961), 699–721.
GHOSH, A.
[1] On Riemann's zeta function—sign changes of $S(T)$, *Recent progress in analytic number theory*, Vol I, 25–46, (Academic Press, London, 1981).
[2] On the Riemann Zeta function – Mean value theorems and the distribution of $|S(T)|$, *J. Number Theory*, 17 (1983), 93–102.
GONEK, S. M.
[1] Analytic properties of zeta and L-functions, Thesis, Univ. Michigan, Ann Arbor, 1979.
[2] Mean values of the Riemann zeta-function and its derivatives, *Invent. Math.* 75 (1984), 123–141.
GOOD, A.
[1] Ein Ω-Resultat für das quadratische Mittel der Riemannschen Zetafunktion auf der kritische Linie, *Invent. Math.* 41 (1977), 233–251.
HAFNER, J. L.
[1] New omega theorems for two classical lattice point problems, *Invent. Math.* 63 (1981), 181–186.
[2] On the average order of a class of arithmetic functions, *J. Number Theory*, 15 (1982), 36–76.

Halász, G.
[1] Über die Mittelwerte multiplikativer zahlentheoretischer Funktionen, *Acta Math. Acad. Sci. Hungar.* 19 (1968), 365–403.
Halász, G., and Turán, P.
[1] On the distribution of Roots of Riemann zeta and allied functions, I, *J. Number Theory*, 1 (1969), 121–137.
Haneke, W.
[1] Verschärfung der Abschätzung von $\zeta(\frac{1}{2}+it)$, *Acta Arith.* 8 (1962–63), 357–430.
Hardy, G. H.
[1] On Dirichlet's divisor problem, *Proc. London Math. Soc.* (2), 15 (1916), 1–25.
Haselgrove, C. B.
[1] A connection between the zeros and the mean values of $\zeta(s)$, *J. London Math. Soc.* 24 (1949), 215–222.
[2] A disproof of a conjecture of Pólya, *Mathematika*, 5 (1958), 141–145.
Heath-Brown, D. R.
[1] The mean square of the Riemann Zeta-function, *Mathematika*, 25 (1978), 177–184.
[2] The twelfth power moment of the Riemann Zeta-function, *Quart. J. Math. Oxford* (2), 29 (1978), 443–462.
[3] Hybrid bounds for Dirichlet L-functions, *Invent. Math.* 47 (1978), 149–170.
[4] The fourth power moment of the Riemann Zeta-function, *Proc. London Math. Soc.* (3), 38 (1979), 385–422.
[5] Simple zeros of the Riemann Zeta-function on the critical line, *Bull. London Math. Soc.* 11 (1979), 17–18.
[6] Zero density estimates for the Riemann Zeta-function and Dirichlet L-functions, *J. London Math. Soc.* (2), 20 (1979), 221–232.
[7] Fractional moments of the Riemann Zeta-function, *J. London Math. Soc.* (2), 24 (1981), 65–78.
[8] Mean values of the Zeta-function and divisor problems, *Recent progress in analytic number theory*, Vol I, 115–119, (Academic Press, London, 1981).
[9] Hybrid bounds for L-functions: a q-analogue of van der Corput's method and a t-analogue of Burgess's method, *Recent progress in analytic number theory*, Vol I, 121–126, (Academic Press, London, 1981).
[10] Gaps between primes, and the pair correlation of zeros of the zeta-function, *Acta Arith.* 41 (1982), 85–99.
[11] The Pjateckiĭ-Šapiro prime number theorem, *J. Number Theory*, 16 (1983), 242–266.
Huxley, M. N.,
[1] On the difference between consecutive primes, *Invent. Math.* 15 (1972), 155–164.
Ingham, A. E.
[1] On two classical lattice point problems, *Proc. Camb. Phil. Soc.* 36 (1940), 131–138.
Ivić, A.
[1] Large values of the error term in the divisor problem, *Invent. Math.* 71 (1983), 513–520.
[2] A zero-density theorem for the Riemann zeta-function, *Trudy Mat. Inst. Steklov.* 163 (1984), 85–89.

[3] *The Riemann zeta-function*, (Wiley-Interscience, New York, 1985).

IWANIEC, H.

[1] Fourier coefficients of cusp forms and the Riemann Zeta-function, Exposé No. 18, *Sémin. Théor. Nombres*, Université Bordeaux, (1979/80).

JURKAT, W., and PEYERIMHOFF, A.

[1] A constructive approach to Kronecker approximations and its applications to the Mertens conjecture, *J. Reine Angew. Math.* 286/287 (1976), 322–340.

JUTILA, M.

[1] Zero-density estimates for L-functions, *Acta Arith.* 32 (1977), 52–62.

[2] Zeros of the zeta-function near the critical line, *Studies in pure mathematics, to the memory of Paul Turán*, 385–394, (Birkhaüser, Basel-Stuttgart, 1982).

[3] Riemann's zeta-function and the divisor problem, *Arkiv för Mat.* 21 (1983), 75–96.

[4] On the value distribution of the zeta-function on the critical line, *Bull. London Math. Soc.* 15 (1983), 513–518.

KARATSUBA, A. A.

[1] Estimates of trigonometric sums by Vinogradov's method, and some applications, *Proc. Steklov. Inst. Math*, 119 (1971), 241–255.

[2] *Principles of analytic number theory*, (Russian), (Izdat. 'Nauka', Moscow, 1975).

KOLESNIK, G.

[1] The improvement of the error term in the divisor problem, *Mat. Zametki*, 6 (1969), 545–554.

[2] On the estimation of certain trigonometric sums, *Acta Arith.* 25 (1973), 7–30.

[3] On the estimation of multiple exponential sums, *Recent progress in analytic number theory*, Vol I, 231–246, (Academic Press, London, 1981).

[4] On the order of $\zeta(\frac{1}{2}+it)$ and $\Delta(R)$, *Pacific J. Math.* 82 (1982), 107–122.

[5] On the method of exponent pairs, *Acta Arith.* 45 (1985), 115–143.

KOROBOV, N. M.

[1] Estimates of trigonometric sums and their applications, *Uspehi Mat. Nauk*, 13 (1958), 185–192.

KUBOTA, T., and LEOPOLDT, H. W.

[1] Eine p-adische Theorie der Zetawerte. I. Einführung der p-adischen Dirichletschen L-funktionen, *J. Reine Angew. Math.* 214/215 (1964), 328–339.

LAVRIK, A. F.

[1] The functional equation for Dirichlet L-functions and the problem of divisors in arithmetic progressions, *Izv. Akad. Nauk SSSR Ser. Mat.* 30 (1966), 433–448.

LAVRIK, A. F., and SOBIROV, A. Š.

[1] On the remainder term in the elementary proof of the prime number theorem, *Dokl. Akad. Nauk SSSR*, 211 (1973), 534–536.

LEVINSON, N.

[1] Ω-theorems for the Riemann zeta-function, *Acta Arith.* 20 (1972), 319–332.

[2] More than one third of the zeros of Riemann's zeta-function are on $\sigma = \frac{1}{2}$, *Adv. Math.* 13 (1974), 383–436.

[3] Zeros of derivative of Riemann's ξ-function, *Bull. Amer. Math. Soc.* 80 (1974), 951–954.

[4] A simplification of the proof that $N_0(T) > \frac{1}{3}N(T)$ for Riemann's zeta-function, *Adv. Math.* 18 (1975), 239–242.

[5] Deduction of semi-optimal mollifier for obtaining lower bounds for $N_0(T)$ for Riemann's zeta-function, *Proc. Nat. Acad. Sci. USA*, 72 (1975), 294–297.

[6] Almost all roots of $\zeta(s) = a$ are arbitrarily close to $\sigma = \frac{1}{2}$, *Proc. Nat. Acad. Sci. USA*, 72 (1975), 1322–1324.

Levinson, N., and Montgomery, H. L.

[1] Zeros of the derivative of the Riemann zeta-function, *Acta Math.* 133 (1974), 49–65.

Lou, S.-T.

[1] A lower bound for the number of zeros of Riemann's zeta-function on $\sigma = \frac{1}{2}$, *Recent progress in analytic number theory*, 319–324, Vol I, (Academic press, London, 1981).

Montgomery, H. L.

[1] *Topics in multiplicative number theory*, Lecture Notes in Math. 227 (Springer, Berlin, 1971).

[2] The pair correlation of zeros of the zeta-function, *Analytic number theory* (Proc. Symp. Pure Math. Vol XXIV), 181–193, (Amer. Math. Soc., Providence, R. I., 1973).

[3] Extreme values of the Riemann zeta-function, *Comment. Math. Helv.* 52 (1977), 511–518.

Montgomery, H. L., and Odlyzko, A. M.

[1] Gaps between zeros of the zeta-function, *Topics in classical number theory*, (Coll. Math. Soc. János Bolyai, 34), 1079–1106, (Budapest, 1981).

Montgomery, H. L., and Vaughan, R. C.

[1] Hilbert's inequality, *J. London Math. Soc.* (2), 8 (1974), 73–82.

Motohashi, Y.

[1] An elementary proof of Vinogradov's zero-free region for the Riemann zeta-function, *Recent progress in analytic number theory*, Vol I, 257–267, (Academic Press, London, 1981).

[2] A note on the approximate functional equation for $\zeta^2(s)$, *Proc. Japan Acad.* 59A (1983), 392–396.

[3] A note on the approximate functional equation for $\zeta^2(s)$. II, *Proc. Japan Acad.* 59A (1983), 469–472.

Mueller, J. H.

[1] On the difference between the consecutive zeros of the Riemann zeta-function, *J. Number Theory*, 14 (1982), 327–331.

[2] On the Riemann Zeta-function $\zeta(s)$ – gaps between sign changes of $S(t)$, *Mathematika*, 29 (1983), 264–269.

Odlyzko, A. M., and te Riele, H. J. J.

[1] Disproof of Mertens conjecture, *J. Reine Angew. Math.* 357 (1985), 138–160.

Ostrowski, A.

[1] Über Dirichletsche Reihen und algebraische Differentialgleichungen, *Math. Zeit.* 8 (1920), 115–143.

PINTZ, J.

[1] On the remainder term of the prime number formula and the zeros of Riemann's zeta-function, *Number theory, Noordwijkerhout* 1983, 186–197, Lecture Notes in Math. 1068, (Springer, Berlin, 1984).

RAMACHANDRA, K.

[1] On the zeros of the Riemann zeta-function and L-series, *Acta Arith.* 34 (1978), 211–218.

[2] Some remarks on a theorem of Montgomery and Vaughan, *J. Number Theory*, 11 (1980), 465–471.

[3] Some remarks on the mean value of the Riemann Zeta-function and other Dirichlet series – II, *Hardy-Ramanujan J.* 3 (1980), 1–24.

[4] Some remarks on the mean value of the Riemann Zeta-function and other Dirichlet series – III, *Ann. Acad. Sci. Fenn. Ser. AI Math.* 5 (1980), 145–158.

[5] Mean-value of the Riemann Zeta-function and other remarks – II, *Trudy Mat. Inst. Steklov.* 163 (1984), 200–204.

RANKIN, R. A.

[1] Van der Corput's method and the theory of exponent pairs, *Quart. J. Math. Oxford* (2), 6 (1955), 147–153.

REICH, A.

[1] Zetafunktion und Differenzen-Differentialgleichungen, *Arch. Math.* (*Basel*), 38 (1982), 226–235.

RICHERT, H.-E.

[1] Verschärfung der Abschätzung beim Dirichiletschen Teilerproblem, *Math. Zeit.* 58 (1953), 204–218.

[2] Einführung in die Theorie der starken Rieszschen Summierbarkeit von Dirichletreihen, *Nachr. Akad. Wiss. Gottingen* (*Math.-Phys.*) *Kl. II*, 1960, 17–75.

[3] Zur Abschätzung der Riemannschen Zetafunktion in der Nähe der Vertikalen $\sigma = 1$, *Math. Ann.* 169 (1967), 97–101.

SELBERG, A.

[1] Discontinuous groups and harmonic analysis, *Proc. Internat. Congr. Mathematicians* (*Stockholm*, 1962), 177–189 (Inst. Mittag-Leffler, Djursholm, 1963).

SPIRA, R.

[1] Zeros of sections of the zeta function. II, *Math. Comp.* 22 (1968), 163–173.

SRINIVASAN, B. R.

[1] Lattice point problems of many-dimensional hyperboloids III, *Math. Ann.* 160 (1965), 280–311.

STEČKIN, S. B.

[1] Mean values of the modulus of a trigonometric sum, *Trudy Mat. Inst. Steklov.* 134 (1975), 283–309, 411.

TATE, J. T.

[1] Fourier analysis in number fields, and Hecke's zeta-functions, *Algebraic Number Theory* (*Proc. Instructional Conf., Brighton*, 1965), 305–347, (Thompson, Washington DC, 1967).

TITCHMARSH, E. C.

[1] *The Zeta-function of Riemann*, Cambr. Tracts in Math. No 26, (Cambridge University Press, 1930).

TONG, K.-C.

[1] On divisor problems III, *Acta Math. Sinica*, 6 (1956), 515–541.

TURGANALIEV, R. T.

[1] The asymptotic formula for fractional mean value moments of the zeta-function of Riemann, *Trudy Mat. Inst. Steklov*. 158 (1981), 203–226.

VINOGRADOV, I. M.

[1] A new estimate for $\zeta(1+it)$, *Izv. Akad. Nauk SSSR, Ser. Mat*. 22 (1958), 161–164.

[2] *Selected works*, (Springer, Berlin, 1985).

VORONIN, M.

[1] On the distribution of nonzero values of the Riemann ζ-function, *Proc. Steklov Inst. Math*. 128 (1972), 153–175.

[2] Theorem on the "universality" of the Riemann Zeta-function, *Math. USSR Izvestija*, 9 (1975), 443–453.

[3] On the zeros of zeta-functions of quadratic forms, *Trudy Mat. Inst. Steklov*. 142 (1976), 135–147.

WALFISZ, A.

[1] *Weylsche Exponentialsummen in der Neueren Zahlentheorie*, (VEB Deutscher Verlag, Berlin, 1963).

WEIL, A.

[1] On the Riemann hypothesis in function-fields, *Proc. Nat. Acad. Sci. USA*, 27 (1941), 345–347.